D0164447

# Computational Methods in Finance

# CHAPMAN & HALL/CRC
Financial Mathematics Series

**Aims and scope:**
The field of financial mathematics forms an ever-expanding slice of the financial sector. This series aims to capture new developments and summarize what is known over the whole spectrum of this field. It will include a broad range of textbooks, reference works and handbooks that are meant to appeal to both academics and practitioners. The inclusion of numerical code and concrete real-world examples is highly encouraged.



**Published Titles**

American-Style Derivatives; Valuation and Computation, *Jerome Detemple*
Analysis, Geometry, and Modeling in Finance: Advanced Methods in Option Pricing, *Pierre Henry-Labordère*
Computational Methods in Finance, *Ali Hirsa*
Credit Risk: Models, Derivatives, and Management, *Niklas Wagner*
Engineering BGM, *Alan Brace*
Financial Modelling with Jump Processes, *Rama Cont and Peter Tankov*
Interest Rate Modeling: Theory and Practice, *Lixin Wu*
Introduction to Credit Risk Modeling, Second Edition, *Christian Bluhm, Ludger Overbeck, and Christoph Wagner*
An Introduction to Exotic Option Pricing, *Peter Buchen*
Introduction to Stochastic Calculus Applied to Finance, Second Edition, *Damien Lamberton and Bernard Lapeyre*
Monte Carlo Methods and Models in Finance and Insurance, *Ralf Korn, Elke Korn, and Gerald Kroisandt*
Monte Carlo Simulation with Applications to Finance, *Hui Wang*
Numerical Methods for Finance, *John A. D. Appleby, David C. Edelman, and John J. H. Miller*
Option Valuation: A First Course in Financial Mathematics, *Hugo D. Junghenn*
Portfolio Optimization and Performance Analysis, *Jean-Luc Prigent*
Quantitative Fund Management, *M. A. H. Dempster, Georg Pflug, and Gautam Mitra*
Risk Analysis in Finance and Insurance, Second Edition, *Alexander Melnikov*
Robust Libor Modelling and Pricing of Derivative Products, *John Schoenmakers*
Stochastic Finance: A Numeraire Approach, *Jan Vecer*
Stochastic Financial Models, *Douglas Kennedy*
Structured Credit Portfolio Analysis, Baskets & CDOs, *Christian Bluhm and Ludger Overbeck*
Understanding Risk: The Theory and Practice of Financial Risk Management, *David Murphy*
Unravelling the Credit Crunch, *David Murphy*


Proposals for the series should be submitted to one of the series editors above or directly to:
**CRC Press, Taylor & Francis Group**
4th, Floor, Albert House
1-4 Singer Street
London EC2A 4BQ
UK

Chapman & Hall/CRC FINANCIAL MATHEMATICS SERIES

# Computational Methods in Finance

Ali Hirsa

CRC Press is an imprint of the
Taylor & Francis Group, an informa business
A CHAPMAN & HALL BOOK

CRC Press
Taylor & Francis Group
6000 Broken Sound Parkway NW, Suite 300
Boca Raton, FL 33487-2742



Printed in the United States of America on acid-free paper
Version Date: 20120518

International Standard Book Number: 978-1-4398-2957-8 (Hardback)



**Library of Congress Cataloging-in-Publication Data**

Hirsa, Ali.
Computational methods in finance / Ali Hirsa.
p. cm. -- (Chapman & Hall/CRC Financial mathematics series)
Includes bibliographical references and index.
ISBN 978-1-4398-2957-8 (hardcover : alk. paper)
1. Derivative securities--Prices--Mathematics. I. Title.

HG6024.A3H57 2013
332.64'57015195--dc23 2012017892

**Visit the Taylor & Francis Web site at**
**http://www.taylorandfrancis.com**

**and the CRC Press Web site at**
**http://www.crcpress.com**

To Kamran Joseph and Tanaz

# Contents

## List of Symbols and Acronyms

| | |
|---|---|
| $B_t$ | money market account at time $t$ starting with \$1 at time 0 and rolling at the instantaneous short rate |
| $i$ | complex number $\sqrt{-1}$ |
| $\mathbb{E}(x)$ | expectation of $x$ under some measure |
| $\mathbb{E}_t(x)$ | expectation of $x$ under some measure conditional on knowing all information up to $t$ |
| f(x) | probability distribution function |
| F(x) | cumulative distribution function |
| $f(t,T)$ | instantaneous forward rate at calendar time $t$ with maturity $T$ |
| $F(t,T,S)$ | simply compounded forward rate for $[T,S]$ prevailing at $t$ |
| $L(t,T)$ | LIBOR rate at calendar time $t$ with maturity $T$ |
| $\mathbb{N}$ | set of natural numbers |
| $\mathbb{P}$ | real-world (physical) measure |
| $\mathbb{P}^T$ | forward measure |
| $\mathbb{P}^{n+1,N}$ | swap measure |
| $p(t,T)$ | price of a zero-coupon at time $t$ maturing at $T$ |
| $P_{k\|k-1}$ | error covariance matrix at time $k$ given observations up to and including time $k-1$ |
| $P_{k\|k}$ | error covariance matrix at time $k$ given observations up to and including time $k$ |
| $\mathbb{Q}$ | risk-neutral measure |
| $\rho$ | correlation |
| $\mathbb{R}$ | set of real numbers |
| $\mathbb{R}^+$ | set of positive real numbers |
| $r_t$ | instantaneous short rate at calendar time $t$ |
| $R(t,T)$ | continuously compounded spot rate with maturity $T$ prevailing at $t$ |
| $R(t,T,S)$ | continuously compounded forward rate for $[T,S]$ prevailing at $t$ |
| $\sigma$ | instantaneous volatility |
| $s(t,T)$ | swap rate at calendar time $t$ with maturity $T$ |
| $S_t$ | underlying price value at time $t$ |
| $Var(x)$ | variance of random variable $x$ |
| $\hat{x}_{k\|k-1}$ | estimate of the state at time $k$ given observations up to and including time $k-1$ |
| $\hat{x}_{k\|k}$ | estimate of the state at time $k$ given observations up to and including time $k$ |
| $\mathbb{Z}$ | set of integers |
| ACT/360 | day count convention — actual days assuming 360 days in a year |
| BVP | boundary value problem |
| CDF | cumulative distribution function |
| CDO | collateralized debt obligation |
| CGF | cumulant generating function |
| CGMY | Carr–Geman–Madan–Yor |
| COS | Fourier-cosine |
| DOC | down-and-out call |
| DOP | down-and-out put |
| FRA | forward rate agreement |
| FFT | fast Fourier transform |
| FrFFT | fractional fast Fourier transform |
| GBM | geometric Brownian motion |
| GBMSA | geometric Brownian motion with stochastic arrival — Heston |
| NIG | normal inverse Gaussian |
| PDE | partial differential equation |
| PDF | probability distribution function |
| PIDE | partial-integro differential equation |
| SAE | sum of the absolute errors |
| SDE | stochastic differential equation |
| SRE | sum of relative errors |
| SSAE | sum of the squares of absolute errors |
| SSRE | sum of the squares of relative errors |
| VG | variance gamma |
| VGSA | variance gamma with stochastic arrival |
| UOC | up-and-out call |
| UOP | up-and-out put |

# *List of Figures*

# List of Tables

# *Preface*

"In order to make any progress, it is necessary to think of approximate techniques, and above all, numerical algorithms ... Once again, what became a major endeavor of mine, the computational solution of complex functional equations, was entered into quite diffidently. I had never been interested in numerical analysis up to that point. Like most mathematicians of my generation, I had been brought up to scorn this utilitarian activity. Numerical solution was considered the last resort of an incompetent mathematician. The opposite, of course, is true. Once working in this area, it is very quickly realized that far more ability and sophistication is required to obtain a numerical solution than to establish the usual existence and uniqueness theorems. It is far more difficult to obtain an effective algorithm than one that stops with a demonstration of validity. A final goal of any scientific theory must be the derivation of numbers." This is an excerpt[1] from *Eye of the Hurricane* [30] on page 185 by Richard Bellman. It seems appropriate to start the preface with this quote considering advances in quantitative finance would have been impossible without utilizing computational/numerical techniques and their impact on the evolution of the field in recent years.

In most applications and physical phenomena, we are in search of a solution that happens to be an approximation of the true solution. As a result, some sort of a computational method/technique or a numerical procedure is a must. In quantitative finance, aside from a few cases with an analytical or a semi-analytical solution, we typically wind up with an approximation as well. As today's financial products have become more complex, quantitative analysts, financial engineers, and others in the financial industry now require robust techniques for numerical solutions. Computational finance has been a field that has been growing tremendously and intricacy of products and markets suggests there will be an even higher demand in the field.

This book is based on lecture notes I have used in my courses at Columbia University and my course at the Courant Institute of New York University. The selection of topics has been influenced by students and market requirements throughout my teaching over the years. Rama Cont, my colleague and friend, suggested to incorporate these notes into a textbook and referred me to the publisher.

My goal has been to write a textbook on computational methods in finance bringing together a full-spectrum of methods and schemes for pricing of derivatives contracts and related products, simulation, model calibration and parameter estimation with many practical examples. This book is intended for first/second year graduate students in the financial engineering or mathematics of finance field as well as practitioners, quants, researchers, technologists implementing models, and those who are interested in the field. My intention has been to keep the book self-contained and stand-alone.

Overall I have been pretty informal about theory.[2] The aim has not been to get into detail on stochastic calculus or martingales pricing as they are not prerequisites for understanding

[1] This quote was brought up to my attention by Michael Johannes, a colleague and friend, of Columbia Business School.

[2] An example of this is the Itô lemma for semi-martingales without defining semi-martingales or the Girsanov theorem without stating the theorem.

the procedures in the book. Yet in some cases it has been unavoidable, and I try to give sufficient explanation so that the reader can proceed without any need to delve into the derivation or the theory behind it.

This book is composed of two parts. The first part of the book describes various methods and techniques for the pricing of derivative contracts and the valuation of a variety of models and processes. In the second part, the book focuses on model calibration, calibration procedure, filtering, and parameter estimation.

Chapter 1 reviews some basic concepts, principally relating to the construction of the characteristic function of stochastic processes. It then shows how the characteristic function can be used to generate the moments of the resulting distribution and some methods used in our derivations of the characteristic functions of different processes. In addition, it reviews various characteristic functions of standard distributions. I then provide a self-contained list of some of the most commonly used stochastic processes that practitioners employ to model assets for derivative pricing applications. However, this list is by no means comprehensive and will certainly not cover every stochastic process used in practice. In describing these processes, I provide as detailed a mathematical description of each process as possible, including the characteristic function for every process, in closed form where available, as well as the stochastic differential equation where a closed form exists. Finally, the chapter contains a basic review of risk-neutral pricing and change of measure. When combined with a model of the stochastic evolution of the underlying asset, this forms the basis for all the derivative pricing algorithms in this book.

Chapters 2–6 cover many computational approaches for pricing derivatives contracts, including (a) transform techniques, (b) the finite difference method for solving partial differential equations and partial-integro differential equations, and (c) Monte Carlo simulation. Chapter 2 presents a range of transform techniques that comprise the fast Fourier transform, fractional fast Fourier transform, the Fourier-cosine (COS) method, and the saddlepoint method. I discuss the pros and cons of each approach and provide plenty of cross comparison. Chapter 3 introduces the finite difference method used for numerically solving partial differential equations. This chapter focuses on the most commonly used finite difference techniques utilized to solve partial differential equations, namely, explicit, implicit, Crank–Nicolson, and multi-step schemes. I discuss stability analysis of those schemes and different structure for the stiffness matrix arising from the discretization of partial differential equations and provide routines for solving the linear equations. A generic approach to derivative approximation by finite differences is also provided. Chapter 4 utilizes finite differences introduced in Chapter 3 to price vanilla and exotic derivatives under models for which a partial differential equation describing derivative prices can be formulated such as the Black–Scholes model and the local volatility models in the one-dimensional case and the Heston stochastic volatility model in the two-dimensional case. I discuss how to implement boundary conditions and exercise boundaries, setting up non-uniform grid points and coordinate transformation as well as dealing with jump conditions. Chapter 5 covers numerical solutions of partial-integro differential equations via finite differences for pricing various different derivative contracts. I look at PIDEs which arise in the pure jump framework, for instance, variance gamma (VG) and CGMY processes.

Not having the characteristic function in closed form, having a fairly complex payoff structure for the derivative contract under consideration, having a non-Markov process, or a high dimensional process or model, we have to utilize Monte Carlo simulation for pricing and valuation as the method of last resort. Chapter 6 covers Monte Carlo simulation. I discuss different sampling methods and sampling from various different distributions. I also go over Monte Carlo integration and numerical integration of stochastic differential equations. The output from simulation is associated with a variance that limits the accuracy of the simulation results. It is the major drawback to simulation and, naturally, various reduction techniques are studied and examined in this chapter. I also delve into simulation of some pure jump processes.

In the second part, the book focuses on essential steps in real-world derivatives pricing and estimation. In Chapter 7, I discuss how to calibrate model parameters so that model prices will be compatible with market prices. Construction of the local volatility surface and calibration of various different models in diffusion or pure-jump framework used for equity, foreign exchange, or interest rate modeling are discussed. The two essential steps in the calibration procedure, namely, the objective function and the optimization methodology are addressed in detail. I also discuss the notation of model risk. Chapter 8, the last chapter of the book covers various filtering techniques and their implementations used on the time series of data to unravel the best parameter set for the model under consideration and provide examples in filtering and parameter estimation of various different models and processes.

The book provides plenty of problems and case studies to help readers and students test their level of understanding in pricing, valuation, scenario analysis, calibration, optimization, and parameter estimation.

I would like to express my gratitude to several people who have influenced me directly or indirectly on this book. I owe a particular debt to my PhD advisor and co-author Dilip B. Madan. Special thanks to my co-authors Peter Carr, Georges Courtadon, and Massoud Heidari. I gained enormously from our many discussions and working together on a variety of different topics. I am thankful to Alireza Javaheri, Michael Johannes, and Nicholas G. Polson; I benefited tremendously on joint work with them regarding filtering and parameter estimation. I learned a great deal from my PhD advisor Howard C. Elman, Ricardo H. Nochetto, R. Bruce (Royal) Kellogg, and Jeffrey Cooper on numerical analysis and scientific computing; without their teaching and guidance I would not have been able to reach the level I am today.

# *Acknowledgments*

I would like to thank all of my students throughout the years at Columbia University and the Courant Institute of New York University for their input on this book. I am thankful to Tony Cheung, Ali Dashti Khavidaky, Helen Chien, Jorgen Berle Christiansen, Peter Derderian, Bernard Hurley, Barry Ichow, Laura Beth Kornstein, Art Kupferman, Jason Lan, Jie Li, Leigh Ronald Maidment, Fabio Niski, Behzad Nouri, Marco Ricciardulli, Seth Jon Rooder, Haohua Wan, Glenn W. Wekony, and Samuel Wolrich for their comments and discussions on this book and/or their help finding errors in the lecture notes from which this book evolved. I would also like to thank Bruno Dupire and his team at Bloomberg L.P. for their feedback and suggestions. I am grateful to Li Bao, Yiming Liu, Xiangji Ma, and Ryan Samson for their assistance producing examples, and checking or extending some of the programming codes to tabulating and plotting results. My sincerest appreciation to Alex Russo, my former student at the Financial Engineering program in the Industrial Engineering and Operations Research Department at Columbia University, for helping me with the structure of the book, editing, constructive comments, and valuable suggestions on this book. All errors are my own responsibility.

# Part I

# Pricing and Valuation

# Chapter 1

## Stochastic Processes and Risk-Neutral Pricing

Derivatives pricing begins with the assumption that the evolution of the underlying asset, be it a stock, commodity, interest rate, or exchange rate, follows some stochastic process. In this chapter, we will review a number of processes that are commonly used to model assets in different markets and explore how derivatives contracts written on these assets can be valued. In describing the many different computational methods which can be used to price derivatives and how they apply under different assumptions of an underlying stochastic process, we will often refer back to this chapter.

We begin this chapter by reviewing some basic probability, principally relating to the construction of the characteristic function of stochastic processes. We review how the characteristic function can be used to generate the moments of the resulting distribution and some methods used in our derivations of the characteristic functions of different processes. In addition, we review various characteristic functions of standard distributions.

Next, we provide a self-contained list of some of the most commonly used stochastic processes that practitioners employ to model assets for derivative pricing applications. However, this list is by no means comprehensive and will certainly not cover every stochastic process used in practice. In describing these processes, we will provide as detailed a mathematical description of each process as possible, including the characteristic function for every process, in closed form where available, as well as the stochastic differential equation (SDE) where a closed form exists.

Finally, this chapter contains a basic review of risk-neutral pricing and change of measure. When combined with a model of the stochastic evolution of the underlying asset, this forms the basis for all the derivative pricing algorithms in this book.

## 1.1 Characteristic Function

This section provides a basic review of the characteristic function of a distribution or a process. These concepts will be essential in our derivation of the characteristic functions of the stochastic processes reviewed in this chapter.

**Definition 1** *Fourier transform and inverse Fourier transform*

For function $f(x)$, its Fourier transform is defined as

$$\Phi(\nu) = \int_{-\infty}^{\infty} e^{i\nu x} f(x) dx \tag{1.1}$$

Having the Fourier transform of a function, $\Phi(\nu)$, the function $f(x)$ can be recovered via inverse Fourier transform

$$f(x) = \frac{1}{2\pi}\int_{-\infty}^{\infty} e^{-i\nu x}\Phi(\nu)d\nu \tag{1.2}$$

**Definition 2** *Characteristic function*

If $f(x)$ is the probability density function (PDF) of a random variable $x$, its Fourier transform is called the *characteristic function*

$$\Phi(\nu) = \int_{-\infty}^{\infty} e^{i\nu x} f(x)dx \tag{1.3}$$

$$= \mathbb{E}(e^{i\nu x}) \tag{1.4}$$

and as before probability density function $f(x)$ can be recovered from its characteristic function via the inverse Fourier transform.

### 1.1.1 Cumulative Distribution Function via Characteristic Function

As shown, the probability density function (PDF) can be recovered from the characteristic function. By integrating the PDF we can recover the cumulative distribution function (CDF). Thus, having characteristic function $\Phi(u) = \int_{-\infty}^{+\infty} e^{iux} f(x)dx$, the probability density function, $f(x)$, can be computed by inverting $\Phi(u)$.

$$f(x) = \frac{1}{2\pi}\int_{-\infty}^{+\infty} e^{-iux}\Phi(u)du$$

and the cumulative distribution function, $F(x)$, can be calculated.

$$\begin{aligned} F(x) &= \int_{-\infty}^{x} f(\eta)d\eta \\ &= \frac{1}{2\pi}\int_{-\infty}^{x}\int_{-\infty}^{\infty} e^{-iu\eta}\Phi(u)dud\eta \end{aligned}$$

However, in most cases the PDF is recovered only in parametric form, not analytically. Thus, recovering the CDF often requires a numerical integration of the parametric form of the PDF. We would prefer to recover cumulative distribution function $F(x)$ directly from the characteristic function $\Phi(u)$ to avoid the need to perform numerical integration twice. To do this we do not use the Fourier transform directly, as this would lead to non-convergence, but instead we use the Fourier transform of $\exp(-\alpha x)F(x)$ where $\exp(-\alpha x)$ is a damping factor with $\alpha > 0$.

$$\int_{-\infty}^{+\infty} e^{iux}e^{-\alpha x}F(x)dx = \int_{-\infty}^{+\infty} e^{-(\alpha - iu)x}F(x)dx$$

Using integration by parts on $\int_{-\infty}^{+\infty} e^{-(\alpha-iu)x}F(x)dx$ and noting that the first term vanishes gives us

$$\begin{aligned} \int_{-\infty}^{+\infty} e^{-(\alpha-iu)x}F(x)dx &= \frac{1}{\alpha - iu}\int_{-\infty}^{+\infty} e^{-(\alpha-iu)x}f(x)dx \\ &= \frac{1}{\alpha - iu}\int_{-\infty}^{+\infty} e^{i(\boxed{u + i\alpha})x}f(x)dx \\ &= \frac{1}{\alpha - iu}\Phi(u + i\alpha) \end{aligned}$$

Therefore

$$\int_{-\infty}^{+\infty} e^{iux} \boxed{e^{-\alpha x} F(x)} dx = \frac{1}{\alpha - iu} \Phi(u + i\alpha)$$

Using the Fourier inversion of this gives us

$$e^{-\alpha x} F(x) = \frac{1}{2\pi} \int_{-\infty}^{\infty} e^{-iux} \frac{1}{\alpha - iu} \Phi(u + i\alpha) du$$

or equivalently

$$F(x) = \frac{1}{2\pi} e^{\alpha x} \int_{-\infty}^{\infty} e^{-iux} \frac{1}{\alpha - iu} \Phi(u + i\alpha) du$$

Thus we can recover the cumulative distribution function directly from the characteristic function using a single numerical integration.

### 1.1.2 Moments of a Random Variable via Characteristic Function

Another useful property of the characteristic function of a distribution is that it allows us to recover an arbitrary number of moments of that distribution. Suppose we have the characteristic function of a random variable $X$ as

$$\phi(u) = \mathbb{E}\left[e^{iuX}\right] \tag{1.5}$$

It is easy to see that the $n^{th}$ derivative of $\phi(u)$ is given by the expression

$$\phi^{(n)}(u) = \mathbb{E}\left[(iX)^n e^{iuX}\right] \tag{1.6}$$

To find its moments, substitute zero for $u$ to obtain

$$\begin{aligned} \phi^{(n)}(0) &= \mathbb{E}\left[(iX)^n e^{i(0)X}\right] \\ &= \mathbb{E}\left[(iX)^n\right] \\ &= i^n \mathbb{E}\left[X^n\right] \end{aligned}$$

Therefore

$$\mathbb{E}\left[X^n\right] = i^{-n} \phi^n(0) \tag{1.7}$$

For example, the first moment of $X$, mean of $X$, is

$$\mathbb{E}\left[X\right] = -i\phi'(0)$$

### 1.1.3 Characteristic Function of Demeaned Random Variables

Assume we are interested in finding the characteristic function of a demeaned random variable $Y = X - \mathbb{E}[X]$ given the characteristic function of $X$. Using the result from 1.1.2, it can be done as follows:

$$\begin{aligned} E\left[e^{iuY}\right] &= E\left[e^{iu(X-E[X])}\right] \\ &= E\left[e^{iu(X+i\phi'(0))}\right] \\ &= e^{-u\phi'(0)} E\left[e^{iuX}\right] \\ &= e^{-u\phi'(0)} \phi(u) \end{aligned}$$

### 1.1.4 Calculating Jensen's Inequality Correction

We can generally express the evolution of an underlying price process $S_t$ using the following geometric law:

$$S_t = S_0 e^{(r-q)t+\omega t+X_t}$$

where $r-q$ is the mean rate of return on the asset under a risk-neutral measure and $X_t$ is the stochastic process of the underlying asset return which may follow any of the stochastic processes discussed in this chapter. We assume we know the characteristic function of the process $\phi(u) = \mathbb{E}(e^{iuX_t})$. The last term, $\omega$, is the so-called *Jensen's inequality correction* to ensure that the mean rate of return on the asset under a risk-neutral measure is $r-q$. As will be shown in the next chapter, in almost all applications of derivatives pricing, we need the characteristic function of the log of the underlying process rather than the characteristic function of the underlying process itself. Using the following derivation we can obtain $\omega$ and also calculate the characteristic function of the log of the underlying process:

$$\begin{aligned}
\Psi(u) &= \mathbb{E}\left(e^{iu\ln S_t}\right) \\
&= \mathbb{E}\left(e^{iu(\ln S_0+(r-q)t+\omega t+X_t)}\right) \\
&= e^{iu(\ln S_0+(r-q)t+\omega t)}\mathbb{E}\left(e^{iuX_t}\right) \\
&= e^{iu(\ln S_0+(r-q)t+\omega t)}\phi(u)
\end{aligned}$$

Substituting $-i$ for $u$ yields

$$\mathbb{E}(S_t) = S_0 e^{(r-q)t} e^{\omega t}\phi(-i).$$

Knowing that under risk-neutral measure we have

$$\mathbb{E}(S_t) = S_0 e^{(r-q)t}$$

comparing two equations implies that

$$e^{\omega t}\phi(-i) = 1$$

or equivalently

$$\omega = -\frac{1}{t}\ln(\phi(-i))$$

and the characteristic function of the log of the underlying process

$$\begin{aligned}
\Psi(u) &= \mathbb{E}\left(e^{iu\ln S_t}\right) \\
&= e^{iu(\ln S_0+(r-q)t)}\frac{\phi(u)}{\phi(-i)}
\end{aligned}$$

### 1.1.5 Calculating the Characteristic Function of the Logarithmic of a Martingale

Assume the random variable $M_t$ is a martingale under some measure and we want to find the characteristic function of the log of $M_t$ under that measure, namely, $\Psi(u) = \mathbb{E}(\exp(iu\ln M_t))$. Observe that if we substitute $-i$ for $u$ we obtain

$$\begin{aligned}
\Psi(-i) &= \mathbb{E}(\exp(i(-i)\ln M_t)) \\
&= \mathbb{E}(M_t) \\
&= M_0
\end{aligned} \tag{1.8}$$

Let $\ln M_t = N_t + A_t$, where $A_t$ and $N_t$ are deterministic and stochastic components of the process, respectively. Assume that the stochastic components of the process are known, but the exact expression for the deterministic component is not known, as is often the case. Moreover, we will assume the characteristic function of the stochastic component, $\Phi_N(u) = \mathbb{E}(\exp(iuN_t))$, is known. We can derive the deterministic component as follows:

$$\begin{aligned}
\Psi(u) &= \mathbb{E}(\exp(iu \ln M_t)) \\
&= \mathbb{E}(\exp(iuN_t + iuA_t)) \\
&= \exp(iuA_t)\mathbb{E}(\exp(iuN_t)) \\
&= \exp(iuA_t)\Phi_N(u)
\end{aligned}$$

where $\Phi_N(u)$ is the characteristic function of $N_t$. Substitute $-i$ for $u$ in $\Psi(u)$ and we get

$$\Psi(-i) = \exp(A_t)\Phi_N(-i) \tag{1.9}$$

Therefore (1.8) and (1.9) imply that

$$M_0 = \Phi_N(-i)\exp(A_t)$$

Solving for $A_t$ we get

$$A_t = \ln \frac{M_0}{\Phi_N(-i)}$$

Substituting for $A_t$ we finally get the following expression for the characteristic function of the logarithmic of $M_t$:

$$\begin{aligned}
\mathbb{E}(\exp(iu \ln M_t)) &= \Phi_N(u)\exp(iuA_t) \\
&= \Phi_N(u)\left(\frac{M_0}{\Phi_N(-i)}\right)^{iu}
\end{aligned}$$

### 1.1.6 Exponential Distribution

The exponential distribution with mean $\lambda$ is the distribution of the time between jumps of a Poisson process with rate $\frac{1}{\lambda}$. It has probability distribution function

$$f(x) = \lambda e^{-\lambda x}, \quad x \geq 0 \tag{1.10}$$

and cumulative distribution function

$$F(x) = 1 - \exp(-x/\theta), \quad x \geq 0 \tag{1.11}$$

Its characteristic function is

$$\phi(u) = \mathbb{E}(e^{iux}) = \int_0^\infty e^{iux}\lambda e^{-\lambda x}$$

This is a complex integral and its solution relies on the knowledge of how to integrate contours on $\mathbb{R}^2$ [124]. The solution can be derived in this case, and the characteristic function is as follows:

$$\phi(u) = \frac{\lambda}{\lambda - iu}$$

### 1.1.7 Gamma Distribution

A gamma random variable has the following probability distribution function:

$$f(x) = \frac{1}{\Gamma(\alpha)}\beta^{\alpha}x^{\alpha-1}e^{-\beta x}$$

where $\alpha$ is the shape parameter and $\beta$ is the scale parameter and we write it as $x \sim \text{gamma}(\alpha, \frac{1}{\beta})$. The characteristic function of a gamma process is obtained by evaluating the following complex integral:

$$\phi(u) = \mathbb{E}(e^{iux}) = \int_0^{\infty} e^{iux}\frac{1}{\Gamma(\alpha)}\beta^{\alpha}x^{\alpha-1}e^{-\beta x}dx$$

As for the exponential distribution, routine methods of complex analysis [124] yield

$$\phi(u) = \left(\frac{\beta}{\beta - iu}\right)^{\alpha} \tag{1.12}$$

This is similar to the result of the exponential distribution, not surprisingly because if $\alpha$ is an integer then $\text{gamma}(\alpha, \frac{1}{\beta})$ represents the sum of $\alpha$ independent exponential random variables, each of which has a mean of $\beta$, which is equivalent to a rate parameter $\frac{1}{\beta}$.

The chi-squared distribution $\chi^2(d)$ is actually a special case of the gamma distribution, $\text{gamma}(\frac{1}{2}d, \frac{1}{2})$, and thus has a characteristic function that can be easily derived from the previous results.

$$\phi(u) = (1 - 2iu)^{-d/2}$$

### 1.1.8 Lévy Processes

The class of the Lévy process consists of all stochastic processes with stationary, independent increments. The Lévy-Khintchine theorem [148] provides a characterization of Lévy processes in terms of the characterization of the underlying process. It states that there exists a measure $\nu$ such that for all $u \in \mathbb{R}$ and $t$ non-negative, the characteristic function of a Lévy process can be written as

$$\mathbb{E}(e^{izX_t}) = \exp(t\phi(z)) \tag{1.13}$$

where

$$\phi(z) = i\gamma z - \frac{\sigma^2 z^2}{2} + \int_{-\infty}^{\infty}(e^{izx} - 1 - izx\mathbb{1}_{|x|\leq 1})d\nu(x) \tag{1.14}$$

where $\sigma$ is non-negative, $\gamma \in \mathbb{R}$ and $\nu$ is a measure on $\mathbb{R}$ such that $\nu(0) = 0$ and

$$\int_{-\infty}^{\infty}(\min(1, x^2)d\nu(x)$$

is bounded.

### 1.1.9 Standard Normal Distribution

One of the most important distributions in finance is the standard normal distribution. It is the main component of a diffusion process and thus is absolutely central to most of the models we will be discussing. If $Z \sim \mathcal{N}(0,1)$, then its characteristic function is calculated as

$$\Phi_Z(\nu) = \mathbb{E}(e^{i\nu Z}) = \int_{-\infty}^{\infty} \frac{1}{\sqrt{2\pi}} \exp\left(i\nu z - \frac{1}{2}z^2\right) dz$$

Following the argument in [124] we instead consider the following integral:

$$\mathbb{E}(e^{sZ}) = \int_{-\infty}^{\infty} \frac{1}{\sqrt{2\pi}} \exp\left(sz - \frac{1}{2}z^2\right) dz$$

Complete the square in the integrand

$$\begin{aligned}
\int_{-\infty}^{\infty} \frac{1}{\sqrt{2\pi}} \exp\left(sz - \frac{1}{2}z^2\right) dz &= \int_{-\infty}^{\infty} \frac{1}{\sqrt{2\pi}} \exp\left(-\frac{1}{2}(z^2 - 2sz)\right) dz \\
&= \int_{-\infty}^{\infty} \frac{1}{\sqrt{2\pi}} \exp\left(-\frac{1}{2}(z-s)^2 + \frac{1}{2}s^2\right) dz \\
&= \frac{1}{\sqrt{2\pi}} e^{\frac{1}{2}s^2} \int_{-\infty}^{\infty} \exp\left(-\frac{1}{2}(z-s)^2\right) dz
\end{aligned}$$

and using the fact that

$$\int_{-\infty}^{\infty} e^{-\frac{1}{2}u^2 du} = \sqrt{2\pi}$$

to get

$$\mathbb{E}(e^{sZ}) = \exp\left(\frac{s^2}{2}\right)$$

As argued in [124] we can substitute $i\nu$ for $s$ by the theory of analytic continuation of functions of a complex variable to get

$$\Phi_Z(\nu) = \mathbb{E}(e^{i\nu Z}) = e^{-\frac{\nu^2}{2}}$$

### 1.1.10 Normal Distribution

A normal random variable with mean $\mu$ and standard deviation $\sigma$ can be constructed from a standard normal variable using $X = \mu + \sigma Z$, so that $X \sim \mathcal{N}(\mu, \sigma^2)$. Thus its characteristic function can be derived as follows:

$$\begin{aligned}
\Phi_X(\nu) &= \mathbb{E}(e^{i\nu X}) \\
&= \mathbb{E}(e^{i\nu(\mu+\sigma Z)}) \\
&= e^{i\nu\mu}\mathbb{E}(e^{i\boxed{\nu\sigma}Z}) \\
&= e^{i\nu\mu}e^{-\frac{(\sigma\nu)^2}{2}} \\
&= e^{i\mu\nu - \frac{\sigma^2\nu^2}{2}}
\end{aligned}$$

Brownian motion $W_t$ is a key component in many models of asset prices. We know that

$$W_t - W_0 = W_t \sim \mathcal{N}(0, t)$$

Therefore, if $X_t = W_t$, its characteristic function is

$$\mathbb{E}(e^{i\nu X_t}) = \mathbb{E}(e^{i\nu W_t}) = \mathbb{E}(e^{i\nu\sqrt{t}Z}) = e^{-\frac{\nu^2 t}{2}} \tag{1.15}$$

## 1.2 Stochastic Models of Asset Prices

To price any derivative, we need to model the statistics of an asset at every point in time which constitutes either a payoff date or a date at which an investment decision needs to be made. Practically, this means describing the evolution of an asset price using a stochastic process, and a number of different stochastic models for asset prices exist to accurately price different derivatives across different markets. In this section we outline a number of the most common models for asset prices.

### 1.2.1 Geometric Brownian Motion — Black–Scholes

One of the oldest and most commonly used models of asset prices in finance is geometric Brownian motion. First proposed by Black and Scholes in 1973, the model's creation was one of the pivotal moments in quantitative finance and is the standard by which most modern derivative pricing models are judged. Its invention helped to create an enormous and liquid market in options and by extension the multi-trillion modern derivatives market. In this section we will give a very brief description of the derivation of the model, as well as its stochastic differential equation and its characteristic function.

#### 1.2.1.1 Stochastic Differential Equation

When modeling asset prices using this model we assume the underlying process, the asset price $S_t$, at time $t$, satisfies the following stochastic differential equation, known as the Black–Scholes SDE:

$$dS_t = (r - q)S_t dt + \sigma S_t dW_t \tag{1.16}$$

where $r$, $q$ and $\sigma$ are a continuous risk-free interest rate, a continuous dividend rate and the instantaneous volatility respectively[1]. This equation models the asset's log returns as growing at a constant rate of $r - q$ and having a volatility of $\sigma$.

By means of Itô's lemma,[2] the solution to the stochastic differential equation is given by

$$\begin{aligned} S_T &= S_0 e^{(r-q-\frac{\sigma^2}{2})T+\sigma W_T} \\ &= S_0 e^{(r-q-\frac{\sigma^2}{2})T+\sigma\sqrt{T}Z} \end{aligned}$$

[1] In case of writing the evolution of the exchange rate under this model we would replace $r$ by $r_d$, the domestic rate, and $q$ by $r_f$, the foreign rate.

[2] Itô's lemma [183] — Assume $X_t$ satisfies

$$dX_t = \mu dt + \sigma dB_t$$

Let $g(t, x) \in C^2([0, \infty) \times \mathbb{R})$. Then

$$Y_t = g(t, X_t)$$

is a stochastic integral and

$$dY_t = \frac{\partial g}{\partial t}(t, X_t)dt + \frac{\partial g}{\partial x}(t, X_t)dX_t + \frac{1}{2}\frac{\partial^2 g}{\partial x^2}(t, X_t)(dX_t)^2$$

where

$$(dt)^2 = dt.dB_t = dB_t.dt = 0, dB_t.dB_t = dt$$

or equivalently

$$s_T = \ln S_T = s_0 + (r - q - \frac{\sigma^2}{2})T + \sigma\sqrt{T}Z$$

where $s_0 = \ln S_0$; therefore

$$s_T \sim \mathcal{N}\Big(s_0 + (r - q - \frac{\sigma^2}{2})T, \sigma^2 T\Big)$$

#### 1.2.1.2 Black–Scholes Partial Differential Equation

The Black–Scholes partial differential equation (1.17) is used to price derivatives whose underlying asset follows geometric Brownian motion. There are various ways to derive the Black–Scholes equation. The report in [164] gives an overview of various derivations of the Black–Scholes equation via (a) standard derivation of the Black–Scholes equation by constructing a replicating portfolio [177], (b) an alternative derivation using the *Capital Asset Pricing Model,* (c) using the return form of *Arbitrage Pricing Theory,* (d) an alternative derivation using *Risk-Neutral Pricing.* The resulting partial differential equation is as follows:

$$\frac{\partial v}{\partial t} + \frac{\sigma^2 S^2}{2}\frac{\partial^2 v}{\partial S^2} + (r - q)S\frac{\partial v}{\partial S} = rv(S, t) \tag{1.17}$$

The closed-form solution to this equation for various derivative contracts is available (for example, see [32] and [202] for European call options).

#### 1.2.1.3 Characteristic Function of the Log of a Geometric Brownian Motion

We already solved the SDE for a geometric Brownian motion to get

$$s_T \sim \mathcal{N}\Big(s_0 + (r - q - \frac{\sigma^2}{2})T, \sigma^2 T\Big) \tag{1.18}$$

Using the characteristic function of a normal random variable, we can easily derive the characteristic function of the log of the asset price as

$$\Phi_{s_T}(\nu) = e^{i(s_0 + (r - q - \frac{\sigma^2}{2})T)\nu - \frac{\sigma^2 \nu^2}{2}T} \tag{1.19}$$

### 1.2.2 Local Volatility Models — Derman and Kani

The Black–Scholes model is still the most widely used benchmark in options pricing, so much so that many markets quote option prices in the model's implied volatility. However, its assumption of a constant volatility for the underlying asset has proven to be incompatible with market prices, leading to the so-called volatility surface of implied volatilities for different option strikes and maturities. While the Black–Scholes model along with the volatility surface are sufficient to price vanilla options, more complex methods are needed to better capture market implied volatilities in order price more complex derivatives. The simplest addition to Black–Scholes is the local volatility model, which relaxes the constant volatility assumption and allows volatility to be a function of both time and the asset price.

#### 1.2.2.1 Stochastic Differential Equation

The stochastic differential equation under the local volatility model is almost exactly the same as the SDE for geometric Brownian motion. The only differences are that volatility is now parameterized on the asset price and time and the drift components are now parameterized by time.

$$dS_t = (r(t) - q(t))S_t dt + \sigma(S_t, t)S_t dW_t \tag{1.20}$$

#### 1.2.2.2 Generalized Black–Scholes Equation

The generalized Black–Scholes partial differential equation (1.21) prices derivatives whose underlying asset follows the local volatility model. It also closely follows the normal Black–Scholes equations, except for the additional parameterization.

$$\frac{\partial v}{\partial t} + \frac{1}{2}\sigma^2(S_t, t)S_t^2 \frac{\partial^2 v}{\partial S_t^2} + (r(t) - q(t))S_t \frac{\partial v}{\partial S_t} = r(t)v(S, t) \tag{1.21}$$

#### 1.2.2.3 Characteristic Function

In general, there is no analytical form available for the characteristic function of the local volatility model because the additional parameterization of the SDE components precludes it.

### 1.2.3 Geometric Brownian Motion with Stochastic Volatility — Heston Model

While the local volatility model can successfully fit the volatility surface of options prices more realistically than the standard Black–Scholes model, the volatility function can be very complex and thus not parsimonious in its use of variables. The effort to model volatility as a non-constant variable without a fully specified volatility function led to the creation of the Heston stochastic volatility model.

#### 1.2.3.1 Heston Stochastic Volatility Model — Stochastic Differential Equation

Under the Heston stochastic volatility model, asset prices $S_t$ follow a stochastic process described by the following set of SDEs:

$$\begin{aligned} dS_t &= rS_t dt + \sqrt{v_t} S_t dW_t^{(1)} \\ dv_t &= \kappa(\theta - v_t)dt + \sigma\sqrt{v_t} dW_t^{(2)} \end{aligned}$$

where the two Brownian components $W_t^{(1)}$ and $W_t^{(2)}$ are correlated with rate $\rho$. The variable $v_t$ represents the mean reverting stochastic volatility, where $\theta$ is the long term variance, $\kappa$ is the mean reversion speed, and $\sigma$ is the volatility of the variance. The presence of the $\sqrt{v_t}$ term in the diffusion component of this equation prevents the volatility from becoming negative by forcing the diffusion component to zero as the volatility approaches zero.

#### 1.2.3.2 Heston Model — Characteristic Function of the Log Asset Price

The characteristic function for the log of the asset price under the Heston stochastic volatility model is given by

$$\begin{aligned} \Phi(u) &= \mathbb{E}(e^{iu \ln S_t}) \\ &= \frac{\exp\{iu \ln S_0 + +iu(r-q)t + \frac{\kappa\theta t(\kappa - i\rho\sigma u)}{\sigma^2}\}}{(\cosh\frac{\gamma t}{2} + \frac{\kappa - i\rho\sigma u}{\gamma}\sinh\frac{\gamma t}{2})^{\frac{2\kappa\theta}{\sigma^2}}} \exp\left\{\frac{-(u^2 + iu)v_0}{\gamma\coth\frac{\gamma t}{2} + \kappa - i\rho\sigma u}\right\} \end{aligned}$$

where $\gamma = \sqrt{\sigma^2(u^2 + iu) + (\kappa - i\rho\sigma u)^2}$ and $S_0$ and $v_0$ are the initial values for the price process and the volatility process, respectively.

There are various different ways of deriving the characteristic function of the Heston model, one of which is manifested in Problem 3 at the end of this chapter. We provide an alternative derivation to its characteristic function below. This derivation is quite generic and can be used to derive the characteristic function of the log of asset prices under various different processes. Derivations similar to the one presented can be seen in [132] and [138].

We define the joint characteristic function of $(x, v)$ at time $t$, $0 < t < T$ as

$$\phi(t, \xi, \varphi) = \mathbb{E}\left[e^{i\xi x_T + i\omega v_T} | x_t = x, v_t = v\right] \tag{1.22}$$

where $(\xi, \varphi)$ are the transform variables. It is conjectured that the characteristic function at time $t = 0$ has a solution of the form

$$\phi(0, \xi, \varphi) = e^{-a(T)-b(T)x-c(T)v}$$

Therefore by the Markov property it must be the case that

$$\phi(t, \xi, \varphi) = e^{-a(T-t)-b(T-t)x-c(T-t)v}$$

The first thing to notice is that evaluating this at $t = T$ gives the following boundary conditions:

$$\begin{aligned}
\phi(T, \xi, \varphi) &= e^{-a(0)-b(0)x-c(0)v} \\
&= \mathbb{E}\left[e^{i\xi x_T + i\omega v_T} | x_T = x, v_T = v\right] \\
&= e^{i\xi x_T + i\omega v_T}
\end{aligned}$$

which implies

$$\begin{aligned}
a(0) &= 0 \\
b(0) &= i\xi \\
c(0) &= -i\varphi
\end{aligned}$$

If we define $G(t)$ to be

$$\mathcal{G}(t) = \phi(t, \xi, \varphi) \tag{1.23}$$

$\mathcal{G}(t)$ is a martingale because it is a conditional expectation of a terminal random variable. Therefore its derivative with respect to $t$, the $dt$ term, must be identically zero.

As stated earlier, for the Heston stochastic volatility model, we assume that $S_t$ evolves according to the following SDE:

$$\begin{aligned}
dS_t &= rS_t dt + \sqrt{v_t} S_t dW_t^{(1)} \\
dv_t &= \kappa(\theta - v_t)dt + \sigma\sqrt{v_t} dW_t^{(2)}
\end{aligned}$$

We define $x_t = \ln S_t = F(t, S_t)$ and apply Itô's lemma to derive $dx_t$.

$$\begin{aligned}
dx_t &= \frac{\partial F}{\partial t} dt + \frac{\partial F}{\partial S_t} dS_t + \frac{1}{2}\frac{\partial^2 F}{\partial S_t^2}(dS_t)^2 \\
&= 0 + \frac{1}{S_t}(rS_t dt + \sqrt{v_t} S_t dW_t^{(1)}) - \frac{1}{2}\frac{1}{S_t^2}(rS_t dt + \sqrt{v_t} S_t dW_t^{(1)})^2 \\
&= r dt + \sqrt{v_t} dW_t^{(1)} - \frac{1}{2} v_t dt \\
&= (r - \frac{1}{2} v_t)dt + \sqrt{v_t} dW_t^{(1)}
\end{aligned}$$

where

$$dv_t = \kappa(\theta - v_t)dt + \sigma\sqrt{v_t}dW_t^{(2)}$$

Thus we have

$$\begin{aligned} dx_t &= (r - \frac{1}{2}v_t)dt + \sqrt{v_t}dW_t^{(1)} \\ dv_t &= \kappa(\theta - v_t)dt + \sigma\sqrt{v_t}dW_t^{(2)} \end{aligned}$$

The goal is to find

$$\mathbb{E}_t^{\mathbb{Q}}\left(e^{iu\ln S_T}\right) = \mathbb{E}_t^{\mathbb{Q}}\left(e^{iux_T}\right)$$

It is clear that $\phi(t, \xi = u, \omega = 0) = \mathbb{E}\left[e^{iux_T}\right]$. Because $\mathcal{G}(t) = \phi(t, \xi, \omega)$ is a martingale, we have

$$\begin{aligned} d\mathcal{G}(t) &= \frac{\partial \mathcal{G}(t)}{\partial t}dt + \ldots \\ &= 0dt + \ldots \end{aligned}$$

because its derivatives with respect to time must be zero and this leads to the expression

$$\begin{aligned} \frac{\partial \mathcal{G}(t)}{\partial t} &= \phi_t + \phi_x(r - \frac{1}{2}v_t) + \phi_v(\kappa(\theta - v_t)) \\ &+ \frac{1}{2}\mathit{Trace}(\phi_{xx}v_t) + \frac{1}{2}\phi_{vv}\sigma^2 v_t + \rho\sigma^2(t)v_t\phi_{xv} = 0 \end{aligned} \tag{1.24}$$

As mentioned, it is conjectured that $\phi(t, \xi, \varphi)$ can be expressed as

$$\phi(t, \xi, \varphi) = e^{-a(T-t)-b(T-t)x-c(T-t)v}$$

and therefore its derivatives would be

$$\begin{aligned} \phi_t &= (-a'(T-t) - b'(T-t)x - c'(T-t)v)\phi \\ \phi_x &= -b(T-t)\phi \\ \phi_v &= -c(T-t)\phi \\ \phi_{xx} &= b^2(T-t)\phi \\ \phi_{vv} &= c^2(T-t)\phi \\ \phi_{vx} &= b(T-t)c(T-t)\phi \end{aligned}$$

Substituting all derivatives into (1.24) we get

$$\begin{aligned} \phi\Big(&-a'(T-t) - b'(T-t)x - c'(T-t)v - b(T-t)(r - \frac{1}{2}v) \\ &-c(T-t)\kappa(\theta - v) + \frac{1}{2}b^2(T-t)v + \frac{1}{2}c^2(T-t)\sigma^2 v + \rho\sigma v b(T-t)c(T-t)\Big) = 0 \end{aligned} \tag{1.25}$$

Since this holds for all $(x, v)$, we get three simpler equations, namely, the Riccati equations, by grouping $x$, $v$, and the remaining terms

$$\begin{cases} a'(T-t) - c(T-t)\kappa\theta - rb(T-t) = 0 \\ b'(T-t) = 0 \\ c'(T-t) + \frac{1}{2}b(\tau) + c(T-t)\kappa + \frac{1}{2}b^2(T-t) + \frac{1}{2}c^2(T-t)\sigma^2 + \rho\sigma b(T-t)c(T-t) = 0 \end{cases}$$

Define $\tau = T - t$ and we get

$$\begin{cases} a'(\tau) - c(\tau)\kappa\theta - rb(\tau) = 0 \\ b'(\tau) = 0 \\ c'(\tau) + \frac{1}{2}b(\tau) + c(\tau)\kappa + \frac{1}{2}b^2(\tau) + \frac{1}{2}c^2(\tau)\sigma^2 + \rho\sigma b(\tau)c(\tau) = 0 \end{cases}$$

The second Riccati equation, $b'(\tau) = 0$, implies that $b(tau)$ is constant and the boundary condition implies

$$b(\tau) = -i\xi$$

Substituting it in the third equation we have

$$\begin{aligned} c'(\tau) &= \frac{1}{2}i\xi - c(\tau)\kappa - \frac{1}{2}(-i\xi)^2 - \frac{1}{2}c^2(\tau)\sigma^2 - \rho\sigma(-i\xi)c(\tau) \\ &= -\frac{1}{2}\sigma^2c^2(\tau) + (i\xi\rho\sigma - \kappa)c(\tau) + (\frac{1}{2}i\xi + \frac{1}{2}\xi^2) \end{aligned}$$

$$\frac{dc(\tau)}{d\tau} = -\frac{1}{2}\sigma^2\left[c^2(\tau) + \frac{2}{\sigma^2}(\kappa - i\xi\rho\sigma)c(\tau) - \frac{i\xi + \xi^2}{\sigma^2}\right]$$

$$\frac{dc(\tau)}{c^2(\tau) + \frac{2}{\sigma^2}(\kappa - i\xi\rho\sigma)c(\tau) - \frac{i\xi+\xi^2}{\sigma^2}} = -\frac{1}{2}\sigma^2 d\tau$$

We solve this equation using partial fractions. The roots of the equation in the denominator

$$c^2(\tau) + \frac{2}{\sigma^2}(\kappa - i\xi\rho\sigma)c(\tau) - \frac{i\xi + \xi^2}{\sigma^2}$$

are

$$\begin{aligned} c_1 &= \frac{\beta - \gamma}{\sigma^2} \\ c_2 &= \frac{\beta + \gamma}{\sigma^2} \end{aligned}$$

where

$$\begin{aligned} \beta &= \kappa - i\xi\rho\sigma \\ \gamma &= \sqrt{(\kappa - i\xi\rho\sigma)^2 + \sigma^2(\xi^2 + i\xi)} \end{aligned}$$

We can find $A$ and $B$ such that

$$\frac{1}{c^2(\tau) + \frac{2}{\sigma^2}(\kappa - i\xi\rho\sigma)c(\tau) - \frac{i\xi+\xi^2}{\sigma^2}} = \frac{A}{c(\tau) - c_1} + \frac{B}{c(\tau) - c_2}$$

We see that $A$ and $B$ should satisfy the following equation:

$$A\left(c(\tau) - \frac{\beta - \gamma}{\sigma^2}\right) + B\left(c(\tau) - \frac{\beta + \gamma}{\sigma^2}\right) = 1$$

Or equivalently

$$\begin{gathered} A + B = 0 \\ -A(\beta - \gamma) - B(\beta + \gamma) = \sigma^2 \end{gathered}$$

and we get $A = \frac{\sigma^2}{2\sigma}$ and $B = -\frac{\sigma^2}{2\sigma}$. Now substituting and integrating we get

$$\begin{aligned}
\int \frac{\frac{\sigma^2}{2\gamma}}{c(\tau) - \frac{\beta+\gamma}{\sigma^2}} dc(\tau) + \int \frac{-\frac{\sigma^2}{2\gamma}}{c(\tau) - \frac{\beta-\gamma}{\sigma^2}} dc(\tau) &= -\frac{1}{2}\sigma^2\tau + constant \\
\frac{\sigma^2}{2\gamma} \ln\left(c(\tau) - \frac{-\beta+\gamma}{\sigma^2}\right) - \frac{\sigma^2}{2\gamma} \ln\left(c(\tau) - \frac{-\beta-\gamma}{\sigma^2}\right) &= -\frac{1}{2}\sigma^2\tau + constant
\end{aligned}$$

or

$$\begin{aligned}
\ln\left(\frac{c(\tau) - \frac{-\beta+\gamma}{\sigma^2}}{c(\tau) - \frac{-\beta-\gamma}{\sigma^2}}\right) &= -\gamma\tau + constant \\
\frac{c(\tau) - \frac{-\beta+\gamma}{\sigma^2}}{c(\tau) - \frac{-\beta-\gamma}{\sigma^2}} &= \alpha e^{-\gamma\tau}
\end{aligned}$$

The boundary condition for $c(\tau)$ is

$$c(\tau = 0) = c(0) = -i\varphi = -i(0) = 0$$

Applying this we obtain

$$\alpha = \frac{0 - \frac{-\beta+\gamma}{\sigma^2}}{0 - \frac{-\beta-\gamma}{\sigma^2}} = \frac{\beta-\gamma}{\beta+\gamma}$$

We substitute this back into 1.26 and solve for $c(\tau)$.

$$c(\tau) = \frac{\gamma-\beta}{\sigma^2} \frac{1-e^{-\gamma\tau}}{1-\alpha e^{-\gamma\tau}}$$

Having $c(\tau)$, we can solve for $a(\tau)$.

$$\begin{aligned}
a'(\tau) &= c(\tau)\kappa\theta + rb(\tau) \\
&= \kappa\theta \frac{\gamma-\beta}{\sigma^2} \frac{1-e^{-\gamma\tau}}{1-\alpha e^{-\gamma\tau}} - i\xi r \\
a(\tau) &= \kappa\theta \frac{\gamma-\beta}{\sigma^2} \int \frac{1-e^{-\gamma\tau}}{1-\alpha e^{-\gamma\tau}} d\tau - i\xi r \int d\tau \\
&= \kappa\theta \frac{\gamma-\beta}{\sigma^2} \left[\int \frac{1}{1-\alpha e^{-\gamma\tau}} d\tau - \int \frac{e^{-\gamma\tau}}{1-\alpha e^{-\gamma\tau}} d\tau\right] - i\xi r \int d\tau \\
&= \kappa\theta \frac{\gamma-\beta}{\sigma^2} \left[\tau + \frac{1}{\gamma}\ln(1-\alpha e^{-\gamma\tau}) - \frac{1}{\alpha\gamma}\ln(1-\alpha e^{-\gamma\tau})\right] - i\xi r\tau + constant \\
&= \kappa\theta \frac{\gamma-\beta}{\sigma^2} \left[\tau + \ln(1-\alpha e^{-\gamma\tau})^{\frac{\alpha-1}{\alpha\gamma}}\right] - iu\xi r\tau + constant
\end{aligned}$$

To find the constant, we know that $a(\tau = 0) = 0$ and that implies

$$a(0) = \kappa\theta \frac{\gamma-\beta}{\sigma^2} \left[0 + \ln(1-\alpha)^{\frac{\alpha-1}{\alpha\gamma}}\right] + constant = 0$$

and we obtain

$$constant = -\kappa\theta \frac{\gamma-\beta}{\sigma^2} \left[\ln(1-\alpha)^{\frac{\alpha-1}{\alpha\gamma}}\right]$$

Substituting it into $a(\tau)$ we get

$$a(\tau) = \kappa\theta\frac{\gamma-\beta}{\sigma^2}\left[\tau + \ln\left(\frac{1-\alpha e^{-\gamma\tau}}{1-\alpha}\right)^{\frac{\alpha-1}{\alpha\gamma}}\right] - iur\tau$$

Now that we have all loadings $a(\tau)$, $b(\tau)$, and $c(\tau)$ explicitly we can calculate the characteristic function

$$\begin{aligned}\mathbb{E}\left(e^{iux_T}\right) &= \phi(t=0,\xi=u,\varphi=0)\\ &= e^{-a(\tau)-b(\tau)x_0-c(\tau)v_0}\end{aligned}$$

where

$$\begin{aligned}a(\tau) &= \kappa\theta\frac{\gamma-\beta}{\sigma^2}\left[\tau + \ln\left(\frac{1-\alpha e^{-\gamma\tau}}{1-\alpha}\right)^{\frac{\alpha-1}{\alpha\gamma}}\right] - iur\tau\\ b(\tau) &= -iu\\ c(\tau) &= \frac{\gamma-\beta}{\sigma^2}\frac{1-e^{-\gamma\tau}}{1-\alpha e^{-\gamma\tau}}\end{aligned}$$

and $\beta = \kappa - iu\rho\sigma$, $\gamma = \sqrt{(\kappa - i\xi\rho\sigma)^2 + \sigma^2(\xi^2 + i\xi)}$, and $\alpha = \frac{\beta-\gamma}{\beta+\gamma}$. Thus the full characteristic function is as follows:

$$\begin{aligned}\mathbb{E}_t\left(e^{iux_T}\right) &= \exp\left\{-\kappa\theta\frac{\gamma-\beta}{\sigma^2}\tau - \ln\left(\frac{1-\alpha e^{-\gamma\tau}}{1-\alpha}\right)^{\kappa\theta\frac{\gamma-\beta}{\sigma^2}\frac{\alpha-1}{\alpha\gamma}}\right\}\\ &\times \exp\left\{iur\tau + iux_0 - \frac{\gamma-\beta}{\sigma^2}\frac{1-e^{-\gamma\tau}}{1-\alpha e^{-\gamma\tau}}v_0\right\}\\ &= \exp\left\{-\kappa\theta\frac{\gamma-\beta}{\sigma^2}\tau - \ln\left(\frac{1-\alpha e^{-\gamma\tau}}{1-\alpha}\right)^{\frac{2\kappa\theta}{\sigma^2}} + iur\tau + iux_0\right\}\\ &\times \exp\left\{-\frac{\gamma-\beta}{\sigma^2}\frac{1-e^{-\gamma\tau}}{1-\alpha e^{-\gamma\tau}}v_0\right\}\\ &= \exp\left\{\kappa\theta\tau\frac{\beta}{\sigma^2} + iux_0 + iur\tau\right\}\\ &\times \exp\left\{-\ln\left(\frac{1-\alpha e^{-\gamma\tau}}{1-\alpha}\right)^{\frac{2\kappa\theta}{\sigma^2}} - \kappa\theta\tau\frac{\gamma}{\sigma^2}\right\}\\ &\times \exp\left\{-\frac{\gamma-\beta}{\sigma^2}\frac{1-e^{-\gamma\tau}}{1-\alpha e^{-\gamma\tau}}v_0\right\}\end{aligned} \tag{1.26}$$

The last term in Equation (1.26) can be simplified as follows:

$$\begin{aligned}
\exp\left\{\frac{\beta-\gamma}{\sigma^2}\frac{1-e^{-\gamma\tau}}{1-\alpha e^{-\gamma\tau}}v_0\right\} &= \exp\left\{\frac{\beta^2-\gamma^2}{\beta+\gamma}\frac{1}{\sigma^2}\frac{e^{\gamma\tau}(1-e^{-\gamma\tau})}{e^{\gamma\tau}(1-\alpha e^{-\gamma\tau})}v_0\right\} \\
&= \exp\left\{-\frac{u^2+iu}{\beta+\gamma}\frac{e^{\gamma\tau}-1}{e^{\gamma\tau}-\alpha}v_0\right\} \\
&= \exp\left\{-(u^2+iu)v_0\frac{e^{\gamma\tau}-1}{e^{\gamma\tau}-\frac{\beta-\gamma}{\beta+\gamma}}\frac{1}{\beta+\gamma}\right\} \\
&= \exp\left\{-(u^2+iu)v_0\frac{e^{\gamma\tau}-1}{\beta(e^{\gamma\tau}-1)+\gamma(e^{\gamma\tau}+1)}\right\} \\
&= \exp\left\{-(u^2+iu)v_0\frac{1}{\frac{\beta(e^{\gamma\tau}-1)+\gamma(e^{\gamma\tau}+1)}{e^{\gamma\tau}-1}}\right\} \\
&= \exp\left\{-(u^2+iu)v_0\frac{1}{\gamma\frac{e^{\gamma\tau}+1}{e^{\gamma\tau}-1}+\beta}\right\} \\
&= \exp\left\{-\frac{(u^2+iu)v_0}{\gamma\coth\frac{\gamma\tau}{2}+\beta}\right\}
\end{aligned}$$

In addition, the second to last term in Equation (1.26) can also be simplified.

$$\begin{aligned}
\exp\left\{-\ln\left(\frac{1-\alpha e^{-\gamma\tau}}{1-\alpha}\right)^{\frac{2\kappa\theta}{\sigma^2}}-\frac{\kappa\theta}{\sigma^2}\gamma\tau\right\} &= \left(e^{\frac{\gamma\tau}{2}}\frac{1-\alpha e^{-\gamma\tau}}{1-\alpha}\right)^{-\frac{2\kappa\theta}{\sigma^2}} \\
&= \left(e^{\frac{\gamma\tau}{2}}\frac{1-\alpha e^{-\gamma\tau}}{\frac{2\gamma}{\gamma+\beta}}\right)^{-\frac{2\kappa\theta}{\sigma^2}} \\
&= \left(\frac{(\gamma+\beta)e^{\frac{\gamma\tau}{2}}}{2\gamma}(1-\alpha e^{-\gamma\tau})\right)^{-\frac{2\kappa\theta}{\sigma^2}} \\
&= \left(\frac{(\gamma+\beta)e^{\frac{\gamma\tau}{2}}+(\gamma-\beta)e^{\frac{-\gamma\tau}{2}}}{2\gamma}\right)^{-\frac{2\kappa\theta}{\sigma^2}} \\
&= \left(\frac{\gamma(e^{\frac{\gamma\tau}{2}}+e^{\frac{\gamma\tau}{2}})+\beta(e^{\frac{\gamma\tau}{2}}-e^{\frac{\gamma\tau}{2}})}{2\gamma}\right)^{-\frac{2\kappa\theta}{\sigma^2}} \\
&= \left(\cosh\frac{\gamma\tau}{2}+\frac{\beta}{\gamma}\sinh\frac{\gamma\tau}{2}\right)^{-\frac{2\kappa\theta}{\sigma^2}}
\end{aligned}$$

Putting them all together we get

$$\mathbb{E}_t\left(e^{iux_T}\right)=\frac{\exp\left\{iu\ln S_0+iu(r-q)\tau+\kappa\theta\tau\frac{\beta}{\sigma^2}\right\}\times\exp\left\{\frac{-(u^2+iu)v_0}{\gamma\coth\frac{\gamma\tau}{2}+\beta}\right\}}{\left(\cosh\frac{\gamma\tau}{2}+\frac{\beta}{\gamma}\sinh\frac{\gamma\tau}{2}\right)^{\frac{2\kappa\theta}{\sigma^2}}}$$

### 1.2.4 Mixing Model — Stochastic Local Volatility (SLV) Model

Stochastic local volatility model is an extension of the local volatility model that incorporates an independent stochastic component to volatility (e.g. [153],[191],[209], and [171]).

The independent stochastic component is modeled by a stochastic process $(V(t), t \geq 0)$ starting at one. In this model the evolution of the stock price is given by

$$dS_t = (r - q)S_t dt + L(S_t, t)V(t)S_t dW_t^1$$

where $(W_t^1, t \geq 0)$ is standard Brownian motion and $L(S_t, t)$ is a deterministic function of the stock price and calendar time which represents average volatility. In [191], they assume $V(t)$ follows a mean-reverting lognormal

$$d\ln V_t = \kappa(\theta(t) - \ln V_t)dt + \lambda dW_t^2$$

where as in Heston $\kappa$ is the rate of mean reversion, $\theta(t)$ is the long-term deterministic drift, and $\lambda$ is the volatility of volatility. Considering that $\sigma^2(S_t, t)$ is interpreted as the average local variance, they put a constraint on the unconditional expectation of $V(t)^2$ to be unity which implies

$$\theta(t) = \frac{\lambda^2}{2\kappa}\left(1 + e^{-2\kappa t}\right)$$

In [209], they consider the following process for $V_t$

$$dV_t = \kappa(\theta(t) - V_t)dt + \lambda V_t dW_t^2$$

One may assume correlation between the Brownian motion driving the stochastic component of volatility and the Brownian motion driving the stock price. However, in [191] they assume zero correlation for simplicity. Note that if $L(S_t, t)$ has no dependence on the stock price, the model is very Heston-like and by letting $\lambda = 0$ the model degenerates to a local volatility model.

### 1.2.5 Geometric Brownian Motion with Mean Reversion — Ornstein–Uhlenbeck Process

While geometric Brownian motion, local volatility, and stochastic volatility models are very popular for modeling assets where the primary concern is modeling the volatility of the underlier, their constant drift assumption is incompatible with market prices in markets where long-term price movements revert to a long-term mean. In markets where mean reversion is a common feature, including interest rates, currency exchange rates, and commodity prices, the Ornstein–Uhlenbeck (OU) process is a popular model.

The OU process is an instance of a Gaussian process that has a bounded variance and admits a stationary probability distribution, in contrast to the Wiener process. The difference between the two is in their *drift* term. For the Wiener process the drift term is constant, whereas for the OU process it is dependent on the current value of the process: if the current value of the process is less than the (long-term) mean, the drift will be positive; if the current value of the process is greater than the (long-term) mean, the drift will be negative. In other words, the mean acts as an equilibrium level for the process. This gives the process its *mean-reverting* characteristics.

#### 1.2.5.1 Ornstein–Uhlenbeck Process — Stochastic Differential Equation

The Ornstein–Uhlenbeck (OU) process is a stochastic process $r_t$ which can be described by the following stochastic differential equation:

$$dr_t = \kappa(\eta - r_t)dt + \lambda dW_t$$

where $\kappa > 0$, $\eta$, and $\lambda > 0$ are model parameters and $W_t$ denotes the Wiener process.

The parameter $\eta$ represents the equilibrium or mean value supported by fundamentals, $\lambda$ the degree of volatility around it caused by shocks, and $\kappa$ the rate by which these shocks dissipate and the variable reverts toward the mean.

#### 1.2.5.2 Vasicek Model

The oldest and most direct application of the OU process in finance is the Vasicek model. Under this model the instantaneous spot interest rate (the short rate) follows an OU process. This model is advantageous in that, unlike geometric Brownian motion, the short rate vacillates around a long-term mean as market rates have done historically. This causes the long-term variance to be bounded, which is also an empirical feature of the interest rate markets; market rates very rarely grow exponentially to very large levels. The disadvantage of using this process for interest rates is that it allows the short rate to become negative, a condition not often seen in the market.

Applying Itô's lemma to $r_t e^{\kappa t}$, we can solve the stochastic differential equation to get

$$r_t = r_0 e^{-\kappa t} + \eta(1 - e^{-\kappa t}) + \sigma e^{-\kappa t} \int_0^t \lambda e^{\kappa s} dW_s \tag{1.27}$$

While in most of the previous cases the characteristic function of the stochastic random variable or its log was most interesting to us, when modeling instantaneous interest rates the integral of the process over time is the most critical component for pricing. If $r(t)$ is the instantaneous interest rate, then the realized interest rate is given by $R(t)$ where

$$R(t) = \int_0^t r(u)du.$$

We can show that $R_t \sim \mathcal{N}(\mu, \sigma^2)$ where

$$\begin{aligned} \mu &= \frac{\eta - r_0}{\kappa}(e^{-\kappa t} - 1) + \eta t \\ \sigma^2 &= \frac{\lambda^2}{2\kappa^2}\left(\frac{4e^{-\kappa t} - e^{-2\kappa t} - 3}{2\kappa} + t\right) \end{aligned}$$

And we know that the characteristic function of a normal random variable, $\mathcal{N}(\mu, \sigma^2)$, is

$$\phi(u) = \mathbb{E}(e^{iuX_t}) \tag{1.28}$$

$$= e^{i\mu u - \frac{\sigma^2 u^2}{2}} \tag{1.29}$$

Substituting $\mu$ and $\sigma$ into the above equation, we get the characteristic function for $R(t)$ as

$$\mathbb{E}(e^{iuR(t)}) = e^{A(t,u) - B(t,u)r(0)}$$

where

$$\begin{aligned} A(t,u) &= (\eta + \frac{\lambda^2}{2\kappa^2} iu)(B(t,u) + iut) - \frac{\lambda^2}{4\kappa} B^2(t,u) \\ B(t,u) &= \left(\frac{e^{-\kappa t} - 1}{\kappa}\right) iu \end{aligned}$$

If we are interested in computing the term structure of bond prices, $P(t,T)$, in the Vasicek model, we can use

$$P(t,T) = \mathbb{E}(e^{-\int_0^T r_t}) = \mathbb{E}(e^{-R_T}) \tag{1.30}$$

which we obtain by simply substituting $-i$ for $u$ in its characteristic function.

## 1.2.6 Cox–Ingersoll–Ross Model

The Cox–Ingersoll–Ross (CIR) model is a modification of the Vasicek model which is meant to maintain all of its advantages while preventing the short interest rate from becoming negative. To do this, a $\sqrt{r_t}$ term is added to the volatility term of the SDE. This causes the volatility to go to zero as our process approaches zero, which prevents the process from resulting in negative interest rates.

### 1.2.6.1 Stochastic Differential Equation

Thus, the stochastic differential equation describing the CIR model is as follows:

$$dr_t = \kappa(\eta - r_t)dt + \lambda\sqrt{r_t}dW_t$$

where $W(t)$ is a Brownian motion, $\eta$ is the long term rate of time change, $\kappa$ is the rate of mean reversion and $\lambda$ is the volatility of the time change.

### 1.2.6.2 Characteristic Function of Integral

As in the Vasicek model, we are interested in the characteristic function of the realized interest rate $R(t)$ where

$$R(t) = \int_0^t r(u)du.$$

It can be shown that the characteristic function for $R(t)$ is

$$\begin{aligned}\mathbb{E}(e^{iuR(t)}) &= \phi(u, t, r(0), \kappa, \eta, \lambda) \\ &= A(t,u)e^{B(t,u)r(0)}\end{aligned}$$

where

$$\begin{aligned}A(t,u) &= \frac{\exp\left(\frac{\kappa^2\eta t}{\lambda^2}\right)}{\left(\cosh(\gamma t/2) + \frac{\kappa}{\gamma}\sinh(\gamma t/2)\right)^{2\kappa\eta/\lambda^2}} \\ B(t,u) &= \frac{2iu}{\kappa + \gamma\coth(\gamma t/2)}\end{aligned}$$

with

$$\gamma = \sqrt{\kappa^2 - 2\lambda^2 iu}$$

As in the Vasicek model, if we are interested in computing the term structure of bond prices in the CIR model, we can use

$$P(t,T) = \mathbb{E}(e^{-\int_0^T r_t}) = \mathbb{E}(e^{-R_T}) \tag{1.31}$$

and simply substitute $-i$ for $u$ in its characteristic function.

## 1.2.7 Variance Gamma Model

All the previous models we have discussed have concentrated on modifying the volatility of the underlying process in order to better capture a dynamic volatility structure or modifying the drift in order to introduce market observed mean reverting behavior. However, real financial markets contain prices and rates which do not move smoothly through time, but in fact jump to different levels instantaneously. The effects of these types of price movements can be seen in the market prices for options. Indeed, the importance of introducing a jump component in modeling stock price dynamics has been noted by experts in the field, who argue that pure diffusion-based models have difficulties in explaining the very steep smile effect in short-dated option prices. Thus a concerted effort has been made to design models which admit price jumps, and Poisson-type jump components in jump diffusion models are designed to address these concerns.

The variance gamma (VG) process is a pure jump process that accounts for high activity, in keeping with the normal distribution, by having an infinite number of jumps in any interval of time. Unlike many other jump models, it is not necessary to introduce a diffusion component for the VG process, as the Black–Scholes model is a parametric special case already and high activity is already accounted for. Unlike normal diffusion, the sum of absolute log price changes is finite for the VG process. Since VG has finite variation, it can be written as the difference of two increasing processes, the first of which accounts for the price increases, while the second explains the price decreases. In the case of the VG process, the two increasing processes that are subtracted to obtained the VG process are themselves gamma processes.

#### 1.2.7.1 Stochastic Differential Equation

The variance gamma process is a three parameter generalization of a Brownian motion as a model for the dynamics of the logarithm of some underlying market variable. The variance gamma process is obtained by evaluating a Brownian motion with a constant drift and constant volatility at a *random time change* given by a gamma process, that is,

$$\begin{aligned} b(t,\sigma,\theta) &= \theta t + \sigma W_t \\ X(t;\sigma,\nu,\theta) &= b(\gamma(t;1,\nu),\sigma,\theta) \\ &= \theta\gamma(t;1,\nu) + \sigma W(\gamma(t;1,\nu)) \end{aligned}$$

Each unit of calendar time may be viewed as having an economically relevant time length given by an independent random variable that has a gamma density with unit mean and positive variance, which we write as $\gamma(t;1,\nu)$. Thus we can view this model as accounting for different levels of trading activity during different time periods. As stated in [54], the economic intuition underlying the stochastic time change approach to stochastic volatility arises from the Brownian scaling property. This property relates changes in scale to changes in time and thus random changes in volatility can alternatively be captured by random changes in time. Thus the stochastic time change of the variance gamma model is an alternative way to represent stochastic volatility in a pure jump process.

Under the variance gamma model the unit period continuously compounded return is normally distributed conditional on the realization of a random process — a random time with a gamma density. The resulting process and associated pricing model provide us with a robust three parameter generalization of the standard Brownian motion model. The log of the asset price process under the variance gamma model is given by

$$\ln S_t = \ln S_0 + (r-q+\omega)t + X(t;\sigma,\nu,\theta)$$

or equivalently

$$S_t = S_0 e^{(r-q+\omega)t + X(t;\sigma,\nu,\theta)}$$

$\omega$ is determined so that

$$\mathbb{E}(S_t) = S_0 e^{(r-q)t}$$

The density of the log asset price under the variance gamma model at time $t$ can be expressed conditional on the realization of gamma time change $g$ as a normal density function. The unconditional density may then be obtained on integrating out $g$.

$$\begin{aligned} f(x;\sigma,\nu,\theta) &= \int_0^\infty \phi(\theta g, \sigma^2 g) \times \text{gamma}(\frac{t}{\nu}, \nu) dg \\ &= \int_0^\infty \frac{1}{\sigma\sqrt{2\pi g}} \exp(-\frac{(x-\theta g)^2}{2\sigma^2 g}) \frac{g^{t/\nu-1} e^{-g/\nu}}{\nu^{t/\nu}\Gamma(t/\nu)} dg \end{aligned}$$

The generalization of this model allows for parameters which control not only the volatility of the Brownian motion, but also (i) *kurtosis* fat tailedness, a symmetric increase in the left and right tail probabilities, relative to the normal for the return distribution and (ii) *skewness* that allows for the asymmetry of the left and right tails of the return density.

An additional attractive feature of VG is that it nests the lognormal density and the Black–Scholes formula as a parametric special case.

#### 1.2.7.2 Characteristic Function

The characteristic function of a VG process can be obtained by first conditioning on the gamma time $g$.

$$\begin{aligned} \mathbb{E}(e^{iuX_t}|g) &= \mathbb{E}\left(e^{iu(\theta g + \sigma W_g)}\right) \\ &= e^{iu\theta g}\mathbb{E}\left(e^{iu\sigma W_g}\right) \\ &= e^{iu\theta g}\mathbb{E}\left(e^{iu\sigma\sqrt{g}Z}\right) \\ &= e^{iu\theta g} e^{-\frac{(u\sigma\sqrt{g})^2}{2}} \\ &= e^{iu\theta g} e^{\frac{-u^2\sigma^2 g}{2}} \\ &= e^{i(u\theta + i\frac{u^2\sigma^2}{2})g} \end{aligned}$$

Now to calculate the characteristic function of a VG process, we have to integrate over $g$.

$$\begin{aligned} \mathbb{E}(e^{iuX_t}) &= \mathbb{E}_g(e^{i(u\theta + i\frac{u^2\sigma^2}{2})g}) \\ &= \int_0^\infty e^{iu\theta g} e^{\frac{-u^2\sigma^2 g}{2}} \frac{g^{t/\nu-1} e^{-g/\nu}}{\nu^{t/\nu}\Gamma(t/\nu)} dg \end{aligned}$$

which is the characteristic function of a gamma process with shape parameter $\frac{t}{\nu}$ and scale parameter $\nu$ evaluated at $u\theta + i\frac{u^2\sigma^2}{2}$. Following expression in Equation (1.12) we obtain

$$\begin{aligned} \mathbb{E}_g\left(e^{i(u\theta + i\frac{u^2\sigma^2}{2})g}\right) &= \left(\frac{\frac{1}{\nu}}{\frac{1}{\nu} - i(u\theta + i\frac{u^2\sigma^2}{2})}\right)^{\frac{t}{\nu}} \\ &= \left(\frac{1}{1 - iu\theta\nu + \frac{u^2\sigma^2\nu}{2}}\right)^{\frac{t}{\nu}} \end{aligned} \tag{1.32}$$

Therefore the characteristic function of the VG process with parameters $\sigma, \nu$, and $\theta$ at time $t$ is

$$\mathbb{E}(e^{iuX(t)}) = \left(\frac{1}{1 - iu\theta\nu + \sigma^2 u^2\nu/2}\right)^{\frac{t}{\nu}} \tag{1.33}$$

In the following chapters we will cover pricing derivatives under the VG model both analytically via transform methods and numerically via partial integro-differential equation solutions depending on the type of the option under consideration. Yet at this point it should be obvious that pricing European options under the VG model involves first conditioning on the random time $g$, then simply using a Black–Scholes type formula to solve for the conditional option value. Thus, the VG European option price, $C(S_0, K, T)$. is obtained on integrating with respect to the gamma density.

$$C(S_0, K, T) = \int_0^\infty \text{Black–Scholes}(S_0, K, g) \frac{g^{t/\nu-1} e^{-g/\nu}}{\nu^{t/\nu} \Gamma(t/\nu)} dg$$

Also, by applying equations (1.13) and (1.14) from the Lévy–Khintchine theorem [148] it can be shown that the Lévy measure for the variance gamma process can be written as $d\nu(x) = k(x)dx$ where $k(x)$ is given by

$$\begin{aligned}
d\nu(x) &= k(x)dx \\
k(x) &= \frac{e^{-\lambda_p x}}{\nu x} \mathbb{1}_{x>0} + \frac{e^{-\lambda_n |x|}}{\nu |x|} \mathbb{1}_{x<0} \\
\lambda_p &= \left( \frac{\theta^2}{\sigma^4} + \frac{2}{\sigma^2 \nu} \right)^{\frac{1}{2}} - \frac{\theta}{\sigma^2} \\
\lambda_n &= \left( \frac{\theta^2}{\sigma^4} + \frac{2}{\sigma^2 \nu} \right)^{\frac{1}{2}} + \frac{\theta}{\sigma^2}
\end{aligned}$$

### 1.2.8 CGMY Model

In this book we consider many different models, some pure diffusion models (e.g., the Black–Scholes model), some pure jump models (e.g., the VG model), and some which combine the two. The CGMY model attempts to accommodate all of these behaviors by introducing a model parameterized in such a way to allow pure diffusion or pure jumps, infinite or finite variation, and infinite or finite arrival rates.

The CGMY process [53] is defined by its Lévy measure, which can be written as $d\nu(x) = k(x)dx$ where $k(x)$ is written as

$$\begin{aligned}
d\nu(x) &= k(x)dx \\
k(x) &= C \frac{e^{-Gx}}{x^{1+Y}} \mathbb{1}_{x>0} + \frac{e^{-M|x|}}{|x|^{1+Y}} \mathbb{1}_{x<0}
\end{aligned}$$

for constants $C > 0$, $G \geq 0$, $M \geq 0$ and $Y < 2$.

We can demonstrate that CGMY generalizes Kou's jump diffusion model [166] ($Y = -1$), and the variance gamma model [175] ($Y = 0$). The CGMY process is a particular case of the Kobol process studied by Boyarchenko and Levendorskii in [36] and Carr, Geman, Madan, and Yor in [54], where constant $C$ is allowed to take on different values on the positive and negative semi axes. The extension to VG is very interesting as it allows for control of the sign of large and small jumps.

By raising $Y$ above zero, one may induce greater activity near zero and less activity further away from zero. There are also some critical values of $Y$ which are of interest: (a) $Y = 1$ separates finite variation $Y < 1$ from $Y > 1$ infinite variation, (b) $Y = 0$ separates finite arrival rate $Y < 0$ from $Y > 0$ infinite arrival rate, (c) $Y = -1$ separates activity concentrated away from zero $Y < -1$ from $Y > -1$ activity concentrated at zero. For $Y > -1$ we have a completely monotone Lévy measure [34].

#### 1.2.8.1 Characteristic Function

The CGMY process model is too general to be described by a single SDE; its description is known only through its characteristic function. The characteristic function of the CGMY process with parameters $C, G, M$, and $Y$ can be computed explicitly as [53]

$$\mathbb{E}\left[e^{iuX_t}\right] = e^{Ct\Gamma(-Y)\left((M-iu)^Y - M^Y + (G+iu)^Y - G^Y\right)}$$

### 1.2.9 Normal Inverse Gaussian Model

The normal inverse Gaussian distribution and process were introduced by Barndorff-Nielsen in [27] and [28]. The process is a time-changed Brownian motion with drift, where the time change is generated via an *inverse* gamma process, in contrast to the VG model, which uses a gamma process. This makes NIG a pure-jump Lévy process with infinite variation, unlike the VG process, which has finite variation. The parameters of this process are the drift and the volatility of the Brownian motion and the variance of the inverse Gaussian distribution whose expectation is assumed to be one. In the limiting case, when the variance is set to zero the NIG process coincides with Brownian motion and the probability density is normal. For other values of the variance, the NIG probability density has nonzero excess kurtosis and skewness similar to variance gamma. Thus in most cases the tails of the NIG distribution decrease more slowly than the normal distribution.

#### 1.2.9.1 Characteristic Function

The characteristic function of the NIG process with parameters $\sigma, \nu$, and $\theta$ is shown to be:

$$\mathbb{E}\left[e^{iuX_t}\right] = e^{\left(\nu - \sigma\sqrt{\frac{\nu^2}{\sigma^2} + \frac{\theta^2}{\sigma^4} - \left(\frac{\theta}{\sigma^2} + iu\right)^2}\right)t}$$

### 1.2.10 Variance Gamma with Stochastic Arrival (VGSA) Model

As discussed in section (1.2.7), the variance gamma model implements stochastic volatility through the use of the stochastic time change. However, the stochastic volatility in the VG model does not allow for volatility clustering, which is a feature of asset prices in many different markets. Volatility clustering can only be achieved in this type of model if random time changes persist, which requires that the rate of time change be mean reverting. The classic example of a mean-reverting positive process is the Cox–Ingersoll–Ross (CIR) process discussed previously. To allow for clustering, the variance gamma with stochastic arrival (VGSA) model was developed in [54]. We construct the VGSA process by taking the VG process, which is a homogeneous Lévy process, and build in stochastic volatility by evaluating it at a continuous time change given by the integral of a Cox–Ingersoll–Ross [82] (CIR) process representing the instantaneous time change. The mean reversion of the CIR process introduces the clustering phenomena often referred to as volatility persistence. This enables us to calibrate to option price surfaces across both strike and maturity simultaneously, unlike the VG model, which we can only calibrate across strike for a fixed maturity. The VGSA process also admits an analytical expression for its characteristic function.

#### 1.2.10.1 Stochastic Differential Equation

As shown earlier, we define the CIR process $y(t)$ as the solution to the stochastic differential equation

$$dy_t = \kappa(\eta - y_t)dt + \lambda\sqrt{y_t}dW_t$$

where $W(t)$ is a Brownian motion, $\eta$ is the long-term rate of time change, $\kappa$ is the rate of mean reversion, and $\lambda$ is the volatility of the time change. The process $y(t)$ is the instantaneous rate of time change and so the time change is given by $Y(t)$ where

$$Y(t) = \int_0^t y(u)du$$

The SDE of the log of the market variable is the same as the VG process with the above time change.

#### 1.2.10.2 Characteristic Function

As shown in Section (1.2.6.2), the characteristic function for the time change $Y(t)$ is given by

$$\begin{aligned} \mathbb{E}(e^{iuY(t)}) &= \phi(u, t, y(0), \kappa, \eta, \lambda) \\ &= A(t, u)e^{B(t,u)y(0)} \end{aligned}$$

where

$$\begin{aligned} A(t, u) &= \frac{\exp\left(\frac{\kappa^2 \eta t}{\lambda^2}\right)}{\left(\cosh(\gamma t/2) + \frac{\kappa}{\gamma}\sinh(\gamma t/2)\right)^{2\kappa\eta/\lambda^2}} \\ B(t, u) &= \frac{2iu}{\kappa + \gamma\coth(\gamma t/2)} \end{aligned}$$

with

$$\gamma = \sqrt{\kappa^2 - 2\lambda^2 iu}$$

The stochastic volatility Lévy process, termed the VGSA process, is defined by

$$Z_{VGSA}(t) = X_{VG}(Y(t); \sigma, \nu, \theta)$$

where $\sigma$, $\nu$, $\theta$, $\kappa$, $\eta$, and $\lambda$ are the six parameters defining the process. Its characteristic function is given by

$$\mathbb{E}(e^{iuZ_{VGSA}(t)}) = \phi(-i\Psi_{VG}(u), t, \frac{1}{\nu}, \kappa, \eta, \lambda)$$

where $\Psi_{VG}$ is the log characteristic function of the variance gamma process at unit time, namely,

$$\Psi_{VG}(u) = -\frac{1}{\nu}\log\left(1 - iu\theta\nu + \sigma^2\nu u^2/2\right)$$

We define the asset price process at time $t$ as follows:

$$S(t) = S(0)\frac{e^{(r-q)t+Z(t)}}{\mathbb{E}[e^{Z(t)}]}$$

We note that

$$\mathbb{E}[e^{Z(t)}] = \phi(-i\Psi_{VG}(-i), t, \frac{1}{\nu}, \kappa, \eta, \lambda)$$

Therefore the characteristic function of the log of the asset price at time $t$ is given by

$$\mathbb{E}[e^{iu \log S_t}] = \exp(iu(\log S_0 + (r-q)t)) \times \frac{\phi(-i\Psi_{VG}(u), t, \frac{1}{\nu}, \kappa, \eta, \lambda)}{\phi(-i\Psi_{VG}(-i), t, \frac{1}{\nu}, \kappa, \eta, \lambda)^{iu}}$$

Thus we have a closed form for the VGSA characteristic function for the log asset price.

---

## 1.3 Valuing Derivatives under Various Measures

### 1.3.1 Pricing under the Risk-Neutral Measure

In the preceding sections we have described a number of different models for asset prices and their various representations. However, valuing derivatives requires more than a model of asset prices. The value of a derivative can be calculated as the expectation of the derivative payoff over all possible asset price paths which affect the payoff. The measure under which this expectation is taken is critical, determining whether the pricing of derivatives is in line with the standard no-arbitrage assumptions present in almost all models of derivative pricing. The fundamental theorem of asset pricing tells us that a complete market is arbitrage free if and only if there exists at least one risk-neutral probability measure. Under this measure all assets have an expected return which is equal to the risk-free rate.

The history of the development of risk-neutral pricing is one that spans decades and largely follows the development of quantitative finance. We will not present a full account of these developments in this text, but refer the reader to [99] and [208] for more exposition. The most general expression of risk-neutral pricing for a derivative whose payoff depends only on the terminal price of the underlying asset can be stated as follows:

$S_T$ is the $T$-time price of the underlying security

$f_T(S) \equiv f(S_T|S_0)$ is the risk-neutral density of $S_T$

$V(S_T)$ is the payoff of a derivative with maturity $T$

$C_T$ is the price of a $T$-maturity derivative with payoff $V(S_T)$

We can express the derivative price using risk-neutral pricing as follows:

$$\begin{aligned} C_T &= e^{-rT}\mathbb{E}^{\mathbb{Q}}[V(S_T)] \\ &= e^{-rT}\int_{-\infty}^{\infty} V(S_T) f_T(S_T) dS_T \end{aligned}$$

The first derivative to be priced in the risk-neutral framework was the European call option and we will illustrate its construction here. Let $\mathbb{Q}$ be the equivalent martingale measure corresponding to taking the cash account, $B_t = e^{\int_0^t r_u du}$, as numeraire. That means under $\mathbb{Q}$ any traded security deflated by $B_t$ is a martingale or equivalently that any security has a return equal to the cash account. This implies a call price at time $t$ with maturity $T$ and strike $K$ is

$$\frac{C_t(K)}{B_t} = \mathbb{E}_t^{\mathbb{Q}}\left(\frac{(S_T - K)^+}{B_T}\right)$$

where $S_T$ is the time-$T$ level of the underlying process. Assuming a constant risk-free interest rate, a call price at time $t$ can be written as

$$C_t(K) = e^{-r(T-t)}\mathbb{E}_t^{\mathbb{Q}}\left((S_T - K)^+\right) \tag{1.34}$$

### 1.3.2 Change of Probability Measure

The risk-neutral measure provides the fundamental link between the no-arbitrage condition in a complete market and the pricing of derivatives. However, for many pricing algorithms it is inconvenient to work in the risk-neutral measure directly. In this case, we apply a change of measure in order to take expectations in a more convenient measure while still remaining consistent with risk-neutral pricing.

Let $\mathbb{Q}$ be a given probability measure and $M_t$ a strictly positive $\mathbb{Q}$-martingale such that $\mathbb{E}^{\mathbb{Q}}[M_t] = 1$ for all $t \in [0\ T]$. We may then define a new equivalent probability measure, $\mathbb{P}$, by defining

$$\mathbb{P}(A) = \mathbb{E}^{\mathbb{Q}}[M_T \mathbb{1}_A] = \int M_T(\omega)\mathbb{1}_A d\mathbb{Q}(\omega) = \int_A M_T(\omega)d\mathbb{Q}(\omega)$$

or in short hand notation $d\mathbb{P} = M_T d\mathbb{Q}$, noting that $\mathbb{P}(\Omega) = 1$. Expectations with respect to $\mathbb{P}$ then satisfy

$$\begin{aligned}
\mathbb{E}^{\mathbb{P}}(X) &= \int X(\omega)d\mathbb{P}(\omega) \\
&= \int X(\omega)M_T(\omega)d\mathbb{Q}(\omega) \\
&= \mathbb{E}^{\mathbb{Q}}[M_T X]
\end{aligned}$$

When we define a change in measure this way, we use the notation $\frac{d\mathbb{P}}{d\mathbb{Q}}$ to refer to $M_T$ so that we often write

$$\mathbb{E}^{\mathbb{P}}(X) = \mathbb{E}^{\mathbb{Q}}\left(\frac{d\mathbb{P}}{d\mathbb{Q}}X\right)$$

The following result explains how to switch between $\mathbb{Q}$ and $\mathbb{P}$ when we are taking conditional expectations.

$$\begin{aligned}
\mathbb{E}_t^{\mathbb{P}}(X) &= \frac{\mathbb{E}_t^{\mathbb{Q}}\left(\frac{d\mathbb{P}}{d\mathbb{Q}}X\right)}{\mathbb{E}_t^{\mathbb{Q}}\left(\frac{d\mathbb{P}}{d\mathbb{Q}}\right)} \\
&= \frac{\mathbb{E}_t^{\mathbb{Q}}\left(\frac{d\mathbb{P}}{d\mathbb{Q}}X\right)}{\mathbb{E}_t^{\mathbb{Q}}(M_T)} \\
&= \frac{\mathbb{E}_t^{\mathbb{Q}}\left(\frac{d\mathbb{P}}{d\mathbb{Q}}X\right)}{M_t}
\end{aligned}$$

given that $M_t$ is a $\mathbb{Q}$-martingale.

### 1.3.3 Pricing under Forward Measure

While the risk-neutral measure is the most common measure used in derivatives pricing, it is not the only one. Derivatives pricing under models with a stochastic interest rate is made tractable when we can eliminate all terms but $(X_T - K)^+$ (for a call option) in the expectation under $\mathbb{Q}$. Thus we need to do some manipulation in order to get rid of any terms inside the expectation but this one and we accomplish this by means of change of measure.

We start with the assumption that $P(t, T)$ is the time $t$ price of a zero-coupon bond maturing at time $T \geq t$ with face value \$1. We assume $B_0 = \$1$ and now use $P(t, T)$ as a numeraire to define a new probability measure; therefore we can write

$$\frac{C_t}{P(t,T)} = \mathbb{E}_t^{\mathbb{P}^T}\left[\frac{C_T}{P(T,T)}\right] \tag{1.35}$$

The new probability measure $\mathbb{P}^T$ we call the $T$-forward probability measure. We can compute the change of measure from the risk-neutral measure to $\mathbb{P}^T$ by noting that

$$\mathbb{E}_0^{\mathbb{Q}}\left[\frac{P(T,T)}{B_T}\right] = \frac{P(0,T)}{B_0}$$

By the fact that $P(T, T) = 1$ and $B_0 = 1$ and $P(0, T)$ is known at time zero we get

$$\mathbb{E}_0^{\mathbb{Q}}\left[\frac{1}{B_T P(0,T)}\right] = 1 \tag{1.36}$$

with $\frac{1}{B_T P(0,T)} > 0$. Therefore we set

$$M_T = \frac{d\mathbb{P}^T}{d\mathbb{Q}} = \frac{1}{B_T P(0,T)}$$

We also note that

$$\mathbb{E}_t^{\mathbb{Q}}\left[\frac{P(T,T)}{B_T}\right] = \frac{P(t,T)}{B_t}$$

and again by the fact that $P(T, T) = 1$ and $P(0, T)$ is known at time zero we have

$$\mathbb{E}_t^{\mathbb{Q}}\left[\frac{1}{P(0,T)B_T}\right] = \frac{P(t,T)}{P(0,T)B_t} \tag{1.37}$$

Now let $C_t$ denote the time $t$ price of a derivative that expires at time $T$. Following the discussion in section (1.3.2) we then have

$$\begin{aligned}
\frac{C_t}{B_t} &= \mathbb{E}_t^{\mathbb{Q}}\left[\frac{C_T}{B_T}\right] \\
C_t &= B_t\mathbb{E}_t^{\mathbb{Q}}\left[\frac{C_T}{B_T}\right] \qquad (1.38)\\
&= B_t\frac{\mathbb{E}_t^{\mathbb{P}^T}\left[\frac{d\mathbb{Q}}{d\mathbb{P}^T}\frac{C_T}{B_T}\right]}{\mathbb{E}_t^{\mathbb{P}^T}\left[\frac{d\mathbb{Q}}{d\mathbb{P}^T}\right]} \\
&= B_t\frac{\mathbb{E}_t^{\mathbb{P}^T}\left[B_T P(0,T)\frac{C_T}{B_T}\right]}{\mathbb{E}_t^{\mathbb{P}^T}\left[B_T P(0,T)\right]} \\
&= B_t P(0,T)\frac{\mathbb{E}_t^{\mathbb{P}^T}\left[C_T\right]}{\mathbb{E}_t^{\mathbb{P}^T}\left[B_T P(0,T)\right]} \\
&= B_t P(0,T)\frac{\mathbb{E}_t^{\mathbb{P}^T}\left[C_T\right]}{\mathbb{E}_t^{\mathbb{Q}}\left[1\right]/\mathbb{E}_t^{\mathbb{Q}}\left[1/B_T P(0,T)\right]} \\
&= P(t,T)\mathbb{E}_t^{\mathbb{P}^T}\left[C_T\right] \qquad (1.39)
\end{aligned}$$

We can now calculate the time-$t$ value of the derivative $C_t$ either through Equation (1.38) or through Equation (1.39) where we use the cash account as numeraire. Computing $C_t$ through (1.38) is our usual method and is often very convenient. When pricing equity derivatives, for example, we usually take interest rates, and hence the cash account, to be deterministic. This means that the factor $\frac{1}{B_T}$ in (1.38) can be taken outside the expectation so only the $\mathbb{Q}$-distribution of $C_T$ is needed to compute $C_t$. When interest rates are stochastic we cannot take the factor $\frac{1}{B_T}$ outside the expectation in (1.38) and we therefore need to find the joint $\mathbb{Q}$-distribution of $(B_T, C_T)$ in order to compute $C_t$. On the other hand, if we use Equation (1.39) to compute $C_t$, then we only need the $\mathbb{P}^T$-distribution of $C_T$, regardless of whether or not interest rates are stochastic.

Working with a univariate distribution is generally much easier than working with a bivariate distribution so if we can easily find the $\mathbb{P}^T$-distribution of $C_T$, then it can often be very advantageous to work with this distribution. The forward measure is therefore particularly useful when studying term structure models.

#### 1.3.3.1 Floorlet/Caplet Price

To demonstrate the utility of the forward measure we will derive the price of a floorlet using expectation under the forward measure, which will illustrate how much more tractable the change of measure makes floorlet pricing.

We know that the forward LIBOR rate can be described in terms of the forward zero coupon bond price as

$$\text{LIBOR}(T,T+h) = \frac{1}{h}\left(\frac{1}{P(T,T+h)} - 1\right)$$

where as before $P(t,T)$ is the zero-coupon bond price at time $t$ with maturity $T$, and $\text{LIBOR}(t,T)$ is the LIBOR rate at time $t$ with period $[t,T]$.

We assume the payment is made in arrears and the notional is $L$, so the time-$t$ value of a floorlet is given by

$$\begin{aligned}
\text{floorlet}_t^i &= L\mathbb{E}_t^{\mathbb{Q}}\left[e^{-\int_t^{T+h} r(s)ds} h\left(k - \text{LIBOR}(T,T+h)\right)^+\right] \\
&= L\mathbb{E}_t^{\mathbb{Q}}\left[e^{-\int_t^{T} r(s)ds} e^{-\int_T^{T+h} r(s)ds} h(k - \text{LIBOR}\left(T,T+h\right))^+\right]
\end{aligned}$$

By the law of iterated expectations, we obtain

$$\begin{aligned}
\text{floorlet}_t^i &= \text{L}\mathbb{E}_t^{\mathbb{Q}}\left\{e^{-\int_t^T r(s)ds}\mathbb{E}_T^{\mathbb{Q}}\left[e^{-\int_T^{T+h} r(s)ds}\right] h\left[k - \frac{1}{h}(\frac{1}{P(T,T+h)} - 1)\right]^+\right\} \\
&= \text{L}\mathbb{E}_t^{\mathbb{Q}}\left\{e^{-\int_t^T r(s)ds}P(T,T+h)\left[hk - \left(\frac{1}{P(T,T+h)} - 1\right)\right]^+\right\} \\
&= \text{L}\mathbb{E}_t^{\mathbb{Q}}\left\{e^{-\int_t^T r(s)ds}\left[(1+hk)P(T,T+h) - 1\right]^+\right\} \\
&= (1+hk)\text{L}\mathbb{E}_t^{\mathbb{Q}}\left\{e^{-\int_t^T r(s)ds}\left[P(T,T+h) - \frac{1}{1+hk}\right]^+\right\}
\end{aligned}$$

Letting $k^* = \frac{1}{1+hk}$, we have

$$\text{floorlet}_t^i = (1+hk)\text{L}\mathbb{E}_t^{\mathbb{Q}}\left\{e^{-\int_t^T r(s)ds}\left[P(T,T+h) - k^*\right]^+\right\} \tag{1.40}$$

Following the previous example, by changing the $\mathbb{Q}$ measure to forward measure $\mathbb{P}^T$ we get

$$\text{floorlet}_t^i = (1+hk)\text{L}P(t,T)\mathbb{E}_t^{\mathbb{P}^T}\left[(P(T,T+h) - k^*)^+\right] \tag{1.41}$$

where $\mathbb{E}_t^{\mathbb{P}^T}[.]$ denotes the expectation under forward measure $\mathbb{P}^T$.

Define the forward discount factor $\kappa_{t,T}$ as

$$\kappa_{t,T} = \frac{P(t,T+h)}{P(t,T)} \tag{1.42}$$

$$= \frac{1}{1+hL(t,T)} \tag{1.43}$$

Then the future forward discount factor $\kappa_{T,T} = P(T,T+h)$, so

$$\text{floorlet}_t^i = (1+hk)\text{L}P(t,T)\mathbb{E}_t^{\mathbb{P}^T}\left[(\kappa_{T,T} - k^*)^+\right] \tag{1.44}$$

And so the expectation represents a call option with underlying $\kappa_{t,T}$ and strike price $k^*$ under forward measure $\mathbb{P}^T$.

### 1.3.4 Pricing under Swap Measure

Another useful measure is the swap measure $\mathbb{P}^{n+1,N}$, which uses as its numeraire $P_{n+1,N}(t)$, the forward zero coupon bond price, and which is very helpful in deriving a tractable solution for swaption pricing, hence its name. To illustrate the use of the swap measure we will construct swaption pricing as a simple vanilla option pricing problem under the swap measure, which greatly simplifies the algorithms which can be used to price swaptions.

The forward par swap rate, $y_{n,N}(t)$, is defined as

$$y_{n,N}(t) = \frac{P(t,T_n) - P(t,T_N)}{\sum_{j=n+1}^{N}\delta P(t,T_j)} = \frac{P(t,T_n) - P(t,T_N)}{P_{n+1,N}(t)}$$

where $P_{n+1,N}(t)$ is called the *present value of a basis point* (PVBP).

A swaption gives the holder the right, but not the obligation, to enter into a particular swap contract. A swaption with option maturity $T_n$ and swap maturity $T_N$ is termed a $T_n \times T_N$-swaption. The total time-swap associated with the swaption is then $T_n + T_N$. A payer swaption gives the holder the right, not the obligation, to enter into a payer swap and can be seen as a call option on a swap rate. The option has the payoff at time $T_n$, the option maturity, of

$$\begin{aligned}
\left[V_{n,N}^{Payer}(T_n)\right]^+ &= [\{1 - P(T_n, T_N)\} - \kappa \sum_{j=n+1}^{N} \delta P(T_n, T_j)]^+ \\
&= [y_{n,N}(T_n)P_{n+1,N}(T_n) - \kappa P_{n+1,N}(T_n)]^+ \\
&= P_{n+1,N}(T_n)\,[y_{n,N}(T_n) - \kappa]^+
\end{aligned}$$

where $\kappa$ denotes the strike rate of the swaption. The second line follows directly from the definition of the forward swap rate. Let $B_t = \exp(\int_0^t r_s ds)$ be the money market account at time $t$. Assuming absence of arbitrage, the value of a payer swaption at time $t < T_n$ denoted by $\mathbf{PS}_t$ can be expressed by the following risk-neutral conditional expectation:

$$\begin{aligned}
\frac{\mathbf{PS}_t}{B_t} &= \mathbb{E}_t^{\mathbb{Q}}\left\{\frac{\left[V_{n,N}^{Payer}(T_n)\right]^+}{B_{T_n}}\right\} \\
\frac{\mathbf{PS}_t}{B_t} &= \mathbb{E}_t^{\mathbb{Q}}\left\{\frac{P_{n+1,N}(T_n)}{B_{T_n}}\,[y_{n,N}(T_n) - K]^+\right\}
\end{aligned}$$

We use $P_{n+1,N}(t)$ as a numeraire to find a new probability measure, $\mathbb{P}^{n+1,N}$, that we call the *swap measure*. Under the swap measure we can show that

$$\mathbf{PS}_t = P_{n+1,N}(t)\mathbb{E}_t^{\mathbb{P}^{n+1,N}}\left\{[y_{n,N}(T_n) - K]^+\right\}$$

Note that under this swap measure the corresponding swap rate, $y_{n,N}(t)$, is a martingale. The change of numeraire shows explicitly why swaptions can be viewed as a call option on swap rates.

---

## 1.4 Types of Derivatives

There are many types of derivatives: European versus American, path-dependent versus non-path-dependent, all with very different payoff structures. We will not attempt to provide a full description of all the different derivative structures in this section, but will instead provide some general guidance on which types of derivatives can be priced under the different methods for derivative pricing discussed in this book. In later sections, when discussing which methods are applicable to each derivative type, we will provide details on why particular methods are well suited to certain derivatives.

In the following chapters we will be presenting three major methods for pricing derivatives, namely, (a) transform methods, (b) numerical solution of partial differential equations (PDEs) and partial integro-differential equations (PIDEs), and (c) Monte Carlo simulation. We will explain how each method can be applied to different models and different products. For example, if we have the characteristic function of the log of the underlying asset, we can use transform methods to price many derivatives depending on their payoff structure.

If the process is Markov but either there is no characteristic function or the derivative price has path dependency, we can use numerical solutions to PDEs/PIDEs for pricing. In case the process is non-Markov or high dimensional, or the derivative price and the payoff has very complex path dependency, then we must use Monte Carlo simulation methods.

---

## Problems

1. Derive the characteristic function for $r_t = \ln(S_t/S_0)$ where there are two possibilities for $r_t$: (a) $a\%$ with probability 0.52 (b) $-a\%$ with probability 0.48.

2. Derive the characteristic function of a normal inverse Gaussian (NIG) process using a similar approach used to derive the characteristic function of the variance gamma process.

3. An alternative and easy way of deriving the characteristic function of the Heston stochastic volatility model is first to

   (a) show that the Heston stochastic volatility model is geometric Brownian motion with stochastic arrival (hence Heston stochastic volatility can be called GBMSA).

   (b) After verifying that, utilize the approach that was used in deriving the characteristic function for VGSA to calculate the characteristic function of the log of the underlying process under Heston stochastic volatility.

4. Having the characteristic function of normal inverse Gaussian from Problem 2, utilize the approach that was used in deriving the characteristic function for VGSA to calculate the characteristic function of the log of the underlying process under normal inverse Gaussian with stochastic arrival (NIGSA).

5. The characteristic function of CGMY is given in Section 1.2.8. Utilize the approach that was used in deriving the characteristic function for VGSA to calculate the characteristic function of the log of the underlying process under CGMYSA.

# Chapter 2

## Derivatives Pricing via Transform Techniques

In this chapter, we will discuss the use of *transform techniques* for pricing derivatives. As discussed in Section 1.1, one of the primary representations of the distribution of prices for a given asset is its characteristic function. The characteristic function of the distribution of asset prices is merely the Fourier transform of its probability distribution function (PDF). Thus its probability distribution function can be recovered from the characteristic function through Fourier inversion. This is particularly important for many classes of models which, as discussed in Section 1.2, have a closed form only in their characteristic function representation. We will outline techniques for pricing derivatives under a variety of different models using transform methods, focusing on fast Fourier transform (FFT) based techniques, fractional fast Fourier transforms, and the recently developed Fourier cosine (COS) method. Finally we will consider the saddlepoint method.

### 2.1 Derivatives Pricing via the Fast Fourier Transform

The first major development in the pricing of derivatives using Fourier techniques was proposed by Carr and Madan [60]. This technique [1] involves first deriving the Fourier transform of the expected value of the derivative under the risk-neutral distribution. We can then express this transform in terms of a known characteristic function and some constants. Finally we can apply the inverse Fourier transform to recover the derivative price.

While this method was a considerable breakthrough in numerical options pricing, like most of the methods discussed in this book, the fast Fourier transform (FFT) pricing method involves a number of trade-offs. This method is very useful as it allows us to efficiently price derivatives under any model with a known characteristic function, which encompasses most of the models discussed in Section 1.2, some of which are only expressible in this form. Also, this method is very fast when using FFT based Fourier inversion, solving derivatives pricing problems in $O(n \ln(n))$ time. Further, this method also allows us to compute not just the desired option price in $O(n \ln(n))$ time, but also the price for options at $n$ different strikes. While there are some restrictions on which option prices are computed for *free*, we are able to extract more information from this method than many others, which is important for calibration.

However, this method cannot be used to price all of the derivatives discussed in Section 1.4. In particular, the method as originally presented is restricted to the pricing of derivatives with European payoffs which are completely path independent. Furthermore, the derivation

[1] Prior to work by Carr and Madan [60], Bakshi et al. [23] had developed a pricing algorithm which involved calculating risk-neutral probabilities $\Pi_1$ and $\Pi_1$ that are recovered from inverting the respective characteristic functions. Pricing in this framework involves two inversions as opposed to one in [60] and calculation of the characteristic functions is not as straightforward as in the case of [60]. Similar treatments can be seen in [134], [29] and [198].

of this method is very dependent on the payoff type, with only two payoffs presented in the original paper. Thus we are restricted to a small, but important, subset of derivative payoffs. Also, to make this method work we need to define a damping factor $\alpha$ whose optimal value must be determined. Finally, this method degrades in accuracy when the option to be priced becomes very far out-of-the-money.

**Models:**

All models for which a characteristic function for the asset price distribution exists.

**Option Types:**

Strictly path independent European options. Restricted set of terminal payoffs.

**Pros**

1. Allows for pricing under any model with a characteristic function
2. Fast, $n$ option prices in $O(n \ln(n))$ time
3. Generates $n$ option prices in a single run

**Cons**

1. Restricted to path independent European options
2. Restricted set of terminal payoffs, each needing to be rederived
3. Requires estimation of proper $\alpha$
4. Inaccurate for highly out-of-the-money options

### 2.1.1 Call Option Pricing via the Fourier Transform

As we observed in Section 1.3, for a security which has a risk-neutral price distribution with a known probability density function we can integrate the payoff via some numerical integration procedure and get its option price. In most cases, we do not know the probability density function analytically or in an integrated form. However, we can often find the characteristic function of an underlying security price or rather the characteristic function of the log of the underlying security price analytically or semi-analytically. It is shown in [60] that if we have the characteristic function analytically, we can efficiently obtain option premiums via the inverse Fourier transform. Following the work in [60], we begin by formulating the option pricing problem for a European call in terms of the density of the log asset price, which allows us to use Fourier transforms to obtain the option premium.

As shown in Section 1.3, many derivative instruments, including vanilla options, caps, floors, and swaptions, can be expressed as a simple call or put option. For that reason, our setup is presented in a very generic form.

We begin with the following definitions. Let

$X_T$ be $T$-time price of the underlying security

$f(X_T) \equiv f(X_T|X_0)$ be the probability density function of $X_T$ under some equivalent martingale measure

$q(x_T) \equiv q(x_T|x_0)$ be the density of the log of the underlying security $x_T = \ln(X_T)$

$k = \ln(K)$ be the log of the strike price

$C_T(k)$ be the price of a $T$-maturity call with strike $K = e^k$

$\Phi(\nu)$ be the characteristic function of the log of the underlying security $x_T$, that is,

$$\Phi(\nu) = \int_{-\infty}^{\infty} e^{i\nu x_T} q(x_T) dx_T$$

The European call option price $C_T(k)$ can be expressed as

$$\begin{aligned} \mathrm{C}\, \mathbb{E}_t\left[(X_T - K)^+\right] &= \mathrm{C} \int_K^{\infty} (X_T - K) f(X_T) dX_T \\ &= \mathrm{C} \int_k^{\infty} (e^{x_T} - e^k) q(x_T) dx_T \\ &= \mathrm{C} \int_k^{\infty} (e^x - e^k) q(x) dx \\ &= C_T(k) \end{aligned}$$

where constant coefficient $C$ depends on the equivalent martingale measure that we are taking expectation under; see Section 1.3. Note that for simplicity we drop the subscript $T$ in the last integral equation. Now that we have expressed the option price $C_T(k)$ in terms of the log price density, we use this representation to calculate the Fourier transform of $C_T(k)$, which we define as $\Psi_T(\nu)$.

$$\begin{aligned} \Psi_T(\nu) &= \int_{-\infty}^{\infty} e^{i\nu k} C_T(k) dk \\ &= \int_{-\infty}^{\infty} e^{i\nu k} \left( \mathrm{C} \int_k^{\infty} (e^x - e^k) q(x) dx \right) dk \\ &= \mathrm{C} \int_{-\infty}^{\infty} \int_{-\infty}^{x} e^{i\nu k} (e^x - e^k) q(x) dk dx \\ &= \mathrm{C} \int_{-\infty}^{\infty} q(x) \left( \int_{-\infty}^{x} e^{i\nu k} (e^x - e^k) dk \right) dx \end{aligned}$$

Here we have used Fubini's theorem to change the order of integration. Now we can evaluate the inner integral

$$\int_{-\infty}^{x} e^{i\nu k} (e^x - e^k) dk = \int_{-\infty}^{x} e^{i\nu k} e^x dk - \int_{-\infty}^{x} e^{i\nu k} e^k dk = e^x \frac{e^{i\nu k}}{i\nu}\bigg|_{-\infty}^{x} - \frac{e^{(i\nu+1)k}}{i\nu + 1}\bigg|_{-\infty}^{x}$$

We see that the first integral does not converge and thus the first term is undefined. As discussed in [60], we reformulate the problem by defining

$$\mathrm{c}_T(k) = e^{\alpha k} C_T(k)$$

the option premium multiplied by an exponential of the strike. This term becomes a damping component in the inner integral which forces convergence and allows the Fourier transform to be calculable. We redefine $\Psi_T(\nu)$ to be the characteristic function of the modified option price $c_T(k)$, and the derivation now becomes

$$\begin{aligned}
\Psi_T(\nu) &= \int_{-\infty}^{\infty} e^{i\nu k} c_T(k) dk \\
&= \int_{-\infty}^{\infty} e^{i\nu k} \left( \mathrm{C} e^{\alpha k} \int_{k}^{\infty} (e^x - e^k) q(x) dx \right) dk \\
&= \mathrm{C} \int_{-\infty}^{\infty} \int_{-\infty}^{x} e^{(\alpha+i\nu)k} (e^x - e^k) q(x) dk dx \\
&= \mathrm{C} \int_{-\infty}^{\infty} q(x) \left( \int_{-\infty}^{x} e^{(\alpha+i\nu)k} (e^x - e^k) dk \right) dx
\end{aligned}$$

However, with the damping factor the inner integral now converges.

$$\begin{aligned}
\int_{-\infty}^{x} e^{\alpha k} e^{i\nu k} (e^x - e^k) dk &= \int_{-\infty}^{x} e^{(\alpha+i\nu)k} (e^x - e^k) dk \\
&= e^x \frac{e^{(\alpha+i\nu)k}}{(\alpha+i\nu)} \bigg|_{-\infty}^{x} - \frac{e^{(\alpha+i\nu+1)k}}{(\alpha+i\nu+1)} \bigg|_{-\infty}^{x}
\end{aligned}$$

Both terms now vanish at negative infinity for $\alpha > 0$, and so we have

$$\begin{aligned}
\int_{-\infty}^{x} e^{(\alpha+i\nu)k} (e^x - e^k) dk &= e^x \frac{e^{(\alpha+i\nu)x}}{(\alpha+i\nu)} - \frac{e^{(\alpha+i\nu+1)x}}{(\alpha+i\nu+1)} \\
&= \frac{e^{(\alpha+i\nu+1)x}}{(\alpha+i\nu)(\alpha+i\nu+1)}
\end{aligned}$$

Now we can compute the characteristic function of the modified option premium using the characteristic of the log asset price $\Phi(\nu)$.

$$\begin{aligned}
\Psi_T(\nu) &= \mathrm{C} \int_{-\infty}^{\infty} q(x) \frac{e^{(\alpha+i\nu+1)x}}{(\alpha+i\nu)(\alpha+i\nu+1)} dx \\
&= \frac{\mathrm{C}}{(\alpha+i\nu)(\alpha+i\nu+1)} \int_{-\infty}^{\infty} e^{(\alpha+i\nu+1)x} q(x) dx \\
&= \frac{\mathrm{C}}{(\alpha+i\nu)(\alpha+i\nu+1)} \int_{-\infty}^{\infty} e^{i(\nu-(\alpha+1)i)x} q(x) dx \\
&= \frac{\mathrm{C}}{(\alpha+i\nu)(\alpha+i\nu+1)} \Phi(\nu - (\alpha+1)i)
\end{aligned}$$

Thus, if we know the characteristic function of the log of an underlying security price, $\Phi(\nu)$, we can calculate $\Psi_T(\nu)$, the Fourier transform of the modified call,

$$\begin{aligned}
\Psi_T(\nu) &= \int_{-\infty}^{\infty} e^{i\nu k} c_T(k) dk && (2.1) \\
&= \frac{\mathrm{C}}{(\alpha+i\nu)(\alpha+i\nu+1)} \Phi(\nu - (\alpha+1)i) && (2.2)
\end{aligned}$$

Because we have the characteristic function of the modified call price

$$\begin{aligned}
\Psi_T(\nu) &= \int_{-\infty}^{\infty} e^{i\nu k} c_T(k) dk && (2.3) \\
&= \int_{-\infty}^{\infty} e^{i\nu k} e^{\alpha k} C_T(k) dk && (2.4)
\end{aligned}$$

we can use the inverse Fourier transform to get

$$C_T(k) = \frac{e^{-\alpha k}}{2\pi} \int_{-\infty}^{\infty} e^{-i\nu k} \Psi_T(\nu) d\nu \tag{2.5}$$

But $C_T(k)$ is a real number, which implies that its Fourier transform $\Psi_T(\nu)$ is even in its real part and odd in its imaginary part. Since we are only concerned with the real part for the option price, we can treat this as an even function and thus we get

$$C_T(k) = \frac{e^{-\alpha k}}{\pi} \int_{0}^{\infty} e^{-i\nu k} \Psi_T(\nu) d\nu \tag{2.6}$$

which is the call option premium. Using Equation (2.2) for the characteristic function and the inverse Fourier transform of $\Psi_T(\nu)$ we can calculate $C_T(k)$.

$$C_T(k) = \frac{e^{-\alpha k}}{\pi} \int_{0}^{\infty} e^{-i\nu k} \Psi_T(\nu) d\nu \tag{2.7}$$

where $\Psi_T(\nu)$ is a known function that will be determined and some suitable parameter $\alpha > 0$.

### 2.1.2 Put Option Pricing via the Fourier Transform

The put option price can also be calculated via the Fourier transform in a similar manner. One might wonder why this formulation is necessary as the put price should be recoverable from the call price and the forward underlier price via put-call parity. However, both put and call options have bid and ask prices, and thus put-call parity does not hold absolutely because there is no single price for the call or put, but in fact a range of possible *true* prices between the bid and ask. Thus a formulation of the put price becomes necessary. If $P_T(k)$ is the price of a $T$-maturity put with strike $K = e^k$, it can be expressed as follows:

$$\begin{aligned}
\mathrm{C}\,\mathbb{E}_t\left[(K - X_T)^+\right] &= \mathrm{C} \int_0^K (K - X_T) f(X_T) dX_T \\
&= \mathrm{C} \int_{-\infty}^{k} (e^k - e^{x_T}) q(x_T) dx_T \\
&= \mathrm{C} \int_{-\infty}^{k} (e^k - e^{x}) q(x) dx \\
&= P_T(k)
\end{aligned}$$

As before, we reformulate the problem by defining

$$p_T(k) = e^{\alpha k} P_T(k)$$

the option premium multiplied by an exponential of the strike. This term becomes a damping component in the inner integral which forces convergence and allows the Fourier transform to be calculable. We redefine $\Psi_T(\nu)$ to be the characteristic function of the modified option price $p_T(k)$, and the derivation now becomes

$$\begin{aligned}
\Psi_T(\nu) &= \int_{-\infty}^{\infty} e^{i\nu k} p_T(k) dk \\
&= \mathrm{C} \int_{-\infty}^{\infty} e^{i\nu k} \left( e^{\alpha k} \int_{-\infty}^{k} (e^k - e^x) q(x) dx \right) dk \\
&= \mathrm{C} \int_{-\infty}^{\infty} \int_{x}^{\infty} e^{(\alpha+i\nu)k} (e^k - e^x) q(x) dk dx \\
&= \mathrm{C} \int_{-\infty}^{\infty} q(x) \left( \int_{x}^{\infty} e^{(\alpha+i\nu)k} (e^k - e^x) dk \right) dx
\end{aligned}$$

With the damping factor the inner integral now converges.

$$\begin{aligned}
\int_{s}^{\infty} e^{\alpha k} e^{i\nu k} (e^k - e^s) dk &= \int_{s}^{\infty} e^{(\alpha+i\nu)k} (e^k - e^s) dk \\
&= \left. \frac{e^{(\alpha+i\nu+1)k}}{(\alpha+i\nu+1)} \right|_s^{\infty} - e^s \left. \frac{e^{(\alpha+i\nu)k}}{(\alpha+i\nu)} \right|_s^{\infty}
\end{aligned}$$

Both terms now vanish at infinity for $\alpha < 0$, and so we have

$$\begin{aligned}
\int_{x}^{\infty} e^{(\alpha+i\nu)k} (e^k - e^x) dk &= -\frac{e^{(\alpha+i\nu+1)x}}{(\alpha+i\nu+1)} + e^x \frac{e^{(\alpha+i\nu)s}}{(\alpha+i\nu)} \\
&= \frac{e^{(\alpha+i\nu+1)x}}{(\alpha+i\nu)(\alpha+i\nu+1)}
\end{aligned}$$

Now we can compute the characteristic function of the modified option premium using the characteristic of the log asset price $\Phi(\nu)$.

$$\begin{aligned}
\Psi_T(\nu) &= \mathrm{C} \int_{-\infty}^{\infty} q(x) \frac{e^{(\alpha+i\nu+1)x}}{(\alpha+i\nu)(\alpha+i\nu+1)} dx \\
&= \frac{\mathrm{C}}{(\alpha+i\nu)(\alpha+i\nu+1)} \int_{-\infty}^{\infty} e^{(\alpha+i\nu+1)x} q(x) dx \\
&= \frac{\mathrm{C}}{(\alpha+i\nu)(\alpha+i\nu+1)} \int_{-\infty}^{\infty} e^{i(\nu-(\alpha+1)i)s} q(x) dx \\
&= \frac{\mathrm{C}}{(\alpha+i\nu)(\alpha+i\nu+1)} \Phi(\nu - (\alpha+1)i)
\end{aligned}$$

Thus, if we know the characteristic function of the log of an underlying security price, $\Phi(\nu)$, we can calculate $\Psi_T(\nu)$, the Fourier transform of the modified put,

$$\Psi_T(\nu) = \int_{-\infty}^{\infty} e^{i\nu k} p_T(k) dk \tag{2.8}$$

$$= \frac{\mathrm{C}}{(\alpha+i\nu)(\alpha+i\nu+1)} \Phi(\nu - (\alpha+1)i) \tag{2.9}$$

Because we have the characteristic function of the modified put price

$$\Psi_T(\nu) = \int_{-\infty}^{\infty} e^{i\nu k} p_T(k) dk \tag{2.10}$$

$$= \int_{-\infty}^{\infty} e^{i\nu k} e^{\alpha k} P_T(k) dk \tag{2.11}$$

we can use the inverse Fourier transform to get

$$P_T(k) = \frac{e^{-\alpha k}}{2\pi} \int_{-\infty}^{\infty} e^{-i\nu k} \Psi_T(\nu) d\nu \tag{2.12}$$

Using the same argument as before, since $P_T(k)$ is a real number, this implies that its Fourier transform $\Psi_T(\nu)$ is odd in its imaginary part and even in its real part. Since we are only concerned with the real part for the option price, we can treat this as an even function and thus we get

$$P_T(k) = \frac{e^{-\alpha k}}{\pi} \int_{0}^{\infty} e^{-i\nu k} \Psi_T(\nu) d\nu \tag{2.13}$$

which is the put option premium. Using Equation (2.9) for the characteristic function and the inverse Fourier transform of $\Psi_T(\nu)$ we can calculate put option premium $P_T(k)$.

$$P_T(k) = \frac{e^{-\alpha k}}{\pi} \int_{0}^{\infty} e^{-i\nu k} \Psi_T(\nu) d\nu \tag{2.14}$$

where $\Psi_T(\nu)$ is a known function that will be determined and some suitable parameter $\alpha < 0$.

### 2.1.3 Evaluating the Pricing Integral

#### 2.1.3.1 Numerical Integration

The Fourier techniques presented thus far give us a method for calculating option prices for models where a closed form PDF is not available but where a closed form characteristic function is. However, we still need to perform the integral to solve for the option premium. It remains to be seen why we would use this method, as we still need to calculate the integral. Note that

$$C_T(k) = \frac{e^{-\alpha k}}{\pi} \int_{0}^{\infty} e^{-i\nu k} \Psi_T(\nu) d\nu \tag{2.15}$$

This integral can be computed easily using simple numerical integration techniques. First we approximate the integral by defining $B$ to be the upper bound for the integration. We can numerically integrate this truncated integral via a simple trapezoidal rule. We let $N$ be the number of equidistant intervals, $\Delta\nu = \frac{B}{N} = \eta$ be the distance between the integration points, and $\nu_j = (j-1)\eta$ be the endpoints for the integration intervals for $j = 1, \dots, N+1$. Applying the trapezoidal rule we get

$$\begin{aligned}
C_T(k) &= \frac{e^{-\alpha k}}{\pi} \int_{0}^{\infty} e^{-i\nu k} \Psi_T(\nu) d\nu \\
&\approx \frac{e^{-\alpha k}}{\pi} \int_{0}^{B} e^{-i\nu k} \Psi_T(\nu) d\nu \\
&\approx \frac{e^{-\alpha k}}{\pi} \big(e^{-i\nu_1 k} \Psi_T(\nu_1) + 2e^{-i\nu_2 k} \Psi_T(\nu_2) + \cdots + 2e^{-i\nu_N k} \Psi_T(\nu_N) \\
&\quad + e^{-i\nu_{N+1} k} \Psi_T(\nu_{N+1})\big) \frac{\eta}{2}
\end{aligned}$$

Since the terms are decaying exponentially, we can just discard the final term to make it suitable for fast Fourier transform. Thus, we end up with

$$C_T(k) \approx \frac{e^{-\alpha k}}{\pi} \sum_{j=1}^{N} e^{-i\nu_j k} \Psi_T(\nu_j) w_j$$

where $w_j = \frac{\eta}{2}(2 - \delta_{j-1})$. For a somewhat more accurate result we could also use Simpson's rule, which would yield

$$C_T(k) \approx \frac{e^{-\alpha k}}{\pi} \sum_{j=1}^{N} e^{-i\nu_j k} \Psi_T(\nu_j) w_j$$

where $w_j = \frac{\eta}{3}(3 + (-1)^j - \delta_{j-N})$. Here

$$\delta_j = \begin{cases} 1 & j = 0 \\ 0 & \text{otherwise} \end{cases}$$

#### 2.1.3.2 Fast Fourier Transform

While the direct integration is sufficient, it is not an efficient method of evaluating the pricing integral. The fast Fourier transform (FFT) algorithm developed by Cooley and Tukey [78] and later extended by many others provides a more efficient algorithm for calculating a set of discrete inverse Fourier transforms with sample points that are powers of two. These transforms take the following form:

$$\omega(m) = \sum_{j=1}^{N} e^{-i\frac{2\pi}{N}(j-1)(m-1)} x(j) \text{ for } m = 1, \ldots, N \tag{2.16}$$

These equations would appear to take $N$ multiplications per inverse transform for a total of $N^2$ multiplications; however, the Cooley–Tukey FFT algorithm can reduce this to $N \log N$ multiplications by using a divide and conquer algorithm to break down discrete Fourier transforms (DFTs). This is crucial for approximating the Fourier integral as this can greatly accelerate the speed at which we can compute option prices under the FFT method.

We can convert our option pricing formula into the FFT form by creating a range of strikes around the strike for which we wish to calculate an accurate option price. A typical case would be an at-the-money option for a particular underlier, and in this case we define the range of (log of) strikes as $k_m = \beta + (m-1)\Delta k = \beta + (m-1)\lambda$, for $m = 1, \ldots, N$ with $\beta = \ln X_0 - \frac{\lambda N}{2}$ which will cause the at-the-money strike to fall in the middle of our range of strikes. For $C_T(k_m)$ we now have

$$\begin{aligned} C_T(k_m) &\approx \frac{e^{-\alpha k_m}}{\pi} \sum_{j=1}^{N} e^{-i\nu_j k_m} \Psi_T(\nu_j) w_j \text{ for } m = 1, \ldots, N \\ &= \frac{e^{-\alpha k_m}}{\pi} \sum_{j=1}^{N} e^{-i(j-1)\eta(m-1)\Delta k} e^{-i\beta\nu_j} \Psi_T(\nu_j) w_j \\ &= \frac{e^{-\alpha k_m}}{\pi} \sum_{j=1}^{N} e^{-i\lambda\eta(j-1)(m-1)} e^{-i\beta\nu_j} \Psi_T(\nu_j) w_j \end{aligned}$$

So, we can see that if we set $\lambda\eta = \frac{2\pi}{N}$ and $x(j) = e^{-i\beta\nu_j}\Psi_T(\nu_j)w_j$ with $\beta = \ln X_0 - \frac{\lambda N}{2}$ we get back the original form of the FFT (2.16). Thus we see that we can generate $N$ option prices using only $O(N \log N)$ multiplications required by the FFT. This entire operation is slower than the $O(N)$ multiplications needed to get a simple option price using direct integration; however, it is rare to only price a single option on an underlier by itself. With the FFT method we have a clear advantage, as the $O(N \log N)$ multiplications, when amortized over $N$ options, are only $O(\log N)$ per option.

However, these $N$ options are not likely to be exactly the $N$ options wanted, say for producing a sensible implied volatility surface with points at market traded strikes. But because the FFT method prices strikes determined by $k_m = \beta + (m-1)\Delta k = \beta + (m-1)\lambda$ we can modify $\eta = \frac{B}{N}$ by modifying $N$, which will in turn change the strikes for which you get option prices using the FFT method. This allows us to extract enough information to interpolate a volatility surface with very small errors in considerably less time than direct integration, which would take $O(N^2)$ for $N$ strikes.

### 2.1.4 Implementation of Fast Fourier Transform

In brief, having the characteristic function of the log of the underlying process $X_t$ that is $\Phi(\nu)$, choose $\eta$ and $N = 2^n$, calculate $\lambda = \frac{2\pi}{N\eta}$, $\nu_j = (j-1)\eta$, and set $\alpha$. Now form vector $\mathbf{x}$.

$$\mathbf{x} = \begin{pmatrix} x_1 \\ x_2 \\ \vdots \\ x_N \end{pmatrix} = \begin{pmatrix} \frac{\eta}{2}\frac{\mathrm{C}}{(\alpha+i\nu_1)(\alpha+i\nu_1+1)}e^{-i(\ln X_0 - \frac{\lambda N}{2})\nu_1}\Phi\left(\nu_1 - (\alpha+1)i\right) \\ \eta\frac{\mathrm{C}}{(\alpha+i\nu_2)(\alpha+i\nu_2+1)}e^{-i(\ln X_0 - \frac{\lambda N}{2})\nu_2}\Phi\left(\nu_2 - (\alpha+1)i\right) \\ \vdots \\ \eta\frac{\mathrm{C}}{(\alpha+i\nu_N)(\alpha+i\nu_N+1)}e^{-i(\ln X_0 - \frac{\lambda N}{2})\nu_N}\Phi\left(\nu_N - (\alpha+1)i\right) \end{pmatrix}$$

where constant coefficient C depends on the equivalent martingale measure that we are taking expectation under; e.g., for $\mathbb{Q}$ we have $\mathrm{C} = e^{-rT}$. Vector $\mathbf{x}$ is the input to the FFT routine, and its output is vector $\mathbf{y}$ of the same size, $\mathbf{y} = \mathrm{fft}(\mathbf{x})$; then call prices at strike $k_m$ for $m = 1, \ldots, N$ are

$$\begin{pmatrix} C_T(k_1) \\ C_T(k_2) \\ \vdots \\ C_T(k_N) \end{pmatrix} = \begin{pmatrix} \frac{e^{-\alpha(\ln X_0 - \frac{N}{2}\lambda)}}{\pi}\mathrm{Re}(y_1) \\ \frac{e^{-\alpha(\ln X_0 - (\frac{N}{2}-1)\lambda)}}{\pi}\mathrm{Re}(y_2) \\ \vdots \\ \frac{e^{-\alpha(\ln X_0 - (\frac{N}{2}-(N-1))\lambda)}}{\pi}\mathrm{Re}(y_N) \end{pmatrix}$$

where $\mathrm{Re}(y_j)$ is the real part of $y_j$.

### 2.1.5 Damping factor $\alpha$

The introduction of the damping factor $\alpha$ made it possible to solve the option pricing problem via a Fourier transform. At a glance it seems that $\alpha$ does not come into the calculation of the integrand as it is hidden in $\Psi_T(\nu)$. We already know that $\alpha$ has to be positive for calls and negative for puts. Theoretically it seems that for any value of $\alpha$ we should get roughly the same results. However, this is not the case. In this section we demonstrate how sensitive the results are to the choice of $\alpha$. We look for a suitable range for its value and illustrate its dependence on the choice of stochastic model. Finally, we run series of empirical studies on three processes, focusing on pricing calls using a positive $\alpha$. The processes we will consider are (a) geometric Brownian motion, (b) the Heston stochastic volatility model, and (c) the variance gamma (VG) model.

For geometric Brownian motion we will use the following parameter set: spot $S_0 = 100$, strike $K = 90$, instantaneous risk-free interest rate $r = 5\%$, maturity $T = 1$ year, and volatility $\sigma = 30\%$. Table 2.1 illustrates Black–Scholes premiums via FFT for various values of $\alpha$, $N$, and $\eta$. The exact Black–Scholes call premium value is 19.6974 for this parameter set. For the Heston stochastic volatility model we will use the following parameter set: spot $S_0 = 100$, strike $K = 90$, risk-free rate $r = 3\%$, maturity $T = 0.5$ years, $\kappa = 2$, $\sigma = 0.5$,

**TABLE 2.1**: Black–Scholes premiums via FFT for various values of $\alpha$, $N$, and $\eta$

| | $\eta = 0.15$ | | | $\eta = 0.1$ | | | $\eta = 0.05$ | | |
|---|---|---|---|---|---|---|---|---|---|
| $\alpha$ | $N = 2^6$ | $2^8$ | $2^{10}$ | $2^6$ | $2^8$ | $2^{10}$ | $2^6$ | $2^8$ | $2^{10}$ |
| 0.01 | 211.907 | 211.91 | 211.91 | 134.048 | 134.054 | 134.054 | 60.4539 | 59.4812 | 59.4812 |
| 0.5 | 19.6922 | 19.6974 | 19.6974 | 19.6264 | 19.6974 | 19.6974 | 19.7932 | 19.6973 | 19.6974 |
| 1 | 19.6911 | 19.6974 | 19.6974 | 19.5632 | 19.6974 | 19.6974 | 18.7819 | 19.6974 | 19.6974 |
| 1.5 | 19.6919 | 19.6974 | 19.6974 | 19.5162 | 19.6974 | 19.6974 | 17.8268 | 19.6974 | 19.6974 |
| 2 | 19.6950 | 19.6974 | 19.6974 | 19.4991 | 19.6974 | 19.6974 | 17.0632 | 19.6975 | 19.6974 |
| 5 | 19.7020 | 19.6974 | 19.6974 | 20.3240 | 19.6974 | 19.6974 | 19.7178 | 19.6968 | 19.6974 |
| 10 | 20.9053 | 19.6974 | 19.6974 | 2.4214 | 19.6974 | 19.6974 | 96.4300 | 19.6968 | 19.6974 |

$\theta = 0.04$, $v_0 = 0.04$, and correlation $\rho = -0.7$. Table 2.2 illustrates Heston premiums via FFT for various values of $\alpha$, $N$, and $\eta$. For reference, the Heston call premium value for this parameter set is 13.4038 as calculated via Monte Carlo simulation. For the variance gamma

**TABLE 2.2**: Heston premiums via FFT for various values of $\alpha$, $N$, and $\eta$

| | $\eta = 0.15$ | | | $\eta = 0.1$ | | | $\eta = 0.05$ | | |
|---|---|---|---|---|---|---|---|---|---|
| $\alpha$ | $N = 2^6$ | $2^8$ | $2^{10}$ | $2^6$ | $2^8$ | $2^{10}$ | $2^6$ | $2^8$ | $2^{10}$ |
| 0.01 | 205.05 | 205.413 | 205.415 | 127.489 | 127.576 | 127.5591 | 55.6651 | 52.745 | 52.9863 |
| 0.5 | 12.7379 | 13.2001 | 13.2023 | 12.8612 | 13.2222 | 13.2023 | 14.8195 | 12.9309 | 13.2025 |
| 1 | 12.6307 | 13.1998 | 13.2023 | 12.5494 | 13.2251 | 13.2023 | 13.5390 | 12.9010 | 13.2026 |
| 1.5 | 12.5201 | 13.1994 | 13.2023 | 12.2137 | 13.2281 | 13.2023 | 12.2191 | 12.8731 | 13.2027 |
| 2 | 12.4092 | 13.1990 | 13.2023 | 11.8667 | 13.2313 | 13.2023 | 10.9782 | 12.8484 | 13.2027 |
| 5 | 11.9182 | 13.1958 | 13.2023 | 10.1369 | 13.2509 | 13.2023 | 6.6138 | 12.8410 | 13.2032 |
| 10 | 12.9279 | 13.1922 | 13.2023 | 10.3712 | 13.2280 | 13.2023 | 5.9025 | 13.9336 | 13.2043 |

model we will use the following parameter set: spot $S_0 = 100$, strike $K = 90$, risk-free rate $r = 10\%$, maturity $T = 1/12$ year, $\sigma = 0.12$, $\theta = -0.14$, and $\nu = 0.2$. Table 2.3 illustrates variance gamma premiums via FFT for various values of $\alpha$, $N$, and $\eta$. The variance gamma call premium value for this parameter set is 10.8288. It is easy to see from Tables 2.1, 2.2,

**TABLE 2.3**: Variance gamma premiums via FFT for various values of $\alpha$, $N$, and $\eta$

| | $\eta = 0.15$ | | | $\eta = 0.1$ | | | $\eta = 0.05$ | | |
|---|---|---|---|---|---|---|---|---|---|
| $\alpha$ | $N = 2^8$ | $2^{10}$ | $2^{12}$ | $2^{18}$ | $2^{10}$ | $2^{12}$ | $2^8$ | $2^{10}$ | $2^{12}$ |
| 0.01 | 203.123 | 203.042 | 203.042 | 125.055 | 125.186 | 125.186 | 49.9824 | 50.6283 | 50.6126 |
| 0.5 | 10.9172 | 10.8286 | 10.8288 | 10.6980 | 10.8293 | 10.8288 | 10.1099 | 10.8444 | 10.8286 |
| 1 | 10.9245 | 10.8285 | 10.8288 | 10.6984 | 10.8293 | 10.8287 | 10.0150 | 10.8443 | 10.8285 |
| 1.5 | 10.9323 | 10.8285 | 10.8288 | 10.6998 | 10.8292 | 10.8287 | 9.9166 | 10.8440 | 10.8285 |
| 2 | 10.9405 | 10.8284 | 10.8288 | 10.7024 | 10.8291 | 10.8287 | 9.8154 | 10.8437 | 10.8282 |
| 5 | 11.0009 | 10.8280 | 10.8288 | 10.7483 | 10.8280 | 10.8287 | 9.2038 | 10.8382 | 10.8274 |
| 10 | 11.1477 | 10.8265 | 10.8287 | 11.0028 | 10.8239 | 10.8285 | 8.5421 | 10.8071 | 10.8223 |

and 2.3 that the results vary significantly depending on $\alpha$ and $N$. For some values of $\alpha$ they either do not make sense or are implausible. To explore why the damping factor $\alpha$

affects the results so much, we plot the integrand for different values of the parameters and examine their behaviors.

In Figure 2.1 we plot the integrand which is the Fourier transform of the call value, namely, $e^{-i\nu k}\Psi_T(\nu)$, for various values of $\alpha$. To see how fast the tail dies out we plot the

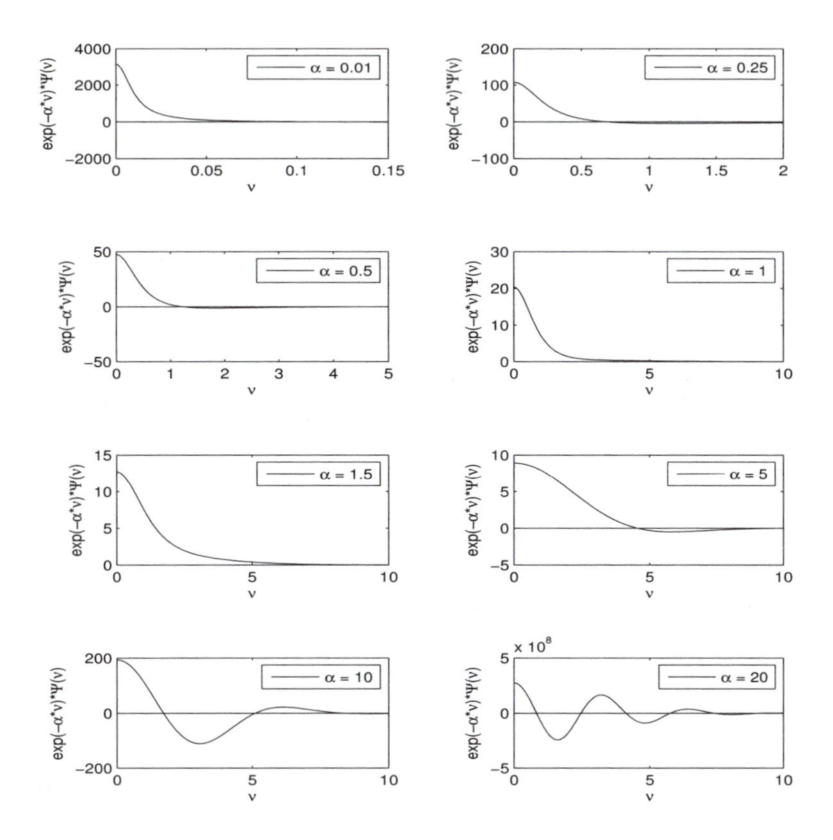


**FIGURE 2.1**: The integrand in geometric Brownian motion for various values of $\alpha$

tail of the integrand for various values of $\alpha$ in Figure 2.2.

We can see from the graphs in Figures 2.1 and 2.2 the integrand oscillates more as $\alpha$ gets larger. The most stable behavior occurs when $\alpha$ is around 1.0, which is consistent with the option pricing results. Also, the tail decays very quickly in each graph, so extending the upper bound of the integral will not have much effect on the value of the integral.

The integrand oscillates quite sharply when $\alpha$ is large, which causes poor results for the integral. But numerically, if we can do the integral more efficiently we should get the same value as we would when using a small $\alpha$. More precisely, say we apply Simpson's rule using a very small $\alpha$ to capture as much detail in the tail as possible. We then need a large $N$ to give us a proper upper bound of the integral because $N \times \eta$ gives us the upper bound of the integral.

To verify this, we take $\alpha = 15$. The graph of the integrand is shown in Figure 2.3 and the graph of its tail is illustrated in Figure 2.4. For this $\alpha$ the function oscillates wildly about zero. In this case, if we take too few intervals the result could be totally unrealistic.

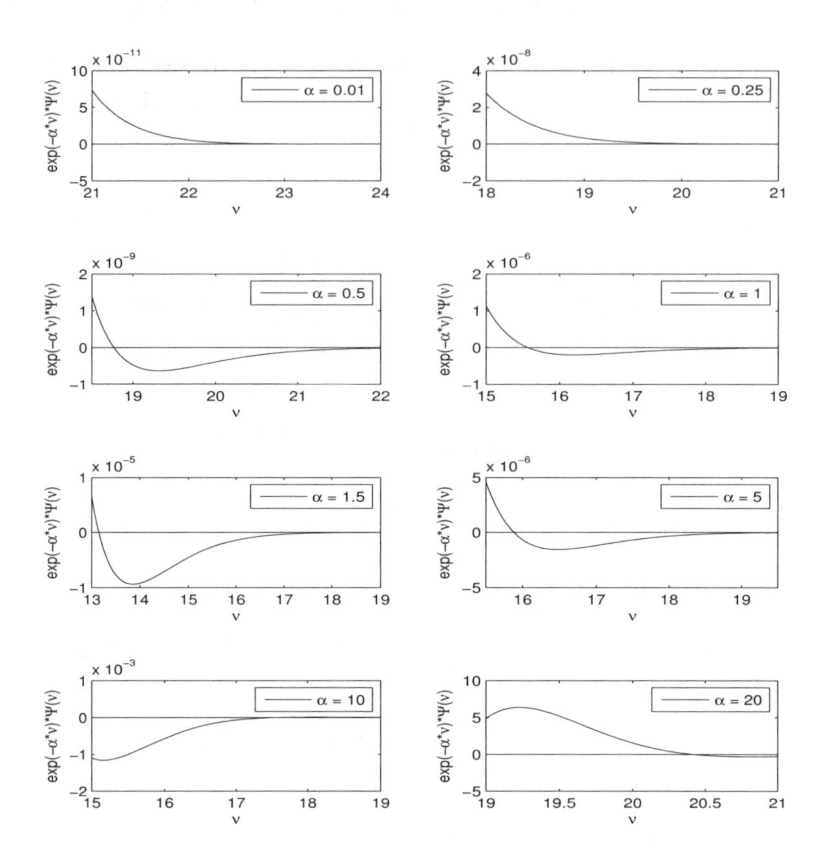


**FIGURE 2.2**: The tail of the integrand in geometric Brownian motion for various values of $\alpha$

Also, we can see a proper upper bound would be larger than 30. We test the results in the following three cases:

- $N = 2^{10}$, $\eta = 0.05$, with the corresponding upper bound of 51.2. Using the Simpson rule, the result is 1302.2, which is way off from the true value expected.
- $N = 2^{12}$, $\eta = 0.01$, with an upper bound of 40.96. The final result is 355.6856. It is getting closer to the true value but still far away from Black–Scholes value.
- $N = 2^{22}$, $\eta = 0.00001$, with an upper bound of 41.943. The final result is 20.2784, which is consistent with the result given by $\alpha = 1$.

Note that if we use the third group of parameters, $\lambda = 0.1498$, which means the log-strike spacing is too large. More precisely, very few strike prices lie in the desired region near the current stock price.

In conclusion, the FFT method will work with a large damping variable $\alpha$ if we use a large enough $N$ and small enough interval $\eta$. However, in this case we will not get a suitable strike range.

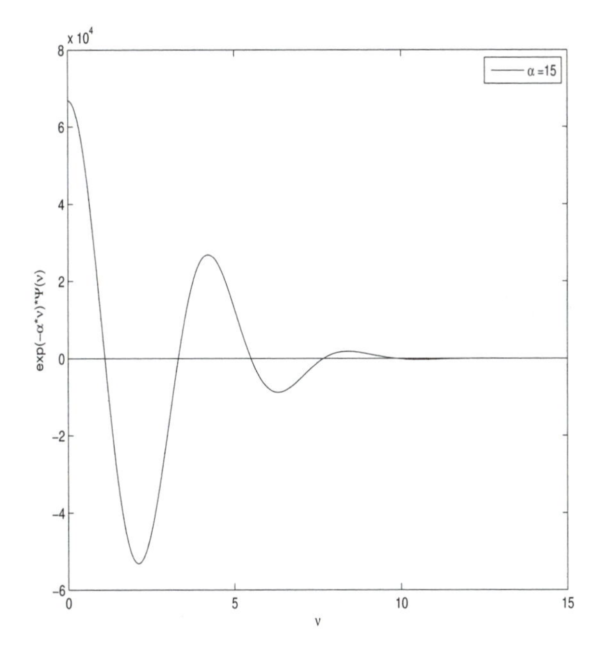


**FIGURE 2.3**: The integrand in geometric Brownian motion for $\alpha = 15$

For the Heston model, we get similar results, as shown in Figures 2.5 and 2.6. Additionally, we can see there is a much larger negative part of the integration than in the GBM model, which means if we use an upper bound for the integration which is too small we will observe higher prices than normal, which is shown when $N$ is $2^8$.

For the VG model, results are consistent with the other two models and are shown in Figures 2.7 and 2.8. We can see that the integrand decays slower when $\alpha$ gets larger, which means for fixed value $\eta$ we need even larger $N$.

Our analyses indicate when using FFT methods for option pricing, results are very sensitive to the choice of $\alpha$, $N$, and $\eta$. FFT methods will work in theory for any damping variable $\alpha$, but one has to be careful as to the choice of $N$ and $\eta$. As shown, the most accurate results are achieved by using a very large $N$ and very small interval $\eta$, but this choice of parameters is not only computationally expensive but also would not provide a useful range of strikes. From the results we can conclude that the optimal range for $\alpha$ is between 1.0 and 1.5.

---

## 2.2 Fractional Fast Fourier Transform

As explained in the previous section, the use of the standard FFT approach dictates the following relationship between $\lambda$ and $\eta$:

$$\lambda\eta = \frac{2\pi}{N} \tag{2.17}$$

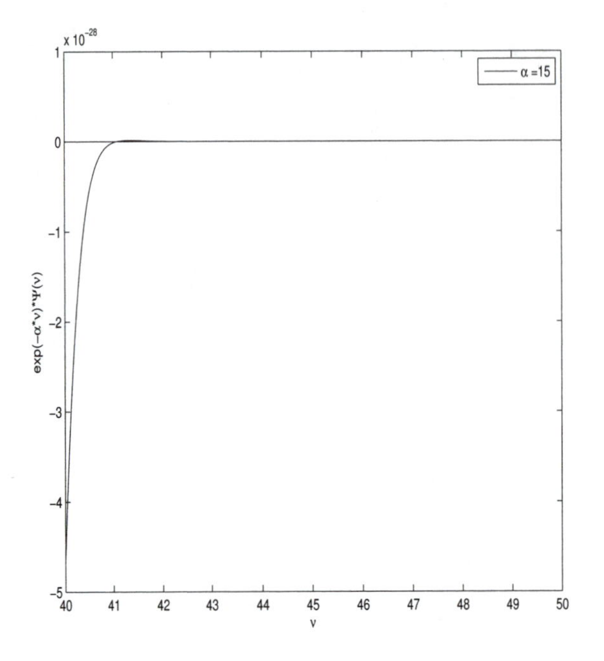


**FIGURE 2.4**: The tail of the integrand in geometric Brownian motion for $\alpha = 15$

with

$$\eta = \frac{B}{N} \tag{2.18}$$

However, we can see that of the four parameters in consideration, $N$, $B$, $\eta$, and $\lambda$, just two can be chosen freely as $\eta$ is determined by $B$ and $N$ and the last one is determined via the constraint (2.17). We further assume that we have a fixed computational budget which dictates a fixed number of integral terms $N$. Given these assumptions we have only two free variables, $B$, the upper bound of the integral, and $\lambda = \Delta k$, the spacing of the $\log(K)$ grid on which we calculate solutions, and they are inversely proportional. So we have an inherent trade-off between the upper bound of the integral, which determines the accuracy of our integration, and the step size in strikes, which will determine if we get relevant pricing information at strikes which are close to traded market strikes. The choice of $B$ will determine how accurate our integral approximation will be; however, if we assume that we want a fixed spacing between integration points $\eta$ to ensure a given degree of accuracy in the integration, we impose a restriction on the $\lambda$ which determines the spacing in the log strikes of the solutions we calculate.

As an example, we fix $\eta$ to ensure a certain degree of local accuracy in the integration, and show the implied upper bound $B$ and the implied log strike spacing $\lambda$ for different values of $N$. These results can be seen in Table 2.4. We observe that it can take quite a large $N$ to generate solutions with strikes which fall close to market traded strikes.

Our results illustrate the relationship between between grid spacing in the integral, the upper bound of the integral, and the log-strike spacing. As mentioned in [66], out of the 4096 option prices that are calculated, roughly 67 ($2 \times \frac{20\%}{0.61\%} + 1 \approx 67$) will fall within the 20% log-strike interval that is relevant for practical applications and thus the remaining option prices calculated are of no practical use. Considering the 4096 option prices calculated in our experiment, were we to price them individually using the FFT algorithm with a log-strike

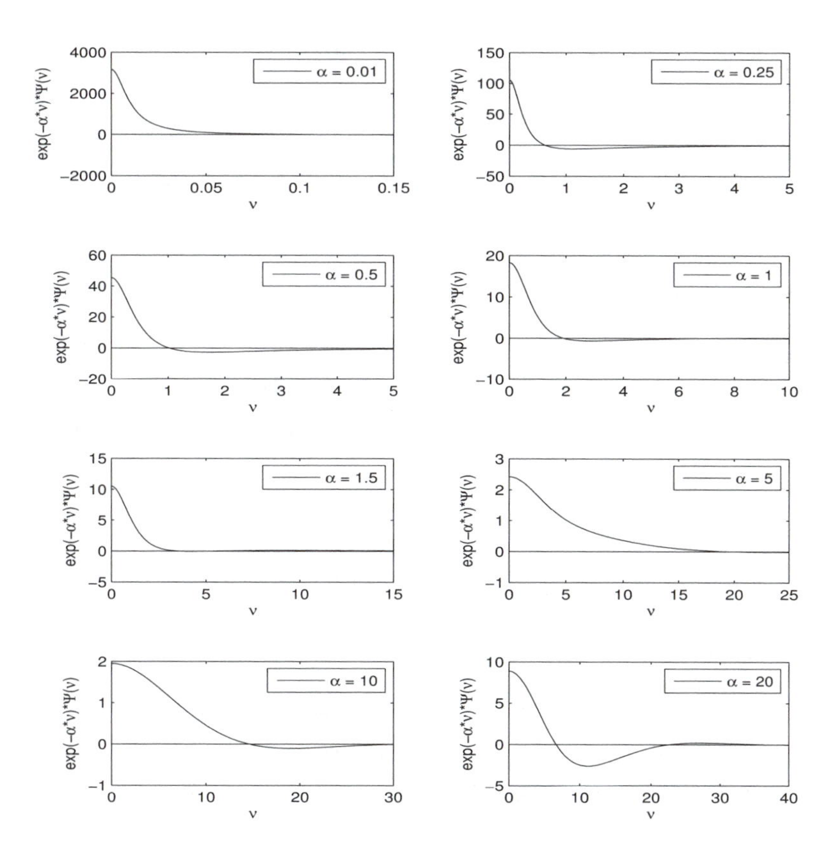


**FIGURE 2.5**: The integrand in the Heston stochastic volatility model for various values of $\alpha$

scale centered around the strike of each option, we would need only a 128– or 256-point grid to get practically accurate prices. In other words, using only 128 or 256 grid points will generally yield acceptably accurate option prices for a given central strike. However, using a 256-point grid in our example would result in $\lambda = 0.0981$ or a log-strike spacing of 9.81%, which is impractical. We would like to eliminate the dependence between the number of terms, $N$, and log-strike spacing, $\lambda$. This would allow us to use a smaller $N$, which will still yield an accurate set of option prices, and to independently choose a log-strike spacing $\lambda$ that is consistent with market traded options. In [66], the author proposes a fractional FFT procedure to achieve this. The fractional FFT procedure computes a sum of the form

$$\sum_{j=1}^{N} e^{-i2\pi\gamma(j-1)(m-1)}x(j) \tag{2.19}$$

for any value of $\gamma$. The standard FFT that we studied in the previous section is a special case for $\gamma = \frac{1}{N}$. The summation in (2.19) can be computed without imposing the constraint in (2.17). This means that the two grid spacings, one for the integral of characteristic function and the other for log-strike prices, can be selected independent of one another.

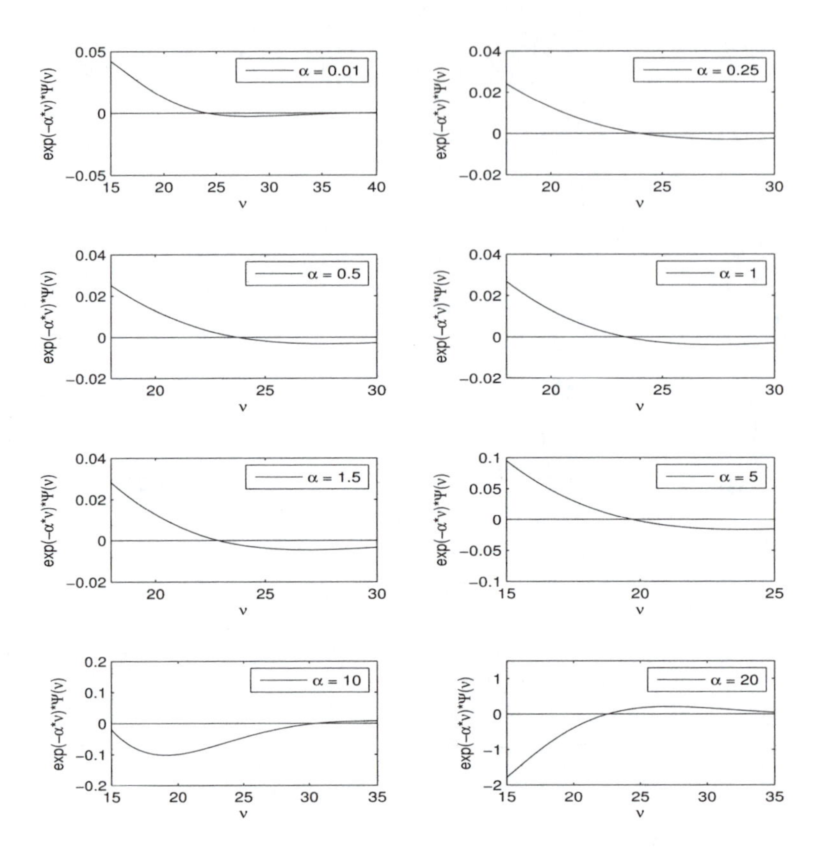


**FIGURE 2.6**: The tail of the integrand in Heston stochastic volatility model for various values of $\alpha$

### 2.2.1 Formation of Fractional FFT

As presented in [21] and [22], in the fractional Fourier transform (FrFFT) we define the $N-$long complex sequence $x$ as

$$G_m(x,\gamma) = \sum_{j=1}^{N} e^{-i2\pi\gamma(j-1)(m-1)} x(j) \tag{2.20}$$

The parameter $\gamma$ in fact may be any complex rational number. The sum can be implemented via three $2N$-point FFT steps. For an $N$-point fractional FFT on the vector $x(j)$, we define the following $2N$-long sequences:

$$\begin{aligned}
y_j &= x_j e^{-i\pi(j-1)^2\gamma} & 1 \leq j \leq N \\
y_j &= 0 & N < j \leq 2N \\
z_j &= e^{i\pi(j-1)^2\gamma} & 1 \leq j \leq N \\
z_j &= e^{i\pi(2N-j)^2\gamma} & N < j \leq 2N
\end{aligned}$$

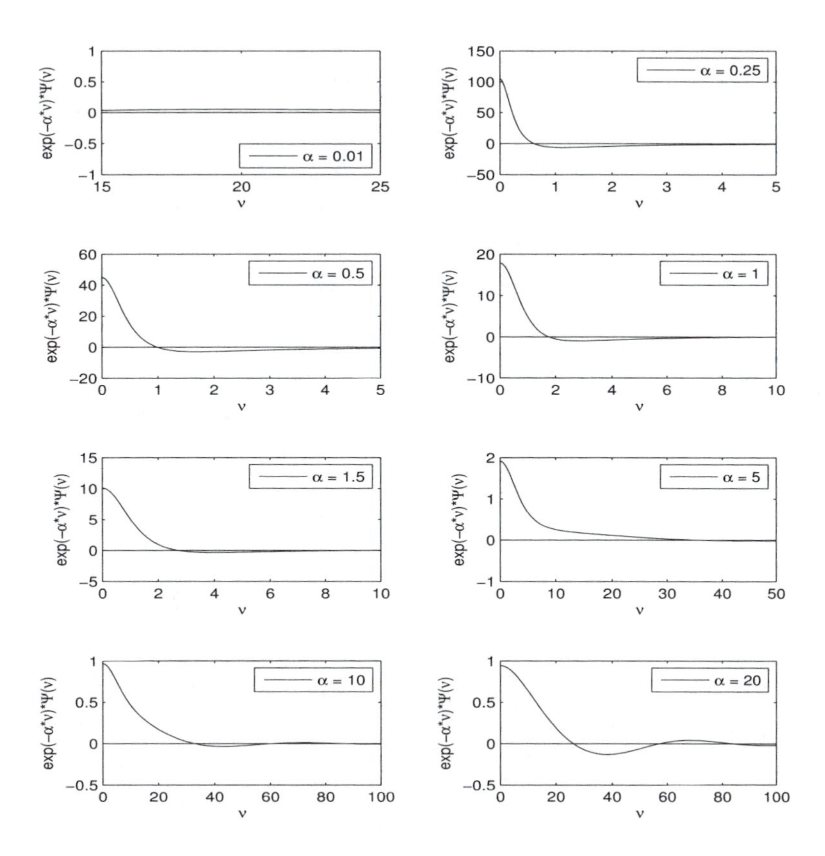


**FIGURE 2.7**: The integrand in variance gamma model for various values of $\alpha$

where $\gamma = \frac{\lambda\eta}{2\pi}$. It is shown in [21] that

$$G_m(x,\gamma) = (e^{-i\pi(m-1)^2\gamma}) \odot \mathrm{D}_m^{-1}(\mathrm{D}(\mathbf{y}) \odot \mathrm{D}(\mathbf{z})) \quad 1 \leq m \leq N$$

where $\odot$ denotes element componentwise vector multiplication and

$$\mathrm{D}(\xi) = \begin{pmatrix} \mathrm{D}_1(\xi) \\ \mathrm{D}_2(\xi) \\ \vdots \\ \mathrm{D}_{2N}(\xi) \end{pmatrix}$$

with

$$\eta_j = \mathrm{D}_j(\xi) = \sum_{m=1}^{2N} \exp\left(-i\frac{2\pi}{2N}(j-1)(m-1)\right)\xi(m) \quad 1 \leq j \leq 2N$$

and

$$\xi_m = \mathrm{D}_m^{-1}(\eta) = \frac{1}{2N}\sum_{j=1}^{2N} \exp\left(i\frac{2\pi}{2N}(j-1)(m-1)\right)\eta(j) \quad 1 \leq m \leq 2N$$

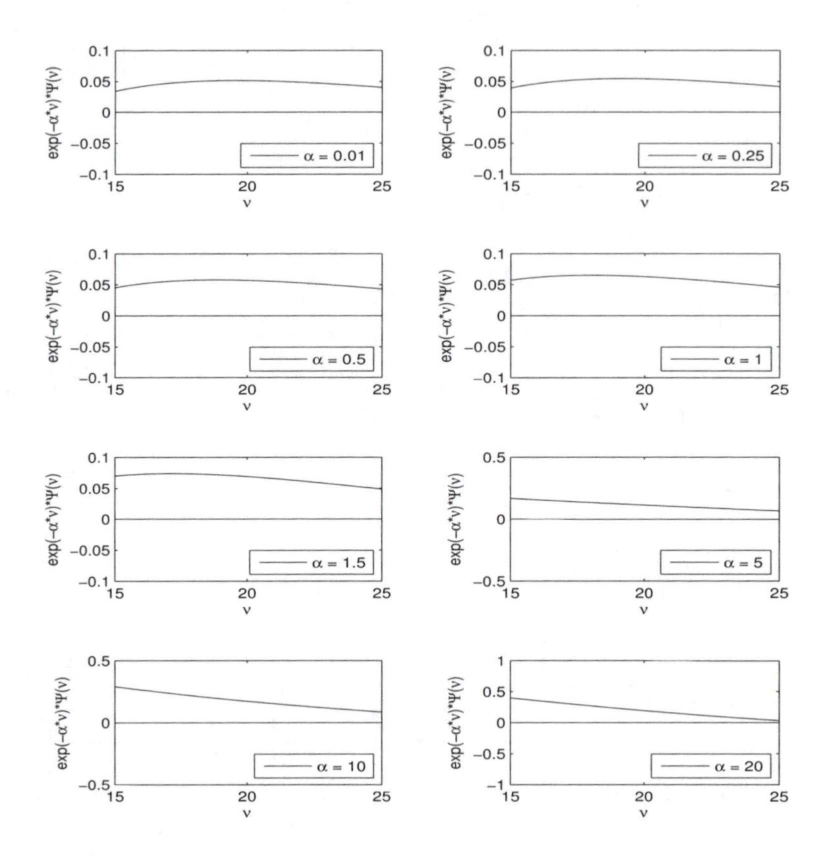


**FIGURE 2.8**: The tail of the integrand in variance gamma model for various values of $\alpha$

The remaining $N$ results of the final inverse discrete Fourier transform are discarded. Note that the exponential quantities do not depend on the actual function that is integrated and therefore can be precomputed and stored.

### 2.2.2 Implementation of Fractional FFT

Having the characteristic function of the log of the underlying process $X_t$ that is $\Phi(\nu)$, we choose $\eta$, $\lambda$ (independently) and $N = 2^n$. Calculate $\gamma = \frac{\eta\lambda}{2\pi}$, $\nu_j = (j-1)\eta$ and set $\alpha$. Form vector $\mathbf{x}$

$$\mathbf{x} = \begin{pmatrix} x_1 \\ x_2 \\ \vdots \\ x_N \end{pmatrix} = \begin{pmatrix} \frac{\eta}{2}\frac{C}{(\alpha+i\nu_1)(\alpha+i\nu_1+1)}e^{-i(\ln X_0-\frac{\lambda N}{2})\nu_1}\Phi\left(\nu_1-(\alpha+1)i\right) \\ \eta\frac{C}{(\alpha+i\nu_2)(\alpha+i\nu_2+1)}e^{-i(\ln X_0-\frac{\lambda N}{2})\nu_2}\Phi\left(\nu_2-(\alpha+1)i\right) \\ \vdots \\ \eta\frac{C}{(\alpha+i\nu_N)(\alpha+i\nu_N+1)}e^{-i(\ln X_0-\frac{\lambda N}{2})\nu_N}\Phi\left(\nu_N-(\alpha+1)i\right) \end{pmatrix}$$

**TABLE 2.4**: Relationship between grid spacing in integral and log-strike spacing

| $n$ | $N=2^n$ | $\eta$ | Upper bound | $\lambda$ | Log-strike spacing |
|---|---|---|---|---|---|
| 7 | 128 | 0.25 | 32 | 0.1963 | 19.0% |
| 9 | 256 | 0.25 | 64 | 0.0981 | 9.81% |
| 9 | 512 | 0.25 | 128 | 0.0491 | 4.91% |
| 10 | 1024 | 0.25 | 256 | 0.0245 | 2.45% |
| 11 | 2048 | 0.25 | 512 | 0.0122 | 1.22% |
| 12 | 4096 | 0.25 | 1024 | 0.0061 | 0.61% |
| 13 | 8192 | 0.25 | 2048 | 0.0031 | 0.30% |

Now form vectors $\mathbf{y}$ and $\mathbf{z}$ as follows:

$$\mathbf{y} = \begin{pmatrix} y_1 \\ y_2 \\ \vdots \\ y_N \\ y_{N+1} \\ y_{N+2} \\ \vdots \\ y_{2N} \end{pmatrix} = \begin{pmatrix} x_1 \\ \exp(-i\pi\gamma)x_2 \\ \vdots \\ \exp(-i\pi\gamma(N-1)^2)x_N \\ 0 \\ 0 \\ \vdots \\ 0 \end{pmatrix} \qquad \mathbf{z} = \begin{pmatrix} z_1 \\ z_2 \\ \vdots \\ z_N \\ z_{N+1} \\ z_{N+2} \\ \vdots \\ z_{2N} \end{pmatrix} = \begin{pmatrix} 1 \\ \exp(i\gamma\pi) \\ \vdots \\ \exp(i\gamma\pi(N-1)^2) \\ \exp(i\gamma\pi(N-1)^2) \\ \exp(i\gamma\pi(N-2)^2) \\ \vdots \\ 1 \end{pmatrix}$$

Vectors $\mathbf{y}$ and $\mathbf{z}$ are the input to the FFT routine, and its output is vectors $\widehat{\mathbf{y}}$ and $\widehat{\mathbf{z}}$ of the same size, respectively. Construct vector $\xi$ by multiplying vectors $\widehat{\mathbf{y}}$ and $\widehat{\mathbf{z}}$ element-wise, that is,

$$\xi = \begin{pmatrix} \xi_1 \\ \xi_2 \\ \vdots \\ \xi_{2N} \end{pmatrix} = \begin{pmatrix} \widehat{y}_1\widehat{z}_1 \\ \widehat{y}_2\widehat{z}_2 \\ \vdots \\ \widehat{y}_{2N}\widehat{z}_{2N} \end{pmatrix}$$

Vector $\xi$ is the input to the inverse FFT (IFFT) routine, and its output is vector $\widehat{\xi}$ of the same size. Using vector $\widehat{\xi}$, call prices at strike $k_m$ for $m = 1, \ldots, N$ are

$$\begin{pmatrix} C_T(k_1) \\ C_T(k_2) \\ \vdots \\ C_T(k_N) \end{pmatrix} = \begin{pmatrix} \frac{e^{-\alpha(\ln X_0 - \frac{N}{2}\lambda)}}{\pi}\mathrm{Re}\left(\widehat{\xi}_1\right) \\ \frac{e^{-\alpha(\ln X_0 - (\frac{N}{2}-1)\lambda)}}{\pi}\mathrm{Re}\left(\exp(-i\pi\gamma)\widehat{\xi}_2\right) \\ \vdots \\ \frac{e^{-\alpha(\ln X_0 - (\frac{N}{2}-(N-1))\lambda)}}{\pi}\mathrm{Re}\left(\exp(-i\pi\gamma(N-1)^2)\widehat{\xi}_N\right) \end{pmatrix}$$

where $\mathrm{Re}(z)$ is the real part of $z$. Note that the last $N$ elements of $\widehat{\xi}$ are never used and are discarded. Considering $\lambda$ and $\eta$ are independent, we can choose $\lambda$ that would yield a range around $\ln X_0$ with desired moneyness (for example, for 25% moneyness we get $\lambda = \frac{2(0.25)}{N}$. Also typically $N = 2^n$ is much smaller than the fast Fourier technique (e.g., $2^7$ as opposed to $2^{14}$).

## 2.3 Derivatives Pricing via the Fourier-Cosine (COS) Method

While the FFT method had a number of very significant advantages, including its ability to price under a model with only a known characteristic function, its speed, and its ability to generate multiple option prices in a single run, we also saw it has a number of significant drawbacks. These drawbacks include a restriction to path independent European options with a restricted subset of terminal payoffs which require the method to be rederived, as well as inaccuracy for highly out-of-the-money options.

In an effort to improve on currently available Fourier-based pricing methods, Fang and Oosterlee developed the COS method in [111]. The FFT method explicitly forms the Fourier transform of the option premium in terms of the characteristic of the log asset price, so the premium can be recovered using Fourier inversion. The COS method takes a different approach, first by representing the probability density function of the log asset price in terms of its Fourier cosine expansion and showing that the coefficients of this expansion can be expressed in terms of the characteristic function of the log asset price. Then this representation is used in the risk-neutral pricing formula, which can be reduced to the sum of an analytically calculable integral and the coefficients which are directly calculable from the characteristic function.

This method improves on the classic FFT method in a number of ways. First, it provides considerable speed improvements over FFT methods. While the COS method does in fact require $O(n)$ multiplications, which is computationally as many as direct integration and more than the amortized $O(n \log n)$ of FFT, the superior performance of the cosine expansion when integrating non-periodic functions reduces the number of terms needed for a certain degree of accuracy by so much that the COS method proves to be faster. Another very important advantage of the COS method is that it completely separates derivation of the cosine expansion coefficients from the terms which depend on the option's terminal payoff. This means that this method can be used to price any European path independent option as long as the term involving the option payoff, a simple cosine integration, can be evaluated analytically. This should be true for almost all payoff structures.

Thus the COS method is a significant improvement on the classic FFT method, but it does come with a few caveats of its own. Similar to the FFT method, it can handle pricing only path independent European options in its original form. Also, the COS method involves approximating the pricing integral by truncating the infinite integral bounds to some finite interval $[a, b]$, so that the COS expansion has a finite number of terms. The optimal value of these bounds needs to be estimated and the resulting prices can be sensitive to one's choice of $[a, b]$. Finally the COS method suffers from similar accuracy problems as the FFT method when trying to price deep out-of-the-money options.

**Models:**

All models for which a characteristic function for the asset price distribution exists

**Option Types:**

Path independent European options with a wide variety of terminal payoffs

**Pros**

1. Allows for pricing under any model with a characteristic function
2. Fastest known Fourier-based method
3. Separates model from payoff, supporting a wide variety of derivatives

**Cons**

1. Restricted to path independent European options
2. Requires estimation of proper $[a, b]$
3. Inaccurate for highly out-of-the-money options

## 2.3.1 COS Method

### 2.3.1.1 Cosine Series Expansion of Arbitrary Functions

The Fourier cosine series expansion of a function $f(\theta)$ on $[0, \pi]$ is

$$\begin{aligned} f(\theta) &= \frac{1}{2}A_0 + \sum_{k=1}^{\infty} A_k \cos(k\theta) \\ &= \overline{\sum}_{k=0}^{\infty} A_k \cos(k\theta) \end{aligned} \tag{2.21}$$

with the Fourier cosine coefficient

$$A_k = \frac{2}{\pi}\int_0^{\pi} f(\theta)\cos(k\theta)d\theta \tag{2.22}$$

where $\overline{\sum}$ indicates the first term in the summation is weighted by one-half.[2] We can extend this definition to arbitrary functions on a finite interval. For functions on any finite interval $[a, b]$, the Fourier cosine series expansion is obtained through the following change of variable that maps $a$ to 0 and $b$ to $\pi$:

$$\theta = \frac{\pi - 0}{b - a}(x - a) = \frac{x - a}{b - a}\pi \tag{2.23}$$

We write $x$ in terms of $\theta$

$$x = \frac{b - a}{\pi}\theta + a \tag{2.24}$$

and substitute it into (2.21) to obtain

$$f(x) = \overline{\sum}_{k=0}^{\infty} A_k \cos\left(k\frac{x - a}{b - a}\pi\right) \tag{2.25}$$

[2]In this section we follow the notes in [111].

with the following Fourier cosine coefficient:

$$A_k = \frac{2}{b-a} \int_a^b f(x) \cos(k\frac{x-a}{b-a}\pi) dx \tag{2.26}$$

#### 2.3.1.2 Cosine Series Coefficients in Terms of Characteristic Function

We already know for a distribution function $f(x)$, its characteristic function is

$$\mathbb{E}(e^{i\nu x}) = \phi(\nu) = \int_{-\infty}^{\infty} e^{i\nu x} f(x) dx$$

By evaluating the characteristic function at $\nu = \frac{k\pi}{b-a}$ we obtain

$$\phi\left(\frac{k\pi}{b-a}\right) = \int_{-\infty}^{\infty} e^{i(\frac{k\pi}{b-a})x} f(x) dx$$

We call the truncated version of this integral $\widehat{\phi}$, that is,

$$\widehat{\phi}\left(\frac{k\pi}{b-a}\right) = \int_a^b e^{i(\frac{k\pi}{b-a})x} f(x) dx \tag{2.27}$$

Multiplying (2.27) by $e^{-i\frac{k\pi a}{b-a}}$ we get

$$\begin{aligned}
\widehat{\phi}\left(\frac{k\pi}{b-a}\right) e^{-i\frac{k\pi a}{b-a}} &= e^{-i\frac{k\pi a}{b-a}} \int_a^b e^{i(\frac{k\pi}{b-a})x} f(x) dx \\
&= \int_a^b e^{ik\pi(\frac{x-a}{b-a})} f(x) dx \\
&= \int_a^b \left(\cos(k\pi(\frac{x-a}{b-a})) + i \sin(k\pi(\frac{x-a}{b-a}))\right) f(x) dx
\end{aligned}$$

and thus we have

$$\text{Re}\left\{\widehat{\phi}\left(\frac{k\pi}{b-a}\right) \exp\left(-i\frac{ka\pi}{b-a}\right)\right\} = \int_a^b \cos\left(k\pi(\frac{x-a}{b-a})\right) f(x) dx \tag{2.28}$$

If we assume $[a, b]$ is chosen such that

$$\widehat{\phi}(\nu) = \int_a^b e^{i\nu x} f(x) dx \approx \int_{-\infty}^{+\infty} e^{i\nu x} f(x) dx = \phi(\nu) \tag{2.29}$$

then comparing (2.26) and (2.28) gives us

$$A_k = \frac{2}{b-a} \text{Re}\left\{\widehat{\phi}\left(\frac{k\pi}{b-a}\right) \exp\left(-i\frac{ka\pi}{b-a}\right)\right\} \tag{2.30}$$

with $A_k \approx F_k$ where

$$F_k = \frac{2}{b-a} \text{Re}\left\{\phi\left(\frac{k\pi}{b-a}\right) \exp\left(-i\frac{ka\pi}{b-a}\right)\right\} \tag{2.31}$$

By substituting $F_k$ for $A_k$ in the Fourier cosine series expansion $f(x)$ on $[a, b]$ we get an approximate cosine series expansion which is a function of the characteristic of $f(x)$.

$$\widehat{f}(x) = \sideset{}{'}\sum_{k=0}^{\infty} F_k \cos(k\frac{x-a}{b-a}\pi) \tag{2.32}$$

By further truncating the summation we get

$$\widetilde{f}(x) = \sideset{}{'}\sum_{k=0}^{N-1} F_k \cos(k\frac{x-a}{b-a}\pi) \tag{2.33}$$

#### 2.3.1.3 COS Option Pricing

Now we show how to use the results of the previous section for options pricing. Let

$x$ be the modeled quantity at $t$, often the log asset price

$y$ be the modeled quantity at $T$, often the log asset price

$f(y|x)$ be the probability density function under the pricing measure

$v(x,t)$ be the value of a path independent European option at $t$

$v(y,T)$ be the option value at $T$, the payoff at expiration

Then the option value at time $t$ can be written as

$$v(x,t) = \mathrm{C}\int_a^b v(y,T)f(y|x)dy$$

for an appropriate value of C. From (2.25) and (2.26) we have

$$\begin{aligned} v(x,t) &= \mathrm{C}\int_a^b v(y,T)\sideset{}{'}\sum_{k=0}^{\infty} A_k \cos(k\frac{y-a}{b-a}\pi)dy \\ &= \mathrm{C}\sideset{}{'}\sum_{k=0}^{\infty} A_k\left(\int_a^b v(y,T)\cos(k\frac{y-a}{b-a}\pi)dy\right) \end{aligned}$$

Define

$$V_k = \frac{2}{b-a}\int_a^b v(y,T)\cos(k\pi\frac{y-a}{b-a})dy \tag{2.34}$$

Then we have

$$v(x,t) = \frac{b-a}{2}\mathrm{C}\sideset{}{'}\sum_{k=0}^{\infty} A_k V_k \tag{2.35}$$

To make this calculable we make another approximation and truncate the integral

$$v(x,t) \approx \frac{b-a}{2}\mathrm{C}\sideset{}{'}\sum_{k=0}^{N-1} A_k V_k \tag{2.36}$$

substitute (2.31), and we get the option premium in terms of the characteristic function of the model and a coefficient $V_k$, which is calculated based on the payoff of the option

$$v(x,t) \approx \mathrm{C}\sideset{}{'}\sum_{k=0}^{N-1} \mathrm{Re}\left\{\phi\left(\frac{k\pi}{b-a};x\right)\exp\left(-ik\pi\frac{a}{b-a}\right)\right\}V_k \tag{2.37}$$

### 2.3.2 COS Option Pricing for Different Payoffs

The option price under the COS method is calculated as

$$\begin{aligned} v(x,t) &\approx \frac{b-a}{2}\mathrm{C}\sideset{}{'}\sum_{k=0}^{N-1} A_k V_k &(2.38) \\ &\approx \mathrm{C}\sideset{}{'}\sum_{k=0}^{N-1} \mathrm{Re}\left\{\phi\left(\frac{k\pi}{b-a};x\right)\exp\left(-ik\pi\frac{a}{b-a}\right)\right\}V_k &(2.39) \end{aligned}$$

All information about the model is wholly contained in the cosine expansion coefficient expressed as a function of the characteristic function. The coefficient $V_k$ contains all the information about the payoff, so we can adapt the COS method to any payoff for which $V_k$ is readily calculable. Since we have

$$V_k = \frac{2}{b-a}\int_a^b v(y,T)\cos(k\pi\frac{y-a}{b-a})dy \tag{2.40}$$

This is true in many cases, and we give some common examples here.

#### 2.3.2.1 Vanilla Option Price under the COS Method

To price vanilla options under the COS method, we first define the following variables:

$X_t$ is the current price of the underlying security

$X_T$ is the $T$-time price of the underlying security

$K$ is the strike of the option

$x = \ln(X_t/K)$

$y = \ln(X_T/K)$

Thus we can express the payoff for vanilla European options as

$$v(y,T) = [\alpha K(e^y - 1)]^+$$

with $\alpha = 1$ for a call and $\alpha = -1$ for a put. In this case $V_k$ has an analytical form. Define

$$\begin{aligned} \chi_k(c,d) &= \int_c^d e^y \cos(k\pi\frac{y-a}{b-a})dy \\ \varphi_k(c,d) &= \int_c^d \cos(k\pi\frac{y-a}{b-a})dy \end{aligned}$$

Then we get their analytical forms

$$\begin{aligned} \chi_k(c,d) = &\frac{1}{1+(\frac{k\pi}{b-a})^2}[\cos(k\pi\frac{d-a}{b-a})e^d - \cos(k\pi\frac{c-a}{b-a})e^c \\ &+\frac{k\pi}{b-a}\sin(k\pi\frac{d-a}{b-a})e^d - \frac{k\pi}{b-a}\sin(k\pi\frac{c-a}{b-a})e^c] \end{aligned} \tag{2.41}$$

and

$$\varphi_k(c,d) = \begin{cases} [\sin(k\pi\frac{d-a}{b-a}) - \sin(k\pi\frac{c-a}{b-a})]\frac{b-a}{k\pi} & k \neq 0 \\ (d-c) & k = 0 \end{cases} \tag{2.42}$$

So for a vanilla call and put we obtain

$$\begin{aligned} V_k^{call} &= \frac{2}{b-a}\int_a^b K(e^y-1)^+\cos(k\pi\frac{y-a}{b-a})dy \\ &= \frac{2}{b-a}K(\chi_k(0,b) - \varphi_k(0,b)) \end{aligned} \tag{2.43}$$

$$\begin{aligned} V_k^{put} &= \frac{2}{b-a}\int_a^b K(1-e^y)^+\cos(k\pi\frac{y-a}{b-a})dy \\ &= \frac{2}{b-a}K(-\chi_k(a,0) + \varphi_k(a,0)) \end{aligned} \tag{2.44}$$

#### 2.3.2.2 Digital Option Price under the COS Method

For a digital option the result is even simpler.

$$\begin{aligned} V_k^{cashcall} &= \frac{2}{b-a}\int_0^b K\cos(k\pi\frac{y-a}{b-a})dy \\ &= \frac{2}{b-a}K\varphi_k(0,b) \end{aligned} \tag{2.45}$$

### 2.3.3 Truncation Range for the COS method

The authors in [111] propose, without any proof, the following formula for the range of the integration $[a,b]$ within the COS method.

$$[a,b] = \left[c_1 - L\sqrt{c_2+\sqrt{c_4}}, c_1 + L\sqrt{c_2+\sqrt{c_4}}\right] \quad with \quad L = 10 \tag{2.46}$$

where $c_n$ denotes the $n$-th cumulant [3] of $x$. In all of our numerical examples we will be using the above formula for range $[a,b]$. Determining the sensitivity of the resultant prices to the choices of $a$ and $b$, as well as justification for this equation, is left as a short case study at the end of the chapter.

### 2.3.4 Numerical Results for the COS Method

In this section we will present some empirical results of the COS method and compare them with the the fractional FFT method. We will present results for the following three models: (a) geometric Brownian motion, (b) Heston stochastic volatility model, and (c) variance gamma model.

#### 2.3.4.1 Geometric Brownian Motion (GBM)

The Black–Scholes model, which models asset prices using geometric Brownian motion, is described in Section 1.2.1 and its characteristic function is described in Section 1.2.1.3. The SDE and characteristic function under GBM are as follows:

$$dS_t = (r-q)S_t dt + \sigma S_t dW_t \tag{2.47}$$

$$\begin{aligned} \phi(\omega) &= \mathbb{E}(e^{i\omega y}) \\ &= \mathbb{E}(e^{i\omega[\ln(\frac{S_0}{K})+(r-q-\frac{\sigma^2}{2})t+\sigma W_T]}) \\ &= e^{i\omega[\ln\frac{S_0}{K}+(r-q-\frac{\sigma^2}{2})T]-\frac{w^2\sigma^2}{2}T} \end{aligned} \tag{2.48}$$

[3] For a random variable $X$ its cumulant generating function is given by

$$G(\omega) = \log\left(\mathbb{E}\left[\exp(\omega X)\right]\right)$$

Having the characteristic function of $X$, $\phi(u)$, then we can write

$$G(\omega) = \log\left(\phi(-i\omega)\right)$$

and its $n$-th cumulant is $n$-th derivatives of the cumulant generating function evaluated at zero, i.e., $c_n = G^{(n)}(0)$. For example,

$$c_1 = G^{(1)}(0) = \frac{-i\phi^{(1)}(0)}{\phi(0)}$$

and the cumulants are

$$c_1 = (r-q)T \tag{2.49}$$
$$c_2 = \sigma^2 T \tag{2.50}$$
$$c_4 = 0 \tag{2.51}$$

We use the following set of parameters for pricing: spot price $S_0 = 100$, risk-free rate $r = 10\%$, time to maturity $T = 0.1$, and volatility $\sigma = 0.25$. Table 2.5 shows the results using the COS and fractional FFT methods with strikes $K = 80, 100, 120$. The reference values, computed using the known Black–Scholes solution, are 20.799, 3.660, and 0.045.

**TABLE 2.5**: COS versus fractional FFT for geometric Brownian motion

| | | COS | | | Fractional FFT | | |
|---|---|---|---|---|---|---|---|
| $K$ | $N$ | Premium | CPU time (msc) | Rel. err. | Premium | CPU time (msc) | Rel. err. |
| 80 | 16 | 20.791 | 0.166 | 3.9e-04 | 16.867 | 2.476 | 1.9e-01 |
| | 64 | 20.799 | 0.184 | 1.3e-06 | 20.858 | 2.631 | -2.8e-03 |
| | 256 | 20.799 | 0.409 | 1.3e-06 | 20.799 | 2.812 | 1.3e-06 |
| 100 | 16 | 3.662 | 0.171 | -5.3e-04 | 7.287 | 2.442 | -9.9e-01 |
| | 64 | 3.660 | 0.191 | -4.2e-07 | 3.858 | 2.677 | -5.4e-02 |
| | 256 | 3.660 | 0.282 | -4.2e-07 | 3.660 | 2.845 | 1.6e-05 |
| 120 | 16 | 0.043 | 0.167 | 3.4e-02 | 2.697 | 2.430 | -6.0e+01 |
| | 64 | 0.045 | 0.203 | 3.1e-07 | -0.102 | 2.642 | 3.3e+00 |
| | 256 | 0.045 | 0.277 | 3.1e-07 | 0.045 | 2.866 | 2.7e-04 |

#### 2.3.4.2 Heston Stochastic Volatility Model

The Heston stochastic volatility model is described in Section 1.2.3 and its characteristic function is described in Section 1.2.3.2. Under this model asset prices are governed by the following SDE:

$$dS_t = (r-q)S_t dt + \sqrt{v_t} S_t dW_t^{(1)}$$
$$dv_t = \kappa(\theta - v_t)dt + \sigma\sqrt{v_t} dW_t^{(2)}$$

where the two Brownian components $W_t^{(1)}$ and $W_t^{(2)}$ are correlated with rate $\rho$. The variable $v_t$ represents the mean reverting stochastic volatility, where $\theta$ is the long-term variance, $\kappa$ is the mean reversion speed, and $\sigma$ is the volatility of the volatility.

The characteristic function for the log of the asset price process is given by

$$\Phi(u) = \mathbb{E}(e^{iu \ln S_t}) = \frac{\exp\left(iu \ln S_0 + iu(r-q)t + \frac{\kappa\theta t(\kappa - i\rho\sigma u)}{\sigma^2}\right)}{\left(\cosh\frac{\gamma t}{2} + \frac{\kappa - i\rho\sigma u}{\gamma}\sinh\frac{\gamma t}{2}\right)^{\frac{2\kappa\theta}{\sigma^2}}} \exp\left\{\frac{-(u^2+iu)v_0}{\gamma\coth\frac{\gamma t}{2} + \kappa - i\rho\sigma u}\right\}$$

where $\gamma = \sqrt{\sigma^2(u^2+iu) + (\kappa - i\rho\sigma u)^2}$, and $S_0$ and $v_0$ are the initial values for the price process and the volatility process, respectively.

**TABLE 2.6**: COS versus fractional FFT for Heston stochastic volatility model

| | | COS | | | Fractional FFT | | |
|---|---|---|---|---|---|---|---|
| $K$ | $N$ | Premium | CPU time (msc) | Rel. err. | Premium | CPU time (msc) | Rel. err. |
| 80 | 16 | 32.582 | 1.792 | -2.1e-05 | 32.224 | 2.592 | 1.1e-02 |
| | 64 | 32.581 | 1.903 | 0.0e+00 | 32.581 | 2.650 | 0.0e+00 |
| | 256 | 32.581 | 2.205 | 0.0e+00 | 32.581 | 2.941 | 0.0e+00 |
| 100 | 16 | 22.578 | 1.915 | -1.2e-02 | 21.553 | 2.535 | 3.4e-02 |
| | 64 | 22.319 | 1.887 | 0.0e+00 | 22.319 | 2.676 | 0.0e+00 |
| | 256 | 22.319 | 2.013 | 0.0e+00 | 22.319 | 3.045 | 0.0e+00 |
| 120 | 16 | 15.192 | 1.814 | -2.6e-02 | 14.296 | 2.454 | 3.4e-02 |
| | 64 | 14.806 | 1.958 | 0.0e+00 | 14.806 | 2.658 | 0.0e+00 |
| | 256 | 14.806 | 2.247 | 0.0e+00 | 14.806 | 3.029 | 0.0e+00 |

The cumulants for the Heston stochastic volatility model are

$$\begin{aligned} c_1 &= rT + (1 - e^{-\kappa T})\frac{\theta - v_0}{2\kappa} - \frac{1}{2}\theta T \\ c_2 &= \frac{1}{8\kappa^3}(\sigma T \kappa e^{-\kappa T}(v_0 - \theta)(8\kappa\rho - 4\sigma) \\ &\quad + \kappa\rho\sigma(1 - e^{-\kappa T})(16\theta - 8v_0) \\ &\quad + 2\theta\kappa T(-4\kappa\rho\sigma + \sigma^2 + 4\kappa^2) \\ &\quad + \sigma^2((\theta - 2v_0)e^{-2\kappa T} + \theta(6e^{-\kappa T} - 7) + 2v_0) \\ &\quad + 8\kappa^2(v_0 - \theta)(1 - e^{-\kappa T})) \end{aligned}$$

Calculating $c_4$ is rather arduous, so instead we use $c_4 = 0$ and assume that $L = 12$ to be conservative due to the choice for $c_4$.

We use the following set of parameters for pricing: spot price $S_0 = 100$, risk-free rate $r = 0\%$, time to maturity $T = 10$, $\lambda = 1.5768$, $\eta = 0.5751$, long-term variance $\theta = 0.0398$, initial variance $v_0 = 0.0175$, and correlation $\rho = -0.5711$. Table 2.6 displays results using the COS and fractional FFT methods with strikes $K = 80, 100, 120$. The reference values are 32.5808, 22.3189, and 14.8058, computed using the fractional FFT method with $N = 2^{15}$ points.

#### 2.3.4.3 Variance Gamma (VG) Model

The variance gamma model is described in Section 1.2.7 and its characteristic function is described in Section 1.2.7.2. This model is a time-changed Brownian motion model described by the following SDE:

$$\begin{aligned} b(t, \sigma, \theta) &= \theta t + \sigma W_t \\ X(t; \sigma, \nu, \theta) &= b(\gamma(t; 1, \nu), \sigma, \theta) \\ &= \theta\gamma(t; 1, \nu) + \sigma W(\gamma(t; 1, \nu)) \end{aligned}$$

**TABLE 2.7**: COS versus fractional FFT for variance gamma

| | | COS | | | Fractional FFT | | |
|---|---|---|---|---|---|---|---|
| $K$ | $N$ | Result | Time (msc) | Rel. Err. | Result | Time (msc) | Rel. err. |
| | 16 | 19.008 | 1.785 | 4.8e-03 | 16.202 | 2.518 | 1.5e-01 |
| 90 | 64 | 19.099 | 1.859 | -2.6e-06 | 19.144 | 2.564 | -2.3e-03 |
| | 256 | 19.099 | 1.979 | -2.6e-06 | 19.099 | 2.952 | -2.6e-06 |
| | 16 | 11.302 | 1.777 | 6.0e-03 | 11.308 | 2.610 | 5.5e-03 |
| 100 | 64 | 11.370 | 1.850 | 0.0e+00 | 11.275 | 2.676 | 8.4e-03 |
| | 256 | 11.370 | 1.957 | 0.0e+00 | 11.370 | 3.151 | 0.0e+00 |
| | 16 | 2.056 | 1.968 | -7.0e-02 | 5.179 | 2.409 | -1.7e+00 |
| 120 | 64 | 1.921 | 2.012 | 0.0e+00 | 2.005 | 2.735 | -4.4e-02 |
| | 256 | 1.921 | 2.041 | 0.0e+00 | 1.921 | 2.916 | 1.0e-05 |

where $\gamma(t; 1, \nu)$ is the gamma distribution of time changes. The characteristic function for the log of the asset price process is given by

$$\mathbb{E}(e^{iuX(t)}) = \left(\frac{1}{1 - iu\theta\nu + \sigma^2 u^2 \nu/2}\right)^{\frac{t}{\nu}} \tag{2.52}$$

and cumulants for the variance gamma model are

$$\begin{aligned} c_1 &= (r+\theta)T \\ c_2 &= (\sigma^2 + v\theta^2)T \\ c_4 &= 3(\sigma^4 v + 2\theta^4 v^3 + 4\sigma^2\theta^2 v^2)T \end{aligned}$$

We use the following set of parameters for pricing: spot price $S_0 = 100$, risk-free rate $r = 10\%$, volatility $\sigma = 0.12$, $\theta = -0.14$, $\nu = 0.2$. Table 2.7 shows the results using the COS and fractional FFT methods with strikes $K = 90, 100, 120$. The reference values are 19.0944, 11.3700, and 1.92123.

#### 2.3.4.4 CGMY Model

CGMY is described in Section 1.2.8 and its characteristic function is described in Section 1.2.8.1. The characteristic function for the log of the asset price process is given by

$$\mathbb{E}\left[e^{iuX_t}\right] = e^{Ct\Gamma(-Y)\left((M-iu)^Y - M^Y + (G+iu)^Y - G^Y\right)}$$

and the cumulants for the CGMY model are

$$\begin{aligned} c_1 &= (r+\omega)T + TC\Gamma(1-Y)(-M^{Y-1} + G^{Y-1}) \\ c_2 &= \chi + CT\Gamma(2-Y)(M^{Y-2} + G^{Y-2}) \\ c_4 &= CT\Gamma(4-Y)(M^{Y-4} + G^{Y-4}) \end{aligned}$$

where

$$\begin{aligned}\omega &= -C\Gamma(-Y)((M-1)^Y) + (G+1)^Y - M^Y - G^Y \\ \chi &= \frac{8C}{((G+M)^2 - (G-M)^2)}\end{aligned}$$

We leave the comparison between the results of the COS and fractional FFT methods under the CGMY model as an exercise.

---

## 2.4 Cosine Method for Path-Dependent Options

The results discussed in the previous sections demonstrate that the COS method is a substantial improvement over previous transform methods. However, in its original form it is still restricted to the pricing of path-independent options. In [112], Oosterlee and Fang present a method for pricing some types of path-dependent options using backwards induction in combination with the COS method. The fundamental cosine method for path-dependent options is the same as for plain vanilla options, using Equation (2.37). The difference between the pricing method for exotic and plain vanilla options is in the calculation of $V_k$. For plain vanilla options, $V_k$ has a straightforward analytical form, as shown earlier. For exotic options, specifically with an early exercise feature, $V_k$ requires recursive backwards calculation and therefore it is time dependent.

### 2.4.1 Bermudan Options

The first class of path-dependent derivatives we will discuss is Bermudian options, which are exercisable at a discrete set of dates prior to their final maturity. We first define some notation before discussing pricing. Let $t_0$ be initial time and $t_1, \ldots, t_M$ be prespecified exercise dates with $t_0 < t_1 < \cdots < t_M = T$, the final maturity, and $\Delta t = t_m - t_{m-1}$. Without a loss of generality it is assumed exercise dates are equidistant. To price a Bermudan option, its value is split into two parts, the continuation value and the immediate exercise payoff.

At time $t_{m-1}$, the value of $v(x, t_{m-1})$ consists of the continuation value and the early exercise payoff value. From Equation (2.37) an approximated continuation value, assuming the option is not exercised in the current period, is

$$c(x, t_{m-1}) \approx \mathrm{C} \sideset{}{'}\sum_{k=0}^{N-1} \mathrm{Re}\left\{\phi\left(\frac{k\pi}{b-a}; y|x\right) \exp\left(-ik\pi\frac{a}{b-a}\right)\right\} V_k(t_m) \tag{2.53}$$

where

$$V_k(t_m) = \frac{2}{b-a}\int_a^b v(y, t_m) \cos\left(k\pi\frac{y-a}{b-a}\right) dy \tag{2.54}$$

and the exercise payoff value is $g(x, t_{m-1})$

$$g(x, t_{m-1}) = [\alpha K(e^x - 1)]^+ \tag{2.55}$$

with $x = \ln(X_{t_{m-1}}/K)$, $y = \ln(X_{t_m}/K)$, and $\alpha = 1$ for a call and $\alpha = -1$ for a put. Therefore the option value at time $t_{m-1}$ is

$$v(x, t_{m-1}) = \max(g(x, t_{m-1}), c(x, t_{m-1})) \tag{2.56}$$

The key point for this method is how to calculate $V_k(t_m)$. This is difficult and requires backwards induction because the option value during the next period $v(y, t_m)$, and thus the coefficient $V_k(t_m)$, is unknown except at expiry. To solve for $V_k(t_m)$ we first define

$$C_k(x_1, x_2, t_m) = \frac{2}{b-a}\int_{x_1}^{x_2} c(x, t_m)\cos(k\pi\frac{x-a}{b-a})dx \tag{2.57}$$

$$G_k(x_1, x_2) = \frac{2}{b-a}\int_{x_1}^{x_2} g(x, t_m)\cos(k\pi\frac{x-a}{b-a})dx \tag{2.58}$$

the coefficients associated with continuation and immediate exercise during period $t_m$. We have the following analytical form for $C_k(x_1, x_2, t_m)$:

$$C_k(x_1, x_2, t_m) = e^{-r\Delta t}\sum_{j=0}^{N-1}{}' \mathrm{Re}\left\{\phi(\frac{k\pi}{b-a})V_j(t_{m+1})M_{k,j}(x_1, x_2)\right\} \tag{2.59}$$

where

$$M_{k,j}(x_1, x_2) = -\frac{i}{\pi}(M^c_{k,j}(x_1, x_2) + M^s_{k,j}(x_1, x_2)) \tag{2.60}$$

and

$$M^c_{k,j}(x_1, x_2) = \begin{cases} \frac{(x_2-x_1)\pi i}{b-a} & k = j = 0 \\ \frac{\exp(i(j+k)\frac{(x_2-a)\pi}{b-a}) - \exp(i(j+k)\frac{(x_1-a)\pi}{b-a})}{j+k} & \text{otherwise} \end{cases} \tag{2.61}$$

$$M^s_{k,j}(x_1, x_2) = \begin{cases} \frac{(x_2-x_1)\pi i}{b-a} & k = j \\ \frac{\exp(i(j-k)\frac{(x_2-a)\pi}{b-a}) - \exp(i(j-k)\frac{(x_1-a)\pi}{b-a})}{j-k} & k \neq j \end{cases} \tag{2.62}$$

For $G_k$ we also have an analytical form

$$G_k(x_1, x_2) = \frac{2}{b-a}\alpha K[\chi_k(x_1, x_2) - \varphi_k(x_1, x_2)] \tag{2.63}$$

where

$$\chi_k(x_1, x_2) = \frac{1}{1 + (\frac{k\pi}{b-a})^2}[\cos(k\pi\frac{x_2-a}{b-a})e^{x_2} - \cos(k\pi\frac{x_1-a}{b-a})e^{x_1} + \frac{k\pi}{b-a}\sin(k\pi\frac{x_2-a}{b-a})e^{x_2} - \frac{k\pi}{b-a}\sin(k\pi\frac{x_1-a}{b-a})e^{x_1}] \tag{2.64}$$

$$\varphi_k(x_1, x_2) = \begin{cases} [\sin(k\pi\frac{x_2-a}{b-a}) - \sin(k\pi\frac{x_1-a}{b-a})]\frac{b-a}{k\pi} & k \neq 0 \\ (x_2 - x_1) & k = 0 \end{cases} \tag{2.65}$$

The calculation of $C_k(x_1, x_2, t_m)$ can be efficiently conducted through fast Fourier transform, as demonstrated in [112].

Now we are able to calculate the coefficients $V_k(t_m)$. First we find $x^*_m$ such that $c(x^*_m, t_m) = g(x^*_m, t_m)$ where $x^*_m$ is the early exercise boundary at time $t_m$, the price at which continuation becomes more profitable than immediate exercise or vice versa. Solving for this boundary price can be done via the Newton–Raphson or the bisection method. We can then calculate the coefficient $V_k(t_m)$ as

$$V_k(t_m) = \begin{cases} C_k(a, x^*_m, t_m) + G_k(x^*_m, b) & \text{for a call} \\ C_k(x^*_m, b, t_m) + G_k(a, x^*_m) & \text{for a put} \end{cases} \tag{2.66}$$

for $m = M-1, M-2, \ldots, 1$. This divides the integral over the next period's prices into the continuation and exercise regions. For the terminal condition we have

$$V_k(t_M) = \begin{cases} G_k(0,b) & \text{for a call} \\ G_k(a,0) & \text{for a put} \end{cases} \tag{2.67}$$

We first start with $V_k(t_M)$ and calculate backwards inductively to get $V_k(t_1)$, with the original option value denoted as

$$v(x,t_0) \approx \mathrm{C} \sum_{k=0}^{N-1}{}' \operatorname{Re}\left\{\phi(\frac{k\pi}{b-a}; y|x) \exp\left(-ik\pi\frac{a}{b-a}\right)\right\} V_k(t_1) \tag{2.68}$$

## 2.4.2 Discretely Monitored Barrier Options

The next class of path-dependent derivatives we will discuss is discretely monitored barrier options, which cease to exist if the underlying price hits a barrier at one of the prespecified dates. There are a number of different types of barrier options, but here we will discuss up-and-out options, which have a payoff of

$$V(X,t) = \max(\alpha(X_T - K), 0)|(X_{t_i} < H) + R|(X_{t_i} >= H) \tag{2.69}$$

where $R$ is the rebate and $X_{t_i}$ is the level of the underlying price at a discretely monitored time. The formula for coefficient $V_k(t_m)$ is again split into a continuation component, assuming that the price did not hit the barrier, and a component indicating the price did hit the barrier,

$$V_k(t_m) = C_k(a,h,t_m) + e^{-r(T-t_{m-1})}\frac{2R}{b-a}\varphi_k(h,b) \tag{2.70}$$

For pricing the option, we start with $V_k(t_M)$ where for $h < 0$

$$V_k(t_M) = \begin{cases} G_k(0,b) + \frac{2R}{b-a}\varphi(a,h) & \text{for a call} \\ G_k(h,0) + \frac{2R}{b-a}\varphi(a,h) & \text{for a put} \end{cases} \tag{2.71}$$

and for $h \geq 0$

$$V_k(t_M) = \begin{cases} G_k(0,h) + \frac{2R}{b-a}\varphi(h,b) & \text{for call} \\ G_k(a,0) + \frac{2R}{b-a}\varphi(h,b) & \text{for put} \end{cases} \tag{2.72}$$

### 2.4.2.1 Numerical Results — COS versus Monte Carlo

Here we provide some numerical results from pricing discretely monitored barrier options using the COS method and Monte Carlo simulation. The models under consideration in this example are the Black–Scholes and variance gamma models. The parameters used for the Black–Scholes model are: spot price $X_0 = 100$, strike $K = 100$, maturity $T = 1$ year, prespecified monitored times $M = 12$, equidistant monthly barrier $H = 120$, rebate $R = 0$, risk-free rate $r = 5\%$ and volatility $\sigma = 0.2$. The parameters used for the VG model are identical except for the VG specific model parameters $\sigma = 0.12$, $\nu = 0.2$, and $\theta = -0.14$.

Results are displayed in Table 2.8. As we observe premiums closely match for each model under both methodologies. However, COS is significantly faster than Monte Carlo simulation.

**TABLE 2.8**: Results on discretely monitored barrier option — COS versus MC

| Method | Model | $N$ | Call | Put | Time (sec) |
|---|---|---|---|---|---|
| COS | BS | $2^6$ | 1.8494 | 5.4846 | 0.1238 |
| MC | BS | $10^6$ | 1.8487 | 5.4794 | 1.0178 |
| COS | VG | $2^6$ | 4.1935 | 3.1572 | 0.1562 |
| MC | VG | $10^6$ | 4.1949 | 3.1557 | 10.1335 |

## 2.5 Saddlepoint Method

The COS and FFT methods have a number of distinct advantages. For instance, they have the ability to price derivatives under a model given only a known characteristic function, they are very fast, and they have the ability to generate multiple option prices in a single run. The COS method in particular has the advantage that it requires very little additional work to rederive for a number of different payoffs (both European and some path-dependent options). However, as we noted previously, both of these methods can be inaccurate for highly out-of-the-money options.

In [61], Carr and Madan suggest an alternative method for pricing options specifically designed to price out-of-the-money options more accurately. This method is called the saddlepoint method and involves expressing the price of an option as the probably of a specially constructed random variable exceeding the log-strike price, then solving for this probability using a modified Lugannani–Rice saddlepoint approximation.

The saddlepoint method offers considerably better accuracy in pricing out-of-the-money options than either the FFT or the COS method. However, the algorithm's accuracy for at the money and in the money options is somewhat lacking compared to these two other methods and like the FFT method its solution must be rederived for each different payoff.

**Models:**

All models for which a cumulant generating function for the asset price distribution exists. By extension any model for which a characteristic function can be derived.

**Option Types:**

Path-independent European options.

**Pros**

1. Allows for very accurate pricing of deep out-of-the-money options

**Cons**

1. Restricted to path-independent European options
2. Requires solution to be rederived for different payoff functions

### 2.5.1 Generalized Lugannani–Rice Approximation

We will begin with a discussion of generalized Lugannani–Rice approximations. These approximations allow us to accurately approximate the probability of a random variable exceeding some level using the cumulant generating function[4] (CGF) of the random variable in question, the CGF of some base random variable typically chosen to resemble the random variable in question up to a shift and scaling transformation, as well as the cumulative distribution function (CDF) and probability density function (PDF) of the base random variable.

The Lugannani–Rice saddlepoint formula presented in [174] has proven to be a remarkably good approximation of the cumulative distribution function of a summation of independent random variables. In its standard form, the Lugannani–Rice approximation for the tail probability of a continuous random variable $X$ is given by

$$P(X \geq y) = 1 - \Phi(\hat{w}) + \phi(\hat{w})\left(\frac{1}{\hat{u}} - \frac{1}{\hat{w}}\right) \tag{2.74}$$

where $\Phi$ and $\phi$ are the cumulative distribution function and probability distribution function of a standard normal distribution

$$\hat{w} = \text{sgn}(\hat{t})\sqrt{2(\hat{t}y - K(\hat{t}))} \tag{2.75}$$

and

$$\hat{u} = \hat{t}\sqrt{K''(\hat{t})} \tag{2.76}$$

where $K(t)$ is the cumulant generating function (CGF) of random variable $X$ and $\hat{t}$ is the unique solution to the saddlepoint equation $K'(t) = y$.

As stated in [219], a central feature of the Lugannani–Rice approximation is for a given $y$ and $K(t)$ we use the transformation from $t$ to $w$ determined by

$$\frac{1}{2}w^2 - \hat{w}w = K(t) - ty \tag{2.77}$$

where $\hat{w}$ is chosen so that the minimum of $\frac{1}{2}w^2 - \hat{w}w$ is equal to the minimum of $K(t) - ty$. The basic idea is to find a transformation which describes the local behavior of the function $K(t) - ty$ over a region containing both $t = \hat{t}$ and $t = 0$ when $\hat{t}$ is small. Such a transformation is

$$\frac{1}{2}(w - \hat{w})^2 = K(t) - ty - K(\hat{t}) + \hat{t}y \tag{2.78}$$

If $\hat{w}$ is chosen such that

$$K(\hat{t}) - \hat{t}y = -\frac{1}{2}\hat{w}^2 \tag{2.79}$$

then this becomes

$$\frac{1}{2}w^2 - \hat{w}w = K(t) - tK'(\hat{t}) \tag{2.80}$$

---

[4]For a random variable $X$ its cumulant generating function is given by

$$G(\omega) = \log\left(\mathbb{E}\left[\exp(\omega X)\right]\right) \tag{2.73}$$

where

$$\hat{w} = \sqrt{2\hat{t}[K'(\hat{t}) - K(\hat{t})]} \tag{2.81}$$

The desired equality is achieved when $\hat{w}$ is given by

$$\hat{u} = 2\hat{t}[K'(\hat{t}) - K(\hat{t})]^{1/2} \tag{2.82}$$

The form of approximation of $\frac{1}{2}w^2 - \hat{w}w$ also explains why the normal CDF and PDF $\Phi$ and $\phi$ appear in the equation, because $\frac{1}{2}w^2$ is the CGF of the standard normal distribution.

However, there is no particular reason why one has to use $\frac{1}{2}w^2$ in the equation; any CGF that is analytic at the origin will work. Suppose $G(w)$ is the CGF of the base distribution. For each $\xi$, $w_\xi$ is the unique solution of the saddlepoint $G'(w) = \xi$. Define $\hat{t}$ as the solution of

$$K'(t) = y \tag{2.83}$$

Then by the same analogy with approximation we find

$$G(w_\xi) - \xi w_\xi = K(\hat{t}) - \hat{t}y \tag{2.84}$$

Now $G(w_\xi) - \xi w_\xi$ is the Legendre–Fenchel transformation of $G$ and so the left-hand side is a concave function of $\xi$. Therefore for fixed $y$ can have at most two solutions in $\xi$.

$$\xi_- = \xi_-(y) < G'(0) < \xi_+ = \xi_+(y) \tag{2.85}$$

Therefore we choose $\hat{\xi} = \xi_-$ if $y < K'(0)$ and $\hat{\xi} = \xi_+$ if $y > K'(0)$. Suppose now that $\Gamma$ and $\gamma$ are the CDF and PDF of the distribution whose CGF is $G$. Then the modified Lugannani–Rice formula for the random variable $X$ takes

$$P(X > y) = 1 - \Gamma(\hat{\xi}) + \gamma(\hat{\xi})\left(\frac{1}{\hat{u}_\xi} - \frac{1}{\omega_\xi}\right) \tag{2.86}$$

where

$$\hat{u} = \hat{t}\left(K''(\hat{t})\right)^{1/2} \tag{2.87}$$

$$\hat{u}_{\hat{\xi}} = \frac{\hat{u}}{G''(w_{\hat{\xi}})^{1/2}} \tag{2.88}$$

This derivation is outlined in [219]. Note that if $G = K$, then $w_{\hat{\xi}} = \hat{u}_{\hat{\xi}} = \hat{t}$ and in this case the approximation is exact.

### 2.5.2 Option Prices as Tail Probabilities

If we are going to use the Lugannani–Rice approximation to improve the pricing of far out-of-the-money options, we must first express options prices in terms of tail probabilities. In this section we will outline how this can be done.

We use $S_t$ to denote the time-$t$ price of the option's underlier and $B_t$ to denote the time-$t$ value of the cash account. We know that under the risk-neutral pricing principle, if we use $B_t$ as the numeraire any tradable security deflated by $B_t$ under risk-neutral measure $\mathbb{Q}$ is a martingale. We define a new measure, the share measure $\mathbb{S}$, where $S_t$ is the numeraire.

Therefore any tradable security deflated by $S_t$ is a martingale under the share measure. This implies that

$$\frac{C_t}{S_t} = \mathbb{E}_t^{\mathbb{S}}\left(\frac{C_T}{S_T}\right) \tag{2.89}$$

For a call option with strike $K$ we can write

$$\frac{C(K)}{S_0} = \mathbb{E}^{\mathbb{S}}\left(\frac{(S_T - K)^+}{S_T}\right) \tag{2.90}$$

and if we assume the security price $S_T$ is always positive, we get

$$\frac{C(K)}{S_0} = \mathbb{E}^{\mathbb{S}}\left(\left(1 - \frac{K}{S_T}\right)^+\right) \tag{2.91}$$

We define $y = \log\left(\frac{S_T}{K}\right)$, which implies $\frac{K}{S_T} = e^{-y}$, and using this definition we can rewrite the normalized call price as follows:

$$\frac{C(K)}{S_0} = \int_0^\infty (1 - e^{-y}) f(y) dy \tag{2.92}$$

where $f(y)$ is the probability density function of $y = \ln(S/K)$ under the share measure. We perform integration by parts to get

$$\begin{aligned}
\frac{C(K)}{S_0} &= \int_0^\infty (1 - e^{-y}) f(y) dy \\
&= (1 - e^{-y}) F(y)\Big|_0^\infty - \int_0^\infty e^{-y} F(y) dy \\
&= (F(y) - F(y) e^{-y})\Big|_0^\infty - \int_0^\infty e^{-y} F(y) dy \\
&= 1 - \int_0^\infty e^{-y} F(y) dy \\
&= \int_0^\infty e^{-y} dy - \int_0^\infty e^{-y} F(y) dy \\
&= \int_0^\infty (1 - F(y)) e^{-y} dy
\end{aligned}$$

and thus we have

$$\frac{C(K)}{S_0} = \int_0^\infty (1 - F(y)) e^{-y} dy$$

For a given $y$, the expression $1 - F(y)$ is the probability that $\ln(S/K)$ is greater than $y$. Considering that $e^{-y}$ is the probability density function of a positive exponential random variable with $\lambda = 1$, the normalized call price is the probability that under the share measure the logarithm of the stock price exceeds the logarithm of strike by an independent exponential variable [61] or equivalently

$$\begin{aligned}
\frac{C(K)}{S_0} &= P(\ln(S/K) > Y) &(2.93)\\
&= P(\ln S - \ln K > Y) &(2.94)\\
&= P(X - Y > \ln K) &(2.95)
\end{aligned}$$

where $X$ is the logarithm of the stock under the share measure, $Y$ is an independent exponential, and $K$ is the strike.

Once we know the cumulant generating function of the random variable $X - Y$, $K(x)$, and its first and second derivatives, $K'(x)$, and $K''(x)$, the saddlepoint method gives us an approximation of the probability $P(X - Y > \ln K)$.

Suppose we have the CGF of the log of the stock price under some risk-neutral model given by $K_0(x)$, then the CGF of the log of the stock price under the share measure less an exponential is

$$K(x) = K_0(x+1) - K_0(1) - \ln(1+x) \tag{2.96}$$

and its first and second derivatives are

$$K'(x) = K_0'(x+1) - \frac{1}{1+x} \tag{2.97}$$

$$K''(x) = K_0''(x+1) + \frac{1}{(1+x)^2} \tag{2.98}$$

### 2.5.3 Lugannani–Rice Approximation for Option Pricing

In the last section we demonstrated that the density which can be used to express the call option price as a tail probability is a Gaussian less an independent exponential random variable. Thus we will consider a Lugannani–Rice base distribution of the same form, using $Z \sim N(0,1)$, a standard Gaussian distribution, and $Y \sim \exp(\lambda)$, an exponential distribution. We define our base distribution as

$$Z + \frac{1}{\lambda} - Y \tag{2.99}$$

The cumulant generating function of this base distribution is

$$\begin{aligned} G(w) &= \frac{w^2}{2} + \frac{w}{\lambda} - \ln\left(\frac{\lambda + w}{\lambda}\right) \\ &= \frac{w^2}{2} + \frac{w}{\lambda} - \ln(\lambda + w) + \ln(\lambda) \end{aligned}$$

And its first two derivatives are

$$\begin{aligned} G'(w) &= w + \frac{1}{\lambda} - \frac{1}{\lambda + w} \\ G''(w) &= 1 + (\frac{1}{\lambda + w})^2 \end{aligned}$$

The CDF and PDF of this distribution are shown in [61] to be

$$\tilde{\Phi}(y) = N(\frac{1}{\lambda} - y) - \exp(\lambda y - 1 + \frac{\lambda^2}{2})N(\frac{1}{\lambda} - y - \lambda) \tag{2.100}$$

$$\tilde{\phi}(y) = n(\frac{1}{\lambda} - y) + \lambda \exp(\lambda y - 1 + \frac{\lambda^2}{2})N(\frac{1}{\lambda} - y - \lambda) \tag{2.101}$$

$$- \exp(\lambda y - 1 + \frac{\lambda^2}{2})n(\frac{1}{\lambda} - y - \lambda) \tag{2.102}$$

where $N(x)$ and $n(x)$ are the cumulative distribution function and the probability distribution function for $\mathcal{N}(0,1)$, respectively. We define $y$ to be the log of the strike price, and the first step of the algorithm is to determine $\hat{t}$ and $\lambda$ by solving

$$K'(\hat{t}) = y \tag{2.103}$$

$$\lambda = \sqrt{K''(\hat{t}+1)} \tag{2.104}$$

We then must solve for $\hat{\xi}$ and $\omega_{\hat{\xi}}$ by solving

$$K(\hat{t}) - \hat{t}y = G(\omega_{\hat{\xi}}) - \hat{\xi}\omega_{\hat{\xi}} \tag{2.105}$$

$$G'(\omega(\hat{\xi})) = \hat{\xi} \tag{2.106}$$

with the Gauss–Fenchel transform

$$\omega_{\hat{\xi}} = \lambda + \frac{c}{2} + \sqrt{\frac{c^2}{4} + 1} \tag{2.107}$$

$$c = \hat{\xi} - \frac{1}{\lambda} + \lambda \tag{2.108}$$

There are two solutions for $\hat{\xi}$, and we choose $\hat{\xi} < G'(0)$ if $y < K'(0)$ and $\hat{\xi} \geq G'(0)$ if $y \geq K'(0)$. If we define

$$\hat{u} = \hat{t}\sqrt{K''(\hat{t})}$$

$$\hat{u}_{\hat{\xi}} = \hat{u}\sqrt{G''(\omega_{\hat{\xi}})}$$

then following the standard Lugannani–Rice approach we can calculate complementary probability as follows:

$$P(X - Y > y) = \tilde{\Phi}(\hat{\xi}) + \tilde{\phi}(\hat{\xi})(\frac{1}{\hat{u}_{\hat{\xi}}} - \frac{1}{\omega_{\hat{\xi}}})$$

### 2.5.4 Implementation of the Saddlepoint Approximation

In this section, we describe a step by step implementation of the saddlepoint approximation. We begin as follows:

- Use the bisection method to solve the equation for $\hat{t}$

$$K'(\hat{t}) = y \tag{2.109}$$

  where $y = \log(K)$ with $K$ being the strike price.

- Solve

$$G'(\omega_{\hat{\xi}}) = \hat{\xi} \tag{2.110}$$

  where we have

$$\omega_{\hat{\xi}} = \lambda + \frac{c}{2} + \sqrt{\frac{c^2}{4} + 1} \tag{2.111}$$

$$c = \hat{\xi} - \frac{1}{\lambda} + \lambda \tag{2.112}$$

- Use the bisection method again with (2.111) and (2.112) to solve

$$K(\hat{t}) - \hat{t}y = G(\omega_{\hat{\xi}}) - \hat{\xi}\omega_{\hat{\xi}} \tag{2.113}$$

  Note that there are two solutions for $\hat{\xi}$, and we choose $\hat{\xi} < G'(0)$ if $y < K'(0)$ and $\hat{\xi} \geq G'(0)$ if $y \geq K'(0)$

- If we define

$$\begin{aligned} \hat{u} &= \hat{t}\sqrt{K''(\hat{t})} \\ \hat{u}_{\hat{\xi}} &= \hat{u}\sqrt{G''(\omega_{\hat{\xi}})} \end{aligned}$$

  then the complementary probability is estimated via

$$P(X > y) = \tilde{\Phi}(\hat{\xi}) + \phi(\hat{\xi})(\frac{1}{\hat{u}_{\hat{\xi}}} - \frac{1}{\omega_{\hat{\xi}}})$$

  where

$$\begin{aligned} \tilde{\Phi}(x) &= N(\frac{1}{\lambda} - x) - \exp(\lambda x - 1 + \frac{\lambda^2}{2})N(\frac{1}{\lambda} - x - \lambda) \\ \phi(x) &= n(\frac{1}{\lambda} - x) + \lambda \exp(\lambda x - 1 + \frac{\lambda^2}{2})N(\frac{1}{\lambda} - x - \lambda) \\ &\quad - \exp(\lambda x - 1 + \frac{\lambda^2}{2})n(\frac{1}{\lambda} - x - \lambda) \end{aligned}$$

  and $n(x)$ and $N(x)$ are the PDF and CDF of the standard normal distribution, respectively.

- Then the call price can be approximated with

$$C = S_0 \times P(X > \ln K) \tag{2.114}$$

  where $X$ is the logarithm of the stock under the share measure less an independent exponential, and $K$ is the strike price.

If we use the base model suggested above, then the formulas for $G(w)$ and $K(x)$ are as follows:

$$G(w) = \frac{w^2}{2} + \frac{w}{\lambda} - \ln(\lambda + w) + \ln \lambda \tag{2.115}$$

$$G'(w) = w + \frac{1}{\lambda} - \frac{1}{\lambda + w} \tag{2.116}$$

$$G''(w) = 1 + (\frac{1}{\lambda + w})^2 \tag{2.117}$$

and

$$K(x) = K_0(x+1) - K_0(1) - \ln(1+x) \tag{2.118}$$

$$K'(x) = K_0'(x+1) - \frac{1}{1+x} \tag{2.119}$$

$$K''(x) = K_0''(x+1) + \frac{1}{(1+x)^2} \tag{2.120}$$

where $K_0$ is the CGF of the log of the stock under the risk-neutral measure, which varies in different models.

In the solution provided we chose

$$\lambda = \sqrt{K''(\hat{t}+1)}$$

### 2.5.5 Numerical Results for Saddlepoint Methods

In this section we present some comparative results utilizing fast Fourier transform, fractional Fourier transform, COS, and saddlepoint techniques for pricing European call options for a variety of strikes under geometric Brownian motion, Heston stochastic volatility, variance gamma, and CGMY models. The tables are meant to provide the reader with some empirical results and to illustrate the accuracy of saddlepoint methods for out-of-the-money options.

#### 2.5.5.1 Geometric Brownian Motion (GBM)

The GBM process follows the following SDE:

$$dS_t = (r-q)S_t dt + \sigma S_t dW_t \tag{2.121}$$

and its CGF and first and second derivatives are

$$\begin{aligned} K_0(x) &= x(\ln S_0 + (r - q - \frac{\sigma^2}{2})t) + \frac{x^2\sigma^2}{2}t && (2.122)\\ K_0'(x) &= \ln(S_0) + (r - q - \frac{\sigma^2}{2})t + \sigma^2 x t && (2.123)\\ K_0''(x) &= \sigma^2 t && (2.124) \end{aligned}$$

We use the following set of parameters for pricing: spot price $S_0 = 100$, risk-free rate $r = 5\%$, time to maturity $T = 1/12$, and volatility $\sigma = 0.25$. Table 2.9 shows the results using the Black–Scholes formula, Fourier cosine (COS), fractional FFT (FrFFT), fast Fourier transform (FFT), and saddlepoint (SP) methods with strike $K$ ranging from 10 to 200.

#### 2.5.5.2 Heston Stochastic Volatility Model

The Heston stochastic volatility model follows the following SDE:

$$\begin{aligned} dS_t &= rS_t dt + \sqrt{v_t} S_t dW_{1t} && (2.125)\\ dv_t &= \kappa(\theta - v_t)dt + \sigma\sqrt{v_t} dW_{2t} && (2.126) \end{aligned}$$

Its CGF is

$$\begin{aligned} K_0(x) &= (\ln S_0 + rt)x + \frac{\kappa\theta t(\kappa - \rho\sigma x)}{\sigma^2} - \frac{2\kappa\theta}{\sigma^2}\ln\alpha(x)\\ &\quad + \frac{(x^2 - x)v_0}{\gamma(x)\alpha(x)}\sinh\frac{\gamma(x)t}{2} && (2.127) \end{aligned}$$

with

$$\begin{aligned} \gamma(x) &= \sqrt{\sigma^2(-x^2+x) + (\kappa - x\rho\sigma)^2} && (2.128)\\ \alpha(x) &= \cosh\frac{\gamma(x)t}{2} + \frac{\kappa - \rho\sigma x}{\gamma(x)}\sinh\frac{\gamma(x)t}{2} && (2.129) \end{aligned}$$

**TABLE 2.9**: GBM European calls for a variety of strikes via various transform techniques

| $K$ | BS | COS | FrFFT | FFT | SP |
|---|---|---|---|---|---|
| 10 | 90.0830 | 94.4112 | 90.0829 | 90.0830 | 90.0832 |
| 20 | 80.1660 | 82.5614 | 80.1660 | 80.1660 | 80.1483 |
| 30 | 70.2490 | 70.2490 | 70.2490 | 70.2490 | 69.0518 |
| 40 | 60.3319 | 60.3319 | 60.3319 | 60.3319 | 60.3529 |
| 50 | 50.4149 | 50.4149 | 50.4149 | 50.4149 | 50.4144 |
| 60 | 40.4979 | 40.4979 | 40.4979 | 40.4979 | 40.4977 |
| 70 | 30.5809 | 30.5809 | 30.5808 | 30.5809 | 30.5809 |
| 80 | 20.6651 | 20.6651 | 20.6650 | 20.6651 | 20.6650 |
| 90 | 10.9147 | 10.9147 | 10.9149 | 10.9147 | 10.9148 |
| 100 | 3.3006 | 3.3006 | 3.3004 | 3.3006 | 3.3004 |
| 110 | 0.4182 | 0.4182 | 0.4182 | 0.4182 | 0.4182 |
| 120 | 0.0207 | 0.0207 | 0.0207 | 0.0207 | 0.0207 |
| 130 | 4.42e-04 | 4.42e-04 | 4.52e-04 | 4.42e-04 | 4.42e-04 |
| 140 | 4.69e-06 | 4.42e-06 | -2.05e-05 | 4.57e-06 | 4.69e-06 |
| 150 | 2.82e-08 | -3.49e-05 | 2.85e-05 | 2.29e-07 | 2.82e-08 |
| 160 | 1.08e-10 | -1.57e-03 | -1.55e-05 | -4.47e-08 | 1.08e-10 |
| 170 | 2.86e-13 | -2.93e-02 | -3.82e-06 | -7.30e-08 | 2.89e-13 |
| 180 | 5.72e-16 | -2.66e-01 | 1.18e-05 | 1.10e-07 | 2.57e-21 |
| 190 | 9.13e-19 | -1.36e+00 | -3.11e-06 | -1.06e-07 | -4.11e-24 |
| 200 | 1.22e-21 | -4.45e+00 | -6.58e-06 | 9.37e-08 | 2.39e-27 |

It should be noted that $\gamma(x)$ and $\alpha(x)$ can be complex and the formula for $K_0(x)$ is still valid. We use numerical differentiation to calculate its first and second derivatives:

$$K_0'(x) = \frac{K_0(x+h) - K_0(x-h)}{2h} \tag{2.130}$$

$$K_0''(x) = \frac{K_0(x+h) - 2K_0(x) + K_0(x-h)}{h^2} \tag{2.131}$$

We use the following set of parameters for pricing: spot price $S_0 = 100$, risk-free rate $r = 3\%$, mean reversion rate $\kappa = 2$, volatility of volatility $\sigma = 0.5$, long-term variance $\theta = 0.04$, initial variance $v_0 = 0.04$, correlation $\rho = -0.7$ and time to maturity $T = 0.5$. Table 2.10 shows the results using Monte Carlo, FFT, FrFFT, COS, and saddlepoint methods with strike $K$ ranging from 10 to 200.

#### 2.5.5.3 Variance Gamma Model

The variance gamma model is described by the following set of equations:

$$\begin{aligned}
\ln(\frac{S_t}{S_0}) &= (r+\omega)t + X_{VG}(t;\sigma,\nu,\theta) \\
X_{VG} &= \theta G(t;\nu) + \sigma W(G(t;\nu)) \\
\omega &= \frac{1}{\nu}\ln(1 - \theta\nu - \frac{\sigma^2\nu}{2})
\end{aligned}$$

**TABLE 2.10**: Heston European calls for a variety of strikes via various transform techniques

| $K$ | MC | COS | FrFFT | FFT | SP |
|---|---|---|---|---|---|
| 10 | 91.5068 | 90.6275 | 90.1489 | 90.1489 | 90.1010 |
| 20 | 81.5068 | 80.3014 | 80.2978 | 80.2977 | 80.1590 |
| 30 | 71.5068 | 70.4467 | 70.4467 | 70.4467 | 69.8742 |
| 40 | 61.5081 | 60.5965 | 60.5967 | 60.5967 | 54.2022 |
| 50 | 51.5170 | 50.7539 | 50.7541 | 50.7541 | 50.3408 |
| 60 | 41.5600 | 40.9446 | 40.9449 | 40.9449 | 40.7358 |
| 70 | 31.7184 | 31.2484 | 31.2486 | 31.2487 | 31.1377 |
| 80 | 22.1918 | 21.8620 | 21.8622 | 21.8622 | 21.8040 |
| 90 | 13.4038 | 13.2020 | 13.2023 | 13.2022 | 13.1275 |
| 100 | 6.1522 | 6.0552 | 6.0555 | 6.0554 | 5.9538 |
| 110 | 1.6659 | 1.6368 | 1.6371 | 1.6371 | 1.6043 |
| 120 | 2.41e-01 | 2.34e-01 | 2.35e-01 | 2.35e-01 | 2.39e-01 |
| 130 | 2.89e-02 | 2.72e-02 | 2.75e-02 | 2.74e-02 | 2.88e-02 |
| 140 | 3.72e-03 | 3.13e-03 | 3.39e-03 | 3.36e-03 | 3.57e-03 |
| 150 | 4.28e-04 | 1.97e-04 | 4.60e-04 | 4.78e-04 | 4.84e-04 |
| 160 | 3.25e-05 | -1.94e-04 | 6.89e-05 | 7.50e-05 | 7.23e-05 |
| 170 | 0.00e+00 | -2.52e-04 | 1.13e-05 | -2.87e-06 | 1.18e-05 |
| 180 | 0.00e+00 | -2.61e-04 | 2.03e-06 | 1.51e-05 | 2.12e-06 |
| 190 | 0.00e+00 | -2.63e-04 | 3.94e-07 | -9.71e-06 | 4.11e-07 |
| 200 | 0.00e+00 | -2.63e-04 | 8.23e-08 | 8.02e-06 | 8.59e-08 |

where $G(t;\nu)$ is the gamma process and $\omega$ is the convexity correction. Under this model the CGF and its first and second derivatives are

$$\begin{aligned}
K_0(x) &= x((r+\omega)t + \ln S_0) - \frac{t}{\nu}\ln(1 - x\theta\nu - \frac{\nu\sigma^2 x^2}{2}) \\
K_0'(x) &= \ln S_0 + (r+\omega)t + t(\frac{\theta + \sigma^2 x}{1 - \theta\nu x - 0.5\nu\sigma^2 x^2}) \\
K_0''(x) &= \frac{t\sigma^2}{1 - \theta\nu x - 0.5\nu\sigma^2 x^2} + t\nu(\frac{\theta + \sigma^2 x}{1 - \theta\nu x - 0.5\nu\sigma^2 x^2})^2
\end{aligned}$$

We use the following set of parameters for pricing: spot price $S_0 = 100$, risk-free rate $r = 10\%$, time to maturity $T = 1/12$, volatility $\sigma = 0.12$, $\theta = -0.14$, and $\nu = 0.2$.

Table 2.11 shows the results using closed-form, FFT, FrFFT, COS, and saddlepoint methods with strike $K$ ranging from 10 to 200.

#### 2.5.5.4 CGMY Model

The CGMY model is defined as

$$S_t = S(0)\exp[(r - q + \omega)t + X_{CGMY}(t; C, G, M, Y)] \tag{2.132}$$

where $X_{CGMY}(t; C, G, M, Y)$ is the CGMY process and $\omega$ is defined as

$$\omega = C\Gamma(-Y)[(M-1)^Y - M^Y + (G+1)^Y - G^Y] \tag{2.133}$$

**TABLE 2.11**: VG European calls for a variety of strikes via various transform techniques

| $K$ | Analytical | COS | FrFFT | FFT | SP |
|---|---|---|---|---|---|
| 10 | 90.0832 | 95.1859 | 90.0830 | 90.0983 | 90.0460 |
| 20 | 80.1660 | 80.8759 | 80.1660 | 80.1687 | 80.0835 |
| 30 | 70.2490 | 70.2490 | 70.2490 | 70.2388 | 70.1157 |
| 40 | 60.3320 | 60.3319 | 60.3319 | 60.3303 | 60.1560 |
| 50 | 50.4149 | 50.4149 | 50.4149 | 50.4010 | 50.2120 |
| 60 | 40.4979 | 40.4979 | 40.4979 | 40.4869 | 40.2881 |
| 70 | 30.5813 | 30.5813 | 30.5813 | 30.5969 | 30.3879 |
| 80 | 20.6702 | 20.6704 | 20.6704 | 20.6617 | 20.5189 |
| 90 | 10.8289 | 10.8289 | 10.8289 | 10.7983 | 10.7156 |
| 100 | 1.8150 | 1.8150 | 1.8151 | 1.7913 | 1.5406 |
| 110 | 0.0195 | 1.94e-02 | 1.95e-02 | 5.29e-02 | 2.26e-02 |
| 120 | 6.9339e-04 | 5.38e-04 | 5.83e-04 | 1.15e-02 | 6.57e-04 |
| 130 | 2.7159e-05 | -1.08e-06 | 2.56e-05 | -3.43e-03 | 2.85e-05 |
| 140 | 5.7237e-06 | -1.25e-04 | 1.89e-07 | -6.44e-03 | 1.63e-06 |
| 150 | 3.90e-08 | -3.71e-04 | -1.73e-06 | 2.86e-03 | 1.16e-07 |
| 160 | 2.14e-09 | -1.53e-03 | 1.45e-06 | 1.47e-03 | 1.00e-08 |
| 170 | 0.00e+00 | -5.19e-03 | 1.37e-06 | -2.84e-03 | 1.01e-09 |
| 180 | 0.00e+00 | -1.73e-02 | -3.27e-07 | 2.61e-03 | 1.16e-10 |
| 190 | 0.00e+00 | -5.47e-02 | 7.88e-07 | -2.05e-03 | 1.51e-11 |
| 200 | 0.00e+00 | -1.68e-01 | -7.31e-07 | 1.65e-03 | 2.19e-12 |

Under this model the CGF and its first and second derivatives are:

$$\begin{aligned}
K_0(x) &= x(\ln S_0 + (r-q+w)T) + TC\Gamma(-Y)\left((M-x)^Y - M^Y + (G+x)^Y - G^Y\right) \\
K_0'(x) &= \ln S_0 + (r-q+w)T + YTC\Gamma(-Y)[-(M-x)^{Y-1} + (G+x)^{Y-1}] \\
K_0''(x) &= Y(Y-1)TC\Gamma(-Y)[(M-x)^{Y-2} + (G+x)^{Y-2}]
\end{aligned}$$

We use the following set of parameters for pricing: spot price $S_0 = 100$, risk-free rate $r = 3\%$, dividend yield $q = 0$, time to maturity $T = 0.5$, $C = 2$, $G = 5$, $M = 10$, and $Y = 0.5$.

Table 2.12 shows the results using Monte Carlo, COS, FrFFT, FFT, and saddlepoint methods with strike $K$ ranging from 10 to 200.

---

## 2.6 Power Option Pricing via the Fourier Transform

As we mentioned in the introduction, one of the caveats of the Fourier method is that it is not only restricted to pricing European options, but it must be rederived for different payoffs and may not exist for an arbitrary terminal payoff. However, there is one payoff for which the derivation of the premium via Fourier techniques is almost identical. This option has the payoff

$$(X_T^p - K)^+ \tag{2.134}$$

and its derivation is almost identical to that of a call option.

We begin by calculating the power call option premium in terms of the distribution of

**TABLE 2.12**: CGMY European calls for a variety of strikes via various transform techniques

| $K$ | COS | FrFFT | FFT | SP |
|---|---|---|---|---|
| 10 | 90.1550 | 90.1489 | 90.1489 | 90.1022 |
| 20 | 80.2889 | 80.8011 | 80.5694 | 80.2066 |
| 30 | 70.4568 | 70.0928 | 70.7195 | 70.3456 |
| 40 | 60.6829 | 59.7980 | 60.9336 | 60.5796 |
| 50 | 51.0318 | 49.9905 | 51.2981 | 50.9989 |
| 60 | 41.7592 | 40.7630 | 41.9856 | 41.7455 |
| 70 | 33.0236 | 32.2781 | 33.2416 | 33.0309 |
| 80 | 25.1175 | 24.7296 | 25.3514 | 25.1301 |
| 90 | 18.2872 | 18.2849 | 18.5797 | 18.3285 |
| 100 | 12.8258 | 13.0302 | 13.0968 | 12.8267 |
| 110 | 8.7340 | 8.9707 | 8.9274 | 8.65796 |
| 120 | 5.6257 | 5.9822 | 5.9417 | 5.68396 |
| 130 | 3.7307 | 3.8926 | 3.9118 | 3.66418 |
| 140 | 2.2610 | 2.4923 | 2.5827 | 2.34064 |
| 150 | 1.5694 | 1.5836 | 1.7335 | 1.49278 |
| 160 | 0.8697 | 1.0064 | 1.1967 | 0.95588 |
| 170 | 0.6637 | 0.6432 | 0.8582 | 0.616932 |
| 180 | 0.4176 | 0.4153 | 0.6437 | 0.402308 |
| 190 | 0.1733 | 0.2714 | 0.5068 | 0.265448 |
| 200 | 0.2730 | 0.1810 | 0.4183 | 0.177337 |

the log asset price.

$$\begin{aligned} C_T^p(K) &= \mathrm{C}\mathbb{E}\left((X_T^p - K)^+\right) \\ &= \mathrm{C}\int_K^\infty (X_T^p - K) f_T(X) dX_T \\ &= \mathrm{C}\int_k^\infty (e^{px} - e^k) q_T(x) dx \end{aligned}$$

We define the modified call premium using a damping factor as before.

$$c_T^p(k) = e^{\alpha k} C_T^p(k)$$

We calculate the characteristic function of the modified option premium, forming an inner integral which can be evaluated analytically.

$$\begin{aligned} \Psi_T^p(\nu) &= \int_{-\infty}^\infty e^{i\nu k} c_T^p(k) dk \\ &= \int_{-\infty}^\infty e^{i\nu k}\left(\mathrm{C}e^{\alpha k}\int_k^\infty (S^p - K) q(s) ds\right) dk \\ &= \int_{-\infty}^\infty e^{i\nu k}\left(\mathrm{C}e^{\alpha k}\int_k^\infty (e^{ps} - e^k) q(s) ds\right) dk \\ &= \mathrm{C}\int_{-\infty}^\infty \int_{-\infty}^s e^{(\alpha+i\nu)k}(e^{ps} - e^k) q(s) dk ds \\ &= \mathrm{C}\int_{-\infty}^\infty q(s)\left(\int_{-\infty}^s e^{(\alpha+i\nu)s}(e^{ps} - e^k) dk\right) ds \end{aligned}$$

We evaluate the inner integral, which converges due to the addition of the damping factor.

$$\begin{aligned}
\int_{-\infty}^{s} e^{(\alpha+i\nu)k}(e^{px} - e^{k})dk &= e^{ps}\frac{e^{(\alpha+i\nu)k}}{(\alpha+i\nu)}\bigg|_{-\infty}^{s} - \frac{e^{(\alpha+i\nu+1)k}}{(\alpha+i\nu+1)}\bigg|_{-\infty}^{s} \\
&= \frac{e^{(\alpha+i\nu+p)s}}{\alpha+i\nu} - \frac{e^{(\alpha+i\nu+1)s}}{\alpha+i\nu+1}
\end{aligned}$$

Finally, we can solve for the characteristic function of the option premium in terms of the characteristic function of the log asset price, giving us

$$\begin{aligned}
\Psi_T^p(\nu) &= \mathrm{C}\int_{-\infty}^{\infty}\left(\frac{e^{(\alpha+i\nu+p)x}}{\alpha+i\nu} - \frac{e^{(\alpha+i\nu+1)s}}{\alpha+i\nu+1}\right)q(s)ds \\
&= \frac{\mathrm{C}}{\alpha+i\nu}\int_{-\infty}^{\infty} e^{(\alpha+i\nu+p)x}q(x)dx - \frac{e^{-rT}}{\alpha+i\nu+1}\int_{-\infty}^{\infty} e^{(\alpha+i\nu+1)x}q(x)dx \\
&= \frac{\mathrm{C}}{\alpha+i\nu}\int_{-\infty}^{\infty} e^{i(\nu-(\alpha+p)i)x}q(x)dx - \frac{e^{-rT}}{\alpha+i\nu+1}\int_{-\infty}^{\infty} e^{i(\nu-(\alpha+1)i)x}q(x)dx \\
&= \frac{\mathrm{C}}{\alpha+i\nu}\int_{-\infty}^{\infty} e^{i(\nu-(\alpha+p)i)x}q(x)dx - \frac{e^{-rT}}{\alpha+i\nu+1}\int_{-\infty}^{\infty} e^{i(\nu-(\alpha+1)i)x}q(x)dx \\
&= \frac{\mathrm{C}}{\alpha+i\nu}\Phi(\nu-(\alpha+p)i) - \frac{\mathrm{C}}{\alpha+i\nu+1}\Phi(\nu-(\alpha+1)i)
\end{aligned}$$

So we have the final expression for the Fourier transform of the modified option premium as

$$\begin{aligned}
\Psi_T^p(\nu) &= \int_{-\infty}^{\infty} e^{i\nu k}c_T(k)dk \\
&= \frac{\mathrm{C}}{\alpha+i\nu}\Phi(\nu-(\alpha+p)i) - \frac{\mathrm{C}}{\alpha+i\nu+1}\Phi(\nu-(\alpha+1)i)
\end{aligned}$$

and one again we can use the equation for the characteristic function and the inverse Fourier transform of $\Psi_T^p(\nu)$ to calculate $C_T^p(k)$, the option premium.

$$C_T^p(k) = \frac{e^{-\alpha k}}{2\pi}\int_{-\infty}^{\infty} e^{-i\nu k}\Psi_T^p(\nu)d\nu \tag{2.135}$$

---

## Problems

1. In fast Fourier transform (FFT), we define the range for log of strikes as $k_m = \beta + (m-1)\Delta k = \beta + (m-1)\lambda$, for $m = 1, \ldots, N$ and some $\beta$. There are many choices for $\beta$. One of which is to set $\beta = \ln S_0 - \frac{\lambda N}{2}$. This choice for $\beta$ would cause at-the-money strike to fall in the middle of our range of strikes where $S_0$ is today spot. For this choice of $\beta$, if one is interested in finding the premium for $k = \log(K)$ one would typically interpolate. If you are interested in finding the premium for a specific strike say $k_0 = \log(K_0)$ without any interpolation, what $\beta$ would you choose? What would be the corresponding index number?

2. The characteristic function of the log of the stock price in the Black–Scholes framework

is:

$$\begin{aligned}\Phi(u) &= \mathbb{E}(e^{iu\ln S_T}) \\ &= \mathbb{E}(e^{ius_T}) \\ &= \exp\left(i(s_0+(r-q-\sigma^2/2)T)u - \frac{1}{2}\sigma^2u^2T\right)\end{aligned}$$

Use the FFT method to price a European call option for the following set of parameters: spot price $S_0 = \$100$, strike price $K = 90$, risk-free interest rate $r = 5.25\%$, dividend rate $q = 2\%$, time to maturity $T = 2$ years, and volatility $\sigma = 25\%$ and compare it with the closed-form call price.

3. In the Heston stochastic volatility model, the stock price follows the following SDE:

$$\begin{aligned}dS_t &= rS_tdt + \sqrt{v_t}S_tdW_t^{(1)} \\ dv_t &= \kappa(\theta - v_t)dt + \sigma\sqrt{v_t}dW_t^{(2)}\end{aligned}$$

and the characteristic function for the log of the stock price process is given by

$$\begin{aligned}\Phi(u) &= \mathbb{E}(e^{iu\ln S_t}) \\ &= \frac{\exp\{\frac{\kappa\theta t(\kappa-i\rho\sigma u)}{\sigma^2} + iutr + iu\ln S_0\}}{(\cosh\frac{\gamma t}{2} + \frac{\kappa-i\rho\sigma u}{\gamma}\sinh\frac{\gamma t}{2})^{\frac{2\kappa\theta}{\sigma^2}}}\exp\left\{-\frac{(u^2+iu)v_0}{\gamma\coth\frac{\gamma t}{2}+\kappa-i\rho\sigma u}\right\},\end{aligned}$$

where $\gamma = \sqrt{\sigma^2(u^2+iu)+(\kappa-i\rho\sigma u)^2}$. Use the FFT method to price a European call using the following parameters: spot price, $S_0 = \$100$, strike price $K = 90$, maturity $T = 1$ year, risk-free rate $r = 5.25\%$, volatility of volatility $\sigma = 30\%$, $\kappa = 1$, $\theta = 0.08$, $\rho = -0.8$, and $\nu_0 = 0.04$.

4. The variance gamma model is described by the following set of equations:

$$\begin{aligned}\ln(\frac{S_t}{S_0}) &= (r-q+\omega)t + X_{VG}(t;\sigma,\nu,\theta) \\ X_{VG} &= \theta G(t;\nu) + \sigma W(G(t;\nu)) \\ \omega &= \frac{1}{\nu}\ln(1-\theta\nu-\frac{\sigma^2\nu}{2})\end{aligned}$$

The characteristic function for the time $t$ level of the VG process is

$$\phi_{X(t)}(u) = \mathbb{E}(e^{iuX(t)}) = \left(\frac{1}{1-iu\theta\nu+\sigma^2u^2\nu/2}\right)^{\frac{t}{\nu}} \tag{2.136}$$

By the definition of risk-neutrality, the price of a European put option with strike $K$ and maturity $T$ is

$$p(S(0);K,t) = e^{-rT}\mathbb{E}_0((K-S(T))^+)$$

For the following set of parameters: spot price $S_0 = \$100$, strike price $K = \$105$, time to maturity $T = 1$ year, risk-free interest rate $r = 4.75\%$, continuous dividend rate $q = 1.25\%$, $\sigma = 25\%$, $\nu = 0.50$, and $\theta = -0.3$, price a European put option via

(a) the FFT technique

(b) simulation

and compare the results.

5. Show that cumulants $c_1$, $c_2$, and $c_4$ for

   (a) the Black–Scholes model are

$$\begin{aligned} c_1 &= (r-q)T \\ c_2 &= \sigma^2 T \\ c_4 &= 0 \end{aligned}$$

   (b) the Heston stochastic volatility model are

$$\begin{aligned} c_1 &= rT + (1.0 - e^{-\kappa T})\frac{\theta - v_0}{2\kappa} - 0.5\theta T \\ c_2 &= \frac{1}{8\kappa^3}(\sigma T\kappa e^{-\kappa T}(v_0 - \theta)(8\kappa\rho - 4\sigma)) \\ &+ \kappa\rho\sigma(1 - e^{-\kappa T}(16\theta - 8v_0)) \\ &+ 2\theta\kappa T(-4\kappa\rho\sigma + \sigma^2 + 4\kappa^2) \\ &+ \sigma^2((\theta - 2v_0)e^{-2\kappa T} + \theta(6e^{-\kappa T} - 7) + 2v_0) \\ &+ 8\kappa^2(v_0 - \theta)(1.0 - e^{-\kappa T}) \end{aligned}$$

   (c) the variance gamma model are

$$\begin{aligned} c_1 &= (r+\theta)T \\ c_2 &= (\sigma^2 + v\theta^2)T \\ c_4 &= 3(\sigma^4 v + 2\theta^4 v^3 + 4\sigma^2\theta^2 v^2)T \end{aligned}$$

   (d) the CGMY model are

$$\begin{aligned} c_1 &= (r+\omega)T + TC\Gamma(1-Y)(-M^{Y-1} + G^{Y-1}) \\ c_2 &= \alpha + CT\Gamma(2-Y)(M^{Y-2} + G^{Y-2}) \\ c_4 &= CT\Gamma(4-Y)(M^{Y-4} + G^{Y-4}) \end{aligned}$$

   where

$$\begin{aligned} \omega &= -C\Gamma(-Y)((M-1)^Y) + (G+1)^Y - M^Y - G^Y \\ \alpha &= \frac{8C}{(G+M)^2 - (G-M)^2} \end{aligned}$$

6. The formula for the range of the integration, $[a, b]$, in the COS method we have been using is

$$[a,b] = \left[c_1 - L\sqrt{c_2 + \sqrt{c_4}}, c_1 + L\sqrt{c_2 + \sqrt{c_4}}\right] \quad \textit{with} \quad L = 10 \qquad (2.137)$$

   where $c_n$ denotes the $n$-th cumulant of $\ln(X_T/K)$. Examine the sensitivity of the results in Tables 2.5, 2.6, and 2.7 to the choices of $a$ and $b$ and compare the results and conclude.

7. For the following set of parameters: spot price $S_0 = \$100$, time to maturity $T = 1$ year, risk-free interest rate of $r = 0.025\%$, continuous dividend rate of $q = 1.25\%$, $\sigma = 25\%$, $\nu = 0.50$, $\theta = -0.3$, and $Y = 0.5$ compare the European put premiums of COS and fractional FFT for the CGMY model for strike range of $K = \$70, \$80, \$90$, and $\$100$.

8. Suppose we have a cumulant generating function of the log of the stock price under the risk-neutral measure for a model given by $K_0(x)$. Show that the cumulant generating function of the log of the stock price under the share measure less an exponential is given by

$$K(x) = K_0(x+1) - K_0(1) - \ln(1+x)$$

# Chapter 3

## Introduction to Finite Differences

In this chapter we will introduce finite difference methods used for numerically solving partial differential equations (PDEs). This chapter will focus on the most commonly used finite difference techniques utilized to solve PDEs, namely, explicit, implicit, Crank–Nicolson, and multi-step schemes. We will then apply these methods to solving the heat equation, which is an excellent example that not only illustrates the critical issues we need to consider when applying finite difference methods, but can in fact be related to the Black–Scholes PDE through variable and coordinate substitution.

Finite differences is one of the oldest and most popular techniques for solving PDEs numerically. It involves discretizing each derivative in the PDE according to the Taylor expansion of some chosen order. The problem domain is then discretized and the discretized PDE is applied to each point in the resulting grid. The technique begins with known solution values at the boundary conditions and then applies the discretized PDE to compute solutions for neighboring points. This step is repeated until we have solved the PDE for every point on the mesh we created on our problem domain. Thus this method involves solving an approximately equal problem over a finite set of points on our problem domain, with the goal of converging to a numerically sufficient solution if we make our discretization small enough.

We should mention that in this book we will not cover finite element method or finite volume method, the other two approaches commonly used in engineering and mathematics, for numerically solving partial differential equations. If not all, a majority of PDEs in finance can be numerically solved via the finite difference method. For that reason, we think it is adequate just to cover the finite difference method. However, at the end of the chapter we will have two simple problems introducing these two methods and showing that for these two special cases they yield very similar results that is comparable to those of the finite difference method. For further reading on finite element with application in finance look at [212] and for finite volume in financial mathematics look at [167] and [220].

### 3.1 Taylor Expansion

The finite difference method relies on the discretization of the derivatives in the PDE which applies under the model used, and derivative discretization begins with use of the Taylor expansion. Thus we will briefly review Taylor expansions and the derivative approximations that can be derived from them.

Assume the function $f(x)$ has infinitely many continuous derivatives, $f \in C^\infty$. If we assume the variable $x$ is deterministic, then its Taylor expansion[1] near $x = a$ is

$$f(x) = f(a) + (x-a)f^{(1)}(a) + \frac{(x-a)^2}{2!}f^{(2)}(a) + \frac{(x-a)^3}{3!}f^{(3)}(a) + \ldots \quad (3.1)$$

$$= \sum_{n=0}^{\infty} \frac{(x-a)^n}{n!} f^{(n)}(a) \quad (3.2)$$

If function $f(x)$ has only $(k+1)$ continuous derivatives, $f \in C^{k+1}$, then we can write its Taylor expansion near $x = a$ as

$$f(x) = f(a) + (x-a)f^{(1)}(a) + \frac{(x-a)^2}{2!}f^{(2)}(a) + \cdots + \frac{(x-a)^k}{k!}f^{(k)}(a) + \frac{(x-a)^{k+1}}{(k+1)!}f^{(k+1)}(\xi)$$

for some $\xi \in (x, a)$. The last term is called the remainder and hereafter it is written in Landau notation as $O((x-a)^{k+1})$.

Below we provide some examples of derivative approximations derived from Taylor expansions.

**Example 1** *Forward and Backward Difference Approximation of First Derivative*

Assume $f \in C^2$ and we are interested in approximating $f^{(1)}(x)$. Using the truncated Taylor expansion, we can write both forward and backward expansion as

$$f(x+h) = f(x) + hf^{(1)}(x) + \frac{h^2}{2!}f^{(2)}(\xi_1) \quad (3.3)$$

$$f(x-h) = f(x) - hf^{(1)}(x) + \frac{h^2}{2!}f^{(2)}(\xi_2) \quad (3.4)$$

Solving for $f^{(1)}(x)$ in (3.3) we see that

$$f^{(1)}(x) = \frac{f(x+h) - f(x)}{h} - \frac{h}{2}f^{(2)}(\xi_1)$$

$$\approx \frac{f(x+h) - f(x)}{h} \quad \text{with} \quad O(h) \quad (3.5)$$

Solving for $f^{(1)}(x)$ in (3.4) we see that

$$f^{(1)}(x) = \frac{f(x) - f(x-h)}{h} + \frac{h}{2}f^{(2)}(\xi_2)$$

$$\approx \frac{f(x) - f(x-h)}{h} \quad \text{with} \quad O(h) \quad (3.6)$$

Difference equations (3.5) and (3.6) are called *forward difference* and *backward difference* approximations of the first derivative, respectively. Both approximations are first order.

**Example 2** *Central Difference Approximation of First Derivative*

[1]Taylor expansion is valid in deterministic calculus; the equivalent of this expansion in stochastic calculus is Itô's Lemma.

Assume $f \in C^3$ and now we are interested in second order approximation of $f^{(1)}(x)$. Using Taylor expansion we can write

$$f(x+h) = f(x) + hf^{(1)}(x) + \frac{h^2}{2!}f^{(2)}(x) + \frac{h^3}{3!}f^{(3)}(\xi_3) \quad (3.7)$$

$$f(x-h) = f(x) - hf^{(1)}(x) + \frac{h^2}{2!}f^{(2)}(x) - \frac{h^3}{3!}f^{(3)}(\xi_4) \quad (3.8)$$

Subtracting (3.8) from (3.7) we obtain

$$f(x+h) - f(x-h) = 2hf^{(1)}(x) + \frac{h^3}{3!}f^{(3)}(\xi_3) + \frac{h^3}{3!}f^{(3)}(\xi_4) \quad (3.9)$$

$$= 2hf^{(1)}(x) + O(h^3) \quad (3.10)$$

Therefore we have

$$f^{(1)}(x) = \frac{f(x+h) - f(x-h)}{2h} + O(h^2)$$

$$\approx \frac{f(x+h) - f(x-h)}{2h} \quad \text{with} \quad O(h^2) \quad (3.11)$$

Difference equation (3.11) is called the *central difference* approximation of the first derivative. This approximation is of order 2.

**Example 3** *Central Difference Approximation of the Second Derivative*

Assume $f \in C^4$ and we are interested in a second order approximation of $f^{(2)}(x)$. Using Taylor expansion we can write

$$f(x+h) = f(x) + hf^{(1)}(x) + \frac{h^2}{2!}f^{(2)}(x) + \frac{h^3}{3!}f^{(3)}(x) + \frac{h^4}{4!}f^{(4)}(\xi_5) \quad (3.12)$$

$$f(x-h) = f(x) - hf^{(1)}(x) + \frac{h^2}{2!}f^{(2)}(x) - \frac{h^3}{3!}f^{(3)}(x) + \frac{h^4}{4!}f^{(4)}(\xi_6) \quad (3.13)$$

By adding (3.12) and (3.13) we obtain

$$f(x+h) + f(x-h) = 2f(x) + 2\frac{h^2}{2!}f^{(2)}(x) + \frac{h^4}{4!}f^{(4)}(\xi_5) + \frac{h^4}{4!}f^{(4)}(\xi_6) \quad (3.14)$$

and solving for $f^{(2)}(x)$ we get

$$f^{(2)}(x) = \frac{f(x-h) - 2f(x) + f(x+h)}{h^2} + O(h^2) \quad (3.15)$$

Difference equation (3.15) is called the *central difference* second derivative approximation of order 2. These examples are very straightforward and cover the majority of cases necessary for applying finite difference techniques. For higher order derivatives and higher order accuracy, deriving the difference equations can become rather cumbersome. In Section 3.4 we review methods for automatically generating coefficients for higher order and higher derivative approximations.

---

## 3.2 Finite Difference Method

In this section we will begin by introducing finite difference methods in the context of solving the heat equation. This equation is somewhat simpler than the Black–Scholes PDE;

however, the Black–Scholes PDE can in fact be transformed into the one-dimensional heat equation through simple variable and coordinate substitution. Thus it contains all of the relevant features necessary to illustrate the finite difference technique.

Consider the one-dimensional heat equation

$$u_\tau - \kappa u_{xx} = 0 \tag{3.16}$$

for $a \leq x \leq b$ and $0 \leq \tau \leq T$, with initial condition

$$u(x,0) = f(x) \tag{3.17}$$

and boundary conditions

$$u(a,\tau) = g(\tau) \tag{3.18}$$

$$u(b,\tau) = h(\tau) \tag{3.19}$$

To apply the finite difference method we must redefine the domain of the problem, going from a continuous domain $D = \{a \leq x \leq b; 0 \leq \tau \leq T\}$ to a discrete grid. The most basic grid construction is an equally spaced grid with $M$ equal sub-intervals on the $\tau$-axis and $N$ equal sub-intervals on the $x$-axis. This results in the following mesh on $[a,b] \times [0,T]$:

$$\bar{D} = \left\{ \begin{array}{lll} x_j = a + (j-1)\Delta x; & \Delta x = \frac{b-a}{N}; & j = 1,\ldots,N+1 \\ \tau_k = 0 + (k-1)\Delta\tau; & \Delta\tau = \frac{T-0}{M}; & k = 1,\ldots,M+1 \end{array} \right\}$$

To illustrate the various finite difference techniques discussed in this chapter we will use the following example grid as shown in Figure 3.1, where bold black lines indicate known initial and boundary conditions. To apply finite difference techniques we will also need to

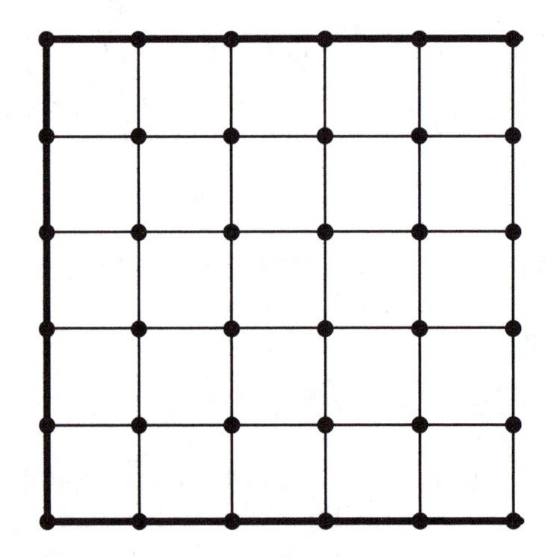

**FIGURE 3.1**: Example grid

write a discrete version of the differential operator

$$\mathcal{L}(u) = u_\tau - \kappa u_{xx} = 0 \tag{3.20}$$

as a difference equation. How we do this determines what type of finite difference technique we are using and the following sections discuss different ways of performing this discretization. In the following sections we let $u_{j,k}$ be the true value of $u(x = x_j, \tau = \tau_k)$ and let $U_{j,k}$ be its discretized approximation.

### 3.2.1 Explicit Discretization

Explicit discretization schemes use only forward difference approximations in time. Let us define all the grid points at a given time $\tau = t$ to be a time slice, consisting of all the solutions on the grid at that time. The explicit discretization schemes allow us to derive the solution to the PDE at each point on the current time slice from points derived at the previous time independent of any points in the current time slice.

To discretize the PDE at the grid point $(x_j, \tau_k)$, which we call the reference point, we use a forward difference approximation for the first term, $\frac{\partial u(x_j,\tau_k)}{\partial \tau}$, and a central difference approximation for the diffusion term, $\frac{\partial^2 u(x_j,\tau_k)}{\partial x^2}$.

The forward difference approximation of the first term is

$$\frac{\partial u(x_j,\tau_k)}{\partial \tau} = \frac{u_{j,k+1} - u_{j,k}}{\Delta\tau} + O(\Delta\tau)$$

and the central difference approximation of the second term is

$$\frac{\partial^2 u(x_j,\tau_k)}{\partial x^2} = \frac{u_{j-1,k} - 2u_{j,k} + u_{j+1,k}}{\Delta x^2} + O(\Delta x^2)$$

So, we obtain the following set of equations at point $(x_j, \tau_k)$:

$$\begin{aligned} u_\tau(x_j,\tau_k) - \kappa u_{xx}(x_j,\tau_k) &= 0 \\ \frac{u_{j,k+1} - u_{j,k}}{\Delta\tau} - \kappa\frac{u_{j-1,k} - 2u_{j,k} + u_{j+1,k}}{\Delta x^2} &= O(\Delta x^2) + O(\Delta\tau) \end{aligned}$$

dropping the orders and use the approximation for $u_{j,k}$ to get

$$\frac{U_{j,k+1} - U_{j,k}}{\Delta\tau} - \kappa\frac{U_{j-1,k} - 2U_{j,k} + U_{j+1,k}}{\Delta x^2} = 0 \tag{3.21}$$

where we can see that the discretization leads to an error on the order of $O(\Delta x^2) + O(\Delta\tau)$.

To solve the PDE at points $(x_j, \tau_k)$ on our problem domain we begin at $\tau = 0$, where the initial conditions determine the solution to the PDE at every grid point in this time slice. In order to solve for the points on the grid where $\tau = \Delta\tau$ we apply the explicit finite difference scheme. Figures 3.2 and 3.3 illustrate this solution technique. In Figure 3.2 we

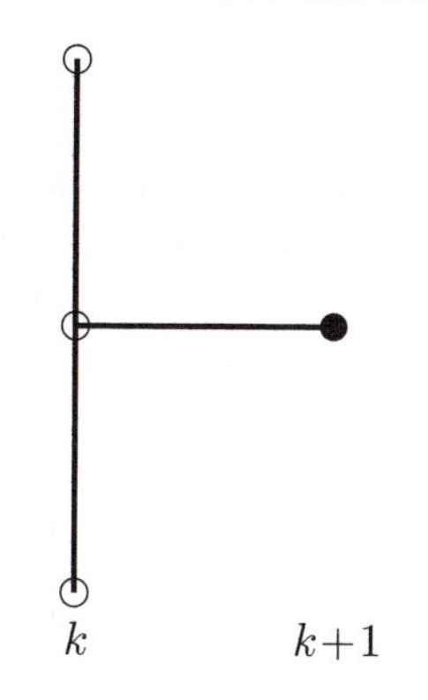


**FIGURE 3.2**: Explicit finite difference stencil

illustrate the explicit finite difference stencil indicating grid points used at time $\tau_k$ in order to calculate points at time $\tau_{k+1}$. In this figure unfilled circles indicate known values and

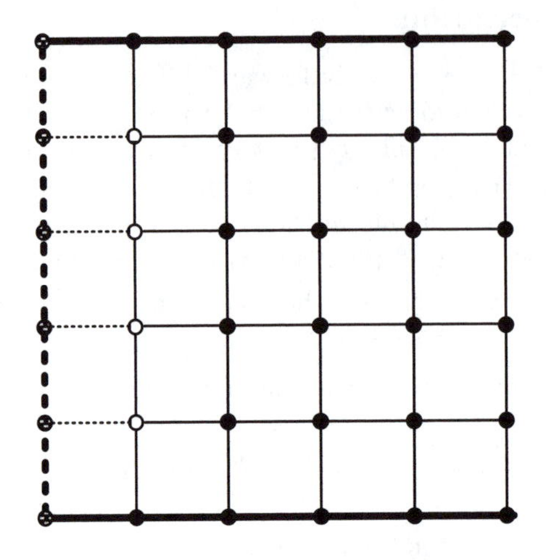

**FIGURE 3.3**: Explicit finite difference grid

disks (filled circles) values we are solving for. In Figure 3.3 bold black lines indicate known boundary conditions, bold dashed lines indicate known initial conditions, unfilled points indicate values used to calculate the solution to the PDE in the first time slice, and filled points indicate points on our grid that are solved for. Notice that when solving for points in the current time slice we are dependent only on values in the previous time period, so these points can be solved for independently.

To solve the PDE at points $(x_j, \tau_k)$ on the entire grid, we simply apply this technique recursively until we get to time $\tau = T$. Typically, we solve for all the points in a time slice simultaneously by rearranging the finite difference equations to solve for each point in terms of grid solutions from the previous time period. In this case, rearranging (3.21) yields

$$U_{j,k+1} - U_{j,k} - \frac{\kappa \Delta\tau}{\Delta x^2}(U_{j-1,k} - 2U_{j,k} + U_{j+1,k}) \quad = \quad 0 \tag{3.22}$$

If we let $\rho = \frac{\kappa\Delta\tau}{\Delta x^2}$ we get

$$U_{j,k+1} = \rho U_{j-1,k} + (1-2\rho)U_{j,k} + \rho U_{j+1,k}, \quad \text{for} \quad 2 \leq j \leq N, \quad 1 \leq k \leq M$$

Note that $j = 1$ and $j = N+1$ correspond to the boundary conditions, $u(x_1, \tau_k) = g(\tau_k), u(x_{N+1}, \tau_k) = h(\tau_k)$, and $k = 1$ corresponds to the initial condition, $u(x_j, \tau_1) = f(x_j) = f_j$. At all of these grid points the solution is already known. The approximate PDE solution for the time slice $\tau_k = (k-1)\Delta\tau$ can be written in vector form as

$$U_k = \begin{pmatrix} U_{2,k} \\ U_{3,k} \\ \vdots \\ U_{N,k} \end{pmatrix}$$

Thus we can write the recursive explicit discretization scheme in matrix form as

$$U_{k+1} = A_{\text{Explicit}} U_k + \rho \begin{pmatrix} U_{1,k} \\ 0 \\ \vdots \\ 0 \\ U_{N+1,k} \end{pmatrix} \tag{3.23}$$

where

$$A_{\text{Explicit}} = \begin{pmatrix} 1-2\rho & \rho & & & \\ \rho & 1-2\rho & \rho & & \\ & \ddots & \ddots & \ddots & \\ & & \rho & 1-2\rho & \rho \\ & & & \rho & 1-2\rho \end{pmatrix}$$

#### 3.2.1.1 Algorithm for the Explicit Scheme

Let the initial condition be

$$U_1 = \begin{pmatrix} U_{2,1} \\ U_{3,1} \\ \vdots \\ U_{N,1} \end{pmatrix} = \begin{pmatrix} f_2 \\ f_3 \\ \vdots \\ f_N \end{pmatrix}$$

For $k = 1 : M$

$$U_{k+1} = A_{\text{Explicit}} U_k + \rho \begin{pmatrix} U_{1,k} \\ 0 \\ \vdots \\ 0 \\ U_{N+1,k} \end{pmatrix}$$

End

The final solution at time $\tau_{M+1} = M\Delta\tau = T$ is

$$U_{M+1} = \begin{pmatrix} U_{2,M+1} \\ U_{3,M+1} \\ \vdots \\ U_{N,M+1} \end{pmatrix}$$

### 3.2.2 Implicit Discretization

Implicit discretization schemes use backward difference approximations in time and central difference approximations within the current time slice. Unlike explicit discretization schemes that allow us to derive the solution to the PDE at each point on the current time slice from only points derived at the previous time, when using implicit discretization schemes we must solve for all points in the current time slice simultaneously as they are interdependent. In practice this means performing a matrix inversion on our coefficient matrix; however, its unique form makes this problem easy to solve.

To perform implicit discretization of (3.16) at the grid point $(x_j, \tau_{k+1})$, we use a backward difference approximation for the first term, $\frac{\partial u(x_j,\tau_{k+1})}{\partial \tau}$, and a central difference approximation for the diffusion term, $\frac{\partial^2 u(x_j,\tau_{k+1})}{\partial x^2}$.

The backward difference approximation for the first term yields

$$\frac{\partial u(x_j, \tau_{k+1})}{\partial \tau} = \frac{u_{j,k+1} - u_{j,k}}{\Delta \tau} + O(\Delta \tau)$$

and the central difference approximation for the second term yields

$$\frac{\partial^2 u(x_j, \tau_{k+1})}{\partial x^2} = \frac{u_{j-1,k+1} - 2u_{j,k+1} + u_{j+1,k+1}}{\Delta x^2} + O(\Delta x^2)$$

So, we obtain the following set of equations at point $(x_j, \tau_{k+1})$:

$$\begin{aligned} u_\tau(x_j, \tau_{k+1}) - \kappa u_{xx}(x_j, \tau_{k+1}) &= 0 \\ \frac{u_{j,k+1} - u_{j,k}}{\Delta \tau} - \kappa \frac{u_{j-1,k+1} - 2u_{j,k+1} + u_{j+1,k+1}}{\Delta x^2} &= O(\Delta x^2) + O(\Delta \tau) \end{aligned}$$

dropping the orders and use the approximation for $u_{j,k}$ to get

$$\frac{U_{j,k+1} - U_{j,k}}{\Delta \tau} - \kappa \frac{U_{j-1,k+1} - 2U_{j,k+1} + U_{j+1,k+1}}{\Delta x^2} = 0 \tag{3.24}$$

where we can see that the discretization leads to an error on the order of $O(\Delta x^2) + O(\Delta \tau)$.

To solve the PDE at points $(x_j, \tau_{k+1})$ on our problem domain, we begin at $\tau = 0$, where the initial conditions determine the solution to the PDE at every grid point in this time slice. In order to solve for the points on the grid for which $\tau = \Delta \tau$ we apply the implicit finite difference scheme. Figures 3.4 and 3.5 illustrate this solution technique. In Figure 3.4 we illustrate the implicit finite difference stencil indicating grid points used at time $\tau_k$ in order to calculate points at time $\tau_{k+1}$. As before, unfilled circles indicate known values and disks (filled circles) values we are solving for.

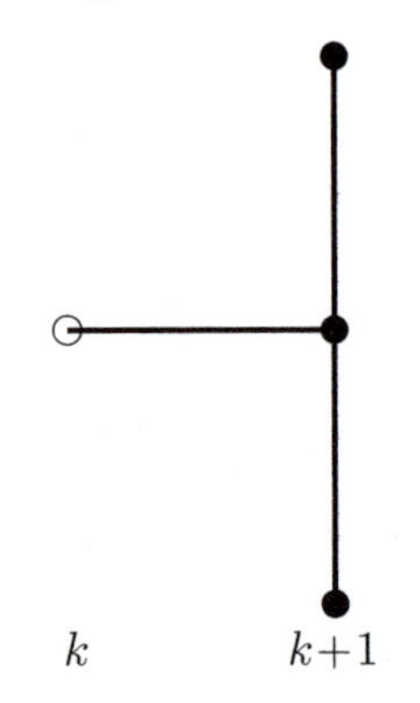


**FIGURE 3.4**: Implicit finite difference stencil

Notice that unlike the explicit finite difference scheme, when solving for points on the current time slice we are dependent on all of the values in the current time slice and thus we must solve for all of these points simultaneously. This is done through matrix inversion of this system of equations.

To solve the PDE at points $(x_j, \tau_{k+1})$ on the entire grid, we simply apply this technique recursively until we get to time $\tau = T$. We solve for all the points in a time slice simultaneously by rearranging the finite difference equations to solve for each point in terms of

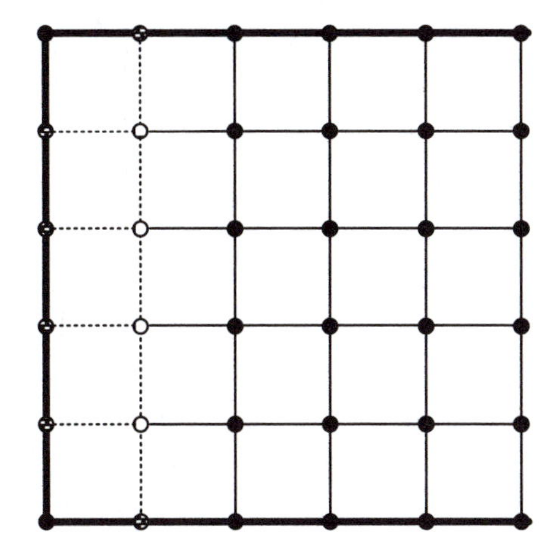

**FIGURE 3.5**: Implicit finite difference grid

grid solutions from the previous time period and from the current time period. In this case, rearranging (3.24) yields

$$U_{j,k+1} - U_{j,k} - \frac{\kappa\Delta\tau}{\Delta x^2}(U_{j-1,k+1} - 2U_{j,k+1} + U_{j+1,k+1}) = 0 \tag{3.25}$$

If we let $\rho = \frac{\kappa\Delta\tau}{\Delta x^2}$ we get

$$-\rho U_{j-1,k+1} + (1+2\rho)U_{j,k+1} - \rho U_{j+1,k+1} = U_{j,k}, \quad \text{for} \quad 2 \leq j \leq N, \quad 1 \leq k \leq M$$

As before, $j = 1$ and $j = N + 1$ correspond to the boundary conditions, $u(1,k) = g(\tau_k), u(N+1,k) = h(\tau_k)$, and $k = 1$ corresponds to the initial conditions, $u_{j,1} = f(x_j) = f_j$. At all of these grid points the solution is already known. The approximate PDE solution for the time slice $\tau = k\Delta\tau$ can be written in matrix form as

$$A_{\text{Implicit}}U_{k+1} = U_k + \rho \begin{pmatrix} U_{1,k+1} \\ 0 \\ \vdots \\ 0 \\ U_{N+1,k+1} \end{pmatrix} \tag{3.26}$$

where

$$A_{\text{Implicit}} = \begin{pmatrix} 1+2\rho & -\rho & & & \\ -\rho & 1+2\rho & -\rho & & \\ & \ddots & \ddots & \ddots & \\ & & -\rho & 1+2\rho & -\rho \\ & & & -\rho & 1+2\rho \end{pmatrix}$$

#### 3.2.2.1 Algorithm for the Implicit Scheme

Let the initial condition be

$$U_1 = \begin{pmatrix} U_{2,1} \\ U_{3,1} \\ \vdots \\ U_{N,1} \end{pmatrix} = \begin{pmatrix} f_2 \\ f_3 \\ \vdots \\ f_N \end{pmatrix}$$

For $k = 1 : M$

$$A_{\text{Implicit}} U_{k+1} = U_k + \rho \begin{pmatrix} U_{1,k+1} \\ 0 \\ \vdots \\ 0 \\ U_{N+1,k+1} \end{pmatrix}$$

End

The final solution at time $T = \tau_{M+1}$ is

$$U_{M+1} = \begin{pmatrix} U_{2,M+1} \\ U_{3,M+1} \\ \vdots \\ U_{N,M+1} \end{pmatrix}$$

Notice that we must solve

$$A_{\text{Implicit}} U_{k+1} = U_k + \rho \begin{pmatrix} U_{1,k+1} \\ 0 \\ \vdots \\ 0 \\ U_{N+1,k+1} \end{pmatrix}$$

for every time step in our implicit algorithm. This necessarily involves inverting the matrix $A_{\text{Implicit}}$ for each step, which for a general matrix would take $O(n^3)$ operations. This would render the algorithm unviable in a practical setting. Fortunately, the matrix $A_{\text{Implicit}}$ is tridiagonal or pentadiagonal, and a modified form of Gaussian elimination can be used to solve a system of equations defined by a tridiagonal or pentadiagonal matrix in $O(n)$ operations. In Section 3.5 we provide pseudo-codes on how to solve matrix equations involving tridiagonal or pentadiagonal matrices. This makes this algorithm not only competitive, but superior to explicit discretization because of its improved stability, which we will discuss later in the chapter.

### 3.2.3 Crank–Nicolson Discretization

The Crank–Nicolson discretization scheme [84] is based on the averaging of the forward and backwards finite difference approximations, allowing for a better order of approximation in time. It uses a central difference approximation in time and the average of two central difference approximations in space. Like implicit discretization, the use of a central difference approximation in the current time slice forces us to solve simultaneously for all points in the current time slice. In practice this means performing a matrix inversion on our coefficient matrix, but again it is a tridiagonal or a pentadiagonal matrix and so numerically easy to solve.

To perform Crank–Nicolson discretization of (3.16) at the grid point $(x_j, \tau_{k+\frac{1}{2}})$, we use a central difference approximation for the first term,

$$\frac{\partial u(x_j, \tau_{k+\frac{1}{2}})}{\partial \tau}$$

and a trapezoidal rule using the average of two central difference approximations for the diffusion term,

$$\frac{\partial^2 u(x_j, \tau_{k+\frac{1}{2}})}{\partial x^2}$$

For the central difference approximation of the first term we refer to (3.11), noting that if we let $x = x - \frac{1}{2}$ and $h = \frac{1}{2}h$ we get the central difference approximation for the first term

$$\frac{\partial u(x_j, \tau_{k+\frac{1}{2}})}{\partial \tau} = \frac{u_{j,k+1} - u_{j,k}}{\Delta \tau} + O(\Delta \tau^2)$$

For the trapezoidal rule approximation for the diffusion term, we average the two central difference formulas around the point $u(x_j, \tau_{k+\frac{1}{2}})$ to get

$$\frac{\partial^2 u(x_j, \tau_{k+\frac{1}{2}})}{\partial x^2} = \frac{u_{j-1,k+1} - 2u_{j,k+1} + u_{j+1,k+1} + u_{j-1,k} - 2u_{j,k} + u_{j+1,k}}{2\Delta x^2} + O(\Delta x^2)$$

So, we obtain the following set of equations at point $(x_j, \tau_{k+\frac{1}{2}})$:

$$u_\tau(x_j, \tau_{k+\frac{1}{2}}) - \kappa u_{xx}(x_j, \tau_{k+\frac{1}{2}}) = 0$$

$$\frac{u_{j,k+1} - u_{j,k}}{\Delta \tau} - \kappa \frac{u_{j-1,k+1} - 2u_{j,k+1} + u_{j+1,k+1} + u_{j-1,k} - 2u_{j,k} + u_{j+1,k}}{2\Delta x^2} = O(\Delta x^2) + O(\Delta \tau^2)$$

dropping the orders and use the approximation for $u_{j,k}$ to get

$$\frac{U_{j,k+1} - U_{j,k}}{\Delta \tau} - \kappa \frac{U_{j-1,k+1} - 2U_{j,k+1} + U_{j+1,k+1} + U_{j-1,k} - 2U_{j,k} + U_{j+1,k}}{2\Delta x^2} = 0 \quad (3.27)$$

where we can see that the discretization leads to an error on the order of $O(\Delta x^2) + O(\Delta \tau^2)$, an improvement over the explicit and implicit methods. Note that the grid point $(x_j, \tau_{k+\frac{1}{2}})$ does not appear in the difference equation (3.27) albeit the entire discretization was based on that grid point.

To solve the PDE at points $(x_j, \tau_{k+1})$ on our problem domain, we begin at $\tau = 0$, where the initial conditions determine the solution to the PDE at every grid point in this time slice. In order to solve for the points on the grid for which $\tau = \Delta \tau$ we apply the Crank–Nicolson finite difference scheme. Figures 3.6 and 3.7 illustrate this solution technique. In Figure 3.6 we display the Crank–Nicolson finite difference stencil. The grid point $(x_j, \tau_{k+\frac{1}{2}})$ is represented as a star (as opposed to a circle) acknowledging it was used in the discretization procedure but does not play any role in the difference equation and is not part of the stencil.

Notice that unlike the explicit finite difference scheme, when solving for points on the current time slice we are again dependent on all of the values in the current time slice, and thus we must solve for all of these points simultaneously, which we do through matrix inversion of this system of equations.

To solve the PDE at points $(x_j, \tau_{k+1})$ on the entire grid, we simply apply this technique recursively until we get to time $\tau = T$. We solve for all the points in a time slice simultaneously by rearranging the finite difference equations to solve for each point in terms of

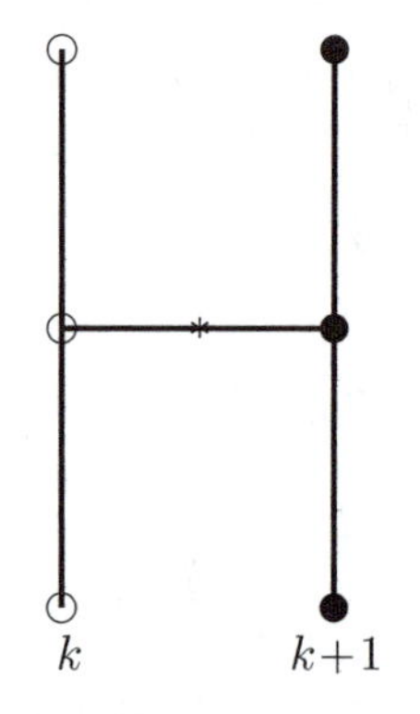


**FIGURE 3.6**: Crank–Nicolson finite difference stencil

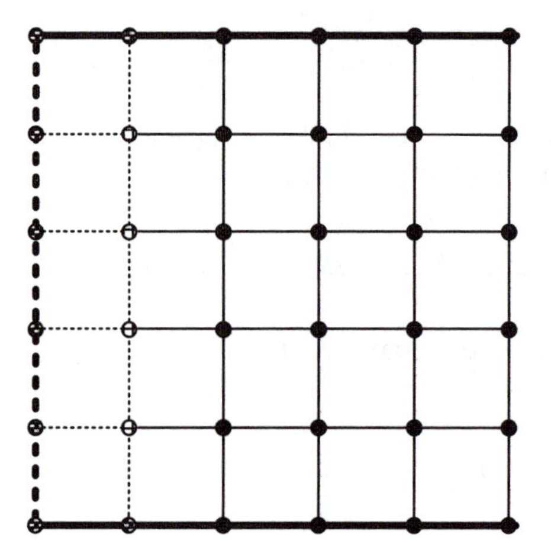

**FIGURE 3.7**: Crank–Nicolson finite difference grid

grid solutions from the previous time period and from the current time period. In this case, rearranging (3.27) yields

$$U_{j,k+1} - U_{j,k} - \frac{\kappa \Delta \tau}{2\Delta x^2}(U_{j-1,k+1} - 2U_{j,k+1} + U_{j+1,k+1} + U_{j-1,k} - 2U_{j,k} + U_{j+1,k}) = 0 \quad (3.28)$$

If we let $\hat{\rho} = \frac{\kappa \Delta \tau}{2\Delta x^2}$ we get

$$U_{j,k+1} - U_{j,k} - \hat{\rho}(U_{j-1,k+1} - 2U_{j,k+1} + U_{j+1,k+1} + U_{j-1,k} - 2U_{j,k} + U_{j+1,k}) = 0$$

We rearrange this equation to move all $\tau_k$ terms to one side, obtaining the following difference equation:

$$-\hat{\rho}U_{j-1,k+1} + (1 + 2\hat{\rho})U_{j,k+1} - \hat{\rho}U_{j+1,k+1} = \hat{\rho}U_{j-1,k} + (1 - 2\hat{\rho})U_{j,k} + \hat{\rho}U_{j+1,k} \quad (3.29)$$

Noting that $\hat{\rho} = \frac{1}{2}\rho$ we can rewrite the above difference equation as

$$-\rho U_{j-1,k+1} + (2 + 2\rho)U_{j,k+1} - \rho U_{j+1,k+1} = \rho U_{j-1,k} + (2 - 2\rho)U_{j,k} + \rho U_{j+1,k} \quad (3.30)$$

As before, $j = 1$ and $j = N + 1$ correspond to the boundary conditions, $u(1, k) =$

$g(\tau_k), u(N+1,k) = h(\tau_k)$, and $k = 1$ corresponds to the initial conditions, $u_{j,1} = f(x_j) = f_j$. At all of these grid points the solution is already known. The approximate PDE solution for the time slice $\tau = k\Delta\tau$ can be written in matrix form as

$$\Big(A_{\text{Implicit}} + I\Big)U_{k+1} = \Big(A_{\text{Explicit}} + I\Big)U_k + \rho \begin{pmatrix} U_{1,k} + U_{1,k+1} \\ 0 \\ \vdots \\ 0 \\ U_{N+1,k} + U_{N+1,k+1} \end{pmatrix} \tag{3.31}$$

We note that adding (3.23) and (3.26) would yield the same results.

#### 3.2.3.1 Algorithm for the Crank–Nicolson Scheme

Let the initial condition be

$$U_1 = \begin{pmatrix} U_{2,1} \\ U_{3,1} \\ \vdots \\ U_{N,1} \end{pmatrix} = \begin{pmatrix} f_2 \\ f_3 \\ \vdots \\ f_N \end{pmatrix}$$

For $k = 1 : M$

$$\Big(A_{\text{Implicit}} + I\Big)U_{k+1} = \Big(A_{\text{Explicit}} + I\Big)U_k + \rho \begin{pmatrix} U_{1,k} + U_{1,k+1} \\ 0 \\ \vdots \\ 0 \\ U_{N+1,k} + U_{N+1,k+1} \end{pmatrix}$$

End

The final solution at time $T = \tau_{M+1}$ is

$$U_{M+1} = \begin{pmatrix} U_{2,M+1} \\ U_{3,M+1} \\ \vdots \\ U_{N,M+1} \end{pmatrix}$$

Notice that we must solve

$$\Big(A_{\text{Implicit}} + I\Big)U_{k+1} = \Big(A_{\text{Explicit}} + I\Big)U_k + \rho \begin{pmatrix} U_{1,k} + U_{1,k+1} \\ 0 \\ \vdots \\ 0 \\ U_{N+1,k} + U_{N+1,k+1} \end{pmatrix}$$

for every step in the Crank–Nicolson algorithm. This again involves a matrix inversion, but the matrix is still tridiagonal and so this only requires $O(n)$ operations. As discussed in [39], the Crank–Nicolson method suffers from problems with oscillating solutions. We will not be exploring this issue here, but instead leave this as an exercise at the end of the chapter for the reader. On how to reduce the Crank–Nicolson oscillations, refer to [184].

### 3.2.4 Multi-Step Scheme

While the explicit, implicit, and Crank–Nicolson schemes remain the most common methods for numerically solving PDEs, we can always use better finite difference approximations for the derivatives in order to improve the accuracy of our method. One such scheme is the multi-step finite difference scheme, which uses a second order backwards approximation in time, using points from two previous periods, in order to achieve an accuracy of $O((\Delta\tau)^2)$ in time. We will use an implicit central difference approximation for the diffusion term, which will again force us to solve simultaneously for all points in the current time slice. But as with the other implicit schemes, the necessary matrix inversion takes only $O(n)$ operations as the coefficient matrix is still tridiagonal.

To perform multi-step discretization of (3.16) at the grid point $(x_j, \tau_{k+2})$, we use backward approximation for the first term of order $O((\Delta\tau)^2)$, $\frac{\partial u(x_j,\tau_{k+2})}{\partial\tau}$, and a central difference approximation for the diffusion term, $\frac{\partial^2 u(x_j,\tau_{k+2})}{\partial x^2}$.

The backward difference approximation for the first term yields

$$\frac{\partial u(x_j,\tau_{k+2})}{\partial\tau} = \frac{u_{j,k} - 4u_{j,k+1} + 3u_{j,k+2}}{2\Delta\tau} + O((\Delta\tau)^2)$$

and the central approximation for the second term yields

$$\frac{\partial^2 u(x_j,\tau_{k+2})}{\partial x^2} = \frac{u_{j-1,k+2} - 2u_{j,k+2} + u_{j+1,k+2}}{\Delta x^2} + O((\Delta x)^2)$$

So, we obtain the following set of equations at point $(x_j, \tau_{k+2})$,

$$\begin{aligned} u_\tau(x_j,\tau_{k+2}) - \kappa u_{xx}(x_j,\tau_{k+2}) &= 0 \\ \frac{u_{j,k} - 4u_{j,k+1} + 3u_{j,k+2}}{2\Delta\tau} - \kappa\frac{u_{j-1,k+2} - 2u_{j,k+2} + u_{j+1,k+2}}{\Delta x^2} &= 0 \\ &+ O((\Delta x)^2) + O((\Delta\tau)^2) \\ \frac{U_{j,k} - 4U_{j,k+1} + 3U_{j,k+2}}{2\Delta\tau} - \kappa\frac{U_{j-1,k+2} - 2U_{j,k+2} + U_{j+1,k+2}}{(\Delta x)^2} &= 0 \end{aligned} \tag{3.32}$$

where we can see that the discretization leads to an error on the order of $O((\Delta x)^2) + O((\Delta\tau)^2)$

Note that we cannot apply the multi-step scheme to the very first time slice, as we only have solutions for the PDE at the immediate previous time step, which is the boundary condition. Thus this multi-step method must have an initial step which generates solutions for the first time slice. In this case we are restricted to using a first order derivative approximation in time, and the implicit method is typically chosen for its greater stability.

Algorithmically this does not pose a problem; however, we have noted previously that the implicit method has an error of $O(\Delta x^2) + O(\Delta\tau)$, which is greater than that of the multi-step method. Also, in the context of options pricing, much of the discontinuity in the option price is centered around the terminal payoff. Thus we must be especially careful that we do not generate errors in this region which will be propagated backwards in time. One obvious solution is to use more granular time steps closer to expiry, which will give us more accuracy in this important region. This topic will be discussed in the next chapter, in the section on adaptive grid points.

To solve the PDE at the reference point $(x_j, \tau_{k+2})$ on our problem domain, we begin at $\tau = 0$, where the initial conditions determine the solution to the PDE at every grid point in this time slice. We include an initialization step for the points on the grid for which $\tau = \Delta\tau$, where we apply the implicit finite difference scheme to solve the PDE at this time slice.

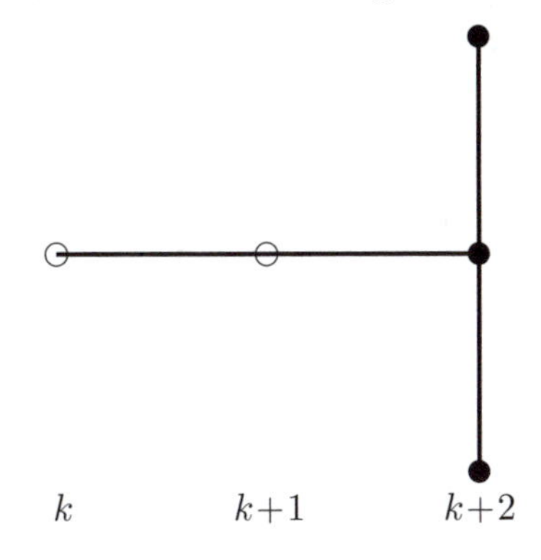


**FIGURE 3.8**: Multi-step finite difference stencil

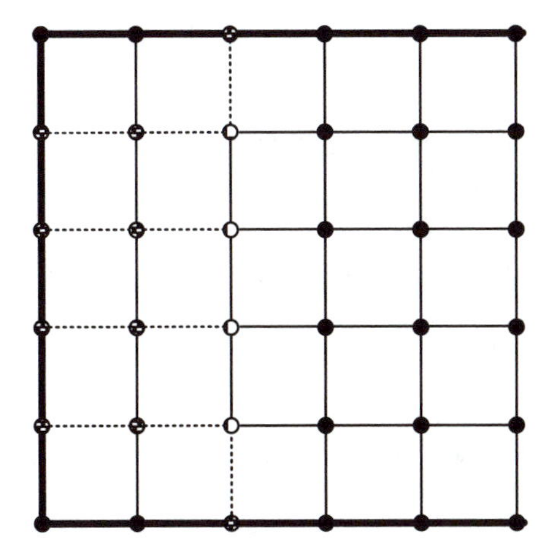

**FIGURE 3.9**: Multi-step finite difference grid

Finally, we recursively apply the multi-step scheme to calculate the solution to the PDE on our entire problem domain. Figures 3.8 and 3.9 illustrate this solution technique.

After we solve the PDE for the time slice at $\tau = \Delta\tau$ during the initialization step, we solve for all the points in the subsequent time slice simultaneously by rearranging the finite difference equations to solve for each point in terms of grid solutions from the previous two time periods and the current time period. In this case rearranging (3.32) yields

$$U_{j,k} - 4U_{j,k+1} + 3U_{j,k+2} - \frac{2\kappa\Delta\tau}{\Delta x^2}(U_{j-1,k+2} - 2U_{j,k+2} + U_{j+1,k+2}) = 0 \quad (3.33)$$

If we let $\hat{\rho} = \frac{2\kappa\Delta\tau}{(\Delta x)^2}$ we get

$$-\hat{\rho}U_{j,k+2} + (3 + 2\hat{\rho})U_{j,k+2} - \hat{\rho}U_{j+1,k+2} = -U_{j,k} + 4U_{j,k+1}, \quad \text{for } 2 \le j \le N, \quad 1 \le k \le M$$

As before, $j = 1$ and $j = N + 1$ correspond to the boundary conditions, $u(1, k) = g(\tau_k), u(N+1, k) = h(\tau_k)$, and $k = 1$ corresponds to the initial conditions, $u_{j,1} = f(x_j) = f_j$. At all of these grid points the solution is already known. The approximate PDE solution for the time slice $\tau = k\Delta\tau$ can be written in matrix form as

$$A_{\text{MS}} U_{k+2} = -U_k + 4U_{k+1} + \hat{\rho} \begin{pmatrix} U_{1,k+2} \\ 0 \\ \vdots \\ 0 \\ U_{N+1,k+2} \end{pmatrix} \tag{3.34}$$

where

$$A_{\text{MS}} = \begin{pmatrix} 3+2\hat{\rho} & -\hat{\rho} & & & \\ -\hat{\rho} & 3+2\hat{\rho} & -\hat{\rho} & & \\ & \ddots & \ddots & \ddots & \\ & & -\hat{\rho} & 3+2\hat{\rho} & -\hat{\rho} \\ & & & -\hat{\rho} & 3+2\hat{\rho} \end{pmatrix}$$

#### 3.2.4.1 Algorithm for the Multi-Step Scheme

Solve for

$$U_2 = \begin{pmatrix} U_{2,2} \\ U_{3,2} \\ \vdots \\ U_{N,2} \end{pmatrix}$$

using the implicit scheme with the initial conditions

For $k = 1 : M$

$$A_{\text{MS}} U_{k+2} = -U_k + 4U_{k+1} + \hat{\rho} \begin{pmatrix} U_{1,k+2} \\ 0 \\ \vdots \\ 0 \\ U_{N+1,k+1} \end{pmatrix}$$

End

The final solution at time $T = \tau_{M+1}$ is

$$U_{M+1} = \begin{pmatrix} U_{2,M+1} \\ U_{3,M+1} \\ \vdots \\ U_{N,M+1} \end{pmatrix}$$

Notice in the above equation we must solve

$$A_{\text{MS}} U_{k+2} = -U_k + 4U_{k+1} + \hat{\rho} \begin{pmatrix} U_{1,k+2} \\ 0 \\ \vdots \\ 0 \\ U_{N+1,k+1} \end{pmatrix}$$

for every step in the multi-step algorithm. This again involves a matrix inversion, but the matrix is still tridiagonal and so this will only take $O(n)$ operations.

## 3.3 Stability Analysis

The stability of a finite difference scheme is of paramount importance when evaluating its effectiveness. A stable scheme will always converge to the true PDE solution as we enhance the granularity of our discretization, decreasing the size of the time and space steps. In this section we will review the definition of stability and the tools necessary to evaluate the stability of a finite difference scheme. We will also evaluate the stability of the explicit, implicit, and Crank–Nicolson schemes.

When discussing the suitability of different finite difference schemes there are a number of important properties we should consider in their evaluation:

- Convergent: A finite difference scheme is said to be convergent if all of its solutions derived from the initial conditions converge point-wise to the corresponding solutions of the original differential equation as $\Delta\tau$ and $\Delta x$ approach zero. In other words, as the $\Delta\tau$ and $\Delta x$ shrink the finite difference solution must eventually converge to the corresponding solution of the original partial differential equation at every point of the problem domain.

- Consistent: A finite difference scheme is said to be consistent with the original partial differential equation if, for any given smooth function $u(x,\tau)$, the results of applying the differential operator to $u(x,\tau)$ approach the results of applying the finite difference equation operating on $U_{j,k}$ as $\Delta\tau$ and $\Delta x$ approach zero. In short, the local truncation error goes to zero as $\Delta\tau$ and $\Delta x$ approach zero. Consistency is proved by invoking Taylor's theorem.

- Stable: A finite difference scheme is said to be stable if the difference between the numerical solution and the exact solution remains bounded as $\Delta\tau$ and $\Delta x$ approach zero.

It is harder to prove a method is stable than to show that it is consistent. Stability can be proven using one of the following approaches:

Eigenvalue analysis of the matrix representation of the finite difference method

Fourier analysis on the grid (von Neumann analysis)

Computing the domain of dependence of the numerical method

The Lax–Richtmeyer equivalence theorem [192] states that *a consistent finite difference scheme for a partial differential equation, for which the initial-value problem is well posed, is convergent if and only if it is stable.* A proof of this theorem is presented in Chapter 10 of [206], but generally this means that stability is equivalent to convergence for finite difference schemes, although this theorem is only valid for linear equations. So, if we can prove a numerical solution to a PDE is stable and thus that its error remains bounded as $\Delta\tau$ approaches zero, then we are guaranteed that the solution converges everywhere on our problem domain.

A PDE is defined by its differential operator and a finite difference scheme depends on the discretization of this operator. Thus, to evaluate the stability of different schemes we must evaluate how they differ in their discretization of the differential operator. We examine the differential operator for the heat equation

$$\mathcal{L}(u) = u_\tau - \kappa u_{xx} = 0 \tag{3.35}$$

and without loss of generality, we assume

$$u(a,\tau) = 0 \tag{3.36}$$
$$u(b,\tau) = 0 \tag{3.37}$$

As discussed in the previous section, the discretization of this operator can be expressed as a linear system of the form

$$U_{k+1} = BU_k \tag{3.38}$$

Specifically, in the three schemes we have examined we have

$B = A_{\text{Explicit}}$ for the explicit scheme

$B = A^{-1}_{\text{Implicit}}$ for the implicit scheme

$B = (A_{\text{Implicit}}+I)^{-1}(A_{\text{Explicit}}+I)$ for the Crank–Nicolson scheme

The essence of the stability condition is that there should be a limit to the extent to which any initial error can be amplified by the numerical solution. Let $U_k$ be the true solution of the linear system and $\hat{U}_k$ be the computed solution. We define the local error term as

$$e_k \equiv \hat{U}_k - U_k \tag{3.39}$$

As discussed in [200], we have the following relationship between the local error and the error propagated by the finite difference scheme:

$$\begin{aligned} Be_k &= B(\hat{U}_k - U_k) \\ &= \hat{U}_{k+1} - U_{k+1} \\ &= e_{k+1} \end{aligned}$$

and by induction we have

$$e_{k+1} = B^k e_1 \tag{3.40}$$

A finite difference method will only be stable if previous errors are reduced by the application of the difference equation; they cannot be propagated indefinitely or grow. This leads us to the stability condition

$$B^k e_1 \to 0 \quad \text{for} \quad k \to \infty \tag{3.41}$$

Let $\lambda_i$ for $i = 1, \ldots, n$ be the eigenvalues of $n \times n$ matrix $A$. Then the spectral radius of $A$ is

$$\rho(A) = \max_i |\lambda_i| \tag{3.42}$$

The spectral radius is closely related to the behavior of the convergence of the power of a matrix.

**Lemma 1** *Let $A \in R^{n\times n}$ be an $n \times n$ matrix and let $\rho(A)$ be its spectral radius. Then*

$$\lim_{k\to\infty} A^k = 0 \text{ if and only if } \rho(A) < 1 \tag{3.43}$$

**Proof 1** *First we show that* $\lim_{k\to\infty} A^k = 0$ *implies* $\rho(A) < 1$*. Let* $\lambda$ *be an eigenvalue of* $A$ *and* $v$ *be its corresponding eigenvector. We know that* $A^k v = \lambda^k v$ *by definition. Since* $\lim_{k\to\infty} A^k = 0$ *we can write*

$$\begin{aligned} 0 &= (\lim_{k\to\infty} A^k)v && (3.44)\\ &= \lim_{k\to\infty} A^k v && (3.45)\\ &= \lim_{k\to\infty} \lambda^k v && (3.46)\\ &= v \lim_{k\to\infty} \lambda^k && (3.47) \end{aligned}$$

*Thus, because* $v$ *is not equal to zero, this implies that* $\lim_{k\to\infty} \lambda^k = 0$*, which implies* $|\lambda| < 1$*. Since this is true for all eigenvalues, we know that* $\rho(A) < 1$*.*

*Now we show that* $\rho(A) < 1$ *implies* $\lim_{k\to\infty} A^k = 0$ *by using* Jordan canonical form. *Any* $n \times n$ *matrix can be put in Jordan canonical form by a similarity transformation*

$$T^{-1}AT = J = \begin{pmatrix} J_1 & & \\ & \ddots & \\ & & J_q \end{pmatrix} \tag{3.48}$$

*where*

$$J_i = \begin{pmatrix} \lambda_i & 1 & & & \\ & \lambda_i & 1 & & \\ & & \ddots & \ddots & \\ & & & \lambda_i & 1 \\ & & & & \lambda_i \end{pmatrix} \in \mathbb{C}^{n_i \times n_i} \tag{3.49}$$

*is called a Jordan block of size* $n_i$ *with eigenvalue* $\lambda_i$*, so* $n = \sum_{i=1}^{q} n_i$*. We can see that* $J$ *is block diagonal and* $J_i$ *is bidiagonal. Rearranging the above equation we get*

$$A = TJT^{-1} \tag{3.50}$$

*and thus it is easy to see that*

$$A^k = TJ^kT^{-1} \tag{3.51}$$

*where*

$$J^k = \begin{pmatrix} J_1^k & & \\ & \ddots & \\ & & J_q^k \end{pmatrix} \tag{3.52}$$

*with*

$$J_i^k = \begin{pmatrix} \lambda_i^k & \binom{k}{1}\lambda_i^{k-1} & \binom{k}{2}\lambda_i^{k-2} & \cdots & \binom{k}{n_i-1}\lambda_i^{k-n_i+1} \\ & \lambda_i^k & \binom{k}{1}\lambda_i^{k-1} & \cdots & \binom{k}{n_i-2}\lambda_i^{k-n_i+2} \\ & & \ddots & \ddots & \vdots \\ & & & \lambda_i^k & \binom{k}{1}\lambda_i^{k-1} \\ & & & & \lambda_i^k \end{pmatrix} \tag{3.53}$$

*Therefore if* $\rho(A) < 1$ *then* $|\lambda_i| < 1$ *for all* $i$ *and that implies all elements of* $J_i^k$ *approach zero as* $k$ *approaches infinity. Therefore* $\lim_{k\to\infty} J^k = 0$ *and thus*

$$\lim_{k\to\infty} A^k = \lim_{k\to\infty} TJ^kT^{-1} = T(\lim_{k\to\infty} J^k)T^{-1} = 0 \tag{3.54}$$

**Lemma 2** *Let $B$ be the following $N \times N$ tridiagonal matrix:*

$$B = \begin{pmatrix} \alpha & \beta & & & \\ \gamma & \alpha & \beta & & \\ & \ddots & \ddots & \ddots & \\ & & \gamma & \alpha & \beta \\ & & & \gamma & \alpha \end{pmatrix}$$

*Then the eigenvalues and eigenvectors of $B$ are*

$$\lambda_B^i = \alpha + 2\beta \left(\frac{\gamma}{\beta}\right)^{1/2} \cos\left(\frac{i\pi}{N+1}\right) \quad i = 1, \ldots, N \tag{3.55}$$

*and*

$$v^{(i)} = \begin{pmatrix} \left(\frac{\gamma}{\beta}\right)^{1/2} \sin\left(\frac{i\pi}{N+1}\right) \\ \left(\frac{\gamma}{\beta}\right)^{2/2} \sin\left(\frac{2i\pi}{N+1}\right) \\ \vdots \\ \left(\frac{\gamma}{\beta}\right)^{N/2} \sin\left(\frac{Ni\pi}{N+1}\right) \end{pmatrix}$$

**Proof 2** *Look at [123] for the proof.*

### 3.3.1 Stability of the Explicit Scheme

The explicit finite difference scheme has a difference operator matrix $B$ which is tridiagonal, therefore we can apply the above lemma to derive its eigenvalues analytically. For the explicit scheme we have $\alpha = 1 - 2\rho$ and $\beta = \gamma = \rho$, so its eigenvalues are

$$\begin{aligned} \lambda^i_{A_{\text{Explicit}}} &= (1 - 2\rho) + 2\rho \cos(\frac{i\pi}{N}) \quad \text{for } i = 1, \ldots, N-1 \\ &= 1 - 2\rho(1 - \cos(\frac{i\pi}{N})) \\ &= 1 - 4\rho \sin^2(\frac{i\pi}{2N}) \end{aligned}$$

We know that this scheme is stable only if $|\lambda^i| < 1$ for all $i$, and we have

$$|1 - 4\rho \sin^2(\frac{i\pi}{2N})| < 1 \Rightarrow 0 < \rho \sin^2(\frac{i\pi}{2N}) < \frac{1}{2} \qquad \text{for } i = 1, \ldots, N-1$$

As the largest sin term is $\sin\left(\frac{(N-1)\pi}{2N}\right) < 1$, we must satisfy $0 < \rho < \frac{1}{2}$ for the explicit scheme to be stable. Considering that $\rho = \frac{\kappa \Delta \tau}{(\Delta x)^2}$, this is equivalent to the following constraint on the time-step size:

$$0 < \Delta \tau < \frac{(\Delta x)^2}{2\kappa}$$

This is the so-called Courant–Friedrichs–Lewy condition (CFL condition [80]). Thus the explicit method is not unconditionally stable, but depends on the relationship between the space and time discretization.

### 3.3.2 Stability of the Implicit Scheme

The implicit finite difference scheme has a difference operator matrix $A^{-1}_{\text{Implicit}}$ which is not tridiagonal. However, $A_{\text{Implicit}}$ is tridiagonal and therefore we can again apply the above lemma to derive its eigenvalues analytically. For the implicit scheme we have $\alpha = 1 + 2\rho$ and $\beta = \gamma = -\rho$, so the eigenvalues of $A^{\text{Implicit}}$ are

$$\begin{aligned}
\lambda^i_{A_{\text{Implicit}}} &= (1+2\rho) + 2(-\rho)\cos(\frac{i\pi}{N}) \quad \text{for } i = 1, \ldots, N-1 \\
&= 1 + 2\rho(1 - \cos(\frac{i\pi}{N})) \\
&= 1 + 4\rho\sin^2(\frac{i\pi}{2N})
\end{aligned}$$

Therefore the eigenvalues of the difference operator matrix are

$$\lambda^i_B = \lambda^i_{A^{-1}_{\text{Implicit}}} = \frac{1}{1 + 4\rho\sin^2\left(\frac{i\pi}{2N}\right)} \quad \text{for } i = 1, \ldots, N$$

and

$$|\lambda^i_B| = \left|\frac{1}{1 + 4\rho\sin^2\left(\frac{i\pi}{2N}\right)}\right| < 1 \quad \text{for } i = 1, \ldots, N-1$$

However, this condition is always true and thus the implicit finite difference scheme is unconditionally stable.

### 3.3.3 Stability of the Crank–Nicolson Scheme

The Crank–Nicolson finite difference scheme has a difference operator matrix $(A_{\text{Implicit}} + I)^{-1}(A_{\text{Explicit}} + I)$. We can see that the eigenvalues of $A_{\text{Implicit}+I}$ are

$$\begin{aligned}
\lambda^i_{A_{\text{Implicit}+I}} &= (2+2\rho) + 2(-\rho)\cos(\frac{i\pi}{N}) \quad \text{for } i = 1, \ldots, N-1 \\
&= 2 + 2\rho(1 - \cos(\frac{i\pi}{N})) \\
&= 2 + 4\rho\sin^2(\frac{i\pi}{2N})
\end{aligned}$$

and the eigenvalues of $A_{\text{Explicit}} + I$ are

$$\begin{aligned}
\lambda^i_{A_{\text{Explicit}+I}} &= (2-2\rho) + 2\rho\cos(\frac{i\pi}{N}) \quad \text{for } i = 1, \ldots, N-1 \\
&= 2 - 2\rho(1 - \cos(\frac{i\pi}{N})) \\
&= 2 - 4\rho\sin^2(\frac{i\pi}{2N})
\end{aligned}$$

Therefore the eigenvalues of $(A_{\text{Implicit}} + I)^{-1}(A_{\text{Explicit}} + I)$ are

$$\lambda^i_B = \frac{1 - 2\rho\sin^2(\frac{i\pi}{2N})}{1 + 2\rho\sin^2(\frac{i\pi}{2N})} \quad \text{for } i = 1, \ldots, N-1$$

and

$$|\lambda^i_B| = \left|\frac{1 - 2\rho\sin^2(\frac{i\pi}{2N})}{1 + 2\rho\sin^2(\frac{i\pi}{2N})}\right| < 1 \quad \text{for } i = 1, \ldots, N-1$$

Thus the Crank–Nicolson finite difference scheme is also unconditionally stable.

### 3.3.4 Stability of the Multi-Step Scheme

It can be shown that the multi-step scheme is also unconditionally stable; however, proof of this is left as an exercise for the reader.

---

## 3.4 Derivative Approximation by Finite Differences: Generic Approach

As mentioned in section (3.1), the most common finite difference formulas are relatively easy to derive. However, once we move to higher derivatives or higher order approximations, deriving the coefficients for the finite difference approximations can become cumbersome. The goal of this section is to develop a method for easily computing an approximation of $f^{(d)}(x)$ with an approximation order of $p$. To do this, we use a method derived in [106].

Assuming $f \in C^{\infty}$, we can write the Taylor series expansion of $f(x+ih)$ as follows:

$$f(x+ih) = \sum_{n=0}^{\infty} \frac{(ih)^n}{n!} f^{(n)}(x) \tag{3.56}$$

for $i \in \mathbb{Z}$ and $h \in \mathbb{R}^+$.

The approximation of $f^{(d)}(x)$ with approximation order $p$ is described by the following equation:

$$f^{(d)}(x) = \sum_{i=i_l}^{i_u} \hat{c}_i f(x+ih) + O(h^p) \tag{3.57}$$

where $\hat{c}_i$ are the unknown coefficients and $i_u$ and $i_l$ are the number of forward and backward terms in our approximation, respectively. If we multiply by $\frac{h^d}{d!}$ and define $c_i = \hat{c}_i \frac{h^d}{d!}$, we get

$$\frac{h^d f^{(d)}(x)}{d!} + O(h^{d+p}) = \sum_{i=i_l}^{i_u} c_i f(x+ih) \tag{3.58}$$

Substituting (3.56) into (3.58) we obtain

$$\begin{aligned}
\frac{h^d f^{(d)}(x)}{d!} + O(h^{d+p}) &= \sum_{i=i_l}^{i_u} c_i f(x+ih) && (3.59)\\
&= \sum_{i=i_l}^{i_u} c_i \left( \sum_{n=0}^{\infty} \frac{(ih)^n}{n!} f^{(n)}(x) \right) && (3.60)\\
&= \sum_{n=0}^{\infty} \left( \sum_{i=i_l}^{i_u} c_i i^n \right) \frac{h^n}{n!} f^{(n)}(x) && (3.61)\\
&= \sum_{n=0}^{d+p-1} \left( \sum_{i=i_l}^{i_u} c_i i^n \right) \frac{h^n}{n!} f^{(n)}(x) + O(h^{d+p}) && (3.62)
\end{aligned}$$

Thus we can see that

$$f^{(d)}(x) = \frac{d!}{h^d} \sum_{n=0}^{d+p-1} \left( \sum_{i=i_l}^{i_u} c_i i^n \right) \frac{h^n}{n!} f^{(n)}(x) \tag{3.63}$$

For this to be true the following constraints should hold:

$$\sum_{i=i_l}^{i_u} c_i i^n = \begin{cases} 1 & n = d \\ 0 & n \neq d \end{cases}$$

**Findings and observations**

- The solution to this equation would be unique if and only if one limits the number of constraints to $d+p$. So we must have $d+p = i_u - i_l + 1$, which restricts the number of terms we can use in our approximation
- In case of a forward difference approximation we set $i_l = 0$ and $i_u = d+p-1$
- In case of a backward difference approximation we set $i_l = -(d+p-1)$ and $i_u = 0$
- In case of a central difference approximation we set $i_l = -\frac{1}{2}(d+p-1)$ and $i_u = \frac{1}{2}(d+p-1)$

**Example 4** *Forward difference approximation of third derivative of second order*

Assume we want to compute the forward difference approximation of $f^{(3)}(x)$ with $O(h^2)$. Thus we have $d = 3$ and $p = 2$, and because we want a forward difference we have $i_l = 0$ and $i_u = 4$. The constraint is then

$$\sum_{i=0}^{4} c_i i^n = \begin{cases} 1 & n = 3 \\ 0 & n \neq 3 \end{cases}$$

In matrix form this is

$$\begin{bmatrix} 0^0 & 1^0 & 2^0 & 3^0 & 4^0 \\ 0^1 & 1^1 & 2^1 & 3^1 & 4^1 \\ 0^2 & 1^2 & 2^2 & 3^2 & 4^2 \\ 0^3 & 1^3 & 2^3 & 3^3 & 4^3 \\ 0^4 & 1^4 & 2^4 & 3^4 & 4^4 \end{bmatrix} \begin{bmatrix} c_0 \\ c_1 \\ c_2 \\ c_3 \\ c_4 \end{bmatrix} = \begin{bmatrix} 0 \\ 0 \\ 0 \\ 1 \\ 0 \end{bmatrix} \Rightarrow \begin{bmatrix} c_0 \\ c_1 \\ c_2 \\ c_3 \\ c_4 \end{bmatrix} = \begin{bmatrix} -5/12 \\ 3/2 \\ -2 \\ 7/6 \\ -1/4 \end{bmatrix}$$

and thus

$$\begin{aligned} f^{(3)}(x) &= \frac{3!}{h^3} \sum_{i=0}^{4} c_i f(x+ih) + O(h^d) \qquad (3.64) \\ &= \frac{-5f(x) + 18f(x+h) - 24f(x+2h) + 14f(x+3h) - 3f(x+4h)}{2h^3} + O(h^2) \end{aligned}$$

**Example 5** *Central difference approximation of second derivative of order* 3

For this example we have $d = 2$ and $p = 3$, for central difference $i_l = -2$, and $i_u = 2$.

$$\sum_{i=-2}^{2} c_i i^n = \begin{cases} 1 & n = 2 \\ 0 & n \neq 2 \end{cases}$$

In matrix form that is

$$\begin{bmatrix} (-2)^0 & (-1)^0 & 0^0 & 1^0 & 2^0 \\ (-2)^1 & (-1)^1 & 0^1 & 1^1 & 2^1 \\ (-2)^2 & (-1)^2 & 0^2 & 1^2 & 2^2 \\ (-2)^3 & (-1)^3 & 0^3 & 1^3 & 2^3 \\ (-2)^4 & (-1)^4 & 0^4 & 1^4 & 2^4 \end{bmatrix} \begin{bmatrix} c_{-2} \\ c_{-1} \\ c_0 \\ c_1 \\ c_2 \end{bmatrix} = \begin{bmatrix} 0 \\ 0 \\ 1 \\ 0 \\ 0 \end{bmatrix} \Rightarrow \begin{bmatrix} c_{-2} \\ c_{-1} \\ c_0 \\ c_1 \\ c_2 \end{bmatrix} = \begin{bmatrix} -1/24 \\ 2/3 \\ -5/4 \\ 2/3 \\ -1/24 \end{bmatrix}$$

Thus

$$
\begin{aligned}
f^{(2)}(x) &= \frac{2!}{h^2}\sum_{i=-2}^{2} c_i f(x+ih) + O(h^3) \qquad (3.65)\\
&= \frac{-f(x-2h)+16f(x-h)-30f(x)+16f(x+h)-f(x+2h)}{12h^2} + O(h^3)
\end{aligned}
$$

## 3.5 Matrix Equations Solver

Nonhomogeneous matrix equations of the form

$$
Ax = b \qquad (3.66)
$$

can be solved by taking the matrix inverse to obtain

$$
x = A^{-1}b
$$

However, numerically this approach is the most expensive and least efficient way of solving the equation and would be of last resort. There are various iterative methods for solving matrix equations (3.66) such as Gaussian elimination, the generalized minimal residual method [195], the quasi-minimal residual method [116], conjugate gradient methods [72], and multi-grid methods [207] just to name few.

By construction, throughout this book, most discretization problems end up to be a linear equation $Ax = b$ where $A$ is a tridiagonal matrix. To get better accuracy by means of techniques that are covered in Section 3.4 we might end up with a pentadiagonal matrix. Here we lay out algorithms for solving these two cases.

### 3.5.1 Tridiagonal Matrix Solver

Assume $A$ is a tridiagonal $N \times N$ matrix as follows:

$$
A = \begin{pmatrix}
d_1 & u_1 & & & \\
l_2 & d_2 & u_2 & & \\
& \ddots & \ddots & \ddots & \\
& & l_{N-1} & d_{N-1} & u_{N-1} \\
& & & l_N & d_N
\end{pmatrix}
$$

and we would like to solve the linear system $Ax = b$ for $x \in \mathbb{R}^N$ where $b$ is an $N \times 1$ vector. We can solve this tridiagonal linear equation by either first making the upper diagonal elements zero and then solving the system or first making the lower diagonal zero and then solving for the system. Here we first make the lower diagonal zero; to do that we use the following steps:

(a) Multiply the first row by $-\frac{l_2}{d_1}$ and add it to the second row to eliminate $l_2$. By doing this we now have $\tilde{d}_2 = d_2 - \frac{l_2}{d_1}u_1$ and $\tilde{b}_2 = b_2 - \frac{l_2}{d_1}b_1$.

(b) Now multiply the second row by $-\frac{l_3}{\tilde{d}_2}$ and add to the third row to eliminate $l_3$. By doing this we now have $\tilde{d}_3 = d_3 - \frac{l_3}{\tilde{d}_2}u_2$ and $\tilde{b}_3 = \tilde{b}_3 - \frac{l_2}{\tilde{d}_2}\tilde{b}_2$.

(c) Repeat until all lower diagonals are eliminated.

Now we have

$$\begin{pmatrix} d_1 & u_1 & & & \\ 0 & \tilde{d}_2 & u_2 & & \\ & \ddots & \ddots & \ddots & \\ & & 0 & \tilde{d}_{N-1} & u_{N-1} \\ & & & 0 & \tilde{d}_N \end{pmatrix} \begin{pmatrix} x_1 \\ x_2 \\ \vdots \\ x_{N-1} \\ x_N \end{pmatrix} = \begin{pmatrix} b_1 \\ \tilde{b}_2 \\ \vdots \\ \tilde{b}_{N-1} \\ \tilde{b}_N \end{pmatrix}$$

Now we can solve for $x$ starting from the $N^{th}$ row:

$$\tilde{d}_N x_N = \tilde{b}_N$$
$$x_N = \frac{\tilde{b}_N}{\tilde{d}_N}$$

Having $x_N$, we can solve for $x_{N-1}$ from the $(N-1)^{th}$ row:

$$\tilde{d}_{N-1} x_{N-1} + u_{N-1} x_N = \tilde{b}_{N-1}$$
$$x_{N-1} = \frac{\tilde{b}_{N-1} - u_{N-1} x_N}{\tilde{d}_{N-1}}$$

and we can repeat this until we solve for $x_1$.

Pseudo-code for this algorithm is as follows:

for $j = 2 : N$

$d_j = d_j - \frac{l_j}{d_{j-1}} u_{j-1}$

$b_j = b_j - \frac{l_j}{d_{j-1}} b_{j-1}$

endfor

$x_N = b_N / d_N$

for $j = N - 1 : 1$

$x_j = (b_j - u_j x_{j+1}) / d_j$

endfor

### 3.5.2 Pentadiagonal Matrix Solver

Assume $A$ is a pentadiagonal $N \times N$ matrix as follows:

$$A = \begin{pmatrix} d_1 & u_1 & v_1 & & & & & & \\ l_2 & d_2 & u_2 & v_2 & & & & & \\ k_3 & l_3 & d_3 & u_3 & v_3 & & & & \\ & & \ddots & \ddots & \ddots & \ddots & \ddots & & \\ & & & & & k_{N-1} & l_{N-1} & d_{N-1} & u_{N-1} \\ & & & & & & k_N & l_N & d_N \end{pmatrix}$$

and we would like to solve the linear system $Ax = b$ for $x \in \mathbb{R}^N$ where $b$ is an $N \times 1$ vector. We can solve this pentadiagonal linear equation by making the lower diagonals zero and then solving for the system. To do that we use the following steps:

(a) Multiply the first row by $-\frac{l_2}{d_1}$ and add it to the second row to eliminate $l_2$; multiply the first row by $-\frac{k_3}{d_1}$ and add it to the third row to eliminate $k_3$. By doing this we overwrite the following entries:

$$\begin{aligned} d_2 &= d_2 - \frac{l_2}{d_1}u_1 \\ u_2 &= u_2 - \frac{l_2}{d_1}v_1 \\ b_2 &= b_2 - \frac{l_2}{d_1}b_1 \\ l_3 &= l_3 - \frac{k_3}{d_1}u_1 \\ d_3 &= d_3 - \frac{k_3}{d_1}v_1 \\ b_3 &= b_3 - \frac{k_3}{d_1}b_1 \end{aligned}$$

(b) To eliminate $l_3$ and $k_4$ multiply the second row by $-\frac{l_3}{d_2}$ and add to the third row to eliminate $l_3$ and multiply the second row by $-\frac{k_4}{d_2}$ and add to the fourth row to eliminate $k_4$. By doing this we overwrite the following entries:

$$\begin{aligned}
d_3 &= d_3 - \frac{l_3}{d_2}u_2 \\
u_3 &= u_3 - \frac{l_3}{d_2}v_2 \\
b_3 &= b_3 - \frac{l_3}{d_2}b_2 \\
l_4 &= l_4 - \frac{k_4}{d_2}u_2 \\
d_4 &= d_4 - \frac{k_4}{d_2}v_2 \\
b_4 &= b_4 - \frac{k_4}{d_2}b_2
\end{aligned}$$

(c) Repeat until all lower diagonals are eliminated.

Now we can solve for $x$ starting from the $N^{th}$ row:

$$d_N x_N = b_N$$
$$x_N = \frac{b_N}{d_N}$$

Having $x_N$ we can solve for $x_{N-1}$ from the $(N-1)^{th}$ row:

$$d_{N-1}x_{N-1} + u_{N-1}x_N = b_{N-1}$$
$$x_{N-1} = \frac{b_{N-1} - u_{N-1}x_N}{\tilde{d}_{N-1}}$$

and we can repeat this until we solve for $x_1$.

Pseudo-code for this algorithm is as follows:

for $j = 2 : N-1$

$d_j = d_j - \frac{l_j}{d_{j-1}}u_{j-1}$

$u_j = u_j - \frac{l_j}{d_{j-1}}v_{j-1}$

$b_j = b_j - \frac{l_j}{d_{j-1}}b_{j-1}$

$l_{j+1} = l_{j+1} - \frac{k_{j+1}}{d_{j-1}}u_{j-1}$

$d_{j+1} = d_{j+1} - \frac{k_{j+1}}{d_{j-1}}v_{j-1}$

$b_{j+1} = b_{j+1} - \frac{k_{j+1}}{d_{j-1}}b_{j-1}$

endfor

$x_N = b_N/d_N$

$x_N = (b_N - u_{N-1}x_N)/d_{N-1}$

for $j = N-2 : 1$

$x_j = (b_j - v_j x_{j+2} - u_j x_{j+1})/d_j$

endfor

## Problems

1. Consider the following one-dimensional boundary value problem (BVP):

$$-u'' = f \quad 0 < x < L \tag{3.67}$$

with boundary conditions:

$$u(0) = u(L) = 0$$

where $f$ is some given function.

(a) Finite difference method — Define the grid point $x_j = 0 + j\Delta x$ where $h = \frac{L-0}{N+1}$ for $j = 0, \ldots, N+1$. Use central difference approximation for the second derivative on this grid to write the differential equation as a difference equation $A\mathbf{u} = \mathbf{f}$ where

$$\mathbf{u} = \begin{pmatrix} u_1 \\ \vdots \\ u_N \end{pmatrix} \quad \mathbf{f} = \begin{pmatrix} f_1 \\ \vdots \\ f_N \end{pmatrix} \tag{3.68}$$

and $A$ is an $N \times N$ tridiagonal matrix. Here $u_j$ is an approximation to $u(x_j)$ and $f_j = f(x_j)$. Write down the entries of matrix $A$.

(b) Finite element method — Multiply both sides of (3.67) by $v \in C^1$ with $v(0) = v(L) = 0$ and integrate and then apply integration by parts to reduce the order of derivatives and to obtain

$$\int_0^L u'(x)v'(x)dx = \int_0^L f(x)v(x)dx \tag{3.69}$$

Now introduce a uniform mesh on $[0, L]$ and define a basis of a piecewise linear function called *hat function* as

$$\phi_j(x) = \begin{cases} 0 & : \quad x \le x_{j-1} \\ \frac{1}{h}(x - x_{j-1}) & : \quad x_{j-1} \le x \le x_j \\ \frac{-1}{h}(x - x_{j+1}) & : \quad x_j \le x \le x_{j+1} \\ 0 & : \quad x \ge x_{j+1} \end{cases}$$

Let $v(x) = \sum_{j=1}^N a_j \phi_j(x)$. We seek a continuous piecewise linear approximate solution of the form $u_h(x) = \sum_{j=1}^N u_j \phi_j(x)$ and require that (3.69) be satisfied for $v = \phi_i$ for $i = 1, \ldots, N$ that is

$$\int_0^L u_h'(x)\phi_i'(x)dx = \int_0^L f(x)\phi_i(x)dx \quad \text{for} \quad i = 1, \ldots, N \tag{3.70}$$

Define

$$a_{ij} = \int_0^L \phi_i'(x)\phi_j'(x)dx$$

and use the trapezoidal rule on the right hand side of (3.70) to show that it can be written as

$$\sum_{j=1}^{N} a_{ij}u_j = hf_i \quad \text{for} \quad i = 1, \ldots, N \tag{3.71}$$

Simplify and write it in matrix form. Show that as in the previous case the boundary value problem can be written as a difference equation $A\mathbf{u} = \mathbf{f}$ which happens to be exactly the same difference equation.

(c) Finite volume method — Define

$$u_j = u(x_j) \sim \frac{1}{h}\int_{j-\frac{1}{2}h}^{j+\frac{1}{2}h} u(x)dx$$

and integrate over $[j - \frac{1}{2}h, j + \frac{1}{2}h]$ to get

$$\int_{j-\frac{1}{2}h}^{j+\frac{1}{2}h} (-u'' - f)dx = 0$$

and define

$$\bar{f}_j = \frac{1}{h}\int_{j-\frac{1}{2}h}^{j+\frac{1}{2}h} f(x)dx$$

Use fundamental theorem of calculus twice to write the boundary value problem as a difference equation $A\mathbf{u} = \bar{\mathbf{f}}$. This is slightly different from the previous two cases nonetheless pretty close.

Difficulty with finite volume method is the fluxes across the faces of volume at $j - \frac{1}{2}h$ and $j + \frac{1}{2}h$ in terms of the integral quantities $u_j$.

2. For the following BVP:

$$-u'' = f \quad 0 < x < L \tag{3.72}$$

with boundary conditions:

$$\begin{aligned} u(0) &= u_0 \\ u(L) &= u_L \end{aligned}$$

Extend the discussion to capture non-zero boundary conditions (just for the finite elements/Galerkin approach).

Hint: Define hat functions at boundaries and extend the derivation.

3. Consider the heat equation

$$\frac{\partial u}{\partial t} - \kappa\frac{\partial^2 u}{\partial x^2} = 0$$

with the following initial condition:

$$u(x, 0) = f(x)$$

where $\kappa > 0$ is a constant. Lay out a grid in the $(x, t)$ plane using points $x_j = j\Delta x$ and $t_k = k\Delta t$, where $\Delta x,\ \Delta t > 0$ are small and $j$ and $k$ are integers. The goal is to calculate $u_{j,k+1}$ such that

$$u_{j,k+1} \approx u(x_j, t_{k+1})$$

and ensure that this approximation converges to the exact solution as $\Delta x,\ \Delta t \to 0$.

(a) Use an explicit finite difference scheme to discretize the PDE and write the difference equation for $u_{j,k+1}$.

(b) We know that the explicit scheme is conditionally stable. To illustrate this, assume that the initial condition is given by $f(x) = \cos(\pi x/\Delta x)$. Use this initial condition to derive the expression for $u_{j,k+1}$ by induction. Considering that the exact solution is bounded, derive a condition that the scheme in (a) becomes stable.

4. Assume $u(x,t)$ solves the heat equation

$$\frac{\partial u}{\partial t} - \kappa \frac{\partial^2 u}{\partial x^2} = 0, \qquad \text{for } 0 < x < L, \;\; t > 0$$

with the following initial condition:

$$u(x,0) = f(x), \qquad \text{for } 0 < x < L$$

and *Neumann* boundary conditions

$$\begin{aligned} \frac{\partial u}{\partial x}(0,t) &= 0 \text{ for all } t \\ \frac{\partial u}{\partial x}(L,t) &= 0 \text{ for all } t \end{aligned}$$

It can be shown that under these boundary conditions

$$\int_0^L u(x,t)dx = \int_0^L f(x)dx \qquad \text{for } t \geq 0$$

For the following initial data

$$f(x) = \begin{cases} 0 & : \; 0 \leq x \leq \frac{L}{4} \\ \frac{4}{L}x - 1 & : \; \frac{L}{4} \leq x \leq \frac{L}{2} \\ -\frac{4}{L}x + 3 & : \; \frac{L}{2} \leq x \leq \frac{3L}{4} \\ 0 & : \; \frac{3L}{4} \leq x \leq L \end{cases}$$

find $u(x,t)$ as $t \to \infty$.

5. Consider the heat equation in Problem 4.

(a) Discretize the PDE (on an arbitrary domain) using the *explicit* scheme and write down the difference equation for an interior grid point and show the truncation error for the scheme is $O(\Delta x^2) + O(\Delta t)$.

(b) Discretize the PDE (on an arbitrary domain) using the *implicit* scheme and write down the difference equation for an interior grid point and show the truncation error for the scheme is $O(\Delta x^2) + O(\Delta t)$.

(a) Now add explicit and implicit schemes obtained in previous two parts to get the *Crank–Nicolson* scheme and write down the difference equation for an interior grid point. Now prove that the truncation error for the scheme is $O(\Delta x^2) + O(\Delta t^2)$. This is an alternative approach to what was done in Section 3.2.3.

No need to bring in boundary conditions or terminal condition into your discretization.

6. Consider the first-order PDE

$$\frac{\partial u}{\partial t} + c\frac{\partial u}{\partial x} = 0$$

with the following initial condition:

$$u(x, 0) = f(x)$$

where $c > 0$ is a constant. Lay out a grid in the $(x, t)$ plane using points $x_j = j\Delta x$ and $t_k = k\Delta t$, where $\Delta x$, $\Delta t > 0$ are small and $j$ and $k$ are integers. The goal is to calculate $u_{j,k+1}$ such that

$$u_{j,k+1} \approx u(x_j, t_{k+1})$$

and ensure that this approximation converges to the exact solution as $\Delta x, \ \Delta t \to 0$.

Use *explicit* finite differences to discretize the PDE and write the difference equation for $u_{j,k+1}$ by

(a) using a *forward difference* approximation to discretize $\frac{\partial u}{\partial x}$.

(b) using a *backward difference* approximation to discretize $\frac{\partial u}{\partial x}$.

Now assume the initial condition $f(x) = \cos(\pi x/\Delta x)$ and show that the scheme in part (a) is unstable no matter how small $\Delta x$ and $\Delta t$ are. Find the condition that would ensure the scheme in part (b) becomes stable.

7. Show that the multi-step scheme used in the discretization of the heat equation is unconditionally stable.

8. Use Taylor series in two variables to show that

$$\frac{\partial^2 u}{\partial x \partial y} \approx \frac{u(x+\Delta x, y+\Delta y) - u(x-\Delta x, y+\Delta y) - u(x+\Delta x, y-\Delta y) + u(x-\Delta x, y-\Delta y)}{4\Delta x \Delta y}$$

and is of order $O(\Delta x^2) + O(\Delta y^2)$.

---

## Case Study

1. Extend the methodology in Section 3.4 to develop a method for computing an approximation of mixed derivatives for multi-variable functions.

# Chapter 4

## Derivative Pricing via Numerical Solutions of PDEs

In this chapter we will discuss the use of the finite difference techniques discussed in Chapter 3 for pricing derivatives under models for which a partial differential equation (PDE) describing derivative prices can be formulated. We start with geometric Brownian motion. As discussed in Section 1.2.1, under geometric Brownian motion the price of an asset follows the following stochastic differential equation (SDE):

$$dS_t = (r - q)S_t dt + \sigma S_t dW_t \tag{4.1}$$

And we know the value of options on that asset satisfy the Black–Scholes partial differential equation

$$\frac{\partial v}{\partial t} + \frac{\sigma^2 S^2}{2}\frac{\partial^2 v}{\partial S^2} + (r - q)S\frac{\partial v}{\partial S} = rv(S, t) \tag{4.2}$$

with terminal and boundary conditions which are dependent on the type of the option we wish to price.

This can be extended to the more general case in which the price of the asset follows the following SDE:

$$dS_t = (r(t) - q(t))S_t dt + \sigma(S_t, t)S_t dW_t \tag{4.3}$$

Here, $r(t)$, and $q(t)$ are deterministic functions of time, and $\sigma(S_t, t)$ is a deterministic volatility function that depends on time and asset price. This constitutes a local volatility model where $\sigma(S_t, t)$ is called *local volatility surface* and $r(t)$ and $q(t)$ are the time dependent term structure of interest rates and dividend rates, respectively. As discussed in Section 1.2.2, local volatility models allow us greater flexibility in pricing, which allows us to calibrate our model more easily to a number of different instruments consistently. For (4.3), the value of the option satisfies the generalized Black–Scholes PDE

$$\frac{\partial v}{\partial t} + \frac{\sigma(S_t, t)^2 S^2}{2}\frac{\partial^2 v}{\partial S^2} + (r(t) - q(t))S\frac{\partial v}{\partial S} = r(t)v(S, t) \tag{4.4}$$

For regular Black–Scholes and the associated local volatility models for which these PDEs apply, we can use standard techniques developed for numerically solving PDEs to price derivatives. This will allow us to price both European and American type options on various different payoff structures with path dependency.

Finite difference techniques have a number of compelling advantages over competing methods. They can be used to price European, Bermudan, and American options. They can be easily adapted to price any derivative with a payoff which is path dependent. Also, because the construction of the mesh on which we solve the problem is arbitrary, these methods admit many mesh constructions which can limit the error of pricing derivatives with payoff properties which cause certain points in our problem domain to become critical to the solution. For example, the mesh on which we apply finite difference methods can be constructed to match the strike of a vanilla option, the exercise times of a Bermudan option, the dividend payment dates and resulting stock price of an underlier, the barrier level of a barrier option, or the exercise boundary of an American put. This allows us to take advantage of our knowledge of the option payoff and critical times and asset prices in order to minimize the errors in our method.

However, finite difference methods also have a number of drawbacks. Obviously they are only applicable to models for which a PDE exists — notably geometric Brownian motion and the local volatility extension. It can be somewhat slower for European options than competing methods, as it requires solving the option price for all time periods up to the current time to expiry, which is more information than is needed to price a European option. Finite difference methods also propagate errors; because of this, errors which occur at some boundary or critical points are propagated to nearby points through the discretized PDE, and this has implications for the conditions under which the finite difference solution is stable. For a higher dimensional PDE, we run into the so-called curse of dimensionality issue that make it difficult and expensive to numerically solve it.

**Models:**
Models for which a PDE describing the option price exists, e.g., geometric Brownian motion, local volatility models.

**Option Types:**
European, Bermudan, and American options with a path-dependent payoff.

**Pros**

1. Handles early-exercise options, Bermudans and Americans
2. Prices path-dependent options with a variety of payoffs
3. Admits a variety of mesh constructions which can be option specific which allow us to minimize propagated errors

**Cons**

1. Restricted to models for which a PDE exists, e.g., geometric Brownian motion and local volatility models
2. Might not be used to price complex path-dependent options
3. Higher dimensional PDE are expensive and difficult to be solved numerically
4. Can be sensitive to boundary conditions, which need to be formulated, and propagates errors from these boundary conditions
5. Required to solve for European option price on full problem domain as the PDE describes the solution only locally

## 4.1 Option Pricing under the Generalized Black–Scholes PDE

Under the generalized Black–Scholes model, extended to include local volatility, term structure of interest rate and dividend rate, we have the following PDE:

$$\begin{cases} \frac{\partial v}{\partial t} + \frac{\sigma(S_t,t)^2 S^2}{2}\frac{\partial^2 v}{\partial S^2} + (r(t) - q(t))S\frac{\partial v}{\partial S} = r(t)v(S,t) \\ v(S,T) = f(S) \quad \text{Terminal condition (payoff function)} \\ \text{Boundary conditions} \end{cases}$$

Clearly, since the maturity of a contract is in the future, in calendar time we have to solve the PDE backwards, and so we use the usual change of variable $\tau = T - t$, expressing all times as time to maturity. Thus we have

$$\begin{aligned} \hat{v}(S,\tau) &= v(S,t) \\ \hat{\sigma}(S,\tau) &= \sigma(S,t) \\ \hat{r}(\tau) &= r(t) \\ \hat{q}(\tau) &= q(t) \end{aligned}$$

and will get

$$\begin{cases} -\frac{\partial \hat{v}}{\partial \tau} + \frac{\hat{\sigma}(S,\tau)^2 S^2}{2}\frac{\partial^2 \hat{v}}{\partial S^2} + (\hat{r}(\tau) - \hat{q}(\tau))S\frac{\partial \hat{v}}{\partial S} = \hat{r}(\tau)\hat{v}(S,\tau) \\ \hat{v}(S,0) = f(S) \quad \text{Initial condition (payoff function)} \\ \text{Boundary conditions} \end{cases}$$

As before, we would like to write a discretized version of the differential operator as a difference equation. First, we must define the domain of the problem, going from a continuous domain $D = \{S_{min} \leq x \leq S_{max};\quad 0 \leq \tau \leq T\}$ to a discrete grid. Without loss of generality we consider $M$ equal sub-intervals in the $\tau$-direction and $N$ equal sub-intervals in the $S$-direction on $[S_{min}\quad S_{max}]$. Thus, we have the following mesh on $[S_{min}, S_{max}] \times [0, T]$:

$$\bar{D} = \left\{ \begin{array}{lll} S_j = S_{min} + (j-1)\Delta S; & \Delta S = \frac{S_{max} - S_{min}}{N}; & j = 1, \ldots, N+1 \\ \tau_k = 0 + (k-1)\Delta\tau; & \Delta\tau = \frac{T-0}{M}; & k = 1, \ldots, M+1 \end{array} \right\}$$

We let $v_{j,k}$ be the approximate value of $\hat{v}(S = S_j, \tau = \tau_k)$, with $\sigma_{j,k} = \hat{\sigma}(S_j, \tau_k)$, $r_k = \hat{r}(\tau_k)$, and $q_k = \hat{q}(\tau_k)$.

Without loss of generality, in all our discretization schemes that follow, we assume boundary values at $S_{min}$ and $S_{max}$ at each time step are known. We will later discuss how to treat boundary conditions for various other cases.

### 4.1.1 Explicit Discretization

Following the procedure outlined in Section 3.2.1, we construct the explicit discretization of the generalized Black–Scholes PDE. Thus to discretize this equation at the grid point $(x_j, \tau_k)$, we use the forward approximation for the theta term, $\frac{\partial \hat{v}}{\partial \tau}$, the central difference equation for the gamma term, $\frac{\partial^2 \hat{v}}{\partial S^2}$, and the first order central difference approximation for the delta term, $\frac{\partial \hat{v}}{\partial S}$.

The forward approximation for the theta term yields

$$\frac{\partial \hat{v}(x_j, \tau_k)}{\partial \tau} = \frac{v_{j,k+1} - v_{j,k}}{\Delta \tau} + O(\Delta \tau)$$

The central approximation for the gamma term yields

$$\frac{\partial^2 \hat{v}(x_j, \tau_k)}{\partial S^2} = \frac{v_{j-1,k} - 2v_{j,k} + v_{j+1,k}}{(\Delta S)^2} + O(\Delta S^2)$$

and the central difference approximation for the delta term yields

$$\frac{\partial \hat{v}(x_j, \tau_k)}{\partial S} = \frac{v_{j+1,k} - v_{j-1,k}}{2\Delta S} + O(\Delta S^2)$$

So, by dropping the error terms, we obtain the following discrete equation at point $(x_j, \tau_k)$:

$$\frac{v_{j,k+1} - v_{j,k}}{\Delta \tau} - \frac{\sigma_{j,k}^2 S_j^2}{2} \left( \frac{v_{j-1,k} - 2v_{j,k} + v_{j+1,k}}{(\Delta S)^2} \right) - (r_k - q_k) S_j \frac{v_{j+1,k} - v_{j-1,k}}{2\Delta S} + r_k v_{j,k} = 0$$

Multiplying it by $\Delta \tau$, we get

$$\begin{aligned} v_{j,k+1} - v_{j,k} - \frac{\sigma_{j,k}^2 S_j^2}{2} \frac{\Delta \tau}{\Delta S^2} (v_{j-1,k} - 2v_{j,k} + v_{j+1,k}) \\ -(r_k - q_k) S_j \frac{\Delta \tau}{2\Delta S} (v_{j+1,k} - v_{j-1,k}) + r_k \Delta \tau v_{j,k} = 0 \end{aligned} \tag{4.5}$$

If we define the constants

$$\alpha_{j,k} = \frac{\sigma_{j,k}^2 S_j^2}{2} \frac{\Delta \tau}{\Delta S^2} \tag{4.6}$$

$$\beta_{j,k} = (r_k - q_k) S_j \frac{\Delta \tau}{2\Delta S} \tag{4.7}$$

we get

$$\begin{aligned} v_{j,k+1} - v_{j,k} - \alpha_{j,k} (v_{j-1,k} - 2v_{j,k} + v_{j+1,k}) \\ -\beta_{j,k} (v_{j+1,k} - v_{j-1,k}) - r_k \Delta \tau v_{j,k} = 0 \end{aligned} \tag{4.8}$$

We can now rearrange (4.8) to get a value in period $k+1$ in terms of values in period $k$. After rearranging the equation, we get

$$v_{j,k+1} = (\beta_{j,k} - \alpha_{j,k}) v_{j-1,k} + (1 - r_k \Delta \tau - 2\alpha_{j,k}) v_{j,k} + (\alpha_{j,k} + \beta_{j,k}) v_{j+1,k} \tag{4.9}$$

and if we define the new constants

$$l_{j,k} = \alpha_{j,k} - \beta_{j,k} \tag{4.10}$$

$$d_{j,k} = 1 - r_k \Delta \tau - 2\alpha_{j,k} \tag{4.11}$$

$$u_{j,k} = \alpha_{j,k} + \beta_{j,k} \tag{4.12}$$

we get the equation

$$v_{j,k+1} = l_{j,k} v_{j-1,k} + d_{j,k} v_{j,k} + u_{j,k} v_{j+1,k} \quad \text{for } j = 2, \ldots, N \tag{4.13}$$

Note that $j = 1$ and $j = N+1$ correspond to the boundary conditions, and $k = 1$ corresponds to the initial conditions, $v_{j,1} = f(S_j) = f_j$. At all of these grid points the solution is already known. The approximate function values at the time slice for $\tau_k = k\Delta\tau$ can be written in vector form as

$$V_k = \begin{pmatrix} v_{2,k} \\ v_{3,k} \\ \vdots \\ v_{N,k} \end{pmatrix}$$

Thus we can write the recursive explicit discretization scheme in matrix form as

$$V_{k+1} = A_k^{\text{Explicit}} V_k + \begin{pmatrix} l_{2,k} v_{1,k} \\ 0 \\ \vdots \\ 0 \\ u_{N,k} v_{N+1,k} \end{pmatrix} \tag{4.14}$$

where

$$A_k^{\text{Explicit}} = \begin{pmatrix} d_{2,k} & u_{2,k} & & & \\ l_{3,k} & d_{3,k} & u_{3,k} & & \\ & \ddots & \ddots & \ddots & \\ & & l_{N-1,k} & d_{N-1,k} & u_{N-1,k} \\ & & & l_{N,k} & d_{N,k} \end{pmatrix} \tag{4.15}$$

### 4.1.2 Implicit Discretization

Following the procedure outlined in Section 3.2.2, we construct the implicit discretization of the generalized Black–Scholes PDE. Thus, to discretize this equation at the grid point $(x_j, \tau_{k+1})$, at time period $k+1$ we use backwards approximation for the theta term, $\frac{\partial \hat{v}}{\partial \tau}$, central difference approximation for the gamma term, $\frac{\partial^2 \hat{v}}{\partial S^2}$, and the first order central difference approximation for the delta term, $\frac{\partial \hat{v}}{\partial S}$.

The backward difference approximation for the theta term yields

$$\frac{\partial \hat{v}(x_j, \tau_{k+1})}{\partial \tau} = \frac{v_{j,k+1} - v_{j,k}}{\Delta\tau} + O(\Delta\tau)$$

The central difference approximation for the gamma term yields

$$\frac{\partial^2 \hat{v}(x_j, \tau_{k+1})}{\partial S^2} = \frac{v_{j-1,k+1} - 2v_{j,k+1} + v_{j+1,k+1}}{(\Delta S)^2} + O(\Delta S^2)$$

and the central difference approximation for the delta term yields

$$\frac{\partial \hat{v}(x_j, \tau_{k+1})}{\partial S} = \frac{v_{j+1,k+1} - v_{j-1,k+1}}{2\Delta S} + O(\Delta S^2)$$

As in the explicit scheme, by dropping the error terms we obtain the following discrete equation at point $(x_j, \tau_{k+1})$:

$$\frac{v_{j,k+1} - v_{j,k}}{\Delta\tau} - \frac{\sigma_{j,k+1}^2 S_j^2}{2}\left(\frac{v_{j-1,k+1} - 2v_{j,k+1} + v_{j+1,k+1}}{(\Delta S)^2}\right)$$
$$-(r_{k+1} - q_{k+1}) S_j \frac{v_{j+1,k+1} - v_{j-1,k+1}}{2\Delta S} + r_{k+1} v_{j,k+1} = 0$$

Multiplying it by $\Delta\tau$, we get

$$v_{j,k+1} - v_{j,k} - \frac{\sigma_{j,k+1}^2 S_j^2}{2}\frac{\Delta\tau}{\Delta S^2}(v_{j-1,k+1} - 2v_{j,k+1} + v_{j+1,k+1})$$
$$-(r_{k+1} - q_{k+1})S_j\frac{\Delta\tau}{2\Delta S}(v_{j+1,k+1} - v_{j-1,k+1}) + r_{k+1}\Delta\tau v_{j,k+1} = 0 \quad (4.16)$$

Using the previous definitions of $\alpha_{j,k}$ and $\beta_{j,k}$ we can now rearrange (4.16) to get

$$(-\alpha_{j,k+1}+\beta_{j,k+1})v_{j-1,k+1}+(1+r_{k+1}\Delta\tau+2\alpha_{j,k+1})v_{j,k+1}+(-\alpha_{j,k+1}-\beta_{j,k+1})v_{j+1,k+1}=v_{j,k}$$

and if we define the new constants

$$\begin{aligned}
\hat{l}_{j,k+1} &= -\alpha_{j,k+1} + \beta_{j,k+1} \\
\hat{d}_{j,k+1} &= 1 + r_{k+1}\Delta\tau + 2\alpha_{j,k+1} \\
\hat{u}_{j,k+1} &= -\alpha_{j,k+1} - \beta_{j,k+1}
\end{aligned}$$

we get the equation

$$\hat{l}_{j,k+1}v_{j-1,k+1} + \hat{d}_{j,k+1}v_{j,k+1} + \hat{u}_{j,k+1}v_{j+1,k+1} = v_{j,k} \quad \text{for } j = 2,\ldots,N \quad (4.17)$$

Note the relationship between $l_{j,k+1}, d_{j,k+1}, u_{j,k+1}$ and $\hat{l}_{j,k+1}, \hat{d}_{j,k+1}, \hat{u}_{j,k+1}$:

$$\begin{aligned}
\hat{l}_{j,k+1} &= -l_{j,k+1} \\
\hat{d}_{j,k+1} &= 2 - d_{j,k+1} \\
\hat{u}_{j,k+1} &= -u_{j,k+1}
\end{aligned}$$

Thus we can write the recursive implicit discretization scheme in matrix form as

$$A_{k+1}^{\text{Implicit}} V_{k+1} = V_k + \begin{pmatrix} -\hat{l}_{2,k+1}v_{1,k+1} \\ 0 \\ \vdots \\ 0 \\ -\hat{u}_{N,k+1}v_{N+1,k+1} \end{pmatrix}$$

where

$$A_{k+1}^{\text{Implicit}} = \begin{pmatrix} \hat{d}_{2,k+1} & \hat{u}_{2,k+1} & & & \\ \hat{l}_{3,k+1} & \hat{d}_{3,k+1} & \hat{u}_{3,k+1} & & \\ & \ddots & \ddots & \ddots & \\ & & \hat{l}_{N-1,k+1} & \hat{d}_{N-1,k+1} & \hat{u}_{N-1,k+1} \\ & & & \hat{l}_{N,k+1} & \hat{d}_{N,k+1} \end{pmatrix} \quad (4.18)$$

### 4.1.3 Crank–Nicolson Discretization

As discussed in Section 3.2.3, the Crank–Nicolson discretization scheme is based on the averaging of the forward and backward finite difference approximations. To perform the Crank–Nicolson discretization of the generalized Black–Scholes PDE we simply add the explicit and implicit schemes (4.13) and (4.17) to obtain

$$(\beta_{j,k} - \alpha_{j,k})v_{j-1,k+1} + (2 + 2\alpha_{j,k} + r(\tau)\Delta\tau)v_{j,k+1} + (-\alpha_{j,k} - \beta_{j,k})v_{j+1,k+1}$$
$$= (\alpha_{j,k} - \beta_{j,k})v_{j-1,k} + (2 + 2\alpha_{j,k} - r(\tau)\Delta\tau)v_{j,k} + (\alpha_{j,k} + \beta_{j,k})v_{j+1,k}$$

or equivalently using the same notations

$$\begin{aligned}
&\hat{l}_{j,k+1}v_{j-1,k+1}+(1+\hat{d}_{j,k+1})v_{j,k+1}+\hat{u}_{j,k+1}v_{j+1,k+1} \\
&\quad = l_{j,k}v_{j-1,k}+(1+d_{j,k})v_{j,k}+u_{j,k}v_{j+1,k} \qquad \text{for } j=2,\ldots,N
\end{aligned} \tag{4.19}$$

where as before

$$\begin{aligned}
l_{j,k} &= \alpha_{j,k}-\beta_{j,k} \\
d_{j,k} &= 1-r_k\Delta\tau-2\alpha_{j,k} \\
u_{j,k} &= \alpha_{j,k}+\beta_{j,k}
\end{aligned}$$

and

$$\begin{aligned}
\hat{l}_{j,k+1} &= -l_{j,k+1} &&= -\alpha_{j,k+1}+\beta_{j,k+1} \\
\hat{d}_{j,k+1} &= 2-d_{j,k+1} &&= 1+r_{k+1}\Delta\tau+2\alpha_{j,k+1} \\
\hat{u}_{j,k+1} &= -u_{j,k+1} &&= -\alpha_{j,k+1}-\beta_{j,k+1}
\end{aligned}$$

Now we can write the recursive Crank–Nicolson discretization scheme in matrix form as

$$(A_k^{\text{Implicit}}+I)V_{k+1}=(A_k^{\text{Explicit}}+I)V_k+\begin{pmatrix} l_{2,k}v_{1,k}-\hat{l}_{2,k+1}v_{1,k+1} \\ 0 \\ \vdots \\ 0 \\ u_{N,k}v_{N+1,k}-\hat{u}_{N,k+1}v_{N+1,k+1} \end{pmatrix}$$

where $A_k^{\text{Explicit}}$ and $A_k^{\text{Implicit}}$ are given in (4.15) and (4.18), respectively.

---

## 4.2 Boundary Conditions and Critical Points

While the previously discussed discretization methods are at the heart of any PDE based pricing algorithm, one important issue not yet discussed in the construction of these pricing techniques is the choice of boundary conditions. The choice of boundary conditions can be critical for ensuring the accuracy of a pricing algorithm, as any errors on the boundaries are propagated through the rest of the mesh through the finite difference scheme. One of the primary advantages of PDE methods is that the mesh on which we construct our solutions can be chosen so as to best fit one's pricing problem, minimizing propagated error. This is especially important for options which have payoff functions which have critical discontinuous points in both time and asset prices. By fitting the mesh in such a way that our grid points are close to or exactly match these critical points, we can improve the accuracy of our pricing algorithm. Some examples include matching the strike of an option, matching the timing and amount of dividend payments, matching barriers for barrier options, and matching exercise time for Bermudan options.

### 4.2.1 Implementing Boundary Conditions

Boundary conditions are the only values in our PDE scheme which are assumed to be (explicitly or implicitly) known; thus the choice of a boundary condition can have a profound effect on the accuracy of computed solutions. For one-dimensional space PDEs, boundary conditions are the value of the derivative at $S_{min}$ and $S_{max}$. The option value at $\tau = 0$ is called the initial condition. In most cases the initial condition is well known, as it is just the payoff of the option that we are pricing:

$$v(S,0) = f(S)$$

where $f(S)$ is the payoff of the derivative at expiration. We distinguish the initial condition from boundary conditions.

Depending on the option, the boundary condition at $S_{min}$ or $S_{max}$ could be explicitly known. For example, for a vanilla call the value of the option at $S_{min}$ at any time can be set to zero, as this is the payoff if the asset has little value, and we often assume it cannot recover value after this occurs. Also for a vanilla put, the boundary condition at $S_{max}$ could be set to zero, as the put option is worthless at a very high price for the underlier.

On the other hand, for a call at $S_{max}$ just simply setting its value to the payoff might underestimate the final value of the option since many options continue to rise in value as the asset price increases. In these cases we make some assumptions about the derivative value at the boundary.

#### 4.2.1.1 Dirichlet Boundary Conditions

A Dirichlet boundary condition is a boundary condition which specifies the derivative value explicitly at the boundary points. A Dirichlet boundary condition can often be found or at least estimated for $S_{min}$ or $S_{max}$ depending on the derivative payoff. For instance, for a call option at $S_{min}$, the derivative payoff could be set to zero, as the option is worthless if $S_{min}$ is zero or small enough. For a put option at $S_{max}$, the derivative payoff could be set to zero if $S_{max}$ is sufficiently large. Also, as we see later for barrier options, the value of the option at the barrier is explicitly known to be rebate if there is any; otherwise it is zero.

We can also approximate a Dirichlet boundary condition for the value of the call at $S_{max}$ and a put at $S_{max}$ by simply setting the option value to be its payoff at the level. However, this should be treated with care, as in a very high volatility regime the value of a call at a high level of the underlier exceeds its payoff.

#### 4.2.1.2 Neumann Boundary Conditions

A Neumann boundary condition is a boundary condition which specifies the partial derivative of the option at the boundary. Neumann boundary conditions can be used at $S_{min}$ and $S_{max}$. This can be advantageous, as the second derivative of the option payoff is often well known at extreme asset values, by typically using

$$\frac{\partial^2 v}{\partial S^2}(S_{min}, \tau) = 0$$

and/or

$$\frac{\partial^2 v}{\partial S^2}(S_{max}, \tau) = 0$$

It is important to note that if we attempt to use the central difference at $S_{min}$ or $S_{max}$, it would result in a grid point outside our grid. If instead we use the forward difference for the left boundary ($S_{min}$), we get

$$\begin{aligned}\frac{\partial^2 v}{\partial S^2}(S_{min}, \tau_{k+1}) &= \frac{\partial^2 v}{\partial S^2}(S_1, \tau_{k+1}) \\ &= \frac{v_{1,k+1} - 2v_{2,k+1} + v_{3,k+1}}{h^2} + O(h)\end{aligned}$$

and as we see, we would not get the second order approximation for the second derivative at the boundary points unless we use four points in our approximation, which will result in a matrix that is not tridiagonal.[1] It is the same scenario if we attempt to use backward difference for the right boundary. Then the question is what to do to get not only a second order approximation but also to preserve the tridiagonal structure of the stiffness matrix without introducing any grid point outside the grid as shown in Figure 4.1. We can argue

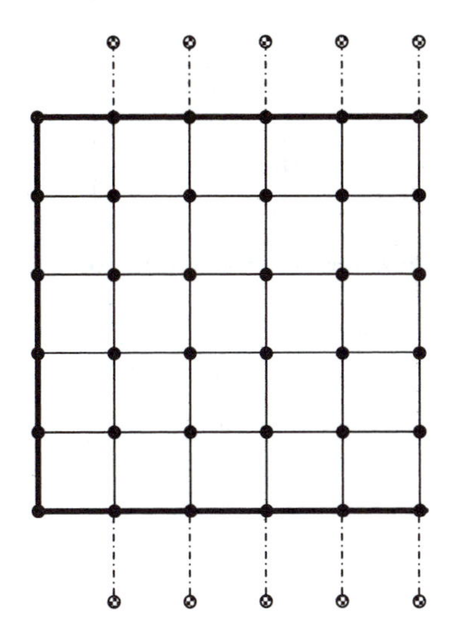

**FIGURE 4.1**: Points outside the grid

that if the second derivative of the option, its gamma, is zero at a boundary, it can be assumed to be zero at the adjacent point as well. Therefore we may consider

$$\frac{\partial^2 v}{\partial S^2}(S_{min} + \Delta S, \tau) = 0$$

and/or

$$\frac{\partial^2 v}{\partial S^2}(S_{max} - \Delta S, \tau) = 0$$

This allows us to use the central difference approximation at a point just inside the boundary conditions and use the central difference approximation; thus it should not add any unwanted truncation error. By doing this we obtain a second order accuracy and preserve the tridiagonal structure of the stiffness matrix. Loosely speaking, for each boundary point we use its neighboring point as the reference point. In Figure 4.2 we illustrate one of the modified reference points for the boundary condition. Using the central difference approximation we get

$$\begin{aligned}\frac{\partial^2 v}{\partial S^2}(S_{min} + \Delta S, \tau_{k+1}) &= \frac{\partial^2 v}{\partial S^2}(S_2, \tau_{k+1}) \\ &= \frac{v_{1,k+1} - 2v_{2,k+1} + v_{3,k+1}}{h^2} + O(h^2)\end{aligned}$$

[1] This would not cause a big obstacle considering that we know how to deal with pentadiagonal matrices efficiently, but what if we can have a second order approximation and preserve the tridiagonal structure of the matrix.

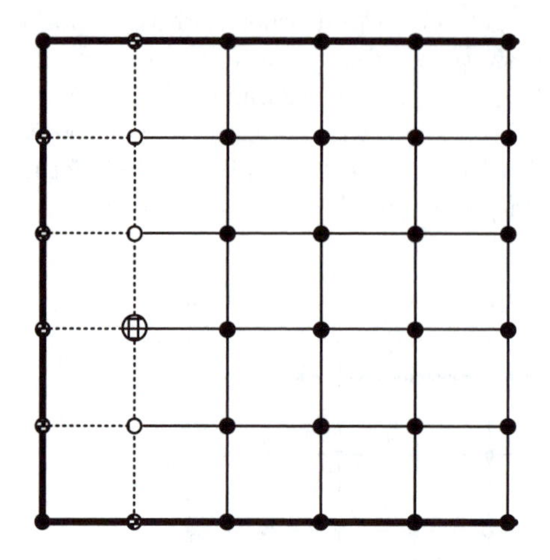

**FIGURE 4.2**: Modified reference points for boundary conditions

Setting it equal to zero we obtain

$$v_{1,k+1} - 2v_{2,k+1} + v_{3,k+1} = 0$$

or equivalently

$$v_{1,k+1} = 2v_{2,k+1} - v_{3,k+1}$$

Substituting it into the difference equation (4.16) for $j = 2$

$$\begin{aligned}
\hat{l}_{j,k+1}v_{1,k+1} + \hat{d}_{j,k+1}v_{2,k+1} + \hat{u}_{j,k+1}v_{3,k+1} &= v_{j,k} \\
\hat{l}_{j,k+1}(2v_{2,k+1} - v_{3,k+1}) + \hat{d}_{j,k+1}v_{2,k+1} + \hat{u}_{j,k+1}v_{3,k+1} &= v_{j,k} \\
(\hat{d}_{j,k+1} + 2\hat{l}_{j,k+1})v_{2,k+1} + (\hat{u}_{j,k+1} - \hat{l}_{j,k+1})v_{3,k+1} &= v_{j,k}
\end{aligned}$$

Using the central difference approximation for the upper boundary condition we get

$$v_{N-1,k+1} - 2v_{N,k+1} + v_{N+1,k+1} = 0$$

or equivalently

$$v_{N+1,k+1} = 2v_{N,k+1} - v_{N-1,k+1}$$

and substituting it into the difference equation (4.17) for $j = N$ we get

$$\begin{aligned}
\hat{l}_{N,k+1}v_{N-1,k+1} + \hat{d}_{N,k+1}v_{N,k+1} + \hat{u}_{N,k+1}v_{N+1,k+1} &= v_{N,k} \\
\hat{l}_{N,k+1}v_{N-1,k+1} + \hat{d}_{N,k+1}v_{N,k+1} + \hat{u}_{N,k+1}(2v_{N,k+1} - v_{N-1,k+1}) &= v_{N,k} \\
(\hat{l}_{N,k+1} - \hat{u}_{N,k+1})v_{N-1,k+1} + (\hat{d}_{N,k+1} + 2\hat{u}_{N,k+1})v_{N,k+1} &= v_{N,k}
\end{aligned}$$

The inclination is that Neumann boundary conditions would be more accurate for the same boundaries, as the second derivative falls off faster than the price, but it would be interesting to see some examples to verify this and perhaps add some rules of thumb or heuristic as to which is the preferred approach. We are not providing any numerical example illustrating the effects of different choices of boundary conditions. However, we set up a short case study on the comparison of using Dirichlet versus Neumann boundary conditions.

### 4.2.2 Implementing Deterministic Jump Conditions

In a number of derivative pricing problems, the asset may undergo a deterministic jump at predetermined times during the life of a derivative contract. One common case of this is discrete dividend payments, which occur at known dates and which in some cases can be assumed to be of known magnitude. Without loss of generality, let us assume there is only one known discrete dividend payment, $D$, at a predetermined time, $t_d$: the following jump condition must be enforced [218]):

$$V(S, t_d^-) = V(S - D, t_d^+)$$

where $t_d^-$ is an instant of time just prior to the discrete dividend payment and $t_d^+$ is an instant after the payment. The times $t_d^-$ and $t_d^+$ refer to calendar times but in the case of a finite difference calculation it is more natural to work in time to maturity $\tau = T - t$. Thus in time to maturity, the jump condition becomes

$$V(S, \tau_d^+) = V(S - D, \tau_d^-)$$

We adjust the time step to coincide with the dividend payment time, $\tau_d$. Therefore, we assume $\tau_d$ happens at the $k^{th}$ time step, that is, $\tau_k = \tau_d$. After the completion of the solution at time $\tau_k$, or to be more precise $\tau_k^-$, we have $V_{i,k}^-$ for all $i$; before proceeding to the solution at time $\tau_{k+1}$, the option values at $S_i$ should be replaced by the option value at $S_i - D$.

To be as accurate as possible, we can adjust the mesh such that the actual time between $\tau_k$ and $\tau_{k+1}$ is very small, and thus minimize the error generated by better representing an instantaneous jump in the asset value.

For now, we consider the case that $D = l\Delta S$ where $l$ is some positive integer. Therefore we will have

$$V_{i,k}^+ = V_{i-l,k}^-$$

Of course $S_i - D$ may not necessarily lie on the finite difference grid, so the option value $V(S, \tau_i^+)$ will be calculated via interpolation. Define

$$\hat{i} = \left\lfloor \frac{S_i - D}{\Delta S} \right\rfloor$$

where $\lfloor x \rfloor$ rounds $x$ to the nearest integer less than or equal to $x$. Then our interpolation scheme is

$$\begin{aligned} V_{i,k}^+ &= V_{\hat{i},k}^- + \frac{V_{\hat{i}+1,k}^- - V_{\hat{i},k}^-}{\Delta S}(S_i - D - \hat{i}\Delta S) \\ &= (1-\alpha)V_{\hat{i},k}^- + \alpha V_{\hat{i}+1,k}^- \end{aligned}$$

with

$$\alpha = \frac{(S_i - D - \hat{i}\Delta S)}{\Delta S}$$

**Pseudo-code for Implementation**

$$\begin{aligned} &\text{for} \quad i = 1 : N \\ &\qquad \hat{i} = \left\lfloor \tfrac{S_i - D}{\Delta S} \right\rfloor \\ &\qquad \alpha = \tfrac{(S_i - D - \hat{i}\Delta S)}{\Delta S} \\ &\qquad V_{i,k}^+ = (1-\alpha)V_{\hat{i},k}^- + \alpha V_{\hat{i}+1,k}^- \\ &\text{end} \end{aligned}$$

We can again minimize the error by adjusting our grid to match the asset price after the dividend price as closely as possible, thus capturing the discontinuity in the price as accurately as possible. What happens when $S_i - D$ lies outside the grid? Add a stencil of an adjusted grid. We could expand this kind of case into a general discussion on ways to fit the solution grid to critical points for the derivative contract.

---

## 4.3 Nonuniform Grid Points

So far we have discussed uniform grid points in both the time and space domains. It might be of interest to have finer grid points near critical prices such as strike or barrier prices, or a coarser grid at locations of less importance. This can be done in two different ways: (a) by laying out the desired grid points which are nonuniform and then discretizing the differential operator that would coincide with those grid points or (b) by applying coordinate transformation that would transfer the original coordinate into a new coordinate and then discretizing the differential operator in the new coordinate. In the second approach we have to rewrite the PDE in the new coordinate. Nonetheless, for discretization purposes we use uniform grid points in new coordinates and after numerically solving it would transfer the numerical solution into the original coordinate which has nonuniform grid points.

The first approach is kind of explicit. First of all there is no need to rewrite the PDE; also if there is a need for various switching regions, going fine to coarse and back to fine grid points should be easily doable. The second approach has an implicit nature to it, in the sense that we have to rewrite the PDE in the new coordinates. However, after we are done with that, we will be dealing with a uniform grid that we are familiar with and there is no reason to rederive the first and second derivative approximation using Taylor expansion at the switching points. subsectionUnequal Sub-intervals Consider grid points $x_0 < x_1 < \cdots < x_N$. Without loss of generality assume there is only one switching point in the grid at $x_i$ such that grid points from $x_0 < x_1 < \cdots < x_i$ are uniform such that $h_1 = x_j - x_{j-1}$ for all $j = 1, \ldots, i$ and $x_i < x_{i+1} < \cdots < x_N$ are uniform such that $h_2 = x_j - x_{j-1}$ for all $j = i+1, \ldots, N$. The interest is to approximate first and second order derivatives $f^{(1)}(x)$ and $f^{(2)}(x)$ at $x_i$. Using Taylor expansion we can write

$$f(x_i - h_1) = f(x_i) - h_1 f^{(1)}(x_i) + \frac{h_1^2}{2!} f^{(2)}(x_i) - \frac{h_1^3}{3!} f^{(3)}(\xi_1) \tag{4.20}$$

$$f(x_i + h_2) = f(x_i) + h_2 f^{(1)}(x_i) + \frac{h_2^2}{2!} f^{(2)}(x_i) + \frac{h_2^3}{3!} f^{(3)}(\xi_2) \tag{4.21}$$

In the case of $h_1 = h_2$ we can simply subtract (4.20) from (4.21) to obtain a central difference for the first derivative. Obviously, this is not the case now. To find an approximation to the first derivative we multiply equations (4.20) and (4.21) by $-h_2^2$ and $+h_1^2$, respectively, and add them to obtain

$$\begin{aligned} -h_2^2 f(x_i - h_1) + h_1^2 f(x_i + h_2) &= (h_1^2 - h_2^2) f(x_i) + (h_1 h_2^2 + h_1^2 h_2) f^{(1)}(x) \\ &+ \frac{h_1^2 h_2^2}{3!}(h_1 f^{(3)}(\xi_1) + h_2 f^{(3)}(\xi_2)) \end{aligned}$$

Therefore we have

$$\begin{aligned} f^{(1)}(x) &= -\frac{h_2}{h_1(h_1+h_2)}f(x_i-h_1)-\frac{h_1-h_2}{h_1h_2}f(x_i)+\frac{h_1}{h_2(h_1+h_2)}f(x_i+h_2) \\ &- \frac{h_1h_2}{3!(h_1+h_2)}(h_1f^{(3)}(\xi_1)+h_2f^{(3)}(\xi_2)) \\ &\approx -\frac{h_2}{h_1(h_1+h_2)}f(x_i-h_1)-\frac{h_1-h_2}{h_1h_2}f(x_i)+\frac{h_1}{h_2(h_1+h_2)}f(x_i+h_2) \quad \text{with} \quad O(h_1h_2) \end{aligned} \tag{4.22}$$

To find an approximation to the second derivative at $x_i$ we multiply equations (4.20) and (4.21) by $h_2$ and $h_1$, respectively, and add them to obtain

$$\begin{aligned} h_2f(x_i-h_1)+h_1f(x_i+h_2) &= (h_1+h_2)f(x_i)+\frac{h_1h_2(h_1+h_2)}{2}f^{(2)}(x) \\ &+ \frac{h_1h_2}{3!}(h_1^2f^{(3)}(\xi_1)+h_2^2f^{(3)}(\xi_2)) \end{aligned}$$

Therefore we have

$$\begin{aligned} f^{(2)}(x) &= \frac{2(h_2f(x_i-h_1)-(h_1+h_2)f(x_i)+h_1f(x_i+h_2))}{h_1h_2(h_1+h_2)} \\ &- \frac{2}{3!(h_1+h_2)}(h_1^2f^{(3)}(\xi_1)+h_2^2f^{(3)}(\xi_2)) \\ &\approx \frac{2(h_2f(x_i-h_1)-(h_1+h_2)f(x_i)+h_1f(x_i+h_2))}{h_1h_2(h_1+h_2)} \quad \text{with} \quad O(h_1+h_2) \end{aligned} \tag{4.23}$$

We see that in non-equal grid points three-point approximation for a second derivative approximation does not yield a second order approximation. To obtain a second order approximation we need to use a four-point approximation. By doing that, however, we would not be able to preserve the tridiagonal structure of the stiffness matrix.

An alternative approach for higher order approximation for non-equal grid points is examined in [203].

### 4.3.1 Coordinate Transformation

As mentioned earlier, in order to have finer grid points near critical prices such as strike or barrier prices, or a coarser grid at locations of less importance, we can apply coordinate transformation [210].

Assume that we are interested in finding a transformation that maps $0 \le \xi \le 1$ to $S_{min} \le S \le S_{max}$, with uniform grid points on $\xi$ and non-uniform grid points on $S$ with concentration around some point $B$ in the domain. There are many ways of doing this, one of which is the following. Let

$$S(\xi) = B + \alpha\sinh(c_1\xi + c_2(1-\xi))$$

With this assumption it is clear that if we want

$$\begin{aligned} S(\xi=0) &= S_{min} \\ S(\xi=1) &= S_{max} \end{aligned}$$

then we should have the following values for $c_1$ and $c_2$

$$
\begin{aligned}
c_1 &= \sinh^{-1}(\frac{S_{max} - B}{\alpha}) \\
c_2 &= \sinh^{-1}(\frac{S_{min} - B}{\alpha})
\end{aligned}
$$

The value of $\xi$ that corresponds to $B$, $S(\xi_B) = B$, is

$$
\xi_B = \frac{c_2}{c_2 - c_1}
$$

To obtain a highly non-uniform grid concentrated around $B$, $\alpha$ should be smaller than $S_{max} - S_{min}$. If we choose $\alpha$ to be greater than $S_{max} - S_{min}$, we get a uniform mesh.

In Figure 4.3 we plot graphs of $S(\xi)$ for various values of $\alpha$ showing concentrations of evaluation points around $B$ when using coordinate transformation. In this example $B = 50$ and, as illustrated for small $\alpha$, points are pretty concentrated around $B$ and for larger $\alpha$ intervals become more equidistant.

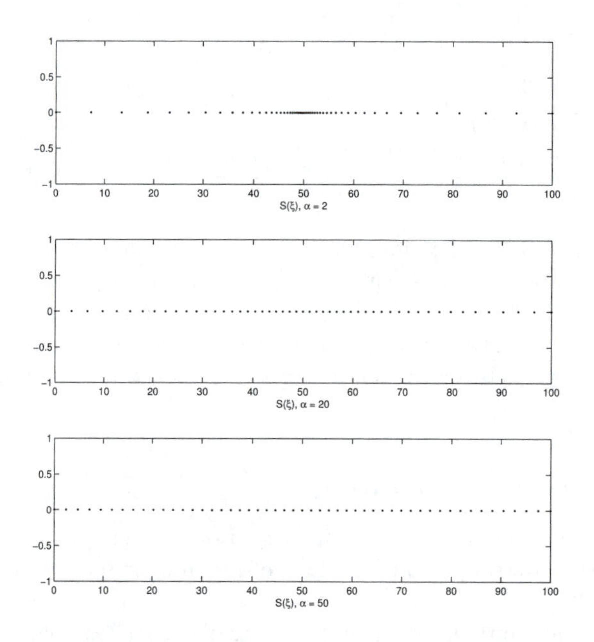


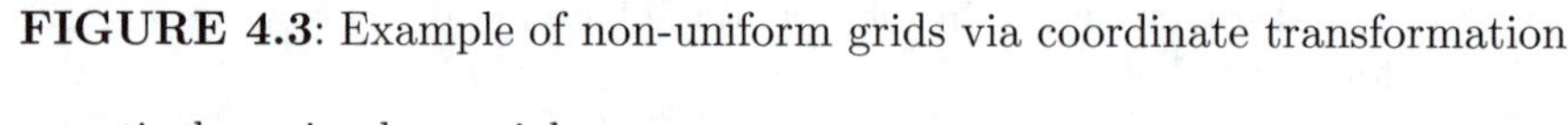

**FIGURE 4.3**: Example of non-uniform grids via coordinate transformation

Two particulary simple special cases are

$$
\begin{aligned}
S(\xi) &= K\sinh(\xi) \\
\frac{\partial S}{\partial \xi} &= J(\xi) = K\cosh(\xi)
\end{aligned}
$$

$$
\begin{aligned}
S(\xi) &= Ke^{\xi} \\
\frac{\partial S}{\partial \xi} &= J(\xi) = Ke^{\xi}
\end{aligned}
$$

We could add an explanation of these special cases and their benefit. Also, maybe a concrete example of using these in a known case, for instance, concentrating points around the strike of a simple option. Is it ever worth it to use coordinate transformation in both directions in order to concentrate the mesh points on a specific time and asset price? What are the numerical differences between this considerably harder coordinate transformation and just using specially created points for the grid while using the original PDE?

#### 4.3.1.1 Black–Scholes PDE after Coordinate Transformation

The Black–Scholes PDE under calendar time and asset price coordinates is

$$\frac{\partial v}{\partial t} + \frac{\sigma^2 S^2}{2}\frac{\partial^2 v}{\partial S^2} + (r-q)S\frac{\partial v}{\partial S} = rv(S,t) \tag{4.24}$$

Let us define

$$\bar{v}(\xi,\tau) = v(S,t)$$

where $\tau = T - t$ time to maturity and $S(\xi) = B + \alpha\sinh(c_1\xi + c_2(1-\xi))$. We are going to see what PDE $\bar{v}(\xi,\tau)$ satisfies. Using the chain rule, we have

$$\frac{\partial v}{\partial S} = \frac{\partial \bar{v}}{\partial \xi}\cdot\frac{\partial \xi}{\partial S} = \frac{\partial \bar{v}}{\partial \xi}\cdot\frac{1}{\frac{\partial S}{\partial \xi}} \tag{4.25}$$

$$\begin{aligned}
\frac{\partial^2 v}{\partial S^2} &= \frac{\partial}{\partial S}\left(\frac{\partial v}{\partial S}\right) = \frac{\partial}{\partial \xi}\left(\frac{\partial v}{\partial S}\right)\cdot\frac{\partial \xi}{\partial S} \\
&= \frac{\partial}{\partial \xi}\left(\frac{\partial \bar{v}}{\partial \xi}\cdot\frac{\partial \xi}{\partial S}\right)\cdot\frac{\partial \xi}{\partial S} = \frac{\partial}{\partial \xi}\left(\frac{\partial \bar{v}}{\partial \xi}\cdot\frac{1}{\frac{\partial S}{\partial \xi}}\right)\cdot\frac{1}{\frac{\partial S}{\partial \xi}} \\
&= \frac{\partial^2 \bar{v}}{\partial \xi^2}\frac{1}{\left(\frac{\partial S}{\partial \xi}\right)^2} - \frac{\partial \bar{v}}{\partial \xi}\cdot\frac{\frac{\partial^2 \xi}{\partial S^2}}{\left(\frac{\partial S}{\partial \xi}\right)^3}
\end{aligned} \tag{4.26}$$

$$\frac{\partial v}{\partial t} = -\frac{\partial \bar{v}}{\partial \tau} \tag{4.27}$$

By substituting (4.25), (4.26), and (4.27) in (4.24) we get

$$-\frac{\partial \bar{v}}{\partial \tau} + \frac{\sigma^2 S^2}{2}\left(\frac{\partial^2 \bar{v}}{\partial \xi^2}\frac{1}{\left(\frac{\partial S}{\partial \xi}\right)^2} - \frac{\partial \bar{v}}{\partial \xi}\cdot\frac{\frac{\partial^2 \xi}{\partial S^2}}{\left(\frac{\partial S}{\partial \xi}\right)^3}\right) + (r-q)S\frac{\partial \bar{v}}{\partial \xi}\cdot\frac{1}{\frac{\partial S}{\partial \xi}} = r\bar{v}$$

$$-\bar{v}_\tau + \frac{\sigma^2 S(\xi)^2}{2}\left(\frac{1}{\frac{\partial S(\xi)}{\partial \xi}}\right)^2 \bar{v}_{\xi\xi} + \left((r-q)S(\xi)\frac{1}{\frac{\partial S(\xi)}{\partial \xi}} - \frac{\sigma^2 S(\xi)^2}{2}\frac{\frac{\partial^2 S(\xi)}{\partial \xi^2}}{\left(\frac{\partial S(\xi)}{\partial \xi}\right)^3}\right)\bar{v}_\xi - r\bar{v}(\xi,\tau) = 0 \tag{4.28}$$

Or simply for the new coordinate $\xi$, we have $S = S(\xi)$ and by using the chain rule we obtain

$$-\frac{\partial \bar{v}}{\partial \tau} + \frac{\sigma^2}{2}\frac{S^2(\xi)}{J(\xi)}\frac{\partial}{\partial \xi}\left(\frac{1}{J(\xi)}\frac{\partial \bar{v}}{\partial \xi}\right) + (r-q)\frac{S(\xi)}{J(\xi)}\frac{\partial \bar{v}}{\partial \xi} - r\bar{v} = 0 \tag{4.29}$$

where

$$J(\xi) = \frac{\partial S(\xi)}{\partial \xi}$$

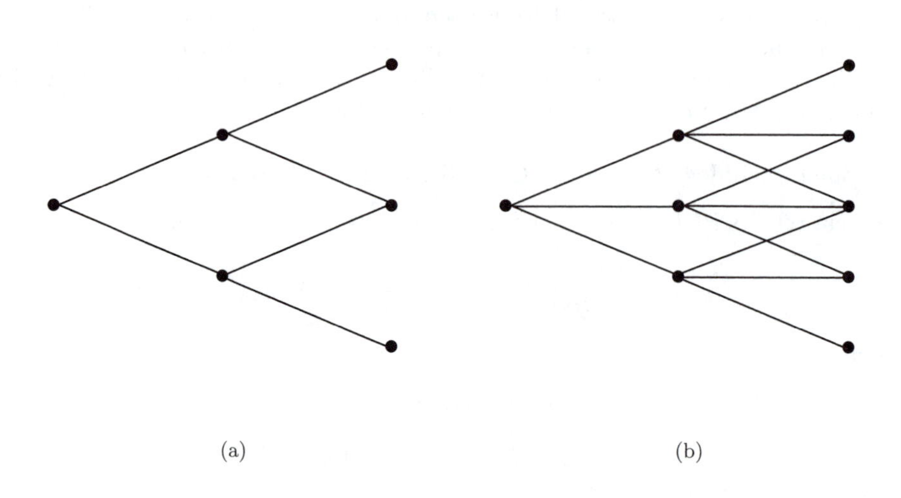


**FIGURE 4.4**: (a) Binomial tree, (b) trinomial tree

## 4.4 Dimension Reduction

One of the earliest computational methods used to price options was the binomial tree methods, developed by Cox, Ross, and Rubinstein. The basic idea of these methods is to model the evolution of the asset using either a binomial or trinomial tree. In Figures 4.4(a) and 4.4(b), we show graphical representations of two-period binomial and trinomial trees respectively. How the points are distributed and how their probabilities are made to be risk-neutral determines which of the many different tree methods we are using. While tree-based methods are still popular, Rubinstein proved that these methods can in fact be directly related to an explicit finite difference scheme and thus inherit its convergence and stability issues. The tree method does, however, have one distinct advantage in that it does not solve for the option price on an entire grid for which we are uninterested. We can, however, use the same dimension reduction technique when solving the option pricing PDE using an explicit method. To do so, we reduce the number of points we solve for at every step, keeping only enough solution points to result in an option price at the current asset level, or possibly slightly more, depending on our needs. We illustrate how it is done schematically in Figure 4.5. Implementing this method is almost identical to implementing the normal explicit finite difference scheme, except the stiffness matrix shrinks in dimension with each step. One other noticeable advantage is that this construction requires us to only consider the boundary conditions at the maturity of the option; all other prices are derived from past prices and thus we eliminate some error prorogation due to boundary conditions. Dimension reduction or domain shrinkage is only applicable to the explicit case.

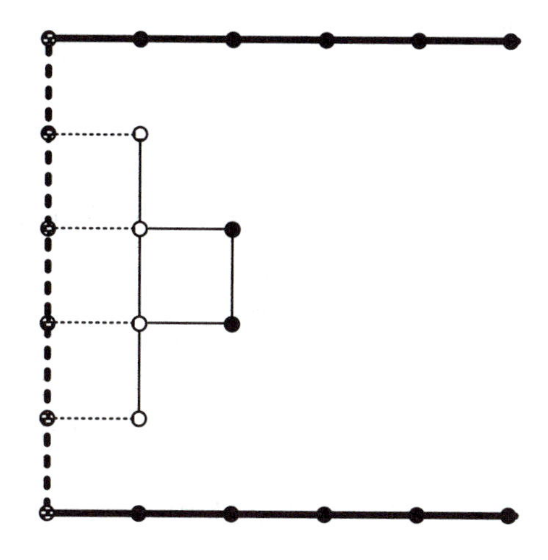

**FIGURE 4.5**: Explicit finite difference with dimension reduction

## 4.5 Pricing Path-Dependent Options in a Diffusion Framework

We will cover some path-dependent options. We start with describing the derivative and cover its boundary conditions and critical points. Then we focus on how to solve it numerically in a diffusion framework.

The previous discussion of numerical techniques for pricing derivatives using PDEs has concentrated only on European options, and up to this point we have eschewed the issue of options with early exercise provisions, i.e., Bermudan and American options. In this section we will discuss a number of different approaches for pricing these options using PDE based methods.

### 4.5.1 Bermudan Options

All of the solution methods discussed thus far have relied on beginning at $\tau = 0$, at which time the payoff of the derivative is known and thus the value is known for any asset price, and walking backwards through time using a finite difference scheme, to generate option prices at a previous time step from results in the current one. This takes the form of a recursive equation taking one of the following forms:

$$V_{k+1} = A_{k+1}^{\text{Explicit}} V_k$$

$$A_{k+1}^{\text{Implicit}} V_{k+1} = V_k$$

$$A_{k+1}^{\text{Implicit}} V_{k+1} = A_{k+1}^{\text{Explicit}} V_k$$

depending on the finite difference scheme we are using.

For example, let us consider an implicit finite difference scheme for pricing European options under the generalized Black–Scholes PDE. We implement such a pricing algorithm using the following pseudo-code

$$\begin{array}{ll} \text{for} & \text{k =1:M} \\ & A_{k+1}^{\text{Implicit}} V_{k+1} = V_k \\ \text{end} & \end{array}$$

The holder of a Bermudan style option can exercise at predetermined dates specified in advance in the contract.

$$V(S_t, t; K, T) = \max_{\tau_e} \left\{ \mathbb{E}[e^{-r(T-t)}(K - S_{\tau_e})^+] \right\} \quad \text{for } \tau_e \in \{t_1, \ldots, t_l\} \tag{4.30}$$

To price Bermudan options, we assume without a loss of generality that we have a grid on which exercise dates coincide with time steps on the grid. Indeed, it is simple to construct a grid where this is true by adjusting the time steps. For illustrative purposes, we modify the grid in Figure 3.1 where dashed lines indicate those time steps that the holder of the option can exercise as depicted in Figure 4.6.

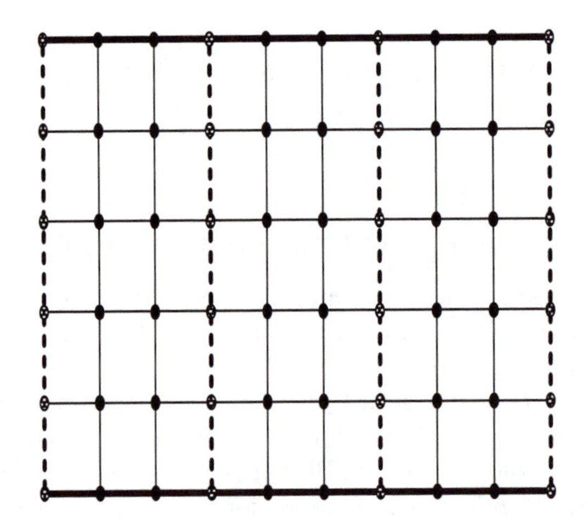

**FIGURE 4.6**: Grid in Bermudan option pricing

Let us suppose that the last exercise date in calendar time, which becomes the first in time-to-maturity since we are going backward in time, coincides with $\tau_k$ for some $k$. Then up to that time step there is no difference between Bermudan and European options since there is no exercise opportunity until we reach $\tau_k$ and so their prices must coincide. When we have iterated from $\tau = 0$ to $\tau_k$ we have valid European option prices $V_k$ for every asset price.

At each asset price level on our grid, we should apply the optimal exercise decision, either choosing not to exercise if the European option value is above the exercise value, or exercising and taking the exercise value if that is advantageous. Thus at every exercise opportunity we correct/adjust the option prices $V_k$ to reflect the optimal exercise condition. For example, if we assume we have a call option we could use the following pseudo-code

$$\begin{array}{l} \text{for } k = 1 : M \\ \quad A_{k+1}^{\text{Implicit}} V_{k+1} = V_k \\ \quad \text{for } i = 1 : N \\ \quad\quad \text{if } V_i^{k+1} < (S_i - K)^+ \end{array}$$

```
            V_i^{k+1} = (S_i - K)^+
        endif
    endfor
endfor
```

### 4.5.2 American Options

Options with American exercise provisions allow the holder to exercise the option at any time during the life of the option, including at maturity. Thus the value of an American option is

$$V(S_t, t; K, T) = \sup_{t \leq \tau \leq T} \left\{ \mathbb{E}_t[e^{-r(T-t)}(K - S_\tau)^+] \right\}$$

where the supremum is taken over all stopping times $\tau$ defined on the probability space with respect to the filtration generated by the stock price. It is shown ([160] and [168]) that for each $t$ there exists a critical stock price $S^*(t)$ such that if $S(t) \leq S^*(t)$ the value of the American put option is the value of immediate exercise or $K - S(t)$ while for $S(t) > S^*(t)$ the value exceeds this immediate exercise value. The curve $S^*(t)$ viewed as a function of time is referred to as the critical exercise boundary while the region

$$\mathcal{C} = \{(S, t) \, | S > S^*(t)\}$$

is called the continuation region. The complement $\mathcal{E}$ of the continuation region is the exercise region. The value of the American put in the exercise region is known and it only remains to determine the value in the continuation region.

In general, the techniques used to price American options via PDE techniques are based on the pricing of Bermudan options, where we evaluate the PDE on successive time slices of the grid and apply corrections of the optimal exercise of the option. This will converge to the American option price as the time step $\Delta\tau$ approaches zero; however, there is always some discretization error involved.

There are a number of different techniques, however, that we can use to implement this correction for optimal exercise of the American option. Some of the different ways of pricing American options under PDE techniques are as follows:

- Bermudan approximation
- Black–Scholes PDE with a synthetic dividend process
- Brennan–Schwartz algorithm

and we will explore these techniques in the following sections.

#### 4.5.2.1 Bermudan Approximation

The most basic method for pricing an American option in a PDE setting is to simply use the technique we outlined for pricing Bermudan options. Simply using the finite difference to solve for the option prices at the next time slice and applying an optimal exercise criterion can determine the true option prices. If we do this at every time step, and make the time step $\Delta\tau$ very small, the resulting Bermudan option price should converge to the American option price as the exercise times become nearly continuous.

#### 4.5.2.2 Black–Scholes PDE with a Synthetic Dividend Process

While the Bermudan approximation will successfully price American options if we make $\Delta\tau$ small enough, it requires us to process every time slice of the mesh a second time to implement optimal exercise conditions. Another approach is to modify the underlying PDE to account for optimal exercise conditions so that the PDE is in fact accurate everywhere on the grid and proceed with our normal discretization using this modified PDE.

The differential operator for the generalized Black–Scholes PDE is

$$\mathcal{L}(v) = -\frac{\partial \hat{v}}{\partial \tau} + \frac{\sigma(S,\tau)^2 S^2}{2}\frac{\partial^2 \hat{v}}{\partial S^2} + (r(\tau) - q(\tau))S\frac{\partial \hat{v}}{\partial S} - r(\tau)\hat{v}(S,\tau) = 0$$

In the case of American options, the optimal exercise policy may be to exercise early depending on the level of the stock. In the case of an American put option, at every time $\tau$ there is a level of the stock price $S^\star(\tau)$ such that at any stock price lower than that level the holder of the option should optimally exercise. This price level is called the critical level and it is time dependent due to the time value of the money. For an American call, at every time $\tau$ there is a level of the stock price $S^*(\tau)$ such that at any stock price higher than $S^*(\tau)$ the holder should optimally exercise. Therefore, these critical values constitute a curve that would divide out the domain to two distinct regions: (a) *exercise region* ($\mathcal{E}$), (b) *continuation region* ($\mathcal{C}$). In the continuation region, the holder of the option does not exercise, and the Black–Scholes PDE holds in the region, i.e., $\mathcal{L}(v) = 0$, in the exercise region on the other hand it does not hold, i.e., $\mathcal{L}(v) \neq 0$ and the holder would exercise and receive $K - S$. Thus in the exercise region $\mathcal{E}$, we have

$$V(S,\tau) = K - S \quad \text{in} \quad \mathcal{E}$$

Knowing the exact value of the option, theta, delta, and gamma of the option can be calculated as well; therefore we have

$$\begin{aligned}
\frac{\partial v}{\partial \tau} &= 0 \\
\frac{\partial v}{\partial S} &= -1 \\
\frac{\partial^2 v}{\partial S^2} &= 0
\end{aligned}$$

So in the exercise region we can substitute these values into $\mathcal{L}(v)$ and compute its value.

$$\begin{aligned}
\mathcal{L}(v) &= -\frac{\partial \hat{v}}{\partial \tau} + \frac{\sigma(S,\tau)^2 S^2}{2}\frac{\partial^2 \hat{v}}{\partial S^2} + (r(\tau) - q(\tau))S\frac{\partial \hat{v}}{\partial S} - r(\tau)\hat{v}(S,\tau) \\
&= 0 + 0 + (r - q)S(-1) - r(K - S) \\
&= qS - rS - rK + rS \\
&= qS - rK
\end{aligned}$$

Therefore, we can say in the entire region the following differential operator applies:

$$\mathcal{L}(v) = \mathbb{1}_{S<S^\star(\tau)}\{qS - rK\}$$

or equivalently

$$\mathcal{L}(v) + \mathbb{1}_{S<S^\star(\tau)}\{rK - qS\} = 0$$

where as before $S^*(\tau)$ is the critical exercise boundary at time $\tau$. This synthetic dividend is consistent with the demonstration by Carr, Jarrow, and Myneni [58] that is one must extract from the American option holder the interest on the strike less the dividend yield for the time the stock spends in the exercise region to get the value back to that of a European option. In our grid, the critical value at time $\tau_k$, $S^*(\tau_k)$, is the smallest value of the stock that the option premium at the value exceeds its payoff; that is,

$$S^*(\tau_k) = \min\{S_i : V_i^k - (K - S_i)^+ > 0\}$$

We assume that $S^*(\tau_0) = S^*(0) = K$. The following pseudo-code implements this scheme and constructs the exercise boundary of critical values at every time.
At $\tau_1$:

$$V_0 = \left[V_i^0\right] \quad \text{and} \quad S(\tau_0 = 0) = K \tag{4.31}$$

$$\hat{l}_{i,1} v_{i-1}^1 + \hat{d}_{j,1} v_j^1 + \hat{u}_{j,1} v_{j+1}^1 = v_i^0 + \mathbb{1}_{S<S(\tau_0)}\{rK - qS_i\} \tag{4.32}$$

and

$$S(\tau_1) = \min\{S_i : V_i^1 - (K - S_i)^+ > 0\} \tag{4.33}$$

and we repeat this procedure to $\tau_M$.

For this scheme to work for a jump framework we need to account for that fact that stock can jump back from the exercise region to the continuation region. Just to be clear here, in our pseudo-code we are explaining the case for the implicit implementation, but theoretically one could use any finite difference scheme with the Heavyside term.

#### 4.5.2.3 Brennan–Schwartz Algorithm

One important shortcoming of the Bermudan approximation method is that the option values for time period $k+1$, $V_{k+1}$ are computed in the implicit method using both the option values at time $k$, $V_k$ and simultaneously all option values at period $k+1$, $V_{k+1}$. These results are then correct to account for optimal exercise, but the continuation region at time $k+1$ still has option values which are influenced by the incorrectly calculated option values in the exercise region which are subsequently corrected. The Brennan–Schwartz algorithm allows us to solve the tridiagonal stiffness matrix in the implicit scheme either bottom-up or top-down, depending where our exercise region is expected to be, and correct for optimal exercise while solving for the $V_{k+1}$ values. This prevents the propagation of incorrect exercise errors into the continuation region.

We know that the implicit finite difference scheme can be implemented with the following equation.

$$A_{k+1}^{\text{Implicit}} V_{k+1} = V_k + r.h.s.$$

Or

$$\begin{pmatrix} \hat{d}_2^{k+1} & \hat{u}_2^{k+1} & & & \\ \hat{l}_3^{k+1} & \hat{d}_3^{k+1} & \hat{u}_3^{k+1} & & \\ & \ddots & \ddots & \ddots & \\ & & \hat{l}_{N-1}^{k+1} & \hat{d}_{N-1}^{k+1} & \hat{u}_{N-1}^{k+1} \\ & & & \hat{l}_N^{k+1} & \hat{d}_N^{k+1} \end{pmatrix} \begin{pmatrix} v_2^{k+1} \\ v_3^{k+1} \\ \vdots \\ v_{N-1}^{k+1} \\ v_N^{k+1} \end{pmatrix} = \begin{pmatrix} v_2^k \\ v_3^k \\ \vdots \\ v_{N-1}^k \\ v_N^k \end{pmatrix} + r.h.s.$$

It is known that we can solve this tridiagonal linear equation by either first making the upper diagonal elements zero and then solving the system or first making the lower diagonal zero and then solving for the system. In a regular tridiagonal system solver, it does not make a difference. However, in the Brennan–Schwartz algorithm it depends on the contract. We wish to apply the optimal exercise criterion while we are solving the tridiagonal system, and so we wish to start solving in the exercise region. Thus for American put options, we make the upper diagonals zero and then solve; for American call options, we first make the lower diagonal zero and then solve.

For American put options, we first make the upper diagonal zero; to do that we use the following steps:

(a) Multiply the $N^{th}$ row (last row) by $-\frac{\hat{u}_{N-1}^{k+1}}{\tilde{d}_N^{k+1}}$ and add to the $(N-1)^{th}$ row; by doing that we eliminate $\hat{u}_{N-1}^{k+1}$

(b) Multiply the $(N-1)^{th}$ row by $-\frac{\hat{u}_{N-2}^{k+1}}{\tilde{d}_{N-1}^{k+1}}$ and add to the $(N-2)^{th}$ row; by doing that we eliminate $\hat{u}_{N-2}^{k+1}$

(c) Repeat until all upper diagonals are eliminated

Now we have

$$\begin{pmatrix} \tilde{d}_2^{k+1} & 0 & & & \\ \hat{l}_3^{k+1} & \tilde{d}_3^{k+1} & 0 & & \\ & \ddots & \ddots & \ddots & \\ & & \hat{l}_{N-1}^{k+1} & \tilde{d}_{N-1}^{k+1} & 0 \\ & & & \hat{l}_N^{k+1} & \hat{d}_N^{k+1} \end{pmatrix} \begin{pmatrix} v_2^{k+1} \\ v_3^{k+1} \\ \vdots \\ v_{N-1}^{k+1} \\ v_N^{k+1} \end{pmatrix} = \begin{pmatrix} v_2^k - \frac{\hat{u}_2^{k+1}}{\tilde{d}_3^{k+1}} v_3^k \\ v_3^k - \frac{\hat{u}_3^{k+1}}{\tilde{d}_4^{k+1}} v_4^k \\ \vdots \\ v_{N-1}^k - \frac{\hat{u}_{N-1}^{k+1}}{\tilde{d}_N^{k+1}} v_N^k \\ v_N^k \end{pmatrix}$$

or equivalently

$$\begin{pmatrix} \tilde{d}_2^{k+1} & 0 & & & \\ \hat{l}_3^{k+1} & \tilde{d}_3^{k+1} & 0 & & \\ & \ddots & \ddots & \ddots & \\ & & \hat{l}_{N-1}^{k+1} & \tilde{d}_{N-1}^{k+1} & 0 \\ & & & \hat{l}_N^{k+1} & \hat{d}_N^{k+1} \end{pmatrix} \begin{pmatrix} v_2^{k+1} \\ v_3^{k+1} \\ \vdots \\ v_{N-1}^{k+1} \\ v_N^{k+1} \end{pmatrix} = \begin{pmatrix} \tilde{v}_2^k \\ \tilde{v}_3^k \\ \vdots \\ \tilde{v}_{N-1}^k \\ v_N^k \end{pmatrix}$$

where $\tilde{v}_j^k = v_j^k - \frac{\hat{u}_j^{k+1}}{\tilde{d}_{j+1}^{k+1}} v_{j+1}^k$ and $\tilde{d}_j^{k+1} = \hat{d}_j^{k+1} - \frac{\hat{u}_j^{k+1}}{\tilde{d}_{j+1}^{k+1}} \hat{l}_{j+1}^{k+1}$ for $j = 2, \ldots, N-1$.
Now we can solve for $V^{k+1}$ starting from the first row.

$$\tilde{d}_2^{k+1} v_2^{k+1} = \tilde{v}_2^k$$
$$v_2^{k+1} = \frac{\tilde{v}_2^k}{\tilde{d}_2^{k+1}}$$

Having $v_2^{k+1}$ we can solve the equation in the second row for $v_3^{k+1}$, that is,

$$\hat{l}_3^{k+1} v_2^{k+1} + \tilde{d}_3^{k+1} v_3^{k+1} = \tilde{v}_3^k$$
$$v_3^{k+1} = \frac{\tilde{v}_3^k - \hat{l}_3^{k+1} v_2^{k+1}}{\tilde{d}_3^{k+1}}$$

and we can repeat this until we solve for $v_N^{k+1}$.

The above algorithm is the procedure for a basic tridiagonal solver. The Brennan–Schwartz algorithm modifies this procedure by taking into the account the exercise value; as one solves for $V^{k+1}$ element by element, we compare the option price to the exercise value and if the option price is smaller than the exercise value correct it by making it equal to the exercise value.

Assuming we solved for the $j^{th}$ element, $v_j^{k+1}$,

if $v_j^{k+1} < (K - S_j)^+$

$$v_j^{k+1} = (K - S_j)^+$$

else

no action is needed

end

and this is repeated for each element.

We provide an example to illustrate the exercise boundary and premiums for an American put for each case. The parameter set used in this example is spot price $S_0 = 100$, strike of $K = 90$, volatility $\sigma = 30\%$, risk-free interest rate $r = 3.0\%$, continuous dividend rate $q = 1.0\%$, and maturity of $T = 0.5$.

Figures 4.7 and 4.8 illustrate the optimal exercise boundary and corresponding premiums for all three cases. At the end of the chapter we present a set of case studies to illustrate

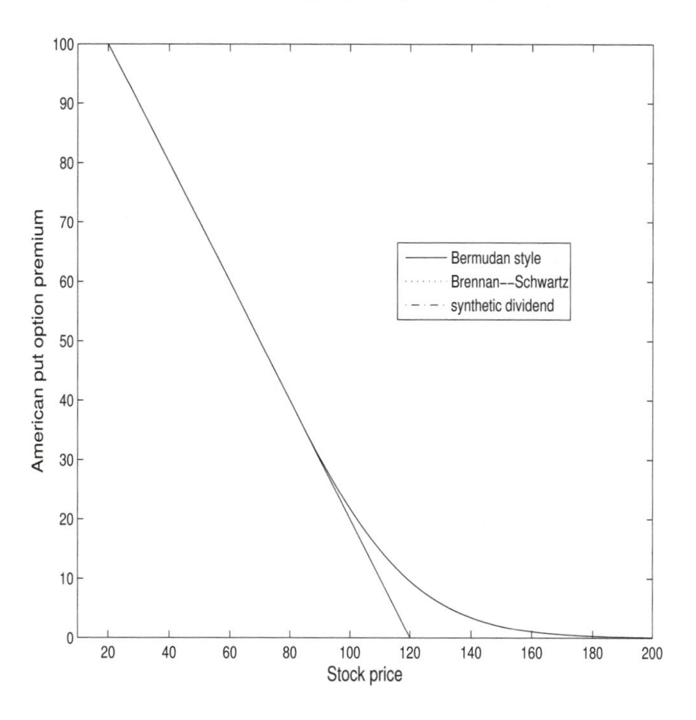


**FIGURE 4.7**: American put premiums

the various levels of efficacy of these techniques.

We are not sure if it is worth to include in this chapter, but it would be interesting to know how many error propagations the Brennan–Schwartz method prevents if we are running case studies.

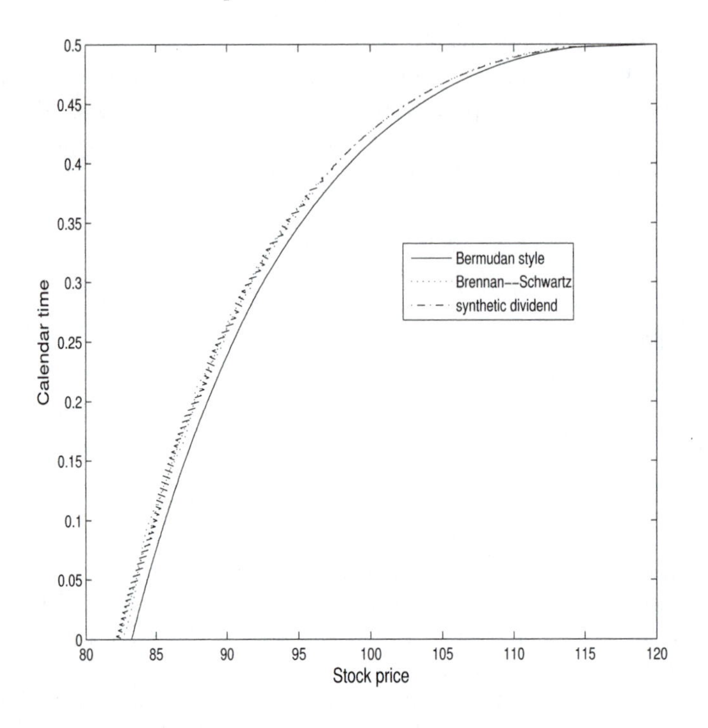


**FIGURE 4.8**: Optimal exercise boundary for an American Put

### 4.5.3 Barrier Options

Barrier options are a family of options that come alive or die when barriers are reached. Two major types of barrier options are:

- knock-in: the option comes alive when the barrier is reached.
- knock-out: the option dies when the barrier is reached.

Option formulas for pricing standard barrier options have been developed [177],[190]. By standard we mean single knock-in barrier options with or without rebate and single knock-out barrier options without rebate or with non-deferred rebate with constant volatility.

$$
\begin{aligned}
A &= \xi S e^{(b-r)T}\Phi(\xi x_1) - \xi X e^{-rT}\Phi(\xi x_1 - \xi\sigma\sqrt{T}) \\
B &= \xi S e^{(b-r)T}\Phi(\xi x_2) - \xi X e^{-rT}\Phi(\xi x_2 - \xi\sigma\sqrt{T}) \\
C &= \xi S e^{(b-r)T}(B/S)^{2(\mu+1)}\Phi(\eta y_1) - \xi X e^{-rT}(B/S)^{2\mu}\Phi(\eta y_1 - \eta\sigma\sqrt{T}) \\
D &= \xi S e^{(b-r)T}(B/S)^{2(\mu+1)}\Phi(\eta y_2) - \xi X e^{-rT}(B/S)^{2\mu}\Phi(\eta y_2 - \eta\sigma\sqrt{T}) \\
E &= R e^{-rT}\left[\Phi(\eta x_2 - \eta\sigma\sqrt{T}) - (B/S)^{2\mu}\Phi(\eta y_2 - \eta\sigma\sqrt{T})\right] \\
F &= R\left[(B/S)^{\mu+\lambda}\Phi(\eta z) + (B/S)^{\mu-\lambda}\Phi(\eta z - 2\eta\sigma\sqrt{T})\right]
\end{aligned}
$$

where

$$
\begin{aligned}
x_1 &= \frac{\log(S/X)}{\sigma\sqrt{T}} + (1+2\mu)\sigma\sqrt{T} \\
x_2 &= \frac{\log(S/B)}{\sigma\sqrt{T}} + (1+2\mu)\sigma\sqrt{T} \\
y_1 &= \frac{\log(B^2/SX)}{\sigma\sqrt{T}} + (1+2\mu)\sigma\sqrt{T} \\
x_2 &= \frac{\log(B/S)}{\sigma\sqrt{T}} + (1+2\mu)\sigma\sqrt{T} \\
z &= \frac{\log(B/S)}{\sigma\sqrt{T}} + \lambda\sigma\sqrt{T} \\
\mu &= \frac{b-\sigma^2/2}{\sigma^2} \\
\lambda &= \sqrt{\mu^2 + \frac{2r}{\sigma^2}} \\
b &= r - q
\end{aligned}
$$

**Knock-In Barrier Options**

Down-and-in call $S > B$, payoff: $\max(S - X, 0)$ if $S \leq B$ before $T$ else $R$ at maturity

$$
\begin{aligned}
c_{di(X>B)} &= C + E & \xi = 1, \eta = 1 \\
c_{di(X<B)} &= A - B + D + E & \xi = 1, \eta = 1
\end{aligned}
$$

Up-and-in call $S < B$, payoff: $\max(S - X, 0)$ if $S \geq B$ before $T$ else $R$ at maturity

$$
\begin{aligned}
c_{ui(X>B)} &= A + E & \xi = 1, \eta = -1 \\
c_{ui(X<B)} &= B - C + D + E & \xi = 1, \eta = -1
\end{aligned}
$$

Down-and-in put $S > B$, payoff: $\max(X - S, 0)$ if $S \leq B$ before $T$ else $R$ at maturity

$$
\begin{aligned}
p_{di(X>B)} &= B - C + D + E & \xi = -1, \eta = 1 \\
p_{di(X<B)} &= A + E & \xi = -1, \eta = 1
\end{aligned}
$$

Up-and-in put $S < B$, payoff: $\max(X - S, 0)$ if $S \geq B$ before $T$ else $R$ at maturity

$$
\begin{aligned}
p_{ui(X>B)} &= A - B + D + E & \xi = -1, \eta = -1 \\
p_{ui(X<B)} &= C + E & \xi = -1, \eta = -1
\end{aligned}
$$

**Knock-Out Barrier Options**

Down-and-out call $S > B$, payoff: $\max(S - X, 0)$ if $S > B$ before $T$ else $R$ at hit

$$
\begin{aligned}
c_{do(X>B)} &= A - C + F & \xi = 1, \eta = 1 \\
c_{do(X<B)} &= B - D + F & \xi = 1, \eta = 1
\end{aligned}
$$

Up-and-out call $S < B$, payoff: $\max(S - X, 0)$ if $S < B$ before $T$ else $R$ at hit

$$
\begin{aligned}
c_{uo(X>B)} &= F & \xi = 1, \eta = -1 \\
c_{uo(X<B)} &= A - B + C - D + F & \xi = 1, \eta = -1
\end{aligned}
$$

Down-and-out put $S > B$, payoff: $\max(X - S, 0)$ if $S > B$ before $T$ else $R$ at hit

$$p_{do(X>B)} = A - B + C - D + F \qquad \xi = -1, \eta = 1$$
$$p_{do(X<B)} = F \qquad \xi = -1, \eta = 1$$

Up-and-out put $S < B$, payoff: $\max(X - S, 0)$ if $S < B$ before $T$ else $R$ at hit

$$p_{uo(X>B)} = B - D + F \qquad \xi = -1, \eta = -1$$
$$p_{uo(X<B)} = A - C + F \qquad \xi = -1, \eta = -1$$

In the case of the deferred rebate knock-out barrier option either down-and-out or up-and-out, we can price the deferred rebate and add it to the price of the barrier without rebate.

In the case of non-constant volatility, there is no analytical solution. Therefore, the only solution would be the numerical solution of the PDE. For numerically solving the PDE we just have to specify boundary conditions and the rest would be the same as before.

#### 4.5.3.1 Single Knock-Out Barrier Options

For up and out calls (UOC) the payoff and boundary conditions are

| | |
|---|---|
| payoff | $V(S,T) = (S-K)^+ \quad \text{for} \quad S \in [0, H)$ |
| boundary conditions | $\lim_{S\downarrow 0} V(S,t) = 0$ or $\lim_{S\downarrow 0} V_{SS}(S,t) = 0$ |
| | $\lim_{S\uparrow H} V(S,t) = \mathrm{R}$ |

where $H$ is the upper barrier level. For the left boundary condition we can apply either Dirichlet or Neumann condition. For the right boundary the only choice is to set it equal to rebate $R$ if there is any, otherwise zero.

For up and out puts (UOP) the payoff and boundary conditions are

| | |
|---|---|
| payoff | $V(S,T) = (K-S)^+ \quad \text{for} \quad S \in [0, H)$ |
| boundary conditions | $\lim_{S\downarrow 0} V_{SS}(S,t) = 0$ or $\lim_{S\downarrow 0} V(S,t) = K - S$ |
| | $\lim_{S\uparrow H} V(S,t) = \mathrm{R}$ |

For down and out calls (DOC) the payoff and boundary conditions are

| | |
|---|---|
| payoff | $V(S,T) = (S-K)^+ \quad \text{for} \quad S \in (L, \infty)$ |
| boundary conditions | $\lim_{S\downarrow L} V(S,t) = \mathrm{R}$ |
| | $\lim_{S\uparrow\infty} V_{SS}(S,t) = 0$ or $\lim_{S\uparrow\infty} V(S,t) = S - K$ |

where $L$ is the lower barrier level. For the right boundary condition we can apply either Dirichlet or Neumann condition. For the left boundary the condition is to set it equal to rebate if there is any otherwise zero.

For down and out puts (DOP) the payoff and boundary conditions are

| | |
|---|---|
| payoff | $V(S,T) = (K-S)^+ \quad \text{for} \quad S \in (L, \infty)$ |
| boundary conditions | $\lim_{S\downarrow L} V(S,t) = \mathrm{R}$ |
| | $\lim_{S\uparrow\infty} V_{SS}(S,t) = 0$ or $\lim_{S\uparrow\infty} V(S,t) = 0$ |

#### 4.5.3.2 Single Knock-In Barrier Options

For premiums of knock-in barrier options we using the parity argument, i.e.. in + out = vanilla, therefore we have

- up and in call (UIC) = vanilla call - UOC
- up and in put (UIP) = vanilla put - UOP

and

- down and in call (DIC) = vanilla call - DOC
- down and in put (DIP) = vanilla put - DOP

From the earlier description, it should be trivial what the boundary conditions are for all these knock-in barrier options. We leave them as an exercise.

#### 4.5.3.3 Double Barrier Options

For a double knock-out call the payoff and boundary conditions are

$$\begin{array}{ll} \text{payoff} & V(S,T) = (S-K)^+ \quad \text{for} \quad S \in (L,H) \\ \text{boundary conditions} & \lim_{S \downarrow L} V(S,t) = R_1 \\ & \lim_{S \uparrow H} V(S,t) = R_2 \end{array}$$

where $R_1$ and $R_2$ are lower and upper rebates, respectively, and if there is none it would be zero. There might be a combination of knock-in and knock-out e.g., knock-out knock-in (KOKI) or knock-out knock-out (KOKO). KOKI has one knock-out barrier and one knock-in barrier. KOKO values options with two knock-out barriers. In KOKI, if the stock price crosses a knock-out barrier before it crosses a knock-in barrier, then the option terminates. If the stock price crosses a knock-in barrier first, then the holder receives either a vanilla or a knock-out option, depending on whether the contract specifies that knock-out is dominant. Knock-out dominance means that the option can be knocked out at any time, even after it is knocked in. In other words, if it knocks in, then it knocks into a knock-out option. If the option is not knock-out dominant, then it knocks into a vanilla option.

The barriers may be constant or may be different in different time intervals. They may be continuous or discrete or a combination of both. In particular, the models can have either continuous or daily discrete monitoring of the barriers. The payoff may have the form of either a standard call or put payoff or digital call or put payoff.

---

## 4.6 Forward PDEs

One can look at option prices $V(S,t;K,T)$ as a four-dimensional problem; dimensions are: (a) spot price space, (b) calendar time space, (c) strike price space, and (d) maturity space. In the case of the Black–Scholes equation, we freeze strike space and maturity space by just picking one point from each space, defining a single option contract, to reduce the pricing problem to a two-dimensional problem. And it happens from spot price space and calendar time space we just need very few points even though we solve for every point and the rest of the points in those spaces are never used. PDEs like the Black–Scholes PDE (4.2) or the generalized Black–Scholes PDE (4.4) whose strike price and maturity are fixed and spot and calendar time are varying are called backward PDEs.

The drawback to backward PDEs is that the solution depends on fixed $K$ and $T$. Therefore we must rerun the scheme separately for each pair of $K$ and $T$, corresponding to each different quoted option. Assuming, as we will see later in Chapter 7 under calibration procedure, we are calibrating a model to all market quoted options across $m$ distinct strikes and $n$ distinct maturities, which requires pricing for $m$ strikes and $n$ maturities that implies running of the pricing scheme $m \times n$ times, for instance an implicit finite difference of the backward Black–Scholes PDEs. The pseudo-code is as follows:

for $i = 1, \ldots, m$
  for $j = 1, \ldots, n$
    set $K_i$ and $T_j$
    $\widehat{V}(K_i, T_k) = \mathbf{G}(S, t; K_i, T_k)$
  endfor
endfor

where $\mathbf{G}$ is a pricing engine/algorithm using some finite differences technique, such as an implicit difference scheme.

### 4.6.1 Vanilla Calls

In an effort to construct a solution which allows us to solve for option prices with different strikes and maturities simultaneously, let us consider again the local volatility geometric Brownian motion stock price process having the following stochastic differential equation:

$$dS_t = (r - q)S_t dt + \sigma(S_t, t)dW(t)$$

where the function $\sigma(S, t)$ is the asset's local volatility function and the corresponding backward PDE is

$$\frac{\partial V}{\partial t} + \frac{1}{2}\sigma^2(S,t)S^2\frac{\partial^2 V}{\partial S^2} + (r(t) - q(t))S\frac{\partial V}{\partial S} = r(t)V(S,t)$$

which we will recognize as the generalized Black–Scholes PDE. In [104], the author derives the associated forward PDE parameterized on strike and maturity (i.e., $K$ and $T$ are varying and $S$ and $t$ are fixed). This forward PDE shown below

$$-\frac{\partial C}{\partial T} + \frac{1}{2}\sigma^2(K,T)K^2\frac{\partial^2 C}{\partial K^2} - [r(T) - q(T)]\, K\frac{\partial C}{\partial K} = q(T)C \tag{4.34}$$

is called the Dupire forward PDE for pricing European vanilla prices. For the forward PDE, if we assume a known local volatility surface, we can calculate option premiums for all strikes and maturities by numerically solving the forward PDE once as opposed to solving the backward PDE for each pair of strikes and maturities. Conversely, having market quotes of option prices $C(K_i, T_k)$, we then can calculate the local volatility surface from market quotes (i.e., calibration) by simply solving (4.34) for $\sigma(K, T)$.

$$\sigma(K,T) = \left(\frac{\frac{\partial C}{\partial T} + [r(T) - q(T)]\, K\frac{\partial C}{\partial K} + q(T)C}{K^2\frac{\partial^2 C}{\partial K^2}}\right)^{1/2}$$

Note that here we assume smooth call prices since in addition to $C(K,T)$ we must also calculate $\frac{\partial C}{\partial T}$, $\frac{\partial C}{\partial K}$, and $\frac{\partial^2 C}{\partial K^2}$. So, we must interpolate/extrapolate from market prices and then apply a smoothing technique. This is not as easy as it looks and this procedure will be discussed in detail in Chapter 7.

### 4.6.2 Down-and-Out Calls

One of the most important motivations and applications of the local volatility model is the ability to more accurately price exotic derivatives as their model prices are fully consistent with the vanilla options market. The last section provided us with a method to efficiently calibrate a local volatility surface from a set of vanilla option prices using the forward PDE for European options. Once we have derived this local volatility surface, the next step is to price more exotic derivatives given the market calibrated local volatility function $\sigma(S,t)$, by either (a) employing a finite difference solution to the underlying partial differential equation in the price of the exotic or (b) by simulating the process. In [57], Carr and Hirsa derive the following forward PDE for down-and-out calls with local volatility.

$$\frac{\sigma^2(K,T)}{2}K^2\frac{\partial^2 D_o^c}{\partial K^2} - [r(T)-q(T)]\,K\frac{\partial D_o^c}{\partial K} - q(T)D_o^c = \frac{\partial D_o^c}{\partial T}$$

with the initial condition

$$D_o^c(K,0) = (S_0-K)^+, \text{ for } K \in [H,\infty), \text{ and } H < S_0$$

and boundary conditions

$$\lim_{K\downarrow H}\frac{\partial^2 D_o^c}{\partial K^2}(K,T) = 0, \;\; T\in[0,\bar{T}]$$

$$\lim_{K\uparrow\infty}\frac{\partial^2 D_o^c}{\partial K^2}(K,T) = 0, \;\; T\in[0,\bar{T}]$$

### 4.6.3 Up-and-Out Calls

In [57], Carr and Hirsa also derive the following forward PDE for up-and-out calls with local volatility.

$$\frac{\sigma^2(K,T)}{2}K^2\frac{\partial^2 U_o^c}{\partial K^2} - [r(T)-q(T)]\,K\frac{\partial U_o^c}{\partial K} - q(T)U_o^c = \frac{\partial U_o^c}{\partial T} + \left[\frac{\sigma^2(H,T)}{2}H^2\frac{\partial^3 U_o^c}{\partial K^3}(H,T)\right](K-H)$$

with initial condition

$$U_o^c(K,0) = (S_0-K)^+, \text{ for } K\in[0,H), \text{ and } S_0 < H$$

and boundary conditions

$$\lim_{K\downarrow 0}\frac{\partial^2 U_o^c}{\partial K^2}(K,T) = 0, \;\; T\in[0,\bar{T}]$$

$$\lim_{K\uparrow H}\frac{\partial^2 U_o^c}{\partial K^2}(K,T) = 0, \;\; T\in[0,\bar{T}]$$

For details on the derivation see [56] and [57]. The schemes discussed earlier can be used to solve these forward PDEs numerically. In case of the forward PDE for an up-and-out call, in order to preserve the tridiagonal structure of the stiffness matrix, we should treat the term

$$\left(\frac{\sigma^2(H,T)}{2}H^2\frac{\partial^3 U_o^c}{\partial K^3}(H,T)\right)(K-H)$$

explicitly. To get a second order approximation for the third derivative we use backward finite difference discretization applying the scheme explained in Section 3.4. For this scheme we use five points.

**Example 6** *Forward versus backward up-and-out call (UOC) premiums*

In this example, we compare UOC premiums by numerically solving both backward and forward PDEs.

The parameter set used in this example is spot price $S_0 = 100$, risk-free interest rate $r = 3.75\%$, continuous dividend rate $q = 2.0\%$, and strike range of $K = 90, 110$, maturity range of $T = 0.25, 0.5, 1.0$. We consider the following local volatility surface:

$$\sigma(K,T) = 0.3e^{-T}\,(100/K)^{0.2}$$

This local volatility surface is plotted in Figure 4.9. In Figure 4.10(a) we display UOC

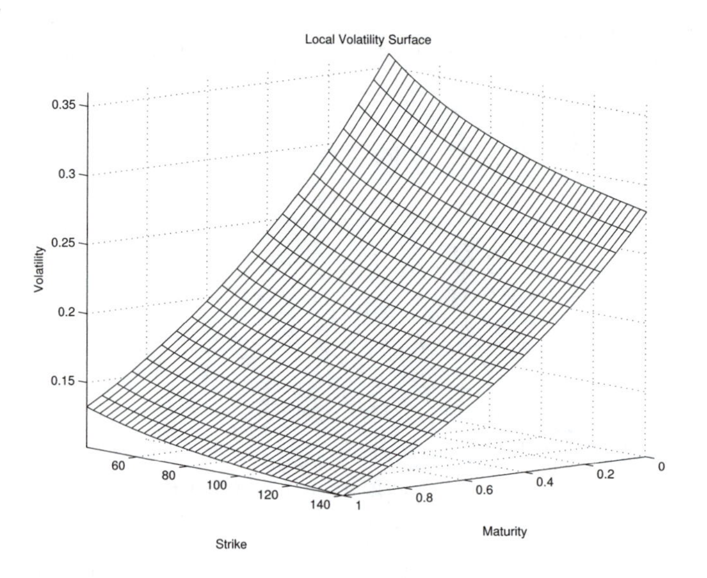


**FIGURE 4.9**: Local volatility surface used for up-and-out calls

premiums for 3-month maturity by solving a backward PDE numerically; the left figure is for a strike of 90 and the right one is for a strike of 110. Out of all those premiums we just pick the one that corresponds to the spot price 100 as pointed out in the figures. That is the drawback with backward PDEs. Figures 4.10(b) and 4.10(c) are the same as Figure 4.10(a) except for 6-month maturity and 12-month maturity, respectively. In Figure 4.11 we display UOC premiums by solving the forward PDE for UOC numerically for all strikes and maturities. From all premiums we pick those that correspond to strikes 90, 110 and maturities 3-month, 6-month, and 12-month as pointed out in the figure.

We see that the premiums from backward and forward PDEs are identical. However, in the case of the forward PDE for UOC, we get the results in one sweep as opposed to backward that we had to solve the backward PDE numerically for each pair of strikes and maturities (in this example we solve it six times, having six pairs of strikes and maturities).

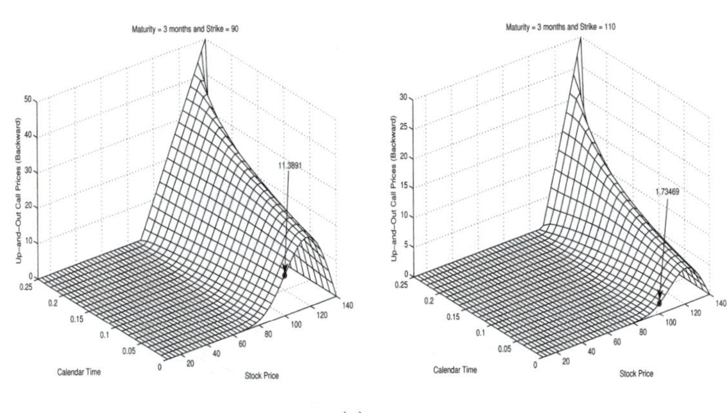


(a)

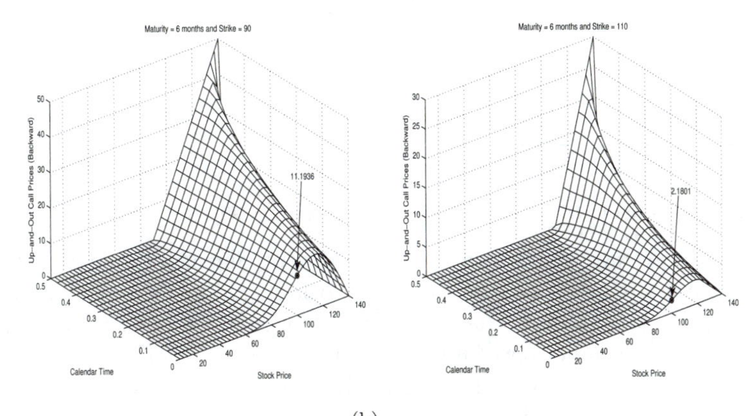

(b)

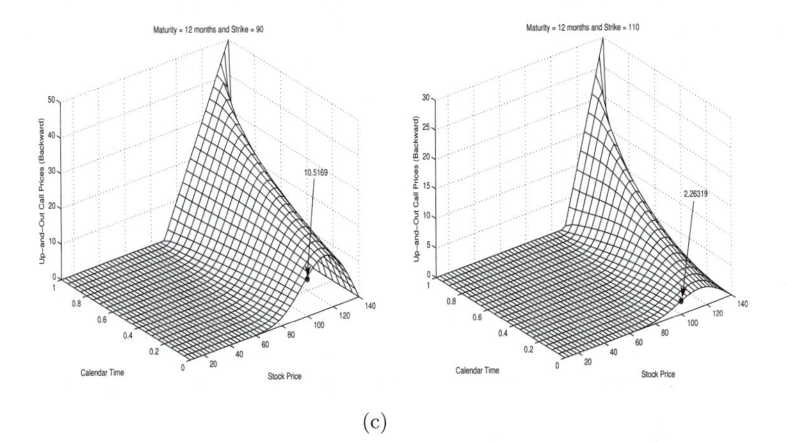


(c)

**FIGURE 4.10**: Up-and-out call prices using a backward PDE

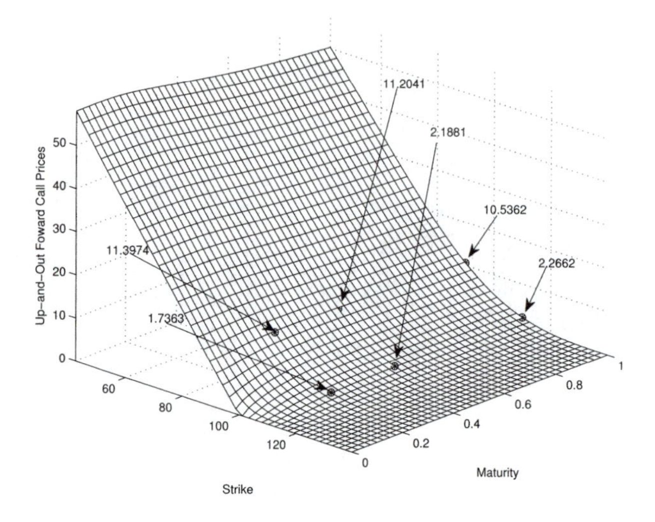


**FIGURE 4.11**: Up-and-out call prices obtained from using a forward PDE

---

## 4.7 Finite Differences in Higher Dimensions

At this point, we have concentrated on options pricing using one-dimensional PDEs.[2] However, the same approach can be extended to higher dimensional spaces, if our model yields a pricing equation in more than one dimension. In any stochastic volatility model the value of the option is a function of underlying price, volatility, and time $v(S, v, t)$ such as Heston stochastic volatility. In models where the interest rate is assumed to be stochastic, for instance interest rate sensitive instruments, the option value is a function of underlying process, interest rate, and time $v(S, r, t)$ such as convertible bond models. The price of an arithmetic average Asian options can be found by solving a PDE in two space dimensions (see Ingersoll [147]). However, Asian options can be reduced to a one-dimensional PDE (e.g. [193] and [214]). In this section we will explore some of the issues arising from the use of PDE techniques in higher dimensions, using the Heston model as our canonical example.

### 4.7.1 Heston Stochastic Volatility Model

One of the most popular two factor stochastic models for asset price evolution is the Heston stochastic volatility model [134], which extends Black–Scholes by allowing volatility to follow a mean reverting stochastic process. Besides being one of the most popular two factor models, the Heston model also admits a closed-form solution, and as such will allow us to assess the accuracy and efficiency of our numerical algorithm.

Under the Heston model, the stock price and stock volatility are modeled using the

[2] In the literature a time variable is not included in the dimension of the numerical solution because the difference equation is presented through a time slice.

following stochastic processes:

$$dS(t) = \mu S dt + \sqrt{v(t)} S dz_1(t) \tag{4.35}$$

$$d\sqrt{v(t)} = -\beta\sqrt{v(t)}dt + \delta dz_2(t) \tag{4.36}$$

where $z_1(t)$ and $z_2(t)$ are correlated Wiener processes, with an instantaneous correlation given by $\rho$. The stochastic process for volatility $\sqrt{v(t)}$ is a Brownian motion with drift; however, it can be transformed, via Itô's lemma, into an Ornstein–Uhlenbeck process for the variance $v(t)$:

$$dv(t) = (\delta^2 - 2\beta v(t))dt + 2\delta\sqrt{v(t)}dz_2(t)$$

$$dv(t) = \kappa(\theta - v(t))dt + \sigma\sqrt{v(t)}dz_2(t)$$

which is the most familiar representation of the model. This is a mean reverting process whose parameters $\kappa$, $\theta$, and $\sigma$ can be interpreted as follows: $\kappa$ is the mean reversion speed, $\theta$ is the long run variance, a nd $\sigma$ is the volatility of the volatility. The proof is not given here; it is fairly trivial via Itô's lemma.

Following arbitrage arguments similar to the ones discussed in Section 1.2.1.2 originating in [32], we can derive the following PDE for the option price $U(S, v, t)$:

$$\frac{1}{2}vS^2\frac{\partial^2 U}{\partial S^2} + \rho\sigma v S\frac{\partial^2 U}{\partial S\partial v} + \frac{1}{2}\sigma^2 v\frac{\partial^2 U}{\partial v^2} + (r-q)S\frac{\partial U}{\partial S} + (\kappa(\theta - v(t)) - \lambda(S,v,t))\frac{\partial U}{\partial v} - rU + \frac{\partial U}{\partial t} = 0 \tag{4.37}$$

where $\lambda(S, v, t)$ is the *price of volatility risk*, the market value assigned to a unit of volatility risk.[3] Under the Heston stochastic volatility model, we make the assumption that the price of volatility risk is proportional to the variance $v$, that is,

$$\lambda(S, v, t) = \lambda v$$

Then the question begs, are there any implications to making the price of volatility risk linear in volatility? and is there any evidence to support this assumption?

The value of $\lambda$, can in practice be estimated using assets which are purely volatility dependent. Using this assumption, we have the following differential equation for the option price:

$$\frac{1}{2}vS^2\frac{\partial^2 U}{\partial S^2} + \rho\sigma v S\frac{\partial^2 U}{\partial S\partial v} + \frac{1}{2}\sigma^2 v\frac{\partial^2 U}{\partial v^2} + (r-q)S\frac{\partial U}{\partial S} + (\kappa(\theta - v) - \lambda v)\frac{\partial U}{\partial v} - rU - \frac{\partial U}{\partial \tau} = 0 \tag{4.38}$$

Estimating $\lambda$ directly can prove to be difficult; however, we can eliminate it by adjusting the measure under which we are pricing. We can eliminate $\lambda$ from our consideration by using risk-neutral pricing probabilities, which results in a new process for the variance:

$$dv(t) = \kappa^\star(\theta^\star - v(t))dt + \sigma\sqrt{v(t)}dz_2(t) \tag{4.39}$$

with

$$\kappa^\star = \kappa + \lambda \text{ and } \theta^\star = \frac{\kappa\theta}{\kappa + \lambda}$$

[3] It would be beneficial to actually extend the arbitrage argument here to give the reader an intuitive feel for what the market price of volatility risk really means, as often in stochastic volatility and jump models this is glossed over, but the economic argument for this follows from dynamic hedging, which makes the Black–Scholes model so relevant. We refer readers to [98], [87], [133], and [35] on price of volatility risk.

Using these constants, the term $\kappa^\star(\theta^\star - v) - \lambda v$ becomes

$$\begin{aligned}\kappa^\star(\theta^\star - v) - \lambda v &= (\kappa+\lambda)((\frac{\kappa\theta}{\kappa+\lambda}) - v) - \lambda v \\ &= \kappa(\theta - v)\end{aligned}$$

and thus we get the following differential equation:

$$\begin{aligned}\frac{1}{2}vS^2\frac{\partial^2 U}{\partial S^2} + \rho\sigma v S\frac{\partial^2 U}{\partial S\partial v} + \frac{1}{2}\sigma^2 v\frac{\partial^2 U}{\partial v^2} + (r-q)S\frac{\partial U}{\partial S}& \\ +\kappa(\theta - v)\frac{\partial U}{\partial v} - rU - \frac{\partial U}{\partial \tau} = 0 & \qquad (4.40)\end{aligned}$$

which does not depend on $\lambda$. Thus by using the risk-neutral measure, we have eliminated the need to estimate $\lambda$ and instead we can estimate the implied $\theta^\star$, $\kappa^\star$ as well as the other model parameters by using option prices.

Using the modified differential equation, we can value options using the standard PDE methods as long as we can formulate the boundary conditions. Thus to calculate the value of a European call option we would solve the PDE in (4.40), subject to the following boundary conditions:

$$\begin{aligned}U(S,v,0) &= (S-K)^+ \\ \lim_{S\downarrow 0}\frac{\partial^2 U}{\partial S^2}(S,v,\tau) &= 0 \\ \lim_{S\uparrow\infty}\frac{\partial^2 U}{\partial S^2}(S,v,\tau) &= 0 \\ (r-q)S\frac{\partial U}{\partial S}(S,0,\tau) + \kappa\theta\frac{\partial U}{\partial v}(S,0,\tau) - rU(S,0,\tau) - \frac{\partial U}{\partial \tau}(S,0,\tau) &= 0 \qquad (4.41) \\ \lim_{v\uparrow\infty} U(S,v,\tau) &= S\end{aligned}$$

where $K$ is the strike price and $\tau$ is time to maturity, $\tau = T - t$.

The first condition is the initial condition that is the payoff. The second and third boundary conditions state the option gamma approaches zero for very small or very large asset prices. [4] The fourth boundary condition enforces the PDE in the case where volatility is zero, and the last boundary condition assumes that the option becomes a stock at infinite variance by the arbitrage argument.

### 4.7.2 Options Pricing under the Heston PDE

Now that we have derived the two dimensional PDE which governs options prices under the Heston stochastic volatility model, we can solve it for pricing derivative contracts using standard techniques.

Before going through discretization we rewrite the PDE in new coordinates by defining

$$\begin{aligned}S(\xi) &= K + \alpha\sinh(c_1\xi + c_2(1-\xi)) \\ v(\eta) &= \beta\sinh(d\eta)\end{aligned}$$

[4] Equivalently we can have Dirichlet conditions for the second and the third one, that is, $U(0,v,\tau) = 0$, which indicates the option is worthless if the asset price goes to zero and $\lim_{S\uparrow\infty} U(S,v,\tau) = S - K$.

where

$$\begin{aligned} c_1 &= \sinh^{-1}(\frac{S_{max} - K}{\alpha}) \\ c_2 &= \sinh^{-1}(\frac{S_{min} - K}{\alpha}) \\ d &= \sinh^{-1}(\frac{v_{max}}{\beta}) \end{aligned}$$

with $\alpha =$ and $\beta =$. Under these transformations we have

$$\begin{aligned} \bar{U}(\xi, \eta, \tau) &= U(S, v, \tau) \\ \frac{\partial U}{\partial S} &= \frac{\partial \bar{U}}{\partial \xi} \frac{1}{\frac{\partial S}{\partial \xi}} \\ \frac{\partial^2 U}{\partial S^2} &= \frac{\partial^2 \bar{U}}{\partial \xi^2} \frac{1}{(\frac{\partial S}{\partial \xi})^2} - \frac{\partial \bar{U}}{\partial \xi} \frac{\frac{\partial^2 S}{\partial \xi^2}}{(\frac{\partial S}{\partial \xi})^3} \\ \frac{\partial U}{\partial v} &= \frac{\partial \bar{U}}{\partial \eta} \frac{1}{\frac{\partial v}{\partial \eta}} \\ \frac{\partial^2 U}{\partial v^2} &= \frac{\partial^2 \bar{U}}{\partial \eta^2} \frac{1}{(\frac{\partial S}{\partial \eta})^2} - \frac{\partial \bar{U}}{\partial \eta} \frac{\frac{\partial^2 v}{\partial \eta^2}}{(\frac{\partial v}{\partial \eta})^3} \\ \frac{\partial^2 U}{\partial S \partial v} &= \frac{\partial^2 \bar{U}}{\partial \xi \partial \eta} \frac{1}{\frac{\partial S}{\partial \xi} \frac{\partial v}{\partial \eta}} \end{aligned}$$

and the PDE under these new coordinates becomes

$$\begin{aligned} \frac{1}{2} v(\eta)\, S^2(\xi) \left( \frac{\partial^2 \bar{U}}{\partial \xi^2} \frac{1}{(\frac{\partial S}{\partial \xi})^2} - \frac{\partial \bar{U}}{\partial \xi} \frac{\frac{\partial^2 S}{\partial \xi^2}}{(\frac{\partial S}{\partial \xi})^3} \right) + \rho \sigma v(\eta) S(\xi) \frac{\partial^2 \bar{U}}{\partial \xi \partial \eta} \frac{1}{\frac{\partial S}{\partial \xi} \frac{\partial v}{\partial \eta}} + \\ \frac{1}{2} \sigma^2 v(\eta) \left( \frac{\partial^2 \bar{U}}{\partial \eta^2} \frac{1}{(\frac{\partial v}{\partial \eta})^2} - \frac{\partial \bar{U}}{\partial \eta} \frac{\frac{\partial^2 v}{\partial \eta^2}}{(\frac{\partial v}{\partial \eta})^3} \right) + (r - q) S(\xi) \frac{\partial \bar{U}}{\partial \xi} \frac{1}{\frac{\partial S}{\partial \xi}} + \\ \kappa(\theta - v(\eta)) \frac{\partial \bar{U}}{\partial \eta} \frac{1}{\frac{\partial v}{\partial \eta}} - r \bar{U} - \frac{\partial \bar{U}}{\partial \tau} = 0 \end{aligned} \quad (4.42)$$

subject to these boundary conditions:

$$\begin{aligned} \bar{U}(\xi, \eta, 0) &= (S(\xi) - K)^+ \\ \frac{1}{(\frac{\partial S}{\partial \xi})^2} \frac{\partial^2 \bar{U}}{\partial \xi^2}(0, \eta, \tau) - \frac{\frac{\partial^2 S}{\partial \xi^2}}{(\frac{\partial S}{\partial \xi})^3} \frac{\partial \bar{U}}{\partial \xi}(0, \eta, \tau) &= 0 \\ \frac{1}{(\frac{\partial S}{\partial \xi})^2} \frac{\partial^2 \bar{U}}{\partial \xi^2}(1, \eta, \tau) - \frac{\frac{\partial^2 v}{\partial \eta^2}}{(\frac{\partial v}{\partial \eta})^3} \frac{\partial \bar{U}}{\partial \xi}(1, \eta, \tau) &= 0 \\ (r-q) S(\xi) \frac{1}{\frac{\partial S}{\partial \xi}} \frac{\partial \bar{U}}{\partial \xi}(\xi, 0, \tau) + \kappa \theta \frac{1}{\frac{\partial v}{\partial \eta}} \frac{\partial \bar{U}}{\partial \eta}(\xi, 0, \tau) - r \bar{U}(\xi, 0, \tau) - \frac{\partial \bar{U}}{\partial \tau}(\xi, 0, \tau) &= 0 \\ \bar{U}(\xi, 1, \tau) &= S(\xi) \end{aligned} \quad (4.43)$$

For the grid, we consider $L$ equal sub-intervals in the $\tau$-direction on $[0, T]$, $N$ equal sub-intervals in the $\xi$-direction on $[0,1]$, and $M$ equal sub-intervals in the $v$-direction on $[0,1]$. Thus, we have the following mesh on $[0,1] \times [0,1] \times [0,T]$:

$$\bar{D} = \left\{ \begin{array}{lll} \xi_i = 0 + (i-1)\Delta\xi; & \Delta\xi = \frac{1}{N}; & i = 1, \ldots, N+1 \\ \eta_j = 0 + (j-1)\Delta\eta; & \Delta\eta = \frac{1}{M}; & j = 1, \ldots, M+1 \\ \tau_k = 0 + (k-1)\Delta\tau; & \Delta\tau = \frac{T-0}{L}; & k = 1, \ldots, L+1 \end{array} \right\}$$

We can use the familiar discretization techniques discussed in the last section to construct either a fully explicit or a fully implicit solution to the finite difference equation under the Heston model. For its inherent advantages of stability, we decided to use a predominantly implicit scheme[5] to numerically solve the Heston PDE. Let $U_{i,j}^k$ be the approximate solution of $\bar{U}(\xi_i, \eta_j, \tau_k)$ on the grid. For t he reference grid point $(xi_i, \eta_j, \tau_{k+1})$ at time slice $\tau_{k+1}$ we have the nine grid points in the stencil indicated below.

$$\begin{array}{ccc} U_{i-1,j+1}^{k+1} & U_{i,j+1}^{k+1} & U_{i+1,j+1}^{k+1} \\ \\ U_{i-1,j}^{k+1} & U_{i,j}^{k+1} & U_{i+1,j}^{k+1} \\ \\ U_{i-1,j-1}^{k+1} & U_{i,j-1}^{k+1} & U_{i+1,j-1}^{k+1} \end{array}$$

Moreover, we assume that numeration of the grid is done from left to right and bottom to top at each time slice.

In our discretization for the partial derivatives of $\bar{U}(\xi_i, \eta_j, \tau_{k+1})$ we use the following approximations: (a) for the first and second derivatives we use the central finite difference approximations, (b) for the cross derivatives there are various different ways of approximation, we use the one which has second order approximation in both $\eta$ and $\xi$. We leave its derivation as an exercise to the reader at the end of the chapter.

$$\begin{aligned}
\frac{\partial^2 \bar{U}}{\partial \xi^2}(\xi_i, \eta_j, \tau_{k+1}) &= \frac{U_{i-1,j}^{k+1} - 2U_{i,j}^{k+1} + U_{i+1,j}^{k+1}}{\Delta\xi^2} + O(\Delta\xi^2) \\
\frac{\partial^2 \bar{U}}{\partial \eta^2}(\xi_i, \eta_j, \tau_{k+1}) &= \frac{U_{i,j-1}^{k+1} - 2U_{i,j}^{k+1} + U_{i,j+1}^{k+1}}{\Delta\eta^2} + O(\Delta\eta^2) \\
\frac{\partial^2 \bar{U}}{\partial \xi \partial \eta}(\xi_i, \eta_j, \tau_{k+1}) &= \frac{U_{i-1,j-1}^{k+1} - U_{i-1,j+1}^{k+1} - U_{i+1,j-1}^{k+1} + U_{i+1,j+1}^{k+1}}{4\Delta\xi\Delta\eta} + O(\Delta\xi^2) + O(\Delta\eta^2) \\
\frac{\partial \bar{U}}{\partial \xi}(\xi_i, \eta_j, \tau_{k+1}) &= \frac{U_{i+1,j}^{k+1} - U_{i-1,j}^{k+1}}{2\Delta\xi} + O(\Delta\xi^2) \\
\frac{\partial \bar{U}}{\partial \eta}(\xi_i, \eta_j, \tau_{k+1}) &= \frac{U_{i,j+1}^{k+1} - U_{i,j-1}^{k+1}}{2\Delta\eta} + O(\Delta\eta^2) \\
\frac{\partial \bar{U}}{\partial \tau}(\xi_i, \eta_j, \tau_{k+1}) &= \frac{U_{i,j}^{k+1} - U_{i,j}^{k}}{\Delta\tau} + O(\Delta\tau)
\end{aligned}$$

Substituting these approximations into the Heston PDE, we obtain the following difference

[5]This is not exactly true as we will be using an implicit-explicit scheme for treatment of one of the boundary conditions.

equation:

$$
\begin{aligned}
&\frac{U_{i,j}^{k+1}-U_{i,j}^{k}}{\Delta\tau}-\frac{1}{2}v(\eta_j)S(\xi_i)^2\frac{1}{(\frac{\partial S}{\partial\xi}(\xi_i))^2}\frac{U_{i-1,j}^{k+1}-2U_{i,j}^{k+1}+U_{i+1,j}^{k+1}}{\Delta\xi^2}\\
&-\rho\sigma v(\eta_j)S(\xi_i)\frac{1}{\frac{\partial S}{\partial\xi}(\xi_i)\frac{\partial v}{\partial\eta}(\eta_j)}\frac{U_{i-1,j-1}^{k+1}-U_{i-1,j+1}^{k+1}+U_{i+1,j-1}^{k+1}+U_{i+1,j+1}^{k+1}}{4\Delta\xi\Delta\eta}\\
&-\frac{1}{2}\sigma^2 v(\eta_j)\frac{1}{(\frac{\partial v}{\partial\eta}(\eta_j))^2}\frac{U_{i,j-1}^{k+1}-2U_{i,j}^{k+1}+U_{i,j+1}^{k+1}}{\Delta\eta^2}\\
&-\left((r-q)S(\xi_i)\frac{1}{\frac{\partial S}{\partial\xi}(\xi_i)}-\frac{1}{2}v(\eta_j)S^2(\xi_i)\frac{\frac{\partial^2 S}{\partial\xi^2}(\xi_i)}{(\frac{\partial S}{\partial\xi}(\xi_i))^3}\right)\frac{U_{i+1,j}^{k+1}-U_{i-1,j}^{k+1}}{2\Delta\xi}\\
&-\left(\kappa(\theta-v(\eta_j))\frac{1}{\frac{\partial v}{\partial\eta}(\eta_j)}-\frac{1}{2}\sigma^2 v(\eta_j)\frac{\frac{\partial^2 v}{\partial\eta^2}(\eta_j)}{(\frac{\partial v}{\partial\eta}(\eta_j))^3}\right)\frac{U_{i,j+1}^{k+1}-U_{i,j-1}^{k+1}}{2\Delta\eta}+rU_{i,j}^{k+1} = 0
\end{aligned}
$$

Multiplying by $\Delta\tau$ and rearranging the terms according to the stencil we obtain

$$
\begin{aligned}
&-\left(\frac{\Delta\tau\rho\sigma}{4\Delta\xi\Delta\eta}v(\eta_j)S(\xi_i)\frac{1}{\frac{\partial S}{\partial\xi}(\xi_i)\frac{\partial v}{\partial\eta}(\eta_j)}\right)U_{i-1,j-1}^{k+1}\\
&-\left(\frac{\sigma^2\Delta\tau}{2(\Delta\eta)^2}v(\eta_j)\frac{1}{(\frac{\partial v}{\partial\eta}(\eta_i))^2}-\frac{\Delta\tau}{2\Delta\eta}\left(\kappa(\theta-v(\eta_j))\frac{1}{\frac{\partial v}{\partial\eta}(\eta_j)}-\frac{1}{2}\sigma^2 v(\eta_j)\frac{\frac{\partial^2 v}{\partial\eta^2}(\eta_j)}{(\frac{\partial v}{\partial\eta}(\eta_j))^3}\right)\right)U_{i,j-1}^{k+1}\\
&+\left(\frac{\Delta\tau\rho\sigma}{4\Delta\xi\Delta\eta}v(\eta_j)S(\xi_i)\frac{1}{\frac{\partial S}{\partial\xi}(\xi_i)\frac{\partial v}{\partial\eta}(\eta_j)}\right)U_{i+1,j-1}^{k+1}\\
&-\left(\frac{\Delta\tau}{(\Delta\xi)^2}v(\eta_j)S(\xi_i)^2\frac{1}{(\frac{\partial S}{\partial\xi}(\xi_i))^2}-\frac{\Delta\tau}{2\Delta\xi}\left((r-q)S(\xi_i)\frac{1}{\frac{\partial S}{\partial\xi}(\xi_i)}-\frac{1}{2}v(\eta_j)S^2(\xi_i)\frac{\frac{\partial^2 S}{\partial\xi^2}(\xi_i)}{(\frac{\partial S}{\partial\xi}(\xi_i))^3}\right)\right)U_{i-1,j}^{k+1}\\
&+\left(1+r\Delta\tau+\frac{\Delta\tau}{(\Delta\xi)^2}v(\eta_j)S(\xi_i)^2\frac{1}{(\frac{\partial S}{\partial\xi}(\xi_i))^2}+\frac{\sigma^2\Delta\tau}{(\Delta\eta)^2}v(\eta_j)\frac{1}{(\frac{\partial v}{\partial\eta}(\eta_i))^2}\right)U_{i,j}^{k+1}\\
&-\left(\frac{\Delta\tau}{(\Delta\xi)^2}v(\eta_j)S(\xi_i)^2\frac{1}{(\frac{\partial S}{\partial\xi}(\xi_i))^2}+\frac{\Delta\tau}{2\Delta\xi}\left((r-q)S(\xi_i)\frac{1}{\frac{\partial S}{\partial\xi}(\xi_i)}-\frac{1}{2}v(\eta_j)S^2(\xi_i)\frac{\frac{\partial^2 S}{\partial\xi^2}(\xi_i)}{(\frac{\partial S}{\partial\xi}(\xi_i))^3}\right)\right)U_{i+1,j}^{k+1}\\
&+\left(\frac{\Delta\tau\rho\sigma}{4\Delta\xi\Delta\eta}v(\eta_j)S(\xi_i)\frac{1}{\frac{\partial S}{\partial\xi}(\xi_i)\frac{\partial v}{\partial\eta}(\eta_j)}\right)U_{i-1,j+1}^{k+1}\\
&-\left(\frac{\sigma^2\Delta\tau}{2(\Delta\eta)^2}v(\eta_j)\frac{1}{(\frac{\partial v}{\partial\eta}(\eta_i))^2}+\frac{\Delta\tau}{2\Delta\eta}\left(\kappa(\theta-v(\eta_j))\frac{1}{\frac{\partial v}{\partial\eta}(\eta_j)}-\frac{1}{2}\sigma^2 v(\eta_j)\frac{\frac{\partial^2 v}{\partial\eta^2}(\eta_j)}{(\frac{\partial v}{\partial\eta}(\eta_j))^3}\right)\right)U_{i,j+1}^{k+1}\\
&-\left(\frac{\Delta\tau\rho\sigma}{4\Delta\xi\Delta\eta}v(\eta_j)S(\xi_i)\frac{1}{\frac{\partial S}{\partial\xi}(\xi_i)\frac{\partial v}{\partial\eta}(\eta_j)}\right)U_{i+1,j+1}^{k+1}=U_{i,j}^{k}
\end{aligned}
$$

Defining the following coefficients for the nine grid points in the stencil.

$$
\begin{aligned}
a_{i,j} &= \frac{\Delta\tau\rho\sigma}{4\Delta\xi\Delta\eta}v(\eta_j)S(\xi_i)\frac{1}{\frac{\partial S}{\partial\xi}(\xi_i)\frac{\partial v}{\partial\eta}(\eta_j)} \\
b_{i,j} &= \frac{\sigma^2\Delta\tau}{2(\Delta\eta)^2}v(\eta_j)\frac{1}{(\frac{\partial v}{\partial\eta}(\eta_i))^2} - \frac{\Delta\tau}{2\Delta\eta}\left(\kappa(\theta - v(\eta_j))\frac{1}{\frac{\partial v}{\partial\eta}(\eta_j)} - \frac{1}{2}\sigma^2 v(\eta_j)\frac{\frac{\partial^2 v}{\partial\eta^2}(\eta_j)}{(\frac{\partial v}{\partial\eta}(\eta_j))^3}\right) \\
c_{i,j} &= \frac{\Delta\tau}{(\Delta\xi)^2}v(\eta_j)S(\xi_i)^2\frac{1}{(\frac{\partial S}{\partial\xi}(\xi_i))^2} - \frac{\Delta\tau}{2\Delta\xi}\left((r-q)S(\xi_i)\frac{1}{\frac{\partial S}{\partial\xi}(\xi_i)} - \frac{1}{2}v(\eta_j)S^2(\xi_i)\frac{\frac{\partial^2 S}{\partial\xi^2}(\xi_i)}{(\frac{\partial S}{\partial\xi}(\xi_i))^3}\right) \\
d_{i,j} &= 1 + r\Delta\tau + \frac{\Delta\tau}{(\Delta\xi)^2}v(\eta_j)S(\xi_i)^2\frac{1}{(\frac{\partial S}{\partial\xi}(\xi_i))^2} + \frac{\sigma^2\Delta\tau}{(\Delta\eta)^2}v(\eta_j)\frac{1}{(\frac{\partial v}{\partial\eta}(\eta_i))^2} \\
e_{i,j} &= \frac{\Delta\tau}{(\Delta\xi)^2}v(\eta_j)S(\xi_i)^2\frac{1}{(\frac{\partial S}{\partial\xi}(\xi_i))^2} + \frac{\Delta\tau}{2\Delta\xi}\left((r-q)S(\xi_i)\frac{1}{\frac{\partial S}{\partial\xi}(\xi_i)} - \frac{1}{2}v(\eta_j)S^2(\xi_i)\frac{\frac{\partial^2 S}{\partial\xi^2}(\xi_i)}{(\frac{\partial S}{\partial\xi}(\xi_i))^3}\right) \\
f_{i,j} &= \frac{\sigma^2\Delta\tau}{2(\Delta\eta)^2}v(\eta_j)\frac{1}{(\frac{\partial v}{\partial\eta}(\eta_i))^2} + \frac{\Delta\tau}{2\Delta\eta}\left(\kappa(\theta - v(\eta_j))\frac{1}{\frac{\partial v}{\partial\eta}(\eta_j)} - \frac{1}{2}\sigma^2 v(\eta_j)\frac{\frac{\partial^2 v}{\partial\eta^2}(\eta_j)}{(\frac{\partial v}{\partial\eta}(\eta_j))^3}\right)
\end{aligned}
$$

We can rewrite this difference equation in a more manageable form:

$$
\begin{aligned}
-a_{i,j}U^{k+1}_{i-1,j-1} - b_{i,j}U^{k+1}_{i,j-1} + a_{i,j}U^{k+1}_{i+1,j-1} - c_{i,j}U^{k+1}_{i-1,j} + d_{i,j}U^{k+1}_{i,j} \\
-e_{i,j}U^{k+1}_{i+1,j} + a_{i,j}U^{k+1}_{i-1,j+1} - f_{i,j}U^{k+1}_{i,j+1} - a_{i,j}U^{k+1}_{i+1,j+1} = U^k_{i,j}
\end{aligned} \tag{4.44}
$$

Writing these difference equations in matrix form we get

$$
AU^{k+1} = U^k + r.h.s. \tag{4.45}
$$

where $A$ is a $(M-1)(N-1) \times (M-1)(N-1)$ block tridiagonal matrix and the solution vector resembles

$$
\mathbf{U^{k+1}} = \begin{pmatrix} U^{k+1}_{2,2} \\ U^{k+1}_{3,2} \\ \vdots \\ U^{k+1}_{N,2} \\ U^{k+1}_{2,3} \\ U^{k+1}_{3,3} \\ \vdots \\ U^{k+1}_{N,3} \\ \vdots \\ U^{k+1}_{2,M} \\ U^{k+1}_{3,M} \\ \vdots \\ U^{k+1}_{N,M} \end{pmatrix}
$$

and the *right-hand side* arises from treatment of some of the boundary conditions. Note that in this setup, indices $i = 1$, $i = N+1$, $j = 1$, $j = M+1$ correspond to boundary points and are not entries of the vector $\mathbf{U^{k+1}}$. Once we have the boundary node values, solving the difference equation is just the problem of solving a $(M-1) \times (N-1)$ system of linear equations in $(M-1) \times (N-1)$ unknowns alternatively written as the above matrix

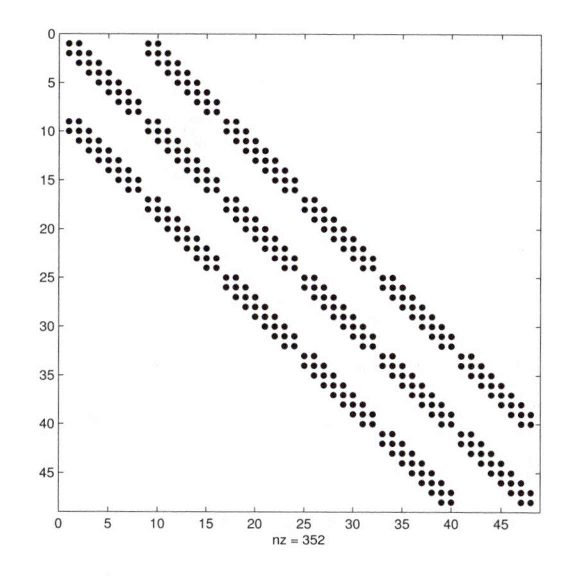


**FIGURE 4.12**: The structure of the sparse stiffness matrix

equation. The sparse structure of $A$ as shown in Figure 4.12 can be explained easily by looking at a stylized version of the mesh. Those rows that contain nine entries correspond to an internal node with no neighboring boundary nodes. A six entry row corresponds to an internal node adjacent to a single boundary (six unknown values, and three given by the boundary conditions). The last scenario is a node neighboring two boundaries, hence, the nearest internal neighbor to a corner node (four unknowns, and five given by the boundary conditions).

#### 4.7.2.1 Implementation of the Boundary Conditions

(a) Boundary condition corresponding to reference point $i=2$ — For this boundary condition we assume gamma is zero (Neumann condition). In $\xi$ coordinate it corresponds to setting the following equation to zero.

$$\frac{1}{(\frac{\partial S}{\partial \xi}(\xi_2))^2}\frac{\partial^2 \bar{U}}{\partial \xi^2}(\xi_2,\eta,\tau) - \frac{\frac{\partial^2 S}{\partial \xi^2}(\xi_2)}{(\frac{\partial S}{\partial \xi}(\xi_2))^3}\frac{\partial \bar{U}}{\partial \xi}(\xi_2,\eta,\tau) = 0 \tag{4.46}$$

We use second order discretization of the first and second derivatives at $\xi_2 = \xi_{min} + \Delta\xi = 0 + \Delta\xi$ to get

$$\frac{1}{(\frac{\partial S}{\partial \xi}(\xi_2))^2}\frac{U_{1,j}^{k+1} - 2U_{2,j}^{k+1} + U_{3,j}^{k+1}}{\Delta\xi^2} - \frac{\frac{\partial^2 S}{\partial \xi^2}(\xi_2)}{(\frac{\partial S}{\partial \xi}(\xi_2))^3}\frac{U_{3,j}^{k+1} - U_{1,j}^{k+1}}{2\Delta\xi} \approx 0 \tag{4.47}$$

which implies

$$\begin{aligned} U_{1,j}^{k+1} &= \frac{\frac{2}{(\frac{\partial S}{\partial \xi}(\xi_2))^2}}{\frac{1}{(\frac{\partial S}{\partial \xi}(\xi_2))^2} + \frac{\Delta\xi}{2}\frac{\frac{\partial^2 S}{\partial \xi^2}(\xi_2)}{(\frac{\partial S}{\partial \xi}(\xi_2))^3}} U_{2,j}^{k+1} - \frac{\frac{1}{(\frac{\partial S}{\partial \xi}(\xi_2))^2} - \frac{\Delta\xi}{2}\frac{\frac{\partial^2 S}{\partial \xi^2}(\xi_2)}{(\frac{\partial S}{\partial \xi}(\xi_2))^3}}{\frac{1}{(\frac{\partial S}{\partial \xi}(\xi_2))^2} + \frac{\Delta\xi}{2}\frac{\frac{\partial^2 S}{\partial \xi^2}(\xi_2)}{(\frac{\partial S}{\partial \xi}(\xi_2))^3}} U_{3,j}^{k+1} \\ &= \underline{l}_2 U_{2,j}^{k+1} + \bar{l}_2 U_{3,j}^{k+1} \end{aligned}$$

where we define

$$\underline{l}_2 = \frac{\frac{2}{(\frac{\partial S}{\partial \xi}(\xi_2))^2}}{\frac{1}{(\frac{\partial S}{\partial \xi}(\xi_2))^2} + \frac{\Delta\xi}{2}\frac{\frac{\partial^2 S}{\partial \xi^2}(\xi_2)}{(\frac{\partial S}{\partial \xi}(\xi_2))^3}} \tag{4.48}$$

$$\bar{l}_2 = -\frac{\frac{1}{(\frac{\partial S}{\partial \xi}(\xi_2))^2} - \frac{\Delta\xi}{2}\frac{\frac{\partial^2 S}{\partial \xi^2}(\xi_2)}{(\frac{\partial S}{\partial \xi}(\xi_2))^3}}{\frac{1}{(\frac{\partial S}{\partial \xi}(\xi_2))^2} + \frac{\Delta\xi}{2}\frac{\frac{\partial^2 S}{\partial \xi^2}(\xi_2)}{(\frac{\partial S}{\partial \xi}(\xi_2))^3}} \tag{4.49}$$

By substituting this discretized boundary condition in (4.44) we obtain

$$\begin{aligned}(-a_{2,j}\underline{l}_2 - b_{2,j})U^{k+1}_{2,j-1} + (-a_{2,j}\bar{l}_2 + a_{2,j})U^{k+1}_{3,j-1} + (-c_{2,j}\underline{l}_2 + d_{2,j})U^{k+1}_{2,j} \\ +(-c_{2,j}\bar{l}_2 - e_{2,j})U^{k+1}_{3,j} + (a_{2,j}\underline{l}_2 - f_{2,j})U^{k+1}_{2,j+1} + (a_{2,j}\bar{l}_2 - a_{2,j})U^{k+1}_{3,j+1} &= U^k_{2,j}\end{aligned}$$

(b) Boundary condition corresponding to reference point $i = N$ — As in case (a), assuming gamma is zero we obtain

$$\frac{1}{(\frac{\partial S}{\partial \xi}(\xi_N))^2}\frac{\partial^2 \bar{U}}{\partial \xi^2}(\xi_N, \eta, \tau) - \frac{\frac{\partial^2 S}{\partial \xi^2}(\xi_N)}{(\frac{\partial S}{\partial \xi}(\xi_N))^3}\frac{\partial \bar{U}}{\partial \xi}(\xi_N, \eta, \tau) = 0 \tag{4.50}$$

We use second order discretization of the first and second derivatives at $\xi_N = \xi_{max} - \Delta\xi$ to get

$$\frac{1}{(\frac{\partial S}{\partial \xi}(\xi_N))^2}\frac{U^{k+1}_{N-1,j} - 2U^{k+1}_{N,j} + U^{k+1}_{N+1,j}}{\Delta\xi^2} - \frac{\frac{\partial^2 S}{\partial \xi^2}(\xi_N)}{(\frac{\partial S}{\partial \xi}(\xi_N))^3}\frac{U^{k+1}_{N+1,j} - U^{k+1}_{N-1,j}}{2\Delta\xi} \approx 0 \tag{4.51}$$

which implies

$$\begin{aligned}U^{k+1}_{N+1,j} &= -\frac{\frac{1}{(\frac{\partial S}{\partial \xi}(\xi_N))^2} + \frac{\Delta\xi}{2}\frac{\frac{\partial^2 S}{\partial \xi^2}(\xi_N)}{(\frac{\partial S}{\partial \xi}(\xi_N))^3}}{\frac{1}{(\frac{\partial S}{\partial \xi}(\xi_N))^2} - \frac{\Delta\xi}{2}\frac{\frac{\partial^2 S}{\partial \xi^2}(\xi_N)}{(\frac{\partial S}{\partial \xi}(\xi_N))^3}}U^{k+1}_{N-1,j} + \frac{\frac{2}{(\frac{\partial S}{\partial \xi}(\xi_N))^2}}{\frac{1}{(\frac{\partial S}{\partial \xi}(\xi_N))^2} - \frac{\Delta\xi}{2}\frac{\frac{\partial^2 S}{\partial \xi^2}(\xi_N)}{(\frac{\partial S}{\partial \xi}(\xi_N))^3}}U^{k+1}_{N,j} \\ &= \underline{r}_N U^{k+1}_{N-1,j} + \bar{r}_N U^{k+1}_{N,j}\end{aligned}$$

where

$$\underline{r}_N = -\frac{\frac{1}{(\frac{\partial S}{\partial \xi}(\xi_N))^2} + \frac{\Delta\xi}{2}\frac{\frac{\partial^2 S}{\partial \xi^2}(\xi_N)}{(\frac{\partial S}{\partial \xi}(\xi_N))^3}}{\frac{1}{(\frac{\partial S}{\partial \xi}(\xi_N))^2} - \frac{\Delta\xi}{2}\frac{\frac{\partial^2 S}{\partial \xi^2}(\xi_N)}{(\frac{\partial S}{\partial \xi}(\xi_N))^3}} \tag{4.52}$$

$$\bar{r}_N = \frac{\frac{2}{(\frac{\partial S}{\partial \xi}(\xi_N))^2}}{\frac{1}{(\frac{\partial S}{\partial \xi}(\xi_N))^2} - \frac{\Delta\xi}{2}\frac{\frac{\partial^2 S}{\partial \xi^2}(\xi_N)}{(\frac{\partial S}{\partial \xi}(\xi_N))^3}} \tag{4.53}$$

By substituting this discretized boundary condition in (4.44) we obtain

$$\begin{aligned}(-a_{N,j} + a_{N,j}\underline{r}_N)U^{k+1}_{N-1,j-1} + (-b_{N,j} + a_{N,j}\bar{r}_N)U^{k+1}_{N,j-1} \\ +(-c_{N,j} - e_{N,j}\underline{r}_N)U^{k+1}_{N-1,j} + (d_{N,j} - e_{N,j}\bar{r}_N)U^{k+1}_{N,j} \\ +(a_{N,j} - a_{N,j}\underline{r}_N)U^{k+1}_{N-1,j+1} + (-f_{N,j} - a_{N,j}\bar{r}_N)U^{k+1}_{N,j+1} &= U^k_{N,j}\end{aligned}$$

(c) Boundary conditions corresponding to $j = M+1$ are Dirichlet boundary conditions and their values are

$$U_{i,N+1}^{k+1} = S(\xi_i)$$

That makes the difference equation

$$-a_{i,M}U_{i-1,M-1}^{k+1} - b_{i,M}U_{i,M-1}^{k+1} + a_{i,M}U_{i+1,M-1}^{k+1} - c_{i,M}U_{i-1,M}^{k+1} + d_{i,M}U_{i,M}^{k+1} - e_{i,M}U_{i+1,M}^{k+1}$$
$$= U_{i,M}^{k} - a_{i,M}S(\xi_{i-1}) + f_{i,M}S(\xi_i) + a_{i,j}S(\xi_{i+1})$$

(d) Boundary values corresponding to $j = 1$, namely, $U_{i,1}^{k+1}$ for $1 \leq i \leq M$ — These boundary nodes are the most complicated ones to evaluate, since it requires us to first solve a first order two-dimensional PDE (4.43) at $\eta = 0$ to find an approximate solution to it. In this case an appropriate discretization is

$$(r-q)\frac{1}{\frac{\partial S}{\partial \xi}(\xi_i)}S(\xi_i)\left(\frac{U_{i+1,1}^{k+1} - U_{i-1,1}^{k+1}}{2\Delta\xi}\right) + \kappa\theta\frac{1}{\frac{\partial v}{\partial \eta}(0)}\frac{-3U_{i,1}^{k} + 4U_{i,2}^{k} - U_{i,3}^{k}}{2\Delta\eta}$$
$$-rU_{i,1}^{k+1} - \frac{U_{i,1}^{k+1} - U_{i,1}^{k}}{\Delta\tau} = 0 \tag{4.54}$$

Note for the partial derivative $\frac{\partial \bar{U}}{\partial \eta}$ we use second order forward discretization which is fully explicit. Thus discretization for this boundary point is not fully implicit. If it were, it would require us to know $U_{i,2}^{k+1}$ and $U_{i,3}^{k+1}$ for $2 \leq i \leq N$, which are nodal values in the interior, hence unknown at time $\tau_{k+1}$. Multiplying by $\Delta\tau$ and gathering terms we can rewrite (4.54) as follows;

$$\alpha_i U_{i-1,1}^{k+1} + (1 + r\Delta\tau)U_{i,1}^{k+1} - \alpha_i U_{i+1,1}^{k+1} = (-3\beta + 1)U_{i,1}^{k} + 4\beta U_{i,2}^{k} - \beta U_{i,3}^{k} \tag{4.55}$$

where

$$\alpha_i = \frac{(r-q)\Delta\tau S(\xi_i)}{2\Delta\xi}\frac{1}{\frac{\partial S}{\partial \xi}(\xi_i)} \tag{4.56}$$

$$\beta = \frac{\kappa\theta\Delta\tau}{2\Delta\eta}\frac{1}{\frac{\partial v}{\partial \eta}(0)} \tag{4.57}$$

with the initial and boundary conditions

$$\begin{aligned} U_{i,1}^{0} &= (S(\xi_i) - K)^{+}, \quad 1 \leq i \leq N \\ U_{1,1}^{k+1} &= \underline{l}_2 U_{2,1}^{k+1} + \bar{l}_2 U_{3,1}^{k+1} \\ U_{N+1,1}^{k+1} &= \underline{r}_N U_{N-1,1}^{k+1} + \bar{r}_N U_{N,1}^{k+1} \end{aligned}$$

where $\underline{l}_2$ and $\bar{l}_2$ are given in (4.48) and (4.49) and $\underline{r}_N$ and $\bar{r}_N$ are given in (4.52) and (4.53). The boundary conditions are written as second order discretization of second derivatives at $\xi_2 = \xi_1 + \Delta\xi$ and $\xi_N = \xi_{N+1} - \Delta\xi$. By substituting those discretized boundary conditions in (4.55) we obtain

$$\begin{aligned} (1 + r\Delta\tau + \alpha_2\underline{l}_2)U_{2,1}^{k+1} + (\alpha_2\bar{l}_2 - \alpha_2)U_{3,1}^{k+1} &= (-3\beta + 1)U_{2,1}^{k} + 4\beta U_{2,2}^{k} - \beta U_{2,3}^{k} \\ (\alpha_N - \alpha_N\underline{r}_N)U_{N-1,1}^{k+1} + (1 + r\Delta\tau - \alpha_N\bar{r}_N)U_{N,1}^{k+1} &= (-3\beta + 1)U_{N,1}^{k} + 4\beta U_{N,2}^{k} - \beta U_{N,3}^{k} \end{aligned}$$

for $i = 2$ and $i = N$, respectively. Laying everything in matrix form we have

$$A\begin{pmatrix} U_{2,1}^{k+1} \\ U_{3,1}^{k+1} \\ \vdots \\ U_{N-1,1}^{k+1} \\ U_{N,1}^{k+1} \end{pmatrix} = \begin{pmatrix} (-3\beta+1)U_{2,1}^{k} + 4\beta U_{2,2}^{k} - \beta U_{2,3}^{k} \\ (-3\beta+1)U_{3,1}^{k} + 4\beta U_{3,2}^{k} - \beta U_{3,3}^{k} \\ \vdots \\ (-3\beta+1)U_{N-1,1}^{k} + 4\beta U_{N-1,2}^{k} - \beta U_{N-1,3}^{k} \\ (-3\beta+1)U_{N,1}^{k} + 4\beta U_{N,2}^{k} - \beta U_{N,3}^{k} \end{pmatrix}$$

where

$$A = \begin{pmatrix} (1+r\Delta\tau+\alpha_2\underline{l}_2) & (\alpha_2\bar{l}_2 - \alpha_2) & & & \\ \alpha_3 & (1+r\Delta\tau) & -\alpha_3 & & \\ & \ddots & \ddots & \ddots & \\ & & \alpha_{N-1} & (1+r\Delta\tau) & -\alpha_{N-1} \\ & & & (\alpha_N - \alpha_N\underline{r}_N) & (1+r\Delta\tau - \alpha_N\bar{r}_N) \end{pmatrix} \tag{4.58}$$

In summary, for each time step it is necessary to solve the following linear system on the $\eta = 0$ boundary for the $U_{i,1}^{k+1}$, $2 \le i \le N$ values, which in turn allows us to solve the original difference equation. Knowing $U_{i,1}^{k+1}$, the original difference equation becomes

$$\begin{aligned} -c_{i,2}U_{i-1,2}^{k+1} + d_{i,2}U_{i,2}^{k+1} - e_{i,2}U_{i+1,2}^{k+1} + a_{i,2}U_{i-1,3}^{k+1} - f_{i,2}U_{i,3}^{k+1} - a_{i,2}U_{i+1,3}^{k+1} \\ = U_{i,j}^{k} + a_{i,2}U_{i-1,1}^{k+1} + b_{i,2}U_{i,1}^{k+1} - a_{i,2}U_{i+1,1}^{k+1} \end{aligned}$$

Now that all boundary conditions are addressed, we should solve linear equation (4.45) at each time step to compute a numerical solution. This fully implicit method is unconditionally stable; however, the solution involves solving a linear system which is block tridiagonal, not strictly tridiagonal like the implicit schemes used in the one dimensional case. An alternative to the above methodology is to use the alternative direction implicit (ADI) method.

### 4.7.3 Alternative Direction Implicit (ADI) Scheme

We would like to solve the two-dimensional PDE using an implicit method; however, the block tridiagonal structure of the resulting matrix makes this expensive. One alternative is to use a partially implicit method which preserves the tridiagonal structure of the matrix for the implicit terms. This construction is called the alternative direction implicit (ADI) scheme. In the ADI scheme, each full time step comprises of two half steps: (a) first doing implicit discretization in $\xi$ and explicit discretization in $\eta$, (b) then doing explicit discretization in $\xi$ and implicit discretization in $\eta$. By doing this we always manage to preserve the tridiagonal structure of the stiffness matrix. Schematically we can say our stencils for each step would be (a) for the first half of the time step we have implicit discretization in $\xi$ and explicit discretization in $\eta$

$$\begin{array}{ccc} U_{i-1,j+1}^{k} & U_{i,j+1}^{k} & U_{i+1,j+1}^{k} \\ U_{i-1,j}^{k+\frac{1}{2}} & U_{i,j}^{k+\frac{1}{2}} & U_{i+1,j}^{k+\frac{1}{2}} \\ U_{i-1,j-1}^{k} & U_{i,j-1}^{k} & U_{i+1,j-1}^{k} \end{array}$$

As before the numeration of the grid is done from left to right and bottom to top. We write $A$ as follows:

$$A = A_1 + A_2$$

and therefore we are now solving the following linear system:

$$A_1 \mathbf{U^{k+\frac{1}{2}}} = A_2 \mathbf{U^k} + r.h.s.$$

and we solve for $\mathbf{U^{k+\frac{1}{2}}}$

$$\mathbf{U^{k+\frac{1}{2}}} = \begin{pmatrix} U_{2,2}^{k+\frac{1}{2}} \\ U_{3,2}^{k+\frac{1}{2}} \\ \vdots \\ U_{N,2}^{k+\frac{1}{2}} \\ U_{2,3}^{k+\frac{1}{2}} \\ U_{3,3}^{k+\frac{1}{2}} \\ \vdots \\ U_{N,3}^{k+\frac{1}{2}} \\ \vdots \\ U_{2,M}^{k+\frac{1}{2}} \\ U_{3,M}^{k+\frac{1}{2}} \\ \vdots \\ U_{N,M}^{k+\frac{1}{2}} \end{pmatrix}$$

(b) for the second half of the time step we have explicit discretization in $S$ and implicit discretization in $\nu$.

$$\begin{matrix} U_{i-1,j+1}^{k+\frac{1}{2}} & U_{i,j+1}^{k+1} & U_{i+1,j+1}^{k+\frac{1}{2}} \\ \\ U_{i-1,j}^{k+\frac{1}{2}} & U_{i,j}^{k+1} & U_{i+1,j}^{k+\frac{1}{2}} \\ \\ U_{i-1,j-1}^{k+\frac{1}{2}} & U_{i,j-1}^{k+1} & U_{i+1,j-1}^{k+\frac{1}{2}} \end{matrix}$$

Now, in order to preserve the tridiagonal structure of the stiffness matrix $A$, the numeration of the grid is done from bottom to top and left to right. We write $\widetilde{A}$ as follows:

$$\widetilde{A} = \widetilde{A}_1 + \widetilde{A}_2$$

and we are now solving the following linear system:

$$\widetilde{A}_1 \widetilde{\mathbf{U}}^{\mathbf{k+1}} = \widetilde{A}_2 \widetilde{\mathbf{U}}^{\mathbf{k+\frac{1}{2}}}$$

where $\widetilde{\mathbf{U}}^{\mathbf{k+\frac{1}{2}}}$ is obtained by reordering $\mathbf{U^{k+\frac{1}{2}}}$. Knowing that in the first half ordering is done left to right and bottom to top and now we should change to bottom to top and left to right, the reordering can be done via the following simple routine:

for $i = 1, \ldots, N-1$
  for $j = 1, \ldots, M-1$
    $\widetilde{U}^{k+\frac{1}{2}}[(i-1)(M-1)+j] = U^{k+\frac{1}{2}}[(j-1)(N-1)+i]$
  endfor
endfor

We solve for $\widetilde{\mathbf{U}}^{\mathbf{k+1}}$:

$$\widetilde{\mathbf{U}}^{\mathbf{k+1}} = \begin{pmatrix} U_{2,2}^{k+1} \\ U_{2,3}^{k+1} \\ \vdots \\ U_{2,M}^{k+1} \\ U_{3,2}^{k+1} \\ U_{3,3}^{k+1} \\ \vdots \\ U_{3,M}^{k+1} \\ \vdots \\ U_{N,2}^{k+1} \\ U_{N,3}^{k+1} \\ \vdots \\ U_{N,M}^{k+1} \end{pmatrix}$$

and reorder back to left to right bottom to top to obtain to $\mathbf{U}^{\mathbf{k+1}}$ for the next time step. This is done via the following routine:

for $i = 1, \ldots, N-1$
  for $j = 1, \ldots, M-1$
    $U^{k+1}[(j-1)(N-1) + i] = \widetilde{U}^{k+1}[(i-1)(M-1)+j]$
  endfor
endfor

This would constitute one full time step in the ADI scheme. It is important to mention that cross derivative $\frac{\partial^2 U}{\partial\xi\partial\eta}$ is treated fully explicitly to make ADI steps possible. This is the Peaceman–Rachford scheme [187]. In the presence of mixed derivatives with high correlation this scheme becomes unstable, as shown by Andersen and Piterbarg [14] and Duffy [102]. Instead we use Douglas–Rachford [97] or Craig–Sneyd [83] for the ADI scheme.

#### 4.7.3.1 Derivation of the Craig–Sneyd Scheme for the Heston PDE

The Craig–Sneyd (CS) scheme is designed to solve a two-dimensional parabolic PDE in the presence of a mixed derivative. The CS scheme can be viewed as a generalization of the Peaceman–Rachford ADI scheme [187]. We first provide some intuition behind such a scheme and we then apply the CS scheme to the Heston PDE. Finally, we implement the CS scheme for both uniform and nonuniform grids for option pricing. Through the derivation, we see why the CS scheme improves numerical stability and accuracy over traditional ADI in dealing with a mixed derivative. The CS scheme is designed to solve the given ODE:

$$\begin{aligned} u'(t) &= F(t, u(t)) \\ F(t, u(t)) &= F_0(t, u(t)) + F_1(t, u(t)) + F_2(t, u(t)) \\ u(0) &= u_0 \end{aligned} \tag{4.59}$$

Following [146], the complete one step CS scheme to get the current time step value $U_n$ from the previous time step value $U_{n-1}$ is as follows:

$$\begin{aligned}
\overline{U}_1 &= \overline{U}_0 + \theta\Delta t(F_1(t_n, \overline{U}_1) - F_1(t_{n-1}, U_{n-1}) && (4.60)\\
\overline{U}_2 &= \overline{U}_1 + \theta\Delta t(F_2(t_n, \overline{U}_2) - F_2(t_{n-1}, U_{n-1}) && (4.61)\\
\widetilde{U}_1 &= \widetilde{U}_0 + \theta\Delta t(F_1(t_n, \widetilde{U}_1) - F_1(t_{n-1}, U_{n-1}) && (4.62)\\
\widetilde{U}_2 &= \widetilde{U}_1 + \theta\Delta t(F_2(t_n, \widetilde{U}_2) - F_2(t_{n-1}, U_{n-1}) && (4.63)\\
U_n &= \widetilde{U}_2 && (4.64)
\end{aligned}$$

where

$$\begin{aligned}
\overline{U}_0 &\triangleq U_{n-1} + \Delta t F(t_{n-1}, U_{n-1}) && (4.65)\\
\widetilde{U}_0 &\triangleq U_0 + \Delta t\lambda(F_0(t_n, U_2) - F_0(t_{n-1}, U_{n-1})) && (4.66)
\end{aligned}$$

We now attempt to give an intuition behind the CS scheme. Given the ODE

$$\begin{aligned}
u'(t) &= F(t, u(t))\\
F(t, u(t)) &= F_0(t, u(t)) + F_1(t, u(t)) + F_2(t, u(t))\\
u(0) &= u_0
\end{aligned}$$

To obtain $u_n$ from $u_{n-1}$, we discretize the above ODE using the $\theta$ method utilizing the iterated splitting scheme [142]

$$\begin{aligned}
\frac{u_1 - u_{n-1}}{\Delta t} &= (1-\theta)(F_1(u_{n-1}) + F_2(u_{n-1})) + \theta(F_1(u_1) + F_2(u_{n-1}))\\
&+ ((1-\lambda)F_0(u_{n-1}) + \lambda F_0(u_{n-1}))\\
\frac{u_2 - u_{n-1}}{\Delta t} &= (1-\theta)(F_1(u_{n-1}) + F_2(u_{n-1})) + \theta(F_1(u_1) + F_2(u_2))\\
&+ ((1-\lambda)F_0(u_{n-1}) + \lambda F_0(u_{n-1}))\\
\frac{\overline{u}_1 - u_{n-1}}{\Delta t} &= (1-\theta)(F_1(u_n) + F_2(u_{n-1})) + \theta(F_1(\overline{u}_1) + F_2(u_{n-1}))\\
&+ ((1-\lambda)F_0(u_{n-1}) + \lambda F_0(u_2))\\
\frac{u_n - u_{n-1}}{\Delta t} &= (1-\theta)(F_1(u_{n-1}) + F_2(u_{n-1})) + \theta(F_1(\overline{u}_1) + F_2(\overline{u}_2))\\
&+ ((1-\lambda)F_0(u_n) + \lambda F_0(u_2))
\end{aligned}$$

We can rewrite the above equations as

$$\begin{aligned}
u_1 &= u_n + \Delta t((1-\theta)F_1(u_n) + \theta F_1(u_1)) + \Delta t(F_2(u_n) + F_0(u_n)) && (4.67)\\
u_2 &= u_n + \Delta t((1-\theta)F_1(u_n) + \theta F_1(u_1)) + \Delta t((1-\theta)F_2(u_n) + \theta F_2(u_2))\\
&+ \Delta t F_0(u_n) && (4.68)\\
\overline{u}_1 &= u_n + \Delta t((1-\theta)F_1(u_n) + \theta F_1(\overline{u}_1)) + \Delta t F_2(u_n)\\
&+ \Delta t((1-\lambda)F_0(u_n) + \lambda F_0(u_2)) && (4.69)\\
\overline{u}_2 &= u_n + \Delta t((1-\theta)F_1(u_n) + \theta F_1(\overline{u}_1)) + \Delta t((1-\theta)F_2(u_n) + \theta F_2(\overline{u}_2))\\
&+ \Delta t((1-\lambda)F_0(u_n) + \lambda F_0(u_2)) && (4.70)
\end{aligned}$$

Equations (4.67), (4.68), (4.69), and (4.70) are the principles behind the CS Scheme. Suppose $U_{n-1}$ is the solution of the ODE (4.59) at time $t = t_{n-1}$ to get the solution at $t = t_n$. Apply (4.67) we get

$$\overline{U}_1 = U_{n-1} + \Delta t(\theta F_1(\overline{U}_1) + (1-\theta)F_1(U_{n-1})) + \Delta t(F_2(U_{n-1}) + F_0(U_{n-1})) \quad (4.71)$$

or equivalently

$$\overline{U}_1 = U_{n-1}+\Delta t\{F_1(U_{n-1})+F_2(U_{n-1})+F_0(U_{n-1})\}+\theta\Delta t(F_1(\overline{U}_1)-F_1(U_{n-1})) \quad (4.72)$$

Define

$$\overline{U}_0 \triangleq U_{n-1} + \Delta t\{F_1(U_{n-1}) + F_2(U_{n-1}) + F_0(U_{n-1})\} = U_{n-1} + \Delta t F(U_{n-1}) \quad (4.73)$$

Therefore

$$\overline{U}_1 = \overline{U}_0 + \theta\Delta t(F_1(U_1) - F_1(U_{n-1})) \quad (4.74)$$

Note that Equation (4.73) justifies Equation (4.65) and Equation (4.74) justifies Equation (4.60). Then we proceed by applying (4.68):

$$\begin{aligned}
\overline{U}_2 &= U_{n-1} + \Delta t(\theta F_1(U_1) + (1-\theta)F_1(U_{n-1})) + \Delta t(\theta F_2(U_2) + (1-\theta)F_2(U_{n-1})) \\
&+ \Delta t F_0(U_{n-1}) \quad (4.75) \\
&= U_{n-1} + \Delta t\{(\theta F_1(U_1) + (1-\theta)F_1(U_{n-1}) + F_2(U_{n-1}) + F_0(U_{n-1})\} \\
&+ \Delta t\theta(F_2(U_2) - F_2(U_{n-1})) \quad (4.76) \\
&= U_{n-1} + U_1 - U_{n-1} + \Delta t\theta(F_2(U_2) - F_2(U_{n-1})) \\
&= U_1 + \Delta t\theta(F_2(U_2) - F_2(U_{n-1})) \quad (4.77)
\end{aligned}$$

Note that the term in Equation (4.76) in big parentheses is $U_1 - U_{n-1}$ via Equations (4.71) and (4.77) justifies Equation (4.61). The following is for justifying Equations (4.62), (4.63), and (4.64), which is similar to the above derivations. Now we proceed by applying (4.69):

$$\begin{aligned}
\widetilde{U}_1 &= U_{n-1} + \Delta t(\theta F_1(\widetilde{U}_1) + (1-\theta)F_1(U_{n-1})) \\
&+ \Delta t F_2(U_{n-1}) + \Delta t(\lambda F_0(U_2) + (1-\lambda)F_0(U_{n-1})) \quad (4.78) \\
&= \{U_{n-1} + \Delta t(F_1(U_{n-1}) + F_2(U_{n-1}) + F_0(U_{n-1})) + \lambda\Delta t(F_0(U_2) - F_0(U_{n-1}))\} \\
&+ \theta\Delta t(F_1(\widetilde{U_1}) - F_1(U_{n-1})) \quad (4.79) \\
&= \widetilde{U}_0 + \Delta t\theta(F_1(\widetilde{U}_1) - F_1(U_{n-1})) \quad (4.80) \\
& \quad (4.81)
\end{aligned}$$

where $\widetilde{U}_0$ is defined as

$$\widetilde{U}_0 \triangleq U_{n-1}+\Delta t(F_1(U_{n-1})+F_2(U_{n-1})+F_0(U_{n-1}))+\lambda\Delta t(F_0(U_2)-F_0(U_{n-1})) \quad (4.82)$$

The intuition behind the above definition of $\widetilde{U}_0$ (4.82) is as follows:

$$\widetilde{U}_0 = U_{n-1} + \Delta t F_1(U_{n-1}) + \Delta t F_2(U_{n-1}) + \Delta t(\lambda F_0(U_2) + (1-\lambda)F_0(U_{n-1})) \quad (4.83)$$

The Euler scheme is used to propagate from $U_{n-1}$ to $\widetilde{U}_0$. For operators $F_1$ and $F_2$ we use old value $U_{n-1}$; for operator $F_0$ we use the average of $\overline{U}_2$ and $U_{n-1}$. The procedure does not involve any implicit scheme for updating $F_1$ and $F_2$ since $F_0$ is always treated explicitly in the CS scheme. Note that the first term in Equation (4.82) in parenthesis is $U_0$ via Equation (4.73) and Equations (4.80) and (4.82) justify (4.66) and (4.62), respectively. Now we proceed by applying (4.70):

$$\begin{aligned}
\widetilde{U}_2 &= U_{n-1} + \Delta t(\theta F_1(\widetilde{U}_1) + (1-\theta)F_1(U_{n-1})) \\
&+ \Delta t(\theta F_2(\widetilde{U}_2) + (1-\theta)F_2(U_{n-1})) + \Delta t(\lambda F_0(U_2) + (1-\lambda)F_0(U_{n-1})) \quad (4.84) \\
&= U_{n-1} + \Delta t\{\theta F_1(\widetilde{U}_1) + (1-\theta)F_1(U_{n-1}) + F_2(U_{n-1}) \\
&+ (\lambda F_0(U_2) + (1-\lambda)F_0(U_{n-1}))\} + \Delta t\theta(F_2(\widetilde{U}_2) - F_2(U_{n-1})) \quad (4.85) \\
&= U_{n-1} + \widetilde{U}_1 - U_{n-1} + \Delta t\theta(F_2(\widetilde{U}_2) - F_2(U_{n-1})) \quad (4.86) \\
&= \widetilde{U}_1 + \Delta t\theta(F_2(\widetilde{U}_2) - F_2(U_{n-1})) \quad (4.87)
\end{aligned}$$

Note that the second term in Equation (4.85) in parentheses is actually $\widetilde{U}_1 - U_{n-1}$ via Equation (4.78). Equation (4.87) justifies Equation (4.63).

### 4.7.4 Heston PDE

$$\frac{\partial u}{\partial \tau} = \frac{1}{2}\nu S^2 \frac{\partial^2 u}{\partial S^2} + (r-q)S\frac{\partial u}{\partial S} + \rho\sigma\nu S\frac{\partial^2 u}{\partial S \partial \nu} + \frac{1}{2}\sigma^2\nu\frac{\partial^2 u}{\partial \nu^2} + \kappa(\theta-\nu)\frac{\partial u}{\partial \nu} - ru$$

Boundary conditions are consistent with what was written earlier in the chapter. The Heston PDE can be rewritten as the ODE as follows:

$$\begin{aligned} U'(t) &= F_0(U,t) + F_1(U,t) + F_2(U,t) \\ U(0) &= U_0 \end{aligned}$$

where

$$\begin{aligned} F_0(u,t) &= \rho\sigma\nu S\frac{\partial^2 u}{\partial S \partial \nu} \\ F_1(u,t) &= \frac{1}{2}\nu S^2\frac{\partial^2 u}{\partial S^2} + (r-q)S\frac{\partial u}{\partial S} - \frac{r}{2}u \\ F_2(u,t) &= \frac{1}{2}\sigma^2\nu\frac{\partial^2 u}{\partial \nu^2} + \kappa(\theta-\nu)\frac{\partial u}{\partial \nu} - \frac{r}{2}u \end{aligned}$$

Applying the CS scheme (4.65-4.64) for the above Heston PDE (ODE) we get a complete CS loop for each time step:

(1) $U_0 = U_{old} + \Delta t F(U_{old})$

(2) $U_1 = (I - \theta\Delta t(A_1 - \frac{r}{2}I))^{-1}(U_0 + \theta\Delta t(\frac{r}{2}U_{old} - A_1 U_{old}))$

(3) $U_2 = (I - \theta\Delta t(A_2 - \frac{r}{2}I))^{-1}(U_1 + \theta\Delta t(\frac{r}{2}U_{old} - A_2 U_{old}))$

(4) $\hat{U}_0 = U_0 + \theta\Delta t A_0(U_2 - U_{old})$

(5) $\hat{U}_1 = (I - \theta\Delta t(A_1 - \frac{r}{2}I))^{-1}(\hat{U}_0 + \theta\Delta t(\frac{r}{2}U_{old} - A_1 U_{old}))$

(6) $\hat{U}_2 = (I - \theta\Delta t(A_2 - \frac{r}{2}I))^{-1}(\hat{U}_1 + \theta\Delta t(\frac{r}{2}U_{old} - A_2 U_{old}))$

(7) $U_{old} = \hat{U}_2$.

Note that $A_0$, $A_1$, and $A_2$ are tridiagonal matrices coming from discretization of $F_0(U,t)$, $F_1(U,t)$, and $F_2(U,t)$, respectively.

### 4.7.5 Numerical Results and Conclusion

For our numerical results we use the following sets of parameters: spot price $S_0 = 1200$, strike price $K = 1200$, risk-free rate $r = 0.25\%$, dividend rate $q = 1.0\%$, $\kappa = 1$, $\theta = 0.15$, correlation $\rho = -0.80$, volatility of variance $\sigma = 0.40$, initial variance $v_0 = 0.15$, $S_{min} = 400$, $S_{max} = 3000$, $v_{min} = 0\%$, $v_{max} = 500\%$. Here are the mesh grid specifications:

Time step spacing

$\Delta t = \frac{T}{400}$

That means $L = 50, 100, 400$ (time intervals) for maturities $T = 0.125, 0.25, 1.0$, respectively.

Stock spacial spacing

$M = 300$

$0 \leq \xi_i \leq 1$, $i = 1, \ldots, M$, uniform grid on interval $[0, 1]$

$S(\xi_i) = K + \alpha \sinh(c_1 \xi_i + c_2(1 - \xi_i))$

For non-uniform grids we set $\alpha = \frac{(K - S_{min})}{3}$

For uniform grids we set $\alpha = S_{max} - S_{min}$

$c_1 = \sinh^{-1}(\frac{S_{max} - K}{\alpha})$

$c_2 = \sinh^{-1}(\frac{S_{min} - K}{\alpha})$

Variance spacing

$N = 100$

$0 \leq \eta_i \leq 1$, $i = 1, \ldots, N$, uniform grid on interval $[0, 1]$

$v(\eta_i) = \beta \sinh(d\eta_i)$

For non-uniform grids we set $\beta = (v_{max} - v_{min})/50$

For uniform grids we set $\beta = v_{max} - v_{min}$

$d = \sinh^{-1}(\frac{v_{max}}{\beta})$

For the FFT method we use $\alpha = 1.2$ and $N = 2^{14}$. Results are summarized in Tables 4.1–4.6. In these tables, we compare premiums from the ADI scheme with those from FFT, implicit scheme and analytical for both uniform and non-uniform grids. We also plot three dimensional plot of ADI finite difference solution in Figures 4.13–4.15. Comparisons of premiums for maturity of 1.5 months are shown in Tables 4.1 and 4.2 for non-uniform and uniform grids respectively. Figure 4.13 depicts the ADI finite difference solution for maturity of 1.5 months.

**TABLE 4.1**: Premiums (cpu time in seconds) comparison for a non-uniform grid on both $S$ and $v$ for maturity $T = 1.5$ months using ADI, implicit scheme, fast Fourier transform and analytical

| $K$ | $T$ | ADI (cpu) | Implicit (cpu) | FFT (cpu) | Analytical (cpu) |
|---|---|---|---|---|---|
| 1200 | 0.125 | 64.256(35.040) | 64.2480(51.085) | 64.351(0.004) | 64.258(0.001) |
| 1250 | 0.125 | 42.367(35.570) | 42.3600(50.151) | 42.601(0.004) | 42.365(0.008) |
| 1300 | 0.125 | 26.246(35.139) | 26.2417(50.844) | 26.355(0.005) | 26.247(0.007) |
| 1350 | 0.125 | 15.193(35.309) | 15.1920(50.506) | 15.361(0.004) | 15.180(0.009) |
| 1400 | 0.125 | 8.153(35.564) | 8.1555(61.744) | 8.223(0.004) | 8.142(0.008) |
| 1450 | 0.125 | 4.031(40.303) | 4.0345(50.450) | 4.135(0.005) | 4.023(0.023) |

Comparisons of premiums for maturity of 3 months are shown in Tables 4.3 and 4.4 for non-uniform and uniform grids respectively. Figure 4.14 depicts the ADI finite difference solution for maturity of 3 months. Comparisons of premiums for maturity of 12 months are shown in Tables 4.5 and 4.6 for non-uniform and uniform grids respectively. Figure 4.15 depicts the ADI finite difference solution for maturity of 12 months. As we see from our

**TABLE 4.2**: Premiums (cpu time in seconds) comparison for a uniform grid on both $S$ and $v$ for maturity $T = 1.5$ months using ADI, implicit scheme, fast Fourier transform and analytical

| $K$ | $T$ | ADI (cpu) | Implicit (cpu) | FFT (cpu) | Analytical (cpu) |
|---|---|---|---|---|---|
| 1200 | 0.125 | 63.573(35.518) | 63.565(511.32) | 64.351(0.004) | 64.258(0.001) |
| 1250 | 0.125 | 41.463(35.292) | 41.455(502.46) | 42.601(0.004) | 42.365(0.008) |
| 1300 | 0.125 | 25.664(35.406) | 25.659(513.03) | 26.355(0.004) | 26.247(0.007) |
| 1350 | 0.125 | 15.164(35.167) | 15.163(502.01) | 15.361(0.004) | 15.180(0.009) |
| 1400 | 0.125 | 8.564(35.766) | 8.565(501.44) | 8.223(0.004) | 8.142(0.008) |
| 1450 | 0.125 | 4.584(35.169) | 4.586(498.88) | 4.135(0.005) | 4.023(0.023) |

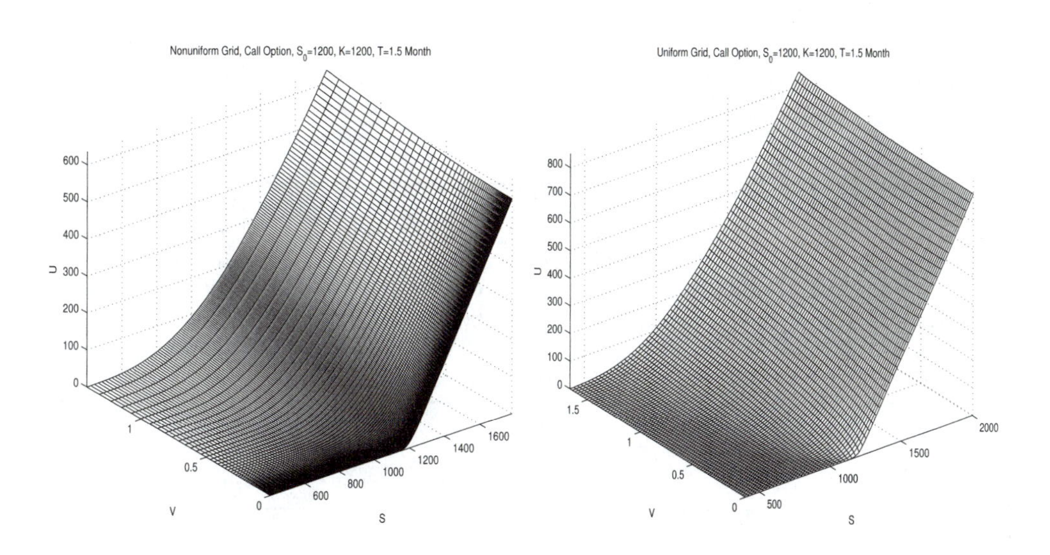


**FIGURE 4.13**: Surface of call price premiums for $S_0 = 1200$, $K = 1200$, $\lambda = 0$, $T = 0.125$

results, both ADI and implicit scheme produce reasonable approximations to the value of the option. To improve the results, we have used non-uniform mesh refinement. There are several ways to do this. We have done by transformation introduced in Section 4.3.1. This transformation simply decreases the number of nodes as $\nu \to \nu_{max}$, and $S \to S_{max}$ (placing a higher density of mesh nodes in the region of high curvature, and hence less in the region of low curvature). This allows for a significantly larger $\nu_{max}$ and $S_{max}$. As expected, we see from the results that ADI scheme is almost twice faster than implicit for the same mesh grid specifications. This is due to the fact the in the implicit scheme we have to solve a linear system which is block tridiagonal.

In general, there could a boundary layer occurring at $\nu = \nu_{max}$. This is understandable since the boundary condition $U = i\Delta S$ for $\nu = \nu_{max}$ is not a good approximation to $U(S, \infty, \tau) = S$, unless the strike price is very small. We would need to increase $\nu$ markedly for this approximate boundary condition to be a good estimate as in our case. This would probably solve the problem of a boundary layer when $\nu = \nu_{max}$.

Due to the choice of $v_0$ and $S_0$ in our case we do not need to interpolate in order to obtain the approximate solution for this case, but in general we can introduce an error into the estimation due to interpolation.

**TABLE 4.3**: Premiums (cpu time in seconds) comparison for a non-uniform grid on both $S$ and $v$ for maturity $T = 3$ months using ADI, implicit scheme, fast Fourier transform and analytical

| $K$ | $T$ | ADI (cpu) | Implicit (cpu) | FFT (cpu) | Analytical (cpu) |
|---|---|---|---|---|---|
| 1200 | 0.25 | 89.599(39.834) | 89.587(504.53) | 89.663(0.003) | 89.603(0.009) |
| 1250 | 0.25 | 66.931(35.917) | 66.920(516.10) | 67.091(0.004) | 66.932(0.037) |
| 1300 | 0.25 | 48.453(36.355) | 48.444(505.67) | 48.537(0.004) | 48.457(0.009) |
| 1350 | 0.25 | 33.915(35.389) | 33.909(532.01) | 34.054(0.003) | 33.907(0.022) |
| 1400 | 0.25 | 22.876(35.860) | 22.873(499.73) | 22.949(0.007) | 22.868(0.002) |
| 1450 | 0.25 | 14.829(35.037) | 14.830(508.10) | 14.959(0.005) | 14.823(0.008) |

**TABLE 4.4**: Premiums (cpu time in seconds) comparison for a uniform grid on both $S$ and $v$ for maturity $T = 3$ months using ADI, implicit scheme, fast Fourier transform and analytical

| $K$ | $T$ | ADI (cpu) | Implicit (cpu) | FFT (cpu) | Analytical (cpu) |
|---|---|---|---|---|---|
| 1200 | 0.25 | 88.657(35.423) | 88.646(58.839) | 89.664(0.004) | 89.603(0.009) |
| 1250 | 0.25 | 65.548(39.981) | 65.537(50.286) | 67.091(0.004) | 66.933(0.037) |
| 1300 | 0.25 | 47.165(35.594) | 47.156(50.163) | 48.537(0.004) | 48.458(0.009) |
| 1350 | 0.25 | 33.137(37.047) | 33.131(50.490) | 34.055(0.004) | 33.908(0.022) |
| 1400 | 0.25 | 22.754(35.175) | 22.752(49.821) | 22.949(0.008) | 22.868(0.001) |
| 1450 | 0.25 | 15.230(35.145) | 15.229(54.356) | 14.959(0.005) | 14.824(0.009) |

## Problems

1. Consider the Black–Scholes PDE

$$\frac{\partial V}{\partial t} + \frac{1}{2}\sigma^2 S^2 \frac{\partial^2 V}{\partial S^2} + (r-q)S\frac{\partial V}{\partial S} = rV$$

Assuming the terminal boundary condition is the payoff of a put option at maturity $T$

$$V(S,T) = \max(K - S, 0)$$

The analytical solution to this PDE at time $t < T$ is the Black–Scholes option pricing formula for a European put, which is given by

$$V(S,t) = Ke^{-r(T-t)}\Phi(-d_2) - Se^{-q(T-t)}\Phi(-d_1)$$

where

$$\begin{aligned} d_1 &= \frac{\ln(S/K) + (r - q + \sigma^2/2)(T-t)}{\sigma\sqrt{T-t}} \\ d_2 &= d_1 - \sigma\sqrt{T-t} \end{aligned}$$

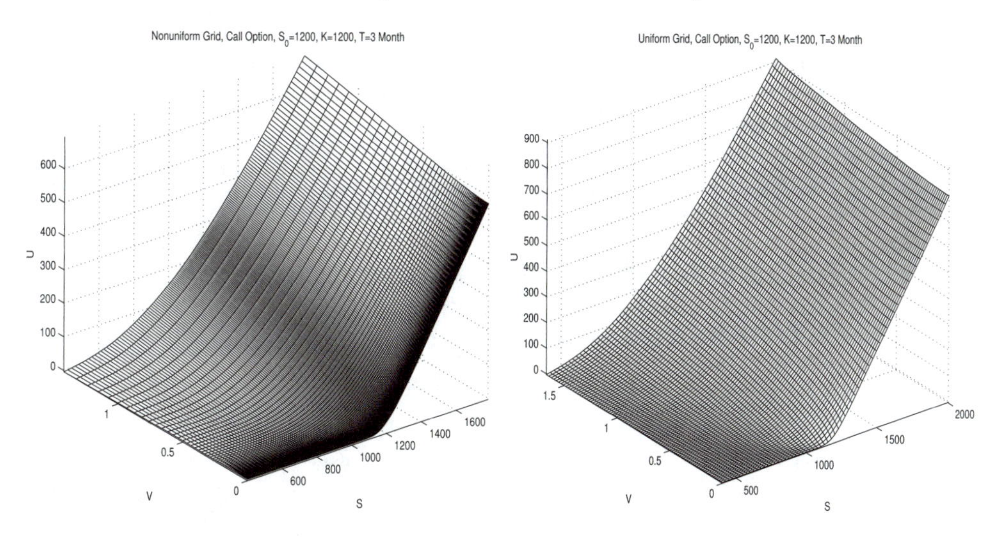


**FIGURE 4.14**: Surface of call price premiums for $S_0 = 1200$, $K = 1200$, $\lambda = 0$, $T = 0.25$

**TABLE 4.5**: Premiums (cpu time in seconds) comparison for a non-uniform grid on both $S$ and $v$ for maturity $T = 12$ months using ADI, implicit scheme, fast Fourier transform and analytical

| $K$ | $T$ | ADI (cpu) | Implicit (cpu) | FFT (cpu) | Analytical (cpu) |
|---|---|---|---|---|---|
| 1200 | 1 | 168.737(217.786) | 168.732(313.129) | 168.781(0.005) | 168.759(0.029) |
| 1250 | 1 | 145.907(175.873) | 145.902(296.699) | 145.981(0.003) | 145.922(0.005) |
| 1300 | 1 | 125.222(179.275) | 125.218(250.785) | 125.268(0.003) | 125.235(0.007) |
| 1350 | 1 | 106.643(174.855) | 106.639(272.039) | 106.714(0.005) | 106.648(0.010) |
| 1400 | 1 | 90.084(185.675) | 90.081(266.624) | 90.130(0.005) | 90.087(0.003) |
| 1450 | 1 | 75.459(173.685) | 75.456(253.322) | 75.546(0.004) | 75.462(0.032) |

Moreover, we have the following analytical expressions for $\Delta$, $\Gamma$, $\kappa$ (Greeks) for a European put option:

$$
\begin{aligned}
\Delta &= e^{-q(T-t)}\left(\Phi(d_1) - 1\right) \\
\Gamma &= \frac{\phi(d_1)e^{-q(T-t)}}{S\sigma\sqrt{T-t}} \\
\kappa &= Se^{-q(T-t)}\phi(d_1)\sqrt{T-t}
\end{aligned}
$$

where $\Phi$ is the standard normal cumulative distribution function and $\phi$ denotes the standard normal probability density function. Using the following pricing parameters, spot price $S_0 = \$100$, strike price $K = \{90, 100, 110\}$, risk-free rate $r = 2\%$, dividend yield $q = 1.5\%$, time to maturity $T = 1$ year, and volatility $\sigma = \{15\%, 30\%, 50\%\}$, solve the Black–Scholes PDE *numerically* to price a European put option by means of

(1) Explicit finite differences

(2) Implicit finite differences

(3) Crank–Nicolson finite differences

**TABLE 4.6**: Premiums (cpu time in seconds) comparison for a uniform grid on both $S$ and $v$ for maturity $T = 12$ months using ADI, implicit scheme, fast Fourier transform and analytical

| $K$ | $T$ | ADI (cpu) | Implicit (cpu) | FFT (cpu) | Analytical (cpu) |
|---|---|---|---|---|---|
| 1200 | 1 | 167.527(182.528) | 167.522(253.134) | 168.780(0.004) | 168.758(0.029) |
| 1250 | 1 | 144.287(174.486) | 144.281(249.799) | 145.981(0.003) | 145.921(0.005) |
| 1300 | 1 | 123.335(174.123) | 123.329(251.339) | 125.268(0.003) | 125.235(0.006) |
| 1350 | 1 | 104.715(178.993) | 104.709(249.550) | 106.713(0.005) | 106.647(0.010) |
| 1400 | 1 | 88.337(186.521) | 88.332(250.629) | 90.130(0.004) | 90.086(0.003) |
| 1450 | 1 | 74.027(186.027) | 74.022(250.657) | 75.545(0.004) | 75.462(0.032) |

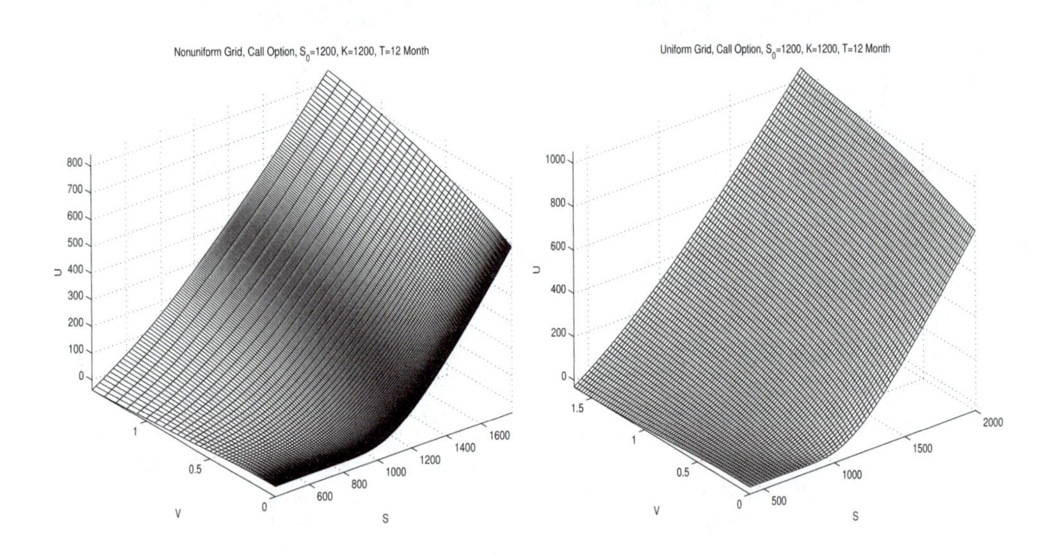


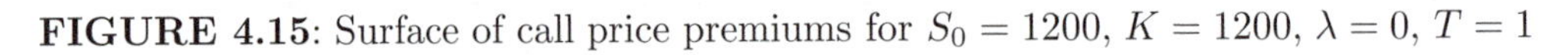
**FIGURE 4.15**: Surface of call price premiums for $S_0 = 1200$, $K = 1200$, $\lambda = 0$, $T = 1$

(4) Multi-step finite differences

Using the following sets of boundary conditions:

I.

$$\begin{aligned}\lim_{S\to 0} V(S,t) &= Ke^{-r(T-t)} - Se^{-q(T-t)} \\ \lim_{S\to\infty} V(S,t) &= 0\end{aligned}$$

II.

$$\begin{aligned}\lim_{S\to 0} \frac{\partial^2 V}{\partial S^2} &= 0 \\ \lim_{S\to\infty} \frac{\partial^2 V}{\partial S^2} &= 0\end{aligned}$$

(a) For each of the eight cases compare the prices and Greeks with the analytical values.

(b) In the case of the explicit finite difference scheme demonstrate that the scheme is conditionally stable and find the condition under which the scheme would be stable.

(c) In the case of the Crank–Nicolson scheme demonstrate that for certain time steps, the solution oscillates in time.

(d) Explain which set of boundary conditions you would prefer and why. Justify your answer by referring to your numerical results.

2. Calculate American put option values in the Black–Scholes framework for the following parameters: spot price, $S_0 = \$100$; strike price $K = \{80, 100, 120\}$; risk-free interest rate, $r = 4.75\%$; dividend rate, $q = 1.75\%$; maturity $T = 1$ year, and volatility, $\sigma = \{15\%, 30\%, 50\%\}$ by:

   (a) Applying the Bermudan approach.

   (b) Solving the following modified Black–Scholes PDE:

$$\frac{\partial V}{\partial t} + \frac{1}{2}\sigma^2 S^2 \frac{\partial^2 V}{\partial S^2} + (r - q)S\frac{\partial V}{\partial S} - rV - \mathbb{1}_{S<S^*(t)}\{qS - rK\} = 0$$

   (c) Applying the Brennan–Schwartz algorithm.

   Compare values and critical stock price curves and conclude.

3. For $K < B$, the forward PDE for up-and-out calls (Carr–Hirsa) is as follows:

$$\frac{\sigma^2(K,T)}{2}K^2\frac{\partial^2 U}{\partial K^2} - [r(T) - q(T)]\, K\frac{\partial U}{\partial K} - q(T)U = \frac{\partial U}{\partial T} + \left[\frac{\sigma^2(B,T)}{2}B^2\frac{\partial^3 U}{\partial K^3}(B,T)\right](K - B)$$

   with initial condition

$$U(K,0) = (S_0 - K)^+, \text{ for } K < B, \text{ and } S_0 < B.$$

   Boundary conditions are

$$U_{KK}(0,T) = 0$$

$$U_{KK}(B,T) = 0$$

   For the following variables up barrier $B = 125$, spot price $S_0 = 100$, the local volatility surface $\sigma(K,T) = 0.5e^{-T}(100/K)^{0.3}$, risk-free rate $r = 4.75\%$, and dividend rate $q = 1.75\%$ fill in the following table:

| Maturity | | $T_1 = 0.25$ | | $T_2 = 0.5$ | | $T_3 = 1.0$ | |
|---|---|---|---|---|---|---|---|
| Barrier | Strike | **Bwd** | *Fwd* | **Bwd** | *Fwd* | **Bwd** | *Fwd* |
| 125 | 110 | | | | | | |
| | 120 | | | | | | |

4. *A fortiori*, the up-and-out call value function $U(S,t)$ solves a backward boundary value problem (BVP), consisting of the backward partial differential equation

$$\frac{\partial U(S,t)}{\partial t} + \frac{\sigma^2(S,t)S^2}{2}\frac{\partial^2 U(S,t)}{\partial S^2} + [r(t) - q(t)]S\frac{\partial U(S,t)}{\partial S} - r(t)U(S,t) = 0$$

subject to the following boundary conditions:

$$\begin{aligned} U(S,T_0) &= (S - K_0)^+, & S &\in [0,B] \\ \lim_{S\downarrow 0} U(S,t) &= 0, & t &\in [0,T_0] \\ \lim_{S\uparrow B} U(S,t) &= 0, & t &\in [0,T_0] \end{aligned}$$

Solve the PDE numerically for the following set of parameters: spot price $S_0 = \$100$, strike price $K = \{105, 115\}$, up barrier $B = 125$, risk-free interest rate $r = 4.75\%$, dividend rate $q = 1.75\%$, maturity $T = 1$ year, and volatility $\sigma = \{15\%, 30\%, 50\%\}$ using

- Uniform mesh points on $S$
- Uniform mesh points on $\xi$ where

$$\xi = \frac{\sinh^{-1}((S-B)/\alpha) - c_2}{c_1 - c_2}$$

where

$$\begin{aligned} c_1 &= \sinh^{-1}\left(\frac{S_{\max} - B}{\alpha}\right) \\ c_2 &= \sinh^{-1}\left(\frac{S_{\min} - B}{\alpha}\right) \\ \alpha &= \frac{S_{\max} - S_{\min}}{20} \end{aligned}$$

Compare approximated values (premium and delta) with the closed-form values.

---

## Case Studies

1. For the following parameters: spot price $S_0 = \$100$, strike price $K = \{90, 100, 110\}$, maturity $T = 0.25$, risk-free interest rate $r = 3.75\%$, volatility of volatility $\sigma = 30\%$, $\kappa = 1$, $\lambda = 0.5$, $\theta = 0.05$, $\rho = -0.75$, and $\nu_0 = 0.05$ apply the following methods to numerically solve the Heston PDE:

   - Fully implicit finite difference scheme
   - Alternative direction implicit (ADI) scheme
   - Fast Fourier transform technique
   - Monte Carlo simulation (with and without variance reduction)

   and compare your results.

2. We would like to investigate the sensitivity of the numerical solution of PDEs to the choice of $S_{min}$, $S_{max}$, and $\Delta S$. Consider the Black–Scholes PDE with the parameter set in problem 1.

| $\Delta S$ | Lower and Upper Boundaries | |
|---|---|---|
| | $S_{min}$ | $S_{max}$ |
| $\frac{1}{100}$ | 0 | 150 |
| | | 200 |
| | | 250 |
| | 25 | 150 |
| | | 200 |
| | | 250 |
| | 50 | 150 |
| | | 200 |
| | | 250 |
| $\frac{1}{25}$ | 0 | 150 |
| | | 200 |
| | | 250 |
| | 25 | 150 |
| | | 200 |
| | | 250 |
| | 50 | 150 |
| | | 200 |
| | | 250 |

| $\Delta S$ | Lower and Upper Boundaries | |
|---|---|---|
| | $S_{min}$ | $S_{max}$ |
| $\frac{1}{10}$ | 0 | 150 |
| | | 200 |
| | | 250 |
| | 25 | 150 |
| | | 200 |
| | | 250 |
| | 50 | 150 |
| | | 200 |
| | | 250 |
| 1 | 0 | 150 |
| | | 200 |
| | | 250 |
| | 25 | 150 |
| | | 200 |
| | | 250 |
| | 50 | 150 |
| | | 200 |
| | | 250 |

Price the Black–Scholes PDE numerically using the implicit scheme for each choice in the table above and compare the numerical results with the close-form solution and draw a conclusion on the sensitivity of results to the choice of $S_{min}$, $S_{max}$, and $\Delta S$.

3. We aim to investigate the effect of higher order discretization on premiums by focusing on a known case. Consider the following parameters: spot price $S_0 = 100$, strike price $K = 90$, risk-free rate $r = 0.25\%$, dividend yield $q = 1.25\%$, time to maturity $T = 1$ year, and volatility $\sigma = 50\%$, $S_{min} = 0$, $S_{max} = 250$, number of grid points in the price direction $N = 1000$ and number of grid points in the time direction $M = 250$. Solve the Black–Scholes PDE *numerically* to price a European put option by means of implicit finite differences using Neumann boundary conditions using

   (a) second order discretization on first and second derivatives which would yield to a tridiagonal stiffness matrix.

   (b) higher order discretization on first and second derivatives which would yield to a pentadiagonal stiffness matrix.

   Compare premiums with the close-form solution and conclude.

4. Is it ever worth using coordinate transformation in order to concentrate the mesh points on a specific price? What are the numerical differences between this considerably harder coordinate transformation and just using specially created points for the grid while using the regular PDE? Here we aim to investigate some special cases and their benefits by looking into some known cases.

   Use the following parameters, spot price $S_0 = 100$, strike price $K = 90$, risk-free rate $r = 0.25\%$, dividend yield $q = 1.25\%$, time to maturity $T = 1$ year, and volatility

$\sigma = 50\%$ and consider $S_{min} = 0$, $S_{max} = 250$, number of grid points in the price direction $N = 1000$ and number of grid points in the time direction $M = 250$. Solve the Black–Scholes PDE *numerically* to price a European put option by means of implicit finite differences using Neumann boundary conditions using

(a) Equidistant subintervals.

(b) Coordinate transformation by concentrating points around the strike.

Compare the prices and Greeks with the close-form solution and conclude.

Now price an up-and-out call with the barrier level of $H = 130$ using

(a) Equidistant subintervals.

(b) Coordinate transformation by concentrating points around the barrier.

Compare the prices and Greeks with the close-form solution and conclude.

# Chapter 5

# Derivative Pricing via Numerical Solutions of PIDEs

A number of authors have recently proposed the use of infinite activity pure jump Lévy processes for the process describing the dynamics of the asset's logarithmic price (Eberlein, Keller, and Prause [105], Barndorff–Nielsen and Shephard [26] and Madan, Carr, and Chang [175]). Further it is argued in Geman, Madan, and Yor [118] that such processes are the norm when it is recognized that time changes with martingale components describe price evolution. At an empirical level, Carr, Geman, Madan and Yor [54] present evidence supporting the view that in the presence of an infinite activity Lévy process one may effectively dispense with a diffusion component.

In this chapter we develop a procedure for pricing options when the underlying asset price dynamics are given by a pure jump infinity activity Lévy process. The method is illustrated for the case of the CGMY process introduced in [53] and discussed in Section 1.2.8. One may easily adapt to the variance gamma process introduced in Madan, Carr, and Chang [175] and other similar processes, including the class of jump diffusion models proposed in Bates [29] or Duffie, Pan, and Singleton [101].

## 5.1 Numerical Solution of PIDEs (a Generic Example)

We first derive a partial-integro differential equation (PIDE) in the value function of the claim, particularly suited to the proposed numerical implementation. We then demonstrate how the PIDE is discretized and develop a numerical scheme for computing an approximate solution to the PIDE. We do not provide an analysis establishing the unconditional stability of our algorithm. Nonetheless, we conjecture that our scheme is consistent, unconditionally stable, and convergent. This claim is based on computational observations and comparisons with the prices of European options using closed forms reported in Madan, Carr, and Chang [175] and [53]. Numerical results and convergence tables presented at the end of this chapter provide further supporting evidence.

An alternative to our approach is presented in [9] and [8]. In [9], the authors prove results on the continuity of the exercise boundary, on the principle of smooth fit (also known as high contact condition introduced by Paul A. Samuelson in [196]), and on the behavior of the exercise boundary near maturity as well. In [9], equations are discretized in space by the collocation method and in time by an explicit backward differentiation formula. They show their discretization is of second-order accuracy. An application of the fast Fourier transform gives the overall amount of work to be $O(MN \log N)$ which represents a fast method.[1]

[1] $N$ is the number of grid point in the space direction and $M$ is the number of grid points in the time direction

To gain some insight into the stability of their resulting discretization of the PIDE, they perform a Von Neumann (Fourier) stability analysis.

### 5.1.1 Derivation of the PIDE

Before we dive straight into the derivation of the PIDE for the CGMY process, we define Lévy processes in general and the Lévy–Khintchine representation. Lévy processes seem to be relatively new to financial engineering and mathematics of finance students and not used extensively in practice. We refer readers to [31] and [185] for further reading on Lévy processes and infinitely divisible distributions. They seem to really be essentials in terms of understanding pure jump and general Lévy models and the decomposition of any Lévy process into a Brownian, small Poisson jump and large Poisson jump component as well as a derivation of the form of the Lévy density. We refer reader to [77] for financial modeling with jump processes which is a good introduction to general Lévy models in finance.

A Lévy process is a stochastic process with stationary independent increments. The Lévy-Khintchine theorem provides a characterization of a Lévy process in terms of the characterization function of the process; that is, there exists a measure $\nu$ such that for all $u \in \mathbb{R}$ and $t$ non-negative

$$\mathbb{E}(e^{iuX_t}) = \exp(t\phi(u)) \tag{5.1}$$

where

$$\phi(u) = i\gamma u - \frac{1}{2}\sigma^2 u^2 + \int_{-\infty}^{+\infty} (e^{iuy} - 1 - iuy\mathbb{1}_{\{|y|\leq 1\}})d\nu(y) \tag{5.2}$$

Here $\gamma$ and $\sigma$ are real numbers, $\nu$ is a measure on $\mathbb{R}$ such that $\nu(\{0\}) = 0$, and $\int_{-\infty}^{+\infty} \min(1, x^2)d\nu(x)$ is bounded. Assume a Lévy process, $\{X_t\}_{t\geq 0}$, of the following form:

$$X_t = (r - q + \mu)t + Z_t \tag{5.3}$$

This process has a drift term controlled by $\mu$ and a pure jump component $\{Z_t\}_{t\geq 0}$. Here we focus on the case that the Lévy measure associated to the pure jump component can be written as $d\nu(y) = k(y)dy$, where $k(y)$ is defined as

$$\begin{aligned}
d\nu(y) &= k(y)dy \\
k(y) &= \frac{e^{-\lambda_p y}}{\nu y^{1+Y}}\mathbb{1}_{y>0} + \frac{e^{-\lambda_n |y|}}{\nu |y|^{1+Y}}\mathbb{1}_{y<0} \\
\lambda_p &= \left(\frac{\theta^2}{\sigma^4} + \frac{2}{\sigma^2\nu}\right)^{\frac{1}{2}} - \frac{\theta}{\sigma^2} \\
\lambda_n &= \left(\frac{\theta^2}{\sigma^4} + \frac{2}{\sigma^2\nu}\right)^{\frac{1}{2}} + \frac{\theta}{\sigma^2}
\end{aligned} \tag{5.4}$$

The variable $Y$ allows for control of the sign of large and small jumps. By raising $Y$ above zero, one may induce greater activity near zero and less activity further away from zero. There are also some critical values of $Y$ of interest.

- $Y = 1$ separates finite variation $Y < 1$ from $Y > 1$ infinite variation
- $Y = 0$ separates finite arrival rate $Y < 0$ from $Y > 0$ infinite arrival rate
- $Y = -1$ separates activity concentrated away from zero $Y < -1$ from $Y > -1$ activity concentrated at zero

Let function $V(S_t, t)$ be the value of derivative security. Applying Itô's lemma for semi-martingales [183] on $e^{r(T-t)}V(S_t, t)$ one gets

$$\begin{aligned}
V(S_T, T) &= e^{rT}V(S_0, 0) + \int_0^T e^{r(T-t)} \frac{\partial V}{\partial S}(S_t, t) dS_t \\
&+ \int_0^T e^{r(T-t)} \int_{-\infty}^{+\infty} \left[ V(S_{t-}e^x, t) - V(S_{t-}, t) - \frac{\partial V}{\partial S}(S_{t-}, t) S_{t-}(e^x - 1) \right] \mu(dx, dt) \\
&+ \int_0^T e^{r(T-t)} \left[ \frac{\partial V}{\partial t}(S_t, t) - rV(S_t, t) \right] dt \\
&= e^{rT}V(S_0, 0) + \int_0^T e^{r(T-t)} \frac{\partial V}{\partial S}(S_t, t) \left[ dS_t - (r - q)S_t dt \right] \\
&+ \int_0^T e^{r(T-t)} \int_{-\infty}^{+\infty} \left[ V(S_{t-}e^x, t) - V(S_{t-}, t) - \frac{\partial V}{\partial S}(S_{t-}, t) S_{t-}(e^x - 1) \right] \mu(dx, dt) \\
&+ \int_0^T e^{r(T-t)} \left[ \frac{\partial V}{\partial t}(S_t, t) + (r - q)S_t \frac{\partial V}{\partial S}(S_t, t) - rV(S_t, t) \right] dt
\end{aligned}$$

where $\mu(dx, dt)$ is the integer valued random measure which counts the number of jumps in any region of space-time. The density $\nu(dy)dt$ is the compensator of $\mu(dx, dt)$ [150]. Add and subtract the following term to the above equation:

$$\int_0^T e^{r(T-t)} \int_{-\infty}^{+\infty} \left[ V(S_{t-}e^y, t) - V(S_{t-}, t) - \frac{\partial V}{\partial S}(S_{t-}, t) S_{t-}(e^y - 1) \right] \nu(dy)dt$$

to get

$$\begin{aligned}
V(S_T, T) &= V(S_0, 0)e^{rT} + \int_0^T e^{r(T-t)} \frac{\partial V}{\partial S}(S_t, t) \left[ dS_t - (r - q)S_t dt \right] \\
&+ \int_0^T e^{r(T-t)} \int_{-\infty}^{+\infty} \left[ V(S_t e^y, t) - V(S_t, t) - \frac{\partial V}{\partial S}(S_t, t) S_t(e^x - 1) \right] \left[ \mu(dx, dt) - \nu(dy)dt \right] \\
&+ \int_0^T e^{r(T-t)} \int_{-\infty}^{+\infty} \left[ V(S_t e^x, t) - V(S_{t-}, t) - \frac{\partial V}{\partial S}(S_t, t) S_{t-}(e^x - 1) \right] \nu(du)dt \\
&+ \int_0^T e^{r(T-t)} \left[ \frac{\partial V}{\partial t}(S_t, t) + (r - q)S_t \frac{\partial V}{\partial S}(S_t, t) - rV(S_t, t) \right] dt
\end{aligned}$$

Now taking the expectation under $\mathbb{Q}$ we will get

$$\begin{aligned}
\mathbb{E}^{\mathbb{Q}} V(S_T, T) &= V(S_0, 0)e^{rT} \\
&+ \int_0^T e^{r(T-t)} \left\{ \int_{-\infty}^{+\infty} \left[ V(S_t e^y, t) - V(S_t, t) - \frac{\partial V}{\partial S}(S_t, t) S_{t-}(e^y - 1) \right] \nu(dy) \right. \\
&+ \left. \frac{\partial V}{\partial t}(S_t, t) + (r - q)S_t \frac{\partial V}{\partial S}(S_t, t) - rV(S_t, t) \right\} dt
\end{aligned}$$

We know that

$$\mathbb{E}^{\mathbb{Q}} \left( V(S_T, T) \right) = V(S_0, 0)e^{rT}$$

Therefore

$$\begin{aligned}
\int_0^T e^{r(T-t)} \left\{ \int_{-\infty}^{+\infty} \left[ V(S_{t-}e^y, t) - V(S_{t-}, t) - \frac{\partial V}{\partial S}(S_{t-}, t) S_{t-}(e^y - 1) \right] \nu(dy) \right. \\
+ \left. \frac{\partial V}{\partial t}(S_t, t) + (r - q)S_t \frac{\partial V}{\partial S}(S_t, t) - rV(S_t, t) \right\} dt &= 0
\end{aligned}$$

Since the integrand is non-negative, that implies

$$\begin{aligned}
&\int_{-\infty}^{+\infty}\left[V(S_{t-}e^y,t)-V(S_{t-},t)-\frac{\partial V}{\partial S}(S_{t-},t)S_{t-}(e^y-1)\right]\nu(dy)\\
+&\ \frac{\partial V}{\partial t}(S_t,t)+(r-q)S_t\frac{\partial V}{\partial S}(S_t,t)-rV(S_t,t)=0
\end{aligned}$$

Note that the PIDE is pretty generic for any Lévy density $\nu(dy)$. Writing $\nu(dy)=k(y)dy$ we get

$$\begin{aligned}
\int_{-\infty}^{\infty}\left[V(S_{t-}e^y,t)-V(S_{t-},t)-\frac{\partial V}{\partial S}(S_{t-},t)S_{t-}(e^y-1)\right]k(y)dy&\\
+\frac{\partial V}{\partial t}(S_t,t)+(r-q)S_t\frac{\partial V}{\partial S}(S_t,t)-rV(S_t,t)&=0 \qquad (5.5)
\end{aligned}$$

which is the partial-integro differential equation (PIDE) we are going to solve numerically for the Lévy density in Equation (5.4). By making the change of variables $x=\ln S$ and $\tau=T-t$ we obtain the following PIDE, as a function of $w(x,\tau)$:

$$\begin{aligned}
\frac{\partial w}{\partial \tau}(x,\tau)-(r-q)\frac{\partial w}{\partial x}(x,\tau)+rw(x,\tau)&\\
-\int_{-\infty}^{\infty}\left[w(x+y,\tau)-w(x,\tau)-\frac{\partial w}{\partial x}(x,\tau)(e^y-1)\right]k(y)dy&=0 \qquad (5.6)
\end{aligned}$$

noting

$$\begin{aligned}
w(x,\tau)&=V(S,t)\\
\frac{\partial^2 w}{\partial x^2}(x,\tau)-\frac{\partial w}{\partial x}(x,\tau)&=S^2\frac{\partial^2 V}{\partial S^2}(S,t)\\
\frac{\partial w}{\partial x}(x,\tau)&=S\frac{\partial V}{\partial S}(S,t)\\
\frac{\partial w}{\partial \tau}(x,\tau)&=-\frac{\partial V}{\partial t}(S,t)\\
w(x+y,\tau)&=V(Se^y,t)
\end{aligned}$$

For European vanilla options, this PIDE must be solved subject to the initial condition

$$w(x,0)=(K-e^x)^+ \qquad (5.7)$$

for a put or

$$w(x,0)=(e^x-K)^+ \qquad (5.8)$$

for a call and the following Neumann boundary conditions:

$$\lim_{x\downarrow-\infty}\frac{\partial^2 w}{\partial x^2}(x,\tau)-\frac{\partial w}{\partial x}(x,\tau)=0\quad\forall\tau \qquad (5.9)$$

$$\lim_{x\uparrow+\infty}\frac{\partial^2 w}{\partial x^2}(x,\tau)-\frac{\partial w}{\partial x}(x,\tau)=0\quad\forall\tau \qquad (5.10)$$

Before starting to discretize the PIDE in (5.6) to set up the difference equation at a grid point, we look into the evaluation of the integral term. There are various approaches to evaluating the integral

$$\int_{-\infty}^{\infty}\left[w(x+y,\tau)-w(x,\tau)-\frac{\partial w}{\partial x}(x,\tau)(e^y-1)\right]k(y)dy$$

The Lévy measure $k(y)dy$ is singular at $y = 0$ but this does not lead to any integrability problems in either the analytical result or the related numerical evaluation. For the purpose of evaluating the integral, the domain of integration is first divided into two regions: (a) $|y| > \epsilon$, (b) $|y| <= \epsilon$ and we write it as

$$\int_{-\infty}^{\infty} \left[ w(x+y,\tau) - w(x,\tau) - \frac{\partial w}{\partial x}(x,\tau)(e^y - 1) \right] k(y)dy \tag{5.11}$$

$$= \int_{|y|<=\epsilon} \left[ w(x+y,\tau) - w(x,\tau) - \frac{\partial w}{\partial x}(x,\tau)(e^y - 1) \right] k(y)dy \tag{5.12}$$

$$+ \int_{|y|>\epsilon} \left[ w(x+y,\tau) - w(x,\tau) - \frac{\partial w}{\partial x}(x,\tau)(e^y - 1) \right] k(y)dy \tag{5.13}$$

Note that the obvious choice for $\epsilon$ is to set it equal to the grid size that is $\Delta x$. The Lévy measure $k(y)dy$ for different processes behaves differently near zero. For instance, the Lévy measure for the CGMY process approaches infinity much faster than the variance gamma process. For this reason we should have the highest possible order of approximation for the integrand near zero, namely,

$$w(x+y,\tau) - w(x,\tau) - \frac{\partial w}{\partial x}(x,\tau)(e^y - 1)$$

(a) For the region $|y| \le \epsilon$ we add and subtract $y\frac{\partial w}{\partial x}(x,\tau)$.

$$\int_{|y|\le\epsilon} \left[ w(x+y,\tau) - w(x,\tau) - \frac{\partial w}{\partial x}(x,\tau)(e^y - 1) \right] k(y)dy$$

$$= \int_{|y|<=\epsilon} \left[ w(x+y,\tau) - w(x,\tau) - y\frac{\partial w}{\partial x}(x,\tau) - \frac{\partial w}{\partial x}(x,\tau)(e^y - 1 - y) \right] k(y)dy \tag{5.14}$$

We can write the following two expansions for $w(x+y,\tau)$ and $e^y$ using the Taylor expansion to get

$$w(x+y,\tau) = w(x,\tau) + y\frac{\partial w}{\partial x}(x,\tau) + \frac{y^2}{2}\frac{\partial^2 w}{\partial x^2}(x,\tau) + O(y^3)$$

and

$$e^y = 1 + y + \frac{y^2}{2} + O(y^3)$$

Inserting these two expansions into Equation (5.14) we can write it as

$$\int_{|y|<=\epsilon} \left[ w(x+y,\tau) - w(x,\tau) - y\frac{\partial w}{\partial x}(x,\tau) - \frac{\partial w}{\partial x}(x,\tau)(e^y - 1 - y) \right] k(y)dy$$

$$= \int_{|y|\le\epsilon} \left[ \frac{y^2}{2}\frac{\partial^2 w}{\partial x^2}(x,\tau) - \frac{y^2}{2}\frac{\partial w}{\partial x}(x,\tau) + O(y^3) \right] k(y)dy$$

$$\approx \int_{|y|\le\epsilon} \left[ \frac{y^2}{2}\frac{\partial^2 w}{\partial x^2}(x,\tau) - \frac{y^2}{2}\frac{\partial w}{\partial x}(x,\tau) \right] k(y)dy \tag{5.15}$$

Define

$$\sigma^2(\epsilon) = \int_{|y|\le\epsilon} y^2 k(y)dy$$

We get the following:

$$\int_{|y|\le\epsilon}\left[w(x+y,\tau)-w(x,\tau)-\frac{\partial w}{\partial x}(x,\tau)(e^y-1)\right]k(y)dy$$
$$\approx \frac{1}{2}\sigma^2(\epsilon)\frac{\partial^2 w}{\partial x^2}(x,\tau)-\frac{1}{2}\sigma^2(\epsilon)\frac{\partial w}{\partial x}(x,\tau) \qquad (5.16)$$

(b) For the region $|y|>\epsilon$

$$\int_{|y|>\epsilon}\left[w(x+y,\tau)-w(x,\tau)-\frac{\partial w}{\partial x}(x,\tau)(e^y-1)\right]k(y)dy$$
$$= \int_{|y|>\epsilon}(w(x+y,\tau)-w(x,\tau))k(y)dy-\frac{\partial w}{\partial x}(x,\tau)\int_{|y|>\epsilon}(e^y-1)k(y)dy$$
$$= \int_{|y|>\epsilon}(w(x+y,\tau)-w(x,\tau))k(y)dy+\frac{\partial w}{\partial x}(x,\tau)\omega(\epsilon) \qquad (5.17)$$

where

$$\omega(\epsilon)=\int_{|y|>\epsilon}(1-e^y)k(y)dy \qquad (5.18)$$

Putting it all back into Equation (5.6) we get

$$\frac{\partial w}{\partial \tau}(x,\tau)-\frac{1}{2}\sigma^2(\epsilon)\frac{\partial^2 w}{\partial x^2}(x,\tau)-\left(r-q+\omega(\epsilon)-\frac{1}{2}\sigma^2(\epsilon)\right)\frac{\partial w}{\partial x}(x,\tau)$$
$$+rw(x,\tau)-\int_{|y|>\epsilon}(w(x+y,\tau)-w(x,\tau))k(y)dy = 0. \qquad (5.19)$$

### 5.1.2 Discretization

In our finite difference discretization of the PIDE, we employ a mixture of two methods in dealing with the integral term ([135], [140]). On the evaluation of this integral, we expand the integrand near its singularity of $y=0$ and treat this part implicitly. The rest of the integral with some exceptions is treated fully explicitly. The differential term of the PIDE is discretized by a fully implicit approach.

Our treatment of the integral term is critical to attaining an unconditionally stable scheme. This observation is also consistent with findings in the diffusion case on noting that the infinitely occurring small jumps essentially behave like a diffusion. The fact that the terms near $y=0$ are treated implicitly is the rationale behind the stability of the scheme. In fact a fully explicit treatment of the integral term would only be conditionally stable. On the other hand, a fully implicit treatment of the integral would be computationally expensive.

For an option with maturity $T$, we consider $M$ equal sub-intervals in the $\tau$-direction. For the $x$-direction we assume $N$ equal sub-intervals on $[x_{\min},x_{\max}]$. Thus, we have the following mesh on $[x_{\min},x_{\max}]\times[0,T]$:

$$D=\left\{\begin{array}{lll} x_i=x_{min}+i\Delta x; & \Delta x=\frac{x_{max}-x_{min}}{N}; & i=0,\ldots,N \\ \tau_j=0+j\Delta\tau; & \Delta\tau=\frac{T-0}{M}; & j=0,\ldots,M \end{array}\right\}$$

A sample grid point on this mesh is $(x_i,\tau_j)\in\mathbb{R}\times\mathbb{R}^+$. Let $w_{i,j}$ be the approximation of

$w(x_i, \tau_j)$ on $D$. Now we start forming the difference equation at the grid point $(x_i, \tau_{j+1})$. As usual the assumption is the discrete values $w_{i,j}$ at time $\tau_j$ are known and we are solving for $w_{i,j+1}$ at time $\tau_{j+1}$. Using a first order finite difference approximation for $\frac{\partial w}{\partial \tau}$, a second order approximation for $\frac{\partial^2 w}{\partial x^2}$, a central difference for $\frac{\partial w}{\partial x}$, and setting $\epsilon$ to $\Delta x$ we obtain the following discrete equation at point $(x_i, \tau_{j+1})$ after dropping the approximation order terms.

$$\begin{aligned}\frac{1}{\Delta\tau}(w_{i,j+1} - w_{i,j}) &- \frac{1}{2(\Delta x)^2}\sigma^2(\Delta x)(w_{i+1,j+1} - 2w_{i,j+1} + w_{i-1,j+1}) \\ -(r - q + \omega(\Delta x) &- \frac{1}{2}\sigma^2(\Delta x))\frac{1}{2\Delta x}(w_{i+1,j+1} - w_{i-1,j+1}) + rw_{i,j+1} \\ &- \int_{|y|>\Delta x} (w(x_i + y, \tau_j) - w(x_i, \tau_j))\, k(y)dy \cong 0\end{aligned}$$

The solution algorithm that is being developed is correct up to a first order in $\Delta\tau$. Equivalently,

$$\begin{aligned}&- \left(\frac{\sigma^2(\Delta x)\Delta\tau}{2\Delta x^2} - \left(r - q + \omega(\Delta x) - \frac{1}{2}\sigma^2(\Delta x)\right)\frac{\Delta\tau}{2\Delta x}\right) w_{i-1,j+1} \\ &+ \left(1 + r\Delta\tau + \sigma^2(\Delta x)\frac{\Delta\tau}{\Delta x^2}\right) w_{i,j+1} \\ &- \left(\frac{\sigma^2(\Delta x)\Delta\tau}{2\Delta x^2} + \left(r - q + \omega(\Delta x) - \frac{1}{2}\sigma^2(\Delta x)\right)\frac{\Delta\tau}{2\Delta x}\right) w_{i+1,j+1} \\ &= w_{i,j} + \Delta\tau \int_{|y|>\Delta x} (w(x_i + y, \tau_j) - w(x_i, \tau_j))\, k(y)dy\end{aligned}$$

or in short we write the difference equation as

$$\begin{aligned}&- B_l w_{i-1,j+1} + (1 + r\Delta\tau + B_l + B_u)w_{i,j+1} - B_u w_{i+1,j+1} \\ &= w_{i,j} + \Delta\tau \int_{|y|>\Delta x} (w(x_i + y, \tau_j) - w(x_i, \tau_j))\, k(y)dy\end{aligned}$$

with

$$\begin{aligned}B_l &= \frac{\sigma^2(\Delta x)\Delta\tau}{2\Delta x^2} - \left(r - q + \omega(\Delta x) - \frac{1}{2}\sigma^2(\Delta x)\right)\frac{\Delta\tau}{2\Delta x} \\ B_u &= \frac{\sigma^2(\Delta x)\Delta\tau}{2\Delta x^2} + \left(r - q + \omega(\Delta x) - \frac{1}{2}\sigma^2(\Delta x)\right)\frac{\Delta\tau}{2\Delta x}\end{aligned}$$

where $w_{i,0} = (K - e^{x_i})^+$.

### 5.1.3 Evaluation of the Integral Term

For numerical evaluation of the integral[2] for $|y| > \Delta x$ we divide it into four sub-intervals and write it as

$$\begin{aligned}\int_{|y|>\Delta x} (w(x_i+y,\tau_j)-w_{i,j})\,k(y)dy &= \int_{-\infty}^{x_0-x_i} (w(x_i+y,\tau_j)-w_{i,j})\,k(y)dy \\ &+ \int_{x_0-x_i}^{-\Delta x} (w(x_i+y,\tau_j)-w_{i,j})\,k(y)dy \\ &+ \int_{+\Delta x}^{x_N-x_i} (w(x_i+y,\tau_j)-w_{i,j})\,k(y)dy \\ &+ \int_{x_N-x_i}^{\infty} (w(x_i+y,\tau_j)-w_{i,j})\,k(y)dy\end{aligned}$$

The rationale is that in the region $|y| > \Delta x$ we have two sub-regions: (a) $y < -\Delta x$, (b) $y > \Delta x$. For $y < -\Delta x$, in order for the quantity $x_i + y$ to be inside the grid we should have $y > x_{min} - x_i$, which is the same as $y > x_0 - x_i = -i\Delta x$. For $y < x_0 - x_i$, the quantity $x_i + y$ would be outside the grid. For $y > \Delta x$, in order for $x_i + y$ to be inside the grid we should have $y < x_{max} - x_i$, which is the same as $y < x_N - x_i = (N-i)\Delta x$. For $y > x_N - x_i$, the quantity $x_i + y$ would be outside the grid.
For $y \in (x_0 - x_i, -\Delta x)$, we do the following:

$$\begin{aligned}\int_{x_0-x_i}^{-\Delta x} (w(x_i+y,\tau_j)-w_{i,j})\,k(y)dy &= \int_{x_0-x_i}^{-\Delta x} (w(x_i+y,\tau_j)-w_{i,j})\,\frac{e^{-\lambda_n|y|}}{\nu|y|^{1+Y}}dy \\ &= \sum_{k=1}^{i-1}\int_{k\Delta x}^{(k+1)\Delta x} (w(x_i-y,\tau_j)-w_{i,j})\,\frac{e^{-\lambda_n y}}{\nu y^{1+Y}}dy\end{aligned}$$

Using linear interpolation on interval $y \in [k\Delta x, (k+1)\Delta x]$, we can write $w(x_i - y, \tau_j)$ as follows:

$$w(x_i-y,\tau_j) \cong w_{i-k,j} + \frac{w_{i-k-1,j}-w_{i-k,j}}{\Delta x}(y-k\Delta x)$$

and therefore we obtain the following:

$$\begin{aligned}&\int_{x_0-x_i}^{-\Delta x} (w(x_i+y,\tau_j)-w_{i,j})\,k(y)dy \\ &= \sum_{k=1}^{i-1}\int_{k\Delta x}^{(k+1)\Delta x}\left(w_{i-k,j}+\frac{w_{i-k-1,j}-w_{i-k,j}}{\Delta x}(y-k\Delta x)-w_{i,j}\right)\frac{e^{-\lambda_n y}}{\nu y^{1+Y}}dy \\ &= \sum_{k=1}^{i-1}\frac{1}{\nu}(w_{i-k,j}-w_{i,j}-k(w_{i-k-1,j}-w_{i-k,j}))\left\{\int_{k\Delta x}^{(k+1)\Delta x}\frac{e^{-\lambda_n y}}{y^{1+Y}}dy\right\} \\ &+ \sum_{k=1}^{i-1}\frac{w_{i-k-1,j}-w_{i-k,j}}{\nu\Delta x}\left(\int_{k\Delta x}^{(k+1)\Delta x}\frac{e^{-\lambda_n y}}{y^{Y}}dy\right)\end{aligned}$$

[2] As stated earlier, an alternative to this approach is to use fast Fourier transform to numerically evaluate the integral as suggested in [9] and [8]. We leave it as a case study at the end of this chapter.

By change of variable we get

$$\begin{aligned}
&\int_{x_0-x_i}^{-\Delta x} (w(x_i+y,\tau_j)-w_{i,j})\,k(y)dy \\
&= \sum_{k=1}^{i-1} \frac{\lambda_n^Y}{\nu}\left(w_{i-k,j}-w_{i,j}-k(w_{i-k-1,j}-w_{i-k,j})\right)\left\{\int_{k\Delta x\lambda_n}^{(k+1)\Delta x\lambda_n} \frac{e^{-z}}{z^{1+Y}}dz\right\} \\
&+ \sum_{k=1}^{i-1} \frac{w_{i-k-1,j}-w_{i-k,j}}{\nu\lambda_n^{1-Y}\Delta x}\left(\int_{k\Delta x\lambda_n}^{(k+1)\Delta x\lambda_n} \frac{e^{-z}}{z^{Y}}dz\right) \\
&= \sum_{k=1}^{i-1} \frac{\lambda_n^Y}{\nu}\left(w_{i-k,j}-w_{i,j}-k(w_{i-k-1,j}-w_{i-k,j})\right)\left\{g_2(k\lambda_n\Delta x)-g_2((k+1)\lambda_n\Delta x)\right\} \\
&+ \sum_{k=1}^{i-1} \frac{w_{i-k-1,j}-w_{i-k,j}}{\lambda_n^{1-Y}\nu\Delta x}\left(g_1(k\Delta x\lambda_n)-g_1((k+1)\Delta x\lambda_n)\right)
\end{aligned}$$

where[3]

$$g_1(\xi) = \int_\xi^\infty \frac{e^{-z}}{z^\alpha}dz \qquad (5.20)$$

$$g_2(\xi) = \int_\xi^\infty \frac{e^{-z}}{z^{\alpha+1}}dz \qquad (5.21)$$

for $0 \le \alpha < 1$.

For the case that $y \in (\Delta x, x_N - x_i)$ we write

$$\int_{\Delta x}^{x_N-x_i} (w(x_i+y,\tau_j)-w_{i,j})\,k(y)dy = \sum_{k=1}^{N-i-1}\int_{k\Delta x}^{(k+1)\Delta x} (w(x_i+y,\tau_j)-w_{i,j})\,\frac{e^{-\lambda_p y}}{\nu y^{1+Y}}dy$$

Similarly, for $y \in [k\Delta x, (k+1)\Delta x]$, a linear approximation yields

$$w(x_i+y,\tau_j) \cong w_{i+k,j} + \frac{w_{i+k+1,j}-w_{i+k,j}}{\Delta x}(y-k\Delta x)$$

and therefore we obtain the following:

$$\begin{aligned}
&\int_{\Delta x}^{x_N-x_i} (w(x_i+y,\tau_j)-w_{i,j})\,k(y)dy \\
&= \sum_{k=1}^{N-i-1}\int_{k\Delta x}^{(k+1)\Delta x}\left(w_{i+k,j}+\frac{w_{i+k+1,j}-w_{i+k,j}}{\Delta x}(y-k\Delta x)-w_{i,j}\right)\frac{e^{-\lambda_p y}}{\nu y^{1+Y}}dy \\
&= \sum_{k=1}^{N-i-1}\frac{1}{\nu}\left(w_{i+k,j}-w_{i,j}-k(w_{i+k+1,j}-w_{i+k,j})\right)\left\{\int_{k\Delta x}^{(k+1)\Delta x}\frac{e^{-\lambda_p y}}{y^{1+Y}}dy\right\} \\
&+ \sum_{k=1}^{N-i-1}\frac{w_{i+k+1,j}-w_{i+k,j}}{\nu\Delta x}\left(\int_{k\Delta x}^{(k+1)\Delta x}\frac{e^{-\lambda_p y}}{y^{Y}}dy\right) \\
&= \sum_{k=1}^{N-i-1}\frac{\lambda_p^Y}{\nu}\left(w_{i+k,j}-w_{i,j}-k(w_{i+k+1,j}-w_{i+k,j})\right)\left\{g_2(k\Delta x\lambda_p)-g_2((k+1)\Delta x\lambda_p)\right\} \\
&+ \sum_{k=1}^{N-i-1}\frac{w_{i+k+1,j}-w_{i+k,j}}{\lambda_p^{1-Y}\nu\Delta x}\left(g_1(k\Delta x\lambda_p)-g_1((k+1)\Delta x\lambda_p)\right)
\end{aligned}$$

[3]At the end of the chapter we explain in detail how to calculate $g_1(\xi)$ and $g_2(\xi)$.

For the region $y \in (-\infty, x_0 - x_i)$

$$\begin{aligned}\int_{-\infty}^{x_0-x_i} \left(w(x_i+y,\tau_j) - w_{i,j}\right) k(y)dy &= \int_{-\infty}^{x_0-x_i} \left(w(x_i+y,\tau_j) - w_{i,j}\right) \frac{e^{-\lambda_n|y|}}{\nu|y|^{1+Y}}dy \\ &= \int_{i\Delta x}^{\infty} \left(w(x_i-y,\tau_j) - w_{i,j}\right) \frac{e^{-\lambda_n y}}{\nu y^{1+Y}}dy\end{aligned}$$

We assume that $w(x_i - y, \tau_j) = K - e^{x_i - y}$ in the interval.[4] Hence

$$\begin{aligned}&\int_{-\infty}^{x_0-x_i} \left(w(x_i+y,\tau_j) - w_{i,j}\right) k(y)dy \\ &= \int_{i\Delta x}^{\infty} \left(K - e^{x_i-y} - w_{i,j}\right) \frac{e^{-\lambda_n y}}{\nu y^{1+Y}}dy \\ &= \frac{1}{\nu}(K - w_{i,j}) \int_{i\Delta x}^{\infty} \frac{e^{-\lambda_n y}}{y^{1+Y}}dy - \frac{1}{\nu}e^{x_i} \int_{i\Delta x}^{\infty} \frac{e^{-(\lambda_n+1)y}}{y^{1+Y}}dy \\ &= \frac{\lambda_n^Y}{\nu}(K - w_{i,j}) g_2(i\Delta x\lambda_n) - \frac{(\lambda_n+1)^Y}{\nu}e^{x_i} g_2(i\Delta x(\lambda_n+1)).\end{aligned}$$

For the region $y \in [(N-i)\Delta x, \infty)$, we assume that $w(x_i + y, \tau_j) = 0$.[5] Thus

$$\begin{aligned}\int_{x_N-x_i}^{\infty} \left(w(x_i+y,\tau_j) - w_{i,j}\right) k(y)dy &= \int_{(N-i)\Delta x}^{\infty} \left(w(x_i+y,\tau_j) - w_{i,j}\right) \frac{e^{-\lambda_p y}}{\nu y^{1+Y}}dy \\ &= -\frac{\lambda_p^Y}{\nu} w_{i,j} g_2((N-i)\Delta x\lambda_p).\end{aligned}$$

### 5.1.4 Difference Equation

Putting together all the terms, we obtain the following difference equation[6] at point $(x_i, \tau_{j+1})$:

$$l_{i,j+1}w_{i-1,j+1} + d_{i,j+1}w_{i,j+1} + u_{i,j+1}w_{i+1,j+1} = w_{i,j} + \frac{\Delta\tau}{\nu}R_{i,j}$$

where

$$\begin{aligned}l_{i,j+1} &= -B_l \\ d_{i,j+1} &= 1 + r\Delta\tau + B_l + B_u + \frac{\Delta\tau}{\nu}\left(\lambda_n^Y g_2(i\Delta x\lambda_n) + \lambda_p^Y g_2((N-i)\Delta x\lambda_p)\right) \\ u_{i,j+1} &= -B_u\end{aligned}$$

[4]We choose $x_0$ small enough such that $w(x_0, \tau_j) = K - e^{x_0}$ for all $j$. Thus it would be true for $w(x_i - y, \tau_j) = K - e^{x_i - y}$ as long as $x_i - y < x_0$. In the case of a European put option, we would assume that $w(x_i - y, \tau_j) = Ke^{-r\tau_j} - e^{x_i-y}e^{-q\tau_j}$.

[5]As explained before, $x_N$ is selected such that $w(x_N, \tau_j) \cong 0$. Therefore, the assumption $w(x_i+y, \tau_j) = 0$ is valid as long as $x_i + y > x_N$.

[6]In the case of $i = 1$ or $i = N - 1$, we impose the boundary conditions.

$$
\begin{aligned}
R_{i,j} &= \sum_{k=1}^{i-1} \lambda_n^Y \left(w_{i-k,j} - w_{i,j} - k(w_{i-k-1,j} - w_{i-k,j})\right) \{g_2(k\Delta x \lambda_n) - g_2((k+1)\Delta x \lambda_n)\} \\
&+ \sum_{k=1}^{i-1} \frac{w_{i-k-1,j} - w_{i-k,j}}{\lambda_n^{1-Y} \Delta x} \left(g_1(k\Delta x \lambda_n) - g_1((k+1)\Delta x \lambda_n)\right) \\
&+ \sum_{k=1}^{N-i-1} \lambda_p^Y \left(w_{i+k,j} - w_{i,j} - k(w_{i+k+1,j} - w_{i+k,j})\right) \{g_2(k\Delta x \lambda_p) - g_2((k+1)\Delta x \lambda_p)\} \\
&+ \sum_{k=1}^{N-i-1} \frac{w_{i+k+1,j} - w_{i+k,j}}{\lambda_p^{1-Y} \Delta x} \left(g_1(k\Delta x \lambda_p) - g_1((k+1)\Delta x \lambda_p)\right) \\
&+ K\lambda_n^Y g_2(i\Delta x \lambda_n) - e^{x_i} (\lambda_n + 1)^Y g_2(i\Delta x(\lambda_n + 1))
\end{aligned}
$$

and as before

$$
\begin{aligned}
B_l &= \frac{\sigma^2(\Delta x)\Delta\tau}{2\Delta x^2} - \left(r - q + \omega(\Delta x) - \frac{1}{2}\sigma^2(\Delta x)\right)\frac{\Delta\tau}{2\Delta x} \\
B_u &= \frac{\sigma^2(\Delta x)\Delta\tau}{2\Delta x^2} + \left(r - q + \omega(\Delta x) - \frac{1}{2}\sigma^2(\Delta x)\right)\frac{\Delta\tau}{2\Delta x}
\end{aligned}
$$

Assuming that at the completion of the time step $\tau_j$ the values $w_{i,j}$ have been computed, we solve a linear system of equations to find the values $w_{i,j+1}$ for all $i$. Notice that in this scheme, the following six vectors (precalculated) are stored:

- $g_1(k\Delta x \lambda_n)$ for $k = 1, \ldots, N$
- $g_1(k\Delta x \lambda_p)$ for $k = 1, \ldots, N$
- $g_2(\lambda_n k\Delta x)$ for $k = 1, \ldots, N$
- $g_2(\lambda_p k\Delta x)$ for $k = 1, \ldots, N$
- $g_2((\lambda_n + 1)k\Delta x)$ for $k = 1, \ldots, N$
- $g_2((\lambda_p - 1)k\Delta x)$ for $k = 1, \ldots, N$

Now, we can evaluate $\sigma^2(\epsilon)$ and $\omega(\epsilon)$ in terms of functions $g_1$ and $g_2$. For $\sigma^2(\epsilon)$ we have

$$
\begin{aligned}
\sigma^2(\epsilon) &= \int_{|y|\le\epsilon} y^2 k(y)dy \\
&= \int_{-\epsilon}^{0} y^2 \frac{e^{-\lambda_n|y|}}{\nu|y|^{1+Y}} dy \\
&+ \int_{0}^{\epsilon} y^2 \frac{e^{-\lambda_p y}}{\nu y^{1+Y}} dy
\end{aligned}
$$

For the first integral, we first do change of variable and then integration by parts and use the definition of $g_1$ to obtain

$$
\begin{aligned}
\int_{-\epsilon}^{0} y^2 \frac{e^{-\lambda_n |y|}}{\nu |y|^{1+Y}} dy &= \int_{0}^{\epsilon} u^2 \frac{e^{-\lambda_n u}}{\nu u^{1+Y}} du \\
&= \frac{1}{\nu} \int_0^{\epsilon} u^{1-Y} e^{-\lambda_n u} du \\
&= \frac{1}{\nu} \lambda_n^{Y-2} \int_0^{\lambda_p \epsilon} z^{1-Y} e^{-z} dz \\
&= \frac{1}{\nu} \lambda_n^{Y-2} \left( -(\lambda_n \epsilon)^{1-Y} e^{-\lambda_n \epsilon} + (1-Y) \int_0^{\lambda_n \epsilon} \frac{e^{-z}}{z^Y} dz \right) \\
&= \frac{1}{\nu} \lambda_n^{Y-2} \left( -(\lambda_n \epsilon)^{1-Y} e^{-\lambda_n \epsilon} + (1-Y)(g_1(0) - g_1(\lambda_n \epsilon)) \right)
\end{aligned}
$$

Similarly for the second integral

$$
\begin{aligned}
\int_{0}^{\epsilon} y^2 \frac{e^{-\lambda_p y}}{\nu y^{1+Y}} dy &= \frac{1}{\nu} \int_0^{\epsilon} y^{1-Y} e^{-\lambda_p y} dy \\
&= \frac{1}{\nu} \lambda_p^{Y-2} \int_0^{\lambda_p \epsilon} z^{1-Y} e^{-z} dz \\
&= \frac{1}{\nu} \lambda_p^{Y-2} \left( -(\lambda_p \epsilon)^{1-Y} e^{-\lambda_p \epsilon} + (1-Y) \int_0^{\lambda_p \epsilon} \frac{e^{-z}}{z^Y} dz \right) \\
&= \frac{1}{\nu} \lambda_p^{Y-2} \left( -(\lambda_p \epsilon)^{1-Y} e^{-\lambda_p \epsilon} + (1-Y)(g_1(0) - g_1(\lambda_p \epsilon)) \right)
\end{aligned}
$$

Therefore

$$
\begin{aligned}
\sigma^2(\epsilon) &= \frac{1}{\nu} \lambda_p^{Y-2} \left( -(\lambda_p \epsilon)^{1-Y} e^{-\lambda_p \epsilon} + (1-Y)(g_1(0) - g_1(\lambda_p \epsilon)) \right) \\
&+ \frac{1}{\nu} \lambda_n^{Y-2} \left( -(\lambda_n \epsilon)^{1-Y} e^{-\lambda_n \epsilon} + (1-Y)(g_1(0) - g_1(\lambda_n \epsilon)) \right)
\end{aligned}
$$

For $\omega(\epsilon)$ we have

$$
\begin{aligned}
\omega(\epsilon) &= \int_{|y|>\epsilon} (1-e^y) k(y) dy \\
&= \int_{-\infty}^{-\epsilon} (1-e^y) \frac{e^{-\lambda_n |y|}}{\nu |y|^{1+Y}} dy + \int_{\epsilon}^{\infty} (1-e^y) \frac{e^{-\lambda_p y}}{\nu y^{1+Y}} dy
\end{aligned}
\tag{5.22}
$$

The first quantity in (5.22) in terms of $g_2$ would be

$$
\begin{aligned}
\int_{\epsilon}^{\infty} (1-e^y) \frac{e^{-\lambda_p y}}{\nu y^{1+Y}} dy &= \frac{1}{\nu} \int_{\epsilon}^{\infty} \frac{e^{-\lambda_p y}}{y^{1+Y}} dy - \frac{1}{\nu} \int_{\epsilon}^{\infty} \frac{e^{-(\lambda_p - 1)y}}{y^{1+Y}} dy \\
&= \frac{\lambda_p^Y}{\nu} \int_{\lambda_p \epsilon}^{\infty} \frac{e^{-z}}{z^{1+Y}} dz - \frac{(\lambda_p - 1)^Y}{\nu} \int_{(\lambda_p - 1)\epsilon}^{\infty} \frac{e^{-z}}{z^{1+Y}} dz \\
&= \frac{\lambda_p^Y}{\nu} g_2(\lambda_p \epsilon) - \frac{(\lambda_p - 1)^Y}{\nu} g_2((\lambda_p - 1)\epsilon)
\end{aligned}
$$

The second quantity in (5.22) in terms of $g_2$ would be

$$\begin{aligned}
\int_{-\infty}^{-\epsilon}(1-e^{y})\frac{e^{-\lambda_n|y|}}{\nu|y|^{1+Y}}dy &= \int_{\epsilon}^{\infty}(1-e^{-x})\frac{e^{-\lambda_n x}}{\nu x^{1+Y}}dx \\
&= \frac{1}{\nu}\int_{\epsilon}^{\infty}\frac{e^{-\lambda_n x}}{x^{1+Y}}dx - \frac{1}{\nu}\int_{\epsilon}^{\infty}\frac{e^{-(\lambda_n+1)x}}{x^{1+Y}}dx \\
&= \frac{\lambda_n^Y}{\nu}\int_{\lambda_n\epsilon}^{\infty}\frac{e^{-z}}{z^{1+Y}}dz - \frac{(\lambda_n+1)^Y}{\nu}\int_{(\lambda_n+1)\epsilon}^{\infty}\frac{e^{-z}}{z^{1+Y}}dz \\
&= \frac{\lambda_n^Y}{\nu}g_2(\lambda_n\epsilon) - \frac{(\lambda_n+1)^Y}{\nu}g_2((\lambda_n+1)\epsilon)
\end{aligned}$$

The first equality follows from setting $x = |y| = -y$. Therefore (5.22) in terms of $g_2$ becomes

$$\omega(\epsilon) = \frac{\lambda_p^Y}{\nu}g_2(\lambda_p\epsilon) - \frac{(\lambda_p-1)^Y}{\nu}g_2((\lambda_p-1)\epsilon) + \frac{\lambda_n^Y}{\nu}g_2(\lambda_n\epsilon) - \frac{(\lambda_n+1)^Y}{\nu}g_2((\lambda_n+1)\epsilon)$$

#### 5.1.4.1 Implementing Neumann Boundary Conditions

Rewriting boundary conditions (5.9) and (5.10) once more

$$\begin{aligned}
\lim_{x\downarrow-\infty}\frac{\partial^2 w}{\partial x^2}(x,\tau) - \frac{\partial w}{\partial x}(x,\tau) &= 0 \quad \forall\tau \\
\lim_{x\uparrow+\infty}\frac{\partial^2 w}{\partial x^2}(x,\tau) - \frac{\partial w}{\partial x}(x,\tau) &= 0 \quad \forall\tau
\end{aligned}$$

Discretization of it yields

$$\frac{w_{i-1,j+1} - 2w_{i,j+1} + w_{i+1,j+1}}{h^2} - \frac{w_{i+1,j+1} - w_{i-1,j+1}}{2h} = 0$$

Or equivalently

$$(1+\frac{h}{2})w_{i-1,j+1} - 2w_{i,j+1} + (1-\frac{h}{2})w_{i+1,j+1} = 0 \tag{5.23}$$

Now in our case, $x_0$ and $x_N$ are boundary points. Applying 5.23 at $i=1$ we can solve for $w_{0,j+1}$, the value at $i=0$, as

$$(1+\frac{h}{2})w_{0,j+1} - 2w_{1,j+1} + (1-\frac{h}{2})w_{2,j+1} = 0$$

and solving for $w_{0,j+1}$ we get

$$w_{0,j+1} = \frac{2}{1+\frac{h}{2}}w_{1,j+1} - \frac{1-\frac{h}{2}}{1+\frac{h}{2}}w_{2,j+1} \tag{5.24}$$

Applying 5.23 at $i=N-1$ we can solve for $w_{N,j+1}$, the value at $i=N$, as

$$(1+\frac{h}{2})w_{N-2,j+1} - 2w_{N-1,j+1} + (1-\frac{h}{2})w_{N,j+1} = 0$$

and solving for $w_{N,j+1}$ we get

$$w_{N,j+1} = -\frac{1+\frac{h}{2}}{1-\frac{h}{2}}w_{N-2,j+1} + \frac{2}{1-\frac{h}{2}}w_{N-1,j+1} \tag{5.25}$$

Assume at $(x_j, \tau_{k+1})$ the difference equation (loosely speaking) looks like

$$l_{i,j+1}w_{i-1,j+1} + d_{i,j+1}w_{i,j+1} + u_{i,j+1}w_{i+1,j+1} = w_{i,j} + r.h.s.$$

where $l_{i,j+1}, d_{i,j+1}, u_{i,j+1}$ are lower-diagonal, diagonal, and upper-diagonal elements of the stiffness matrix, respectively.
Then for $j = 1$ we get

$$l_{1,j+1}w_{0,j+1} + d_{1,j+1}w_{1,j+1} + u_{1,j+1}w_{2,j+1} = w_{1,j} + r.h.s.$$

Substituting (5.24) we obtain

$$l_{1,j+1}\left(\frac{2}{1+\frac{h}{2}}w_{1,j+1} - \frac{1-\frac{h}{2}}{1+\frac{h}{2}}w_{2,j+1}\right) + d_{1,j+1}w_{1,j+1} + u_{1,j+1}w_{2,j+1} = w_{1,j} + r.h.s.$$

and gathering terms we get

$$\left(\frac{2}{1+\frac{h}{2}}l_{1,j+1} + d_{1,j+1}\right)w_{1,j+1} + \left(u_{1,j+1} - \frac{1-\frac{h}{2}}{1+\frac{h}{2}}l_{1,j+1}\right)w_{2,j+1} = w_{1,k} + r.h.s.$$

For $j = N-1$ we have

$$l_{N-1,j+1}w_{N-2,j+1} + d_{N-1,j+1}w_{N-1,j+1} + u_{N-1,j+1}w_{N,j+1} = w_{N-1,k} + r.h.s.$$

Substituting (5.25) we obtain

$$\begin{aligned} l_{N-1,j+1}w_{N-2,j+1} + d_{N-1,j+1}w_{N-1,j+1} + u_{N-1,j+1}\left(-\frac{1+\frac{h}{2}}{1-\frac{h}{2}}w_{N-2,j+1} + \frac{2}{1-\frac{h}{2}}w_{N-1,j+1}\right) \\ = \quad w_{N-1,k} + r.h.s. \end{aligned}$$

and gathering terms we get

$$\begin{aligned} \left(l_{N-1,j+1} - \frac{1+\frac{h}{2}}{1-\frac{h}{2}}u_{N-1,j+1}\right)w_{N-2,j+1} + \left(d_{N-1,j+1} + \frac{2}{1-\frac{h}{2}}u_{N-1,j+1}\right)w_{N-1,j+1} \\ = \quad w_{N-1,k} + r.h.s. \end{aligned}$$

---

## 5.2 American Options

As in the diffusion framework, we can price an American option by

a. applying the Bermudan approach at each time step

b. applying the Brennan–Schwartz algorithm

c. or the synthetic dividend process

Here we just explain the last approach. As previously discussed, the PIDE in (5.19) holds in the continuation region. In the exercise region, we know the value function is $w(x,t) = K - e^x$. Now, we can extend the PIDE to the entire region by first applying the infinitesimal generator to this known value function in the exercise region. That yields the equation $\mathcal{L}w = \delta(x)$ where in particular we write that

$$\begin{aligned}\mathcal{L}w &= \frac{\partial w}{\partial \tau}(x,\tau) - (r-q)\frac{\partial w}{\partial x}(x,\tau) + rw(x,\tau) \\ &\quad - \int_{-\infty}^{\infty}\left[w(x+y,\tau) - w(x,\tau) - \frac{\partial w}{\partial x}(x,\tau)(e^y - 1)\right]k(y)dy \\ &= \delta(x)\end{aligned}$$

The function $\delta(x)$ is often called the *dividend process*. This is best seen using the fact that $w(x,\tau) = K - e^x$ for $x \leq x(\tau)$. Therefore in the exercise region we have

$$\begin{aligned}w(x,\tau) &= K - e^x \\ \frac{\partial w}{\partial \tau}(x,\tau) &= 0 \\ \frac{\partial w}{\partial x}(x,\tau) &= -e^x\end{aligned}$$

Before substituting those values we should note that $w(x+y,\tau)$ is not known for the case that $x + y > x(\tau)$, which means we are in the continuation region. Therefore we divide the integral into two regions: (a) $x + y \leq x(\tau)$, and (b) $x + y > x(\tau)$. Or equivalently considering that $y$ in the integral runs from $-\infty$ to $+\infty$, we write it as (a) $y \leq x(\tau) - x$, and (b) $y > x(\tau) - x$. For the case that $y \leq x(\tau) - x$, the integrand vanishes.

$$\begin{aligned}w(x+y,\tau) - w(x,\tau) - \frac{\partial w}{\partial x}(x,\tau)(e^y - 1) &= (K - e^{x+y}) - (K - e^x) - (-e^x)(e^y - 1) \\ &= 0\end{aligned}$$

For the case that $y > x(\tau) - x$, the integrand becomes

$$\begin{aligned}w(x+y,\tau) - w(x,\tau) - \frac{\partial w}{\partial x}(x,\tau)(e^y - 1) &= w(x+y,\tau) - (K - e^x) - (-e^x)(e^y - 1) \\ &= w(x+y,\tau) - (K - e^{x+y})\end{aligned}$$

and hence in the exercise region we get the following for the dividend process:

$$\begin{aligned}\delta(x) &= 0 - (r-q)(-e^x) + r(K - e^x) \\ &\quad - \int_{x(\tau)-x}^{\infty}\left[w(x+y,\tau) - (K - e^{x+y})\right]k(y)dy \\ &= rK - qe^x - \int_{x(\tau)-x}^{\infty}\left[w(x+y,\tau) - (K - e^{x+y})\right]k(y)dy \qquad (5.26)\end{aligned}$$

This is consistent with the demonstration by Carr, Jarrow, and Myneni [58]. Namely, one must extract from the American option holder the interest on the strike less the dividend yield for the time the stock spends in the exercise region to get the value back to that of a European option. For a jump process, this amount is further reduced by the expected shortfall that the stop-loss-start-gain strategy may experience on account of jumping back into the continuation region, as explained further in Gukhal [125].

Substituting the dividend definition of Equation (5.26) back into the PIDE we obtain the PIDE in $w(x,t)$ over the entire region as

$$\begin{aligned} &\frac{\partial w}{\partial \tau}(x,\tau) - (r-q)\frac{\partial w}{\partial x}(x,\tau) + rw(x,\tau) \\ &- \int_{-\infty}^{\infty} \left[ w(x+y,\tau) - w(x,\tau) - \frac{\partial w}{\partial x}(x,\tau)(e^y - 1) \right] k(y)dy \\ -\mathbb{1}_{x<x(\tau)} &\left\{ rK - qe^x - \int_{x(\tau)-x}^{\infty} \left[ w(x+y,\tau) - (K - e^{x+y}) \right] k(y)dy \right\} = 0 \end{aligned} \tag{5.27}$$

and adhering to what was done in the previous section we get the following difference equation

$$\begin{aligned} &-B_l w_{i-1,j+1} + (1 + r\Delta\tau + B_l + B_u) w_{i,j+1} - B_u w_{i+1,j+1} \\ &= w_{i,j} + \Delta\tau \int_{|y|>\Delta x} \left( w(x_i + y, \tau_j) - w(x_i, \tau_j) \right) k(y)dy \\ &\quad + \Delta\tau \; \mathbb{1}_{x_i < x(\tau_j)} \left\{ rK - qe^{x_i} - \int_{x(\tau_j)-x_i}^{\infty} \left[ w(x_i + y, \tau_j) - (K - e^{x_i+y}) \right] k(y)dy \right\} \end{aligned}$$

where $w_{i,0} = (K - e^{x_i})^+$, $x(\tau_0) = K$ and

$$x(\tau_j) = \min_{x_i} \{ x_i : w(x_i, \tau_j) - (K - e^{x_i})^+ > 0 \} \quad \text{for} \quad j = 1, \ldots, M$$

with $B_l$ and $B_u$ defined as before. As shown, at the first time step, $\tau_0$, the exercise boundary is the strike price. At time $\tau_{j+1}$, for $j = 1, \ldots, M$, the precomputed exercise boundary at the previous time step, namely, $x(\tau_j)$, is used. After solving for the discrete values $w_{i,j+1}$, at time $\tau_{j+1}$, the smallest $x_i$ with the property that the associated value $w_{i,j+1}$ exceeds the intrinsic value of $(K - e^{x_i})^+$ defines the critical boundary $x(\tau_{j+1})$ for the next time step. This procedure introduces an error that is first order in $\Delta\tau$.

As stated in [9], it is known from the classical Black-Scholes situation [25], [161], and [218] that the exercise boundary behaves differently, depending on whether the risk-free interest rate $r$ is less or greater than the dividend rate $q$. In the diffusion framework, the exercise boundary approaches $\frac{r}{q}K$, for $q \leq r$, and strike price $K$ for $q > r$. This fact is different for VG and CGMY. It is proved in [9] that the exercise boundary tends to $K$ for $q > r + \varpi$, where $\varpi$ depends on the model parameters, and is a strictly positive number. If the opposite inequality occurs, they numerically show that the boundary tends to the zero of the *dividend process*. We leave this as a case study at the end of this chapter.

### 5.2.1 Heaviside Term – Synthetic Dividend Process

The integral inside the Heaviside term would be treated in the same manner explained earlier. Therefore, we obtain

$$\begin{aligned}
&\int_{x(\tau_j)-x_i}^{\infty} \left[w(x_i+y,\tau_j) - (K - e^{x_i+y})\right] k(y)dy \\
&= \sum_{k=l-i}^{N-i-1} \int_{k\Delta x}^{(k+1)\Delta x} \left(w_{i+k,j} + \frac{w_{i+k+1,j} - w_{i+k,j}}{\Delta x}(y - k\Delta x)\right) \frac{e^{-\lambda_p y}}{\nu y^{1+Y}} dy \\
&- \frac{1}{\nu}\left\{K \int_{x(\tau_j)-x_i}^{\infty} \frac{e^{-\lambda_p y}}{y^{1+Y}} dy - e^{x_i} \int_{x(\tau_j)-x_i}^{\infty} \frac{e^{-(\lambda_p-1)y}}{y^{1+Y}} dy\right\} \\
&= \sum_{k=l-i}^{N-i-1} \frac{1}{\nu}\left(w_{i+k,j} - k(w_{i+k+1,j} - w_{i+k,j})\right)\left(\int_{k\Delta x}^{(k+1)\Delta x} \frac{e^{-\lambda_p y}}{y^{1+Y}} dy\right) \\
&+ \sum_{k=l-i}^{N-i-1} \frac{w_{i+k+1,j} - w_{i+k,j}}{\nu \Delta x}\left(\int_{k\Delta x}^{(k+1)\Delta x} \frac{e^{-\lambda_p y}}{y^{Y}} dy\right) \\
&- \frac{1}{\nu}\left\{K \int_{x(\tau_j)-x_i}^{\infty} \frac{e^{-\lambda_p y}}{y^{1+Y}} dy - e^{x_i} \int_{x(\tau_j)-x_i}^{\infty} \frac{e^{-(\lambda_p-1)y}}{y^{1+Y}} dy\right\} \\
&= \sum_{k=l-i}^{N-i-1} \frac{\lambda_p^Y}{\nu}\left(w_{i+k,j} - k(w_{i+k+1,j} - w_{i+k,j})\right)\left(g_2(k\Delta x\lambda_p) - g_2((k+1)\Delta x\lambda_p)\right) \\
&+ \sum_{k=l-i}^{N-i-1} \frac{w_{i+k+1,j} - w_{i+k,j}}{\nu \lambda_p^{1-Y}\Delta x}\left(g_1(k\Delta x\lambda_p) - g_1((k+1)\Delta x\lambda_p)\right) \\
&- \frac{\lambda_p^Y}{\nu} K g_2((l-i)\Delta x\lambda_p) + \frac{(\lambda_p-1)^Y}{\nu} e^{x_i} g_2((l-i)\Delta x(\lambda_p - 1))
\end{aligned}$$

Putting together all the terms, we obtain the following difference equation[7] at point $(x_i, \tau_{j+1})$:

$$l_{i,j+1}w_{i-1,j+1} + d_{i,j+1}w_{i,j+1} + u_{i,j+1}w_{i+1,j+1} = w_{i,j} + \frac{\Delta\tau}{\nu}R_{i,j} + \Delta\tau \mathbb{1}_{x_i<x(\tau_j)}H_{i,j}$$

where

$$\begin{aligned}
l_{i,j+1} &= -B_l \\
d_{i,j+1} &= 1 + r\Delta\tau + B_l + B_u + \frac{\Delta\tau}{\nu}\left(\lambda_n^Y g_2(i\Delta x\lambda_n) + \lambda_p^Y g_2((N-i)\Delta x\lambda_p)\right) \\
u_{i,j+1} &= -B_u
\end{aligned}$$

[7]In the case of $i = 1$ or $i = N - 1$, we impose the boundary conditions.

$$
\begin{aligned}
R_{i,j} &= \sum_{k=1}^{i-1} \lambda_n^Y \left(w_{i-k,j} - w_{i,j} - k(w_{i-k-1,j} - w_{i-k,j})\right) \{g_2(k\Delta x \lambda_n) - g_2((k+1)\Delta x \lambda_n)\} \\
&+ \sum_{k=1}^{i-1} \frac{w_{i-k-1,j} - w_{i-k,j}}{\lambda_n^{1-Y}\Delta x} \left(g_1(k\Delta x \lambda_n) - g_1((k+1)\Delta x \lambda_n)\right) \\
&+ \sum_{k=1}^{N-i-1} \lambda_p^Y \left(w_{i+k,j} - w_{i,j} - k(w_{i+k+1,j} - w_{i+k,j})\right) \{g_2(k\Delta x \lambda_p) - g_2((k+1)\Delta x \lambda_p)\} \\
&+ \sum_{k=1}^{N-i-1} \frac{w_{i+k+1,j} - w_{i+k,j}}{\lambda_p^{1-Y}\Delta x} \left(g_1(k\Delta x \lambda_p) - g_1((k+1)\Delta x \lambda_p)\right) \\
&+ K\lambda_n^Y g_2(i\Delta x \lambda_n) - e^{x_i}(\lambda_n + 1)^Y g_2(i\Delta x(\lambda_n + 1))
\end{aligned}
$$

$$
\begin{aligned}
H_{i,j} &= rK - qe^{x_i} \\
&- \sum_{k=l-i}^{N-i-1} \frac{\lambda_p^Y}{\nu} \left(w_{i+k,j} - k(w_{i+k+1,j} - w_{i+k,j})\right) \left(g_2(k\Delta x \lambda_p) - g_2((k+1)\Delta x \lambda_p)\right) \\
&- \sum_{k=l-i}^{N-i-1} \frac{w_{i+k+1,j} - w_{i+k,j}}{\lambda_p^{1-Y}\nu\Delta x} \left(g_1(k\Delta x \lambda_p) - g_1((k+1)\Delta x \lambda_p)\right) \\
&+ \frac{\lambda_p^Y}{\nu} K g_2((l-i)\Delta x \lambda_p) - \frac{(\lambda_p - 1)^Y}{\nu} e^{x_i} g_2((l-i)\Delta x(\lambda_p - 1))
\end{aligned}
$$

and as before

$$
\begin{aligned}
B_l &= \frac{\sigma^2(\Delta x)\Delta\tau}{2\Delta x^2} - \left(r - q + \omega(\Delta x) - \frac{1}{2}\sigma^2(\Delta x)\right)\frac{\Delta\tau}{2\Delta x} \\
B_u &= \frac{\sigma^2(\Delta x)\Delta\tau}{2\Delta x^2} + \left(r - q + \omega(\Delta x) - \frac{1}{2}\sigma^2(\Delta x)\right)\frac{\Delta\tau}{2\Delta x}
\end{aligned}
$$

Assuming that at the completion of the time step $\tau_j$ the values $w_{i,j}$ have been computed, we solve a linear system of equations to find the values $w_{i,j+1}$ for all $i$.

### 5.2.2 Numerical Experiments

This section contains numerical results for American option pricing under the variance gamma model. The VG parameters employed in our study of American option pricing are obtained by calibrating the European option pricing model to market data separately for each maturity. The prices used in the calibration are those of all exchange traded strikes lying within 20% of the forward price on either side. The criterion for selection of the parameters is the minimization over the parameter space, $(\sigma,\nu,\theta)$, of the root mean square percentage error on an equally weighted basis between market prices, $\text{MarketPrice}(K_i,T)$, and model prices, $\text{VG}(S_0,K_i,r,q,T;\sigma,\nu,\theta)$. Specifically

$$
z = \sqrt{\frac{1}{m}\sum_{i=1}^{m}\left(\ln(\text{MarketPrice}(K_i,T)) - \ln(\text{VG}(S_0,K_i,r,q,T;\sigma,\nu,\theta))\right)^2}
$$

The VG prices are calculated using the closed-form solution[8] for European options.

[8]See [175] for the VG closed-form solution for European options.

Market prices used in parameter estimation are for out-of-the-money options on account of their relative liquidity. More exactly, for strikes below the forward price we use put prices and call prices for strikes above the forward price. The following parameters were obtained from calibrating to S&P 500 options on data for June 30, 1999:

**TABLE 5.1**: Calibrated VG parameters for the S&P 500 on June 30, 1999

| $T$ | $r$ | $q$ | $\sigma$ | $\nu$ | $\theta$ |
|---|---|---|---|---|---|
| 0.13972 | 0.0533 | 0.011 | 0.17875 | 0.13317 | -0.30649 |
| 0.21643 | 0.0536 | 0.012 | 0.18500 | 0.22460 | -0.28837 |
| 0.46575 | 0.0549 | 0.011 | 0.19071 | 0.49083 | -0.28113 |
| 0.56164 | 0.0541 | 0.012 | 0.20722 | 0.50215 | -0.22898 |

The S&P 500 spot price on June 30, 1999 was 1369.41. As we see in Table 5.1, as maturity gets larger the annualized kurtosis parameter $\nu$ increases and annualized skewness $\theta$ decreases. The increase in $\nu$ is slower than the increase in maturity and this is consistent with an approach to normality, though at a rate slower than would be the case if $\nu$ were constant or falling. The decrease in $\theta$ is also broadly consistent with the approach to normality. Table 5.2 contains the Black–Scholes implied volatility for these option prices. We observe a significant skewness in these implied volatilities with a drop of 10 volatility points over the specified strike range.

We first show some convergence results demonstrating numerical stability. In Table 5.3, we illustrate some examples of convergence as $\Delta\tau$ and $\Delta x$ approach zero. For all the options in the table, the maturity is $T = 0.56164$ and the corresponding parameters for this maturity are shown in Table 5.1. The results in Table 5.3 support the claim that the scheme is stable and convergent. For reasons of space, only a ratio of $N/M = 2$ is shown. Computations of other ratios give similar results and lead to the conjecture that the scheme is unconditionally stable. This conjecture is solely based on our computational experience.

Table 5.4 contains the early exercise premiums from pricing American options under the variance gamma and geometric Brownian motion dynamics, respectively. The American option prices for the geometric Brownian motion model were obtained at the implied volatility for the option reported in Table 5.2. We observe that across all strikes and maturities, the VG early exercise premiums dominate those from geometric Brownian motion.

**TABLE 5.2**: Implied volatility for S&P 500 options on June 30, 1999

| Strike | $T = 0.13972$ | $T = 0.21643$ | $T = 0.46575$ | $T = 0.56164$ |
|---|---|---|---|---|
| 1200 | 0.2675 | 0.2737 | 0.2801 | 0.2868 |
| 1220 | 0.2592 | 0.2662 | 0.2743 | 0.2818 |
| 1240 | 0.2508 | 0.2587 | 0.2686 | 0.2768 |
| 1260 | 0.2422 | 0.2509 | 0.2629 | 0.2718 |
| 1280 | 0.2334 | 0.2431 | 0.2571 | 0.2667 |
| 1300 | 0.2244 | 0.2351 | 0.2513 | 0.2616 |
| 1320 | 0.2152 | 0.227 | 0.2455 | 0.2565 |
| 1340 | 0.2057 | 0.2187 | 0.2396 | 0.2514 |
| 1360 | 0.1961 | 0.2102 | 0.2337 | 0.2462 |
| 1380 | 0.1863 | 0.2016 | 0.2277 | 0.2409 |
| 1400 | 0.1767 | 0.193 | 0.2217 | 0.2357 |

**TABLE 5.3**: Convergence results for maturity $T = 0.56164$ with various strike prices as $M$ and $N$ increase

| $N$ | $M$ | Strike price | | | |
|---|---|---|---|---|---|
| | | 1200 | 1260 | 1320 | 1380 |
| 500 | 250 | 35.5615 | 48.8537 | 66.0738 | 88.1319 |
| 1000 | 500 | 35.5318 | 48.8051 | 66.0078 | 88.0179 |
| 2000 | 1000 | 35.5307 | 48.7982 | 65.9926 | 87.9922 |
| 4000 | 2000 | 35.5301 | 48.7976 | 65.9908 | 87.9911 |

**TABLE 5.4**: Early exercise premiums for variance gamma (VG) and geometric Brownian motion (GBM)

| Maturity | $T_1 = 0.13972$ | | $T_2 = 0.21643$ | | $T_3 = 0.46575$ | | $T_4 = 0.56164$ | |
|---|---|---|---|---|---|---|---|---|
| Strike | GBM | VG | GBM | VG | GBM | VG | GBM | VG |
| 1200 | 0.052 | 0.025 | 0.036 | 0.063 | 0.443 | 0.539 | 0.828 | 1.041 |
| 1220 | 0.026 | 0.070 | 0.123 | 0.124 | 0.647 | 0.677 | 0.710 | 1.250 |
| 1240 | 0.010 | 0.117 | 0.165 | 0.191 | 0.569 | 0.835 | 1.131 | 1.489 |
| 1260 | 0.065 | 0.159 | 0.081 | 0.253 | 0.918 | 0.997 | 1.192 | 1.742 |
| 1280 | 0.113 | 0.196 | 0.251 | 0.322 | 0.856 | 1.185 | 1.483 | 2.023 |
| 1300 | 0.164 | 0.239 | 0.359 | 0.392 | 1.298 | 1.388 | 1.756 | 2.339 |
| 1320 | 0.231 | 0.428 | 0.415 | 0.602 | 1.318 | 1.681 | 1.966 | 2.741 |
| 1340 | 0.328 | 0.608 | 0.465 | 0.791 | 1.856 | 2.008 | 2.462 | 3.178 |
| 1360 | 0.462 | 0.869 | 0.745 | 1.078 | 1.999 | 2.392 | 2.686 | 3.687 |
| 1380 | 0.630 | 1.103 | 1.069 | 1.374 | 2.664 | 2.795 | 3.400 | 4.241 |
| 1400 | 1.066 | 1.483 | 1.475 | 1.678 | 3.112 | 3.291 | 3.771 | 4.861 |

This suggests that the traditional practice of adding geometric Brownian motion based American option premia inferred from the implied volatility of a European price quote to get an American option price is biased downward with respect to the true American option price of the underlying VG dynamics for the stock price. The differences can be substantial and for example for the 1320 strike with maturity 0.2164 it is about 1/3 of the American option value under geometric Brownian motion. We are led to conclude that even though the pricing of American options under the right underlying dynamics may be difficult, it is important from the perspective of correctly accounting for the values of these instruments. We also compare the exercise boundary for VG and GBM for strike $K = 1300$ and maturity $T = 0.56164$ in Figure 5.1. This example exhibits that a smaller continuation region may be associated with an earlier exercise. The exact timing is difficult to comment on as differences in the underlying dynamics enter into the issues of passage times to these boundaries.

## 5.3 PIDE Solutions for Lévy Processes

The prices of options under models described by Lévy processes can compute via numerical solutions of partial integro-differential equations that are similar to the one illustrated

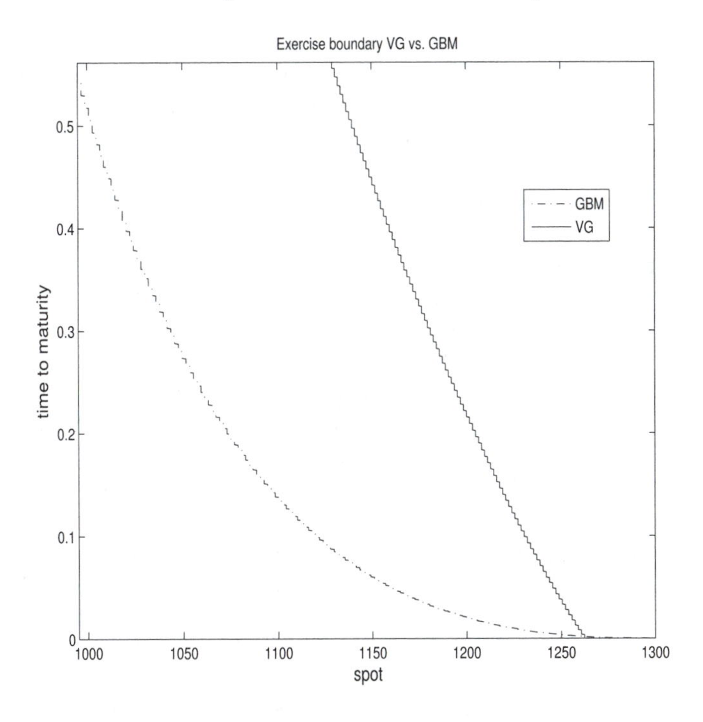


**FIGURE 5.1**: Optimal exercise boundary for an American put

here for the case of the CGMY process. The methods developed here are therefore applicable to a wide class of models. In particular our combination of analytical and numerical approaches to the singularity at zero for infinite activity Lévy processes should prove useful in many contexts.

## 5.4 Forward PIDEs

### 5.4.1 American Options

Since the original development of forward equations for European options in continuous models, several extensions have been proposed for both using these techniques under different model assumptions and pricing exotics by means of PDE under the standard local volatility model. For example, Esser and Schlag [110] develop forward equations for European options written on the forward price rather than the spot price. Forward equations for European options in jump diffusion models were developed in Andersen and Andreasen [12] and extended by Andreasen and Carr [17]. It is straightforward to develop the relevant forward equations for European binary options or for European power options by differentiating or integrating the forward equation for standard European options. Buraschi and Dumas [49] develop forward equations for compound options. However, their definition of a compound option is non-standard in that the critical stock price is specified in the contract. In contrast to the PDEs determined by others, their evolution equation is an ordinary differential equation whose sole independent variable is the intermediate maturity date.

Given the close relationship between compound options and American options, it seems plausible that there might be a forward equation for American options. The development of such an equation has important practical implications since all listed options on individual stocks are American-style. The Dupire equation cannot be used to infer the volatility function from market prices of American options, nor can it be used to efficiently value a collection of American options of differing strikes and maturities. This problem is addressed for American calls on stocks paying discrete dividends in Buraschi and Dumas [49] and it is also considered in a lattice setting in Chriss [67].

In [56], Carr and Hirsa direct their attention to the more difficult problem of pricing continuously exercisable American puts in continuous time models. To do so, they depart from the diffusive models which characterize most of the previous research on forward equations in continuous time. To capture the volatility smile, they assume that prices jump rather than assuming that the instantaneous volatility is a function of stock price and time. Dumas, Fleming, and Whaley [103] find little empirical support for the Dupire model, whereas there is a long history of empirical support for jump-diffusion models. Three recent papers documenting support for such models are [15], [53], and [62]. In particular, they assume that the returns on the underlying asset have stationary independent increments, or in other words that the log price is a Lévy process. Besides the [32] model, their framework includes as special cases the variance gamma (VG) model of Madan, Carr, and Chang [175], the CGMY model of Carr, Geman, Madan, and Yor [53], the finite moment logstable model of Carr and Wu [62], the Merton [178] and Kou [166] jump diffusion models, and the hyperbolic models of Eberlein, Keller, and Prause [105]. In all of these models except Black–Scholes, the existence of a jump component implies that the backward and forward equations contain an integral in addition to the usual partial derivatives. Despite the computational complications introduced by this term, Carr and Hirsa [56] use finite differences to solve both of these fundamental partial-integro differential equations (PIDEs). They demonstrate that American option values in the diffusion extended VG option pricing model are very similar when using either the forward PIDE or the traditional backward approach. For details on the application of finite differences to valuing American options in the VG model, see [140].

The approach to determining the forward equation for American options in [56] begins with the well-known backward equation and then exploits the symmetries which essentially define Lévy processes. In the process of developing the forward equation, the authors also determine two hybrid equations which are more computationally efficient when one is interested in the variation of prices or Greeks across strike or maturity at a fixed time, for example at market close. They also hold in greater generality and depending on the problem at hand, can have large computational advantages over the backward or forward equations when the model has already been calibrated.

The first of these hybrid equations has the stock price and maturity as independent variables. The numerical solution of this hybrid equation is an alternative to the backward equation in producing a spot slide, which shows how American option prices vary with the initial spot price of the underlying asset. If one is interested in understanding how this spot slide varies with maturity, then this hybrid equation is much more efficient than the backward equation. This hybrid equation also has important implications for path-dependent options such as cliquets whose payoff directly depends on the particular level reached by an intermediate stock price.

The second hybrid equation has strike price and calendar time as independent variables. The numerical solution of this hybrid equation is an alternative to the forward equation in producing an implied volatility smile at a fixed maturity. If one is interested in understanding how the model predicts that this smile will change over time, then this hybrid equation is much more computationally efficient than the forward equation. This second hybrid equation also allows parameters to have a term structure, whereas our forward equation does not. Note, however, that implied volatility can have a term or strike structure in the Lévy setting.

Hence, if one needs to efficiently value a collection of American options of different strikes in the time-dependent Black–Scholes model, then it is far more efficient to solve this hybrid equation than to use the standard backward equation.

In [56], Carr and Hirsa derive the following forward PIDE for pricing American puts for any semi-martingale process with Lévy density $\hat{k}(y)$.

$$\frac{\partial P(K,T;s,t)}{\partial T} - \frac{\sigma^2}{2}K^2\frac{\partial^2 P(K,T;s,t)}{\partial K^2} + (r-q)K\frac{\partial P(K,T;s,t)}{\partial K} + qP(K,T;s,t)$$
$$-\int_{-\infty}^{+\infty}\left[P(Ke^{-y},T;s,t) - P(K,T;s,t) - \frac{\partial P(K,T;s,t)}{\partial K}K(e^{-y}-1)\right]e^y\nu(y)dy$$
$$-\mathbb{1}_{K>\bar{K}(s,t;T)}\left\{rK - qs - \int_{\ln(K/\bar{K}(s,t;T))}^{\infty}\left[P(Ke^{-y},T;s,t) - (Ke^{-y}-s)\right]e^y\hat{k}(y)dy\right\} = 0$$

It is easy to show

$$\int_{-\infty}^{+\infty}(e^{-y}-1)e^y\hat{k}(y)dy = \omega \tag{5.28}$$

Therefore, we get (dropping $s$ and $t$ for simplicity)

$$\frac{\partial P(K,T)}{\partial T} - \frac{\sigma^2}{2}K^2\frac{\partial^2 P(K,T)}{\partial K^2} + (r-q+\omega)K\frac{\partial P(K,T)}{\partial K} + qP(K,T)$$
$$-\int_{-\infty}^{+\infty}\left((P(Ke^{-y},T) - P(K,T)\right)e^y\hat{k}(y)dy$$
$$-\mathbb{1}_{K>\bar{K}(T)}\left\{rK - qs - \int_{\ln(K/\bar{K}(T))}^{\infty}\left[P(Ke^{-y},T) - (Ke^{-y}-s)\right]e^y\hat{k}(y)\right\} = 0$$

By making the change of variable $x = \ln K$ we obtain, noting

$$\begin{aligned}
p(x,T) &= P(K,T) \\
\frac{\partial p}{\partial x}(x,T) &= K\frac{\partial P}{\partial K}(K,T) \\
\frac{\partial^2 p}{\partial x^2}(x,T) - \frac{\partial p}{\partial x}(x,T) &= K^2\frac{\partial^2 P}{\partial K^2}(K,T) \\
p(x-y,T) &= P(Ke^{-y},T)
\end{aligned}$$

the following PIDE, as a function of $p(x,T)$:

$$\frac{\partial p}{\partial T}(x,T) - \frac{\sigma^2}{2}\frac{\partial^2 p(x,T)}{\partial x^2} + (r-q+\frac{\sigma^2}{2}+\omega)\frac{\partial p}{\partial x}(x,T) + qp(x,T)$$
$$-\int_{-\infty}^{+\infty}\left(p(x-y,T) - p(x,T)\right)\hat{k}(y)(y)dy$$
$$-\mathbb{1}_{x>\bar{x}(T)}\left\{re^x - qs - \int_{x-\bar{x}(T)}^{\infty}\left(p(x-y,T) - (e^{x-y}-s)\right)\hat{k}(y)dy\right\} = 0$$

where

$$\begin{aligned}
\hat{k}(y) &= \frac{e^{-\hat{\lambda}_p y}}{\nu y}\mathbb{1}_{y>0} + \frac{e^{-\hat{\lambda}_n|y|}}{\nu|y|}\mathbb{1}_{y<0} \\
\hat{\lambda}_p &= \left(\frac{\theta^2}{\sigma^4} + \frac{2}{\sigma^2\nu}\right)^{\frac{1}{2}} - \frac{\theta}{\sigma^2} - 1 \\
\hat{\lambda}_n &= \left(\frac{\theta^2}{\sigma^4} + \frac{2}{\sigma^2\nu}\right)^{\frac{1}{2}} + \frac{\theta}{\sigma^2} + 1
\end{aligned}$$

This PIDE should be solved subject to the initial condition

$$p(x,0) = (e^x - s)^+$$

and boundary conditions

$$\frac{\partial^2 p}{\partial x^2}(-\infty,T) - \frac{\partial p}{\partial x}(-\infty,T) = 0 \quad \forall T \tag{5.29}$$

$$\frac{\partial^2 p}{\partial x^2}(+\infty,T) - \frac{\partial p}{\partial x}(+\infty,T) = 0 \quad \forall T \tag{5.30}$$

### 5.4.2 Down-and-Out and Up-and-Out Calls

In a different paper [57], the authors derive the forward evolution equations for up-and-out and down-and-out call options when the log price is a Lévy process. In this setting the forward equations are again partial-integro differential equations (PIDEs).

We assume the standard model of perfect capital markets, continuous trading, and no arbitrage opportunities. When a pure discount bond is used as numeraire, then it is well known that no arbitrage implies that there exists a probability measure $\mathbb{Q}$ under which all non-dividend-paying asset prices are martingales. Under this measure we assume that a stock price $S_t$ obeys the following stochastic differential equation:

$$dS_t = [r(t) - q(t)]S_{t-}dt + a(S_{t-},t)dW_t + \int_{-\infty}^{\infty} S_{t-}(e^x - 1)[\mu(dx,dt) - \nu(x,t)dxdt] \tag{5.31}$$

for all $t \in [0,\bar{T}]$, where the initial stock price, $S_0 > 0$, is known, and $\bar{T}$ is some arbitrarily distant horizon. The process is Markov in itself since the coefficients of the stock price process at time $t$ depend on the path only through $S_{t-}$, which is the pre jump price at $t$. Thus, the dynamics are fully determined by the drift function $b(S,t) \equiv [r(t) - q(t)]S$, the (normal) volatility function $a(S,t)$, and the jump compensation function $\nu(x,t)$. The term $dW_t$ denotes increments of a standard Brownian motion (SBM) $W_t$ defined on the time set $[0,\bar{T}]$ and on a complete probability space $(\Omega,\mathcal{F},Q)$. The random measure $\mu(dx,dt)$ counts the number of jumps of size $x$ in the log price at time $t$. The function $\{\nu(x,t), x \in \mathbb{R}, t \in [0,\bar{T}]\}$ is used to compensate the jump process $J_t \equiv \int_0^t \int_{-\infty}^{\infty} S_{t-}(e^x - 1)\mu(dx,ds)$, so that the last term in (5.31) is the increment of a $\mathbb{Q}$ jump martingale. The function $\nu(x,t)$ must have the following properties:

1. $\nu(0,t) = 0$
2. $\int_{-\infty}^{\infty}(x^2 \wedge 1)\nu(x,t)dx < \infty, \quad t \in [0,\bar{T}]$

Thus, each price change is the sum of the increment in a general diffusion process with proportional drift and the increment in a pure jump martingale, where the latter is an additive process in the log price. We restrict the function $a(S,t)$ so that the spot price is always non-negative and absorbing at the origin. A sufficient condition for keeping the stock price away from the origin is to bound the lognormal volatility. In particular, we set

$$a(0,t) = 0$$

Hence, (5.31) describes a continuous-time Markov model for the spot price dynamics, which is both arbitrage-free and consistent with limited liability. Aside from the Markov property, the main restrictions inherent in (5.31) are the standard assumptions that interest rates, dividend yields, and the compensator do not depend on the spot price.

For down-and-out calls they find the forward PIDE to be

$$\begin{aligned}
&\frac{\partial}{\partial T} D_o^c(K,T) \\
&= -q(T)D_o^c(K,T) - [r(T)-q(T)]K\frac{\partial}{\partial K}D_o^c(K,T) + \frac{a^2(K,T)}{2}\frac{\partial^2}{\partial K^2}D_o^c(K,T) \\
&+ \int_{-\infty}^{\infty}\left[D_o^c(Ke^{-x},T) - D_o^c(K,T) - \frac{\partial}{\partial K}D_o^c(K,T)K(e^{-x}-1)\right]e^x\nu(x,T)dx \qquad (5.32)
\end{aligned}$$

This PIDE holds on the domain $K \geq L, T \in [0,\bar{T}]$ where $L$ is the barrier level. For a down-and-out call, the initial condition is

$$D_o^c(K,0) = (S_0 - K)^+, \qquad K \geq L \qquad (5.33)$$

Since a down-and-out call behaves like a standard call as its strike approaches infinity, we have

$$\lim_{K\uparrow\infty} D_o^c(K,T) = \lim_{K\uparrow\infty}\frac{\partial}{\partial K}D_o^c(K,T) = \lim_{K\uparrow\infty}\frac{\partial^2}{\partial K^2}D_o^c(K,T) = 0, \quad T\in[0,\bar{T}] \qquad (5.34)$$

For a lower boundary condition, note that a down-and-out call on a stock with the dynamics in (5.31) has the same value prior to knocking out as a down-and-out call on a stock which absorbs at $L$. The second derivative of this latter call gives the r-discounted risk-neutral probability density for the event that the stock price has survived to at least $T$ and is in the interval $(K, K+dK)$. Now it is well known that the appropriate boundary condition for an absorbing process is that this PDF vanishes on the boundary. Hence

$$\lim_{K\downarrow L}\frac{\partial^2}{\partial K^2}D_o^c(K,T) = 0, \quad T\in[0,\bar{T}] \qquad (5.35)$$

Evaluating (5.32) at $K = L$ and substituting in (5.35) implies

$$\begin{aligned}
\frac{\partial D_o^c(L,T)}{\partial T} &= \int_{-\infty}^{\infty}\left[D_o^c(Le^{-x},T) - D_o^c(L,T) - \frac{\partial}{\partial K}D_o^c(L,T)L(e^{-x}-1)\right]e^x\nu(x,T)dx \\
&- [r(T)-q(T)]L\frac{\partial}{\partial K}D_o^c(L,T) - q(T)D_o^c(K,T), \quad T\in[0,\bar{T}] \qquad (5.36)
\end{aligned}$$

This is a Robin condition as it involves the value and both its first partial derivatives along the boundary (in some contexts, the generalized Neumann boundary conditions are also referred to as the Robin boundary conditions).

They find the forward PIDE for up-and-out call to be

$$\begin{aligned}
\frac{\partial U_o^c(K,T)}{\partial T} &= \int_{-\infty}^{\infty}[U_o^c(Ke^{-x},T) - U_o^c(K,T) - \frac{\partial U_o^c(K,T)}{\partial K}K(e^{-x}-1)]e^x\nu(x,T)dx \\
&+ \frac{a^2(K,T)}{2}\frac{\partial^2 U_o^c(K,T)}{\partial K^2} - [r(T)-q(T)]K\frac{\partial U_o^c(K,T)}{\partial K}(K,T) - q(T)U_o^c(K,T) \\
&- \left[\int_0^{\infty}\frac{\partial U_o^c(He^{-x},T)}{\partial K}\nu(x,T)dx + \frac{a^2(H,T)}{2}\frac{\partial^3 U_o^c(H,T)}{\partial K^3}\right](K-H) \\
&- \int_0^{\infty}U_o^c(He^{-x},T)e^x\nu(x,T)dx \qquad (5.37)
\end{aligned}$$

for $K\in(0,H), T\in[0,\bar{T}]$ and for $H > S_0$. Recall that for an up-and-out call, the initial condition is

$$u(K,0) = (S_0 - K)^+, \quad K\in[0,H) \qquad (5.38)$$

and for $H > S_0$. For boundary conditions, we use the following:

$$\lim_{K\downarrow 0} \frac{\partial^2}{\partial K^2} U_o^c(K,T) = 0, \quad T \in [0, \bar{T}] \tag{5.39}$$

$$\lim_{K\uparrow H} \frac{\partial^2}{\partial K^2} U_o^c(K,T) = 0, \quad T \in [0, \bar{T}] \tag{5.40}$$

To interpret the PIDE in 5.37 financially, first note that if an investor buys a calendar spread of up-and-out calls, then the initial cost is given by the left-hand side. The first term on the right-hand side arises only from paths which survive to $T$ and then cross $K$. It can be shown that this first term is the initial value of a path-dependent claim that pays the overshoots of the strike at maturity. The second term on the right-hand side arises only from paths which survive to maturity and finish at strike $K$. Consider the infinite position in the later maturing call at time $t = T$ if the option survives until then. This position will have infinite time value when $S_T = K$ and zero value otherwise. The greater is the local variance rate at $S_T = K$, the greater is this conditional time value and the more valuable is this position initially. The next two terms arise only from paths which survive to $T$ and finish above strike $K$. They capture the additional carrying costs of stock and bond which are embedded in the time value of the later maturing call. The operator given by the first four terms on the right-hand side also represents the present value of benefits obtained at maturity $T$ when an investor buys a calendar spread of standard or down-and-out calls. In contrast, the last two terms in Equation 5.37 have no counterpart for calendar spreads in standard or down-and-out calls.

**Example 7** *Forward versus backward up-and-out call (UOC) premiums*

We employ the methodology cover in this chapter and in [56] and [140] to numerically solve the backward and forward PIDEs for up-and-out calls. For our numerical examples, we consider $\nu(x)dx$ to be the *Lévy density* for the VG process in the following form:

$$\nu(x) = \frac{e^{-\lambda_p x}}{\nu x} \text{ for } x > 0 \quad \text{and} \quad \nu(x) = \frac{e^{-\lambda_n |x|}}{\nu |x|} \text{ for } x < 0$$

and

$$\lambda_p = \left(\frac{\theta^2}{\sigma^4} + \frac{2}{\sigma^2 \nu}\right)^{\frac{1}{2}} - \frac{\theta}{\sigma^2} \qquad \lambda_n = \left(\frac{\theta^2}{\sigma^4} + \frac{2}{\sigma^2 \nu}\right)^{\frac{1}{2}} + \frac{\theta}{\sigma^2}$$

where $\sigma$, $\nu$, and $\theta$ are VG parameters. The parameter set for our numerical experiments is spot $S_0$=100, risk-free rate $r = 3.75\%$, dividend rate $q = 2.0\%$, and VG parameters $\sigma = 0.3$, $\nu = 0.25$, $\theta = -0.3$ and strike range $K = 90, 110$, maturity range $T = 0.25, 0.5, 1.0$. In this example, we compare UOC premiums by numerically solving both backward and forward PIDEs.

In Figure 5.2(a) we display UOC premiums for 3-month maturity by solving the backward PIDE numerically, the left figure is for a strike of 90 and the right one is for a strike 110. Out of all those premiums we just pick the one that corresponds to the spot price 100 as pointed out in the figures. That is the drawback with backward PDEs. Figures 5.2(b) and 5.2(c) are the same as Figure 5.2(a) except for 6-month maturity and 12-month maturity respectively. In Figure 5.3 we display UOC premiums by solving the forward PIDE for UOC numerically for all strikes and maturities. From all premiums we pick those that correspond to strikes $90, 110$ and maturities 3-month, 6-month, and 12-month as pointed out in the figure. We see that the premiums from backward and forward PIDEs are identical. However, in case of forward PIDE for UOC, we get the results in one sweep as opposed to backward that we had to solve the backward PIDE numerically for each pair of strike and maturity (in this example we solve it six times, having six pairs of strike and maturity).

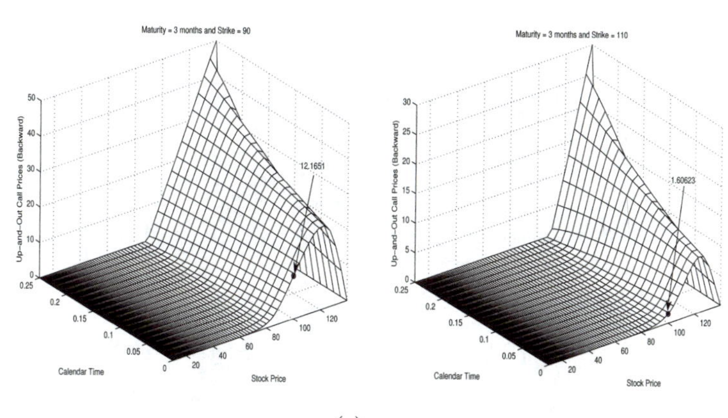

(a)

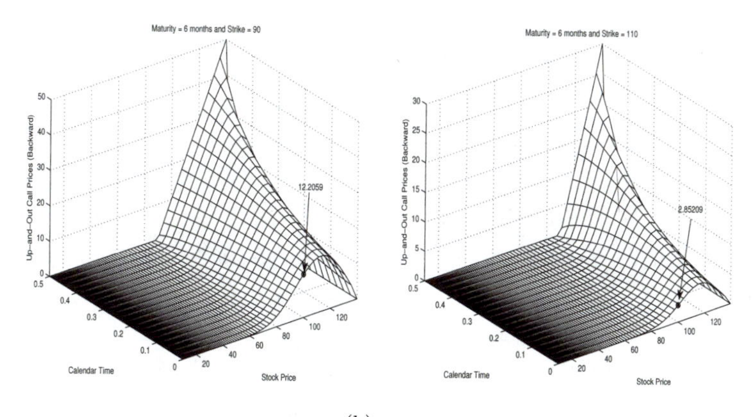

(b)

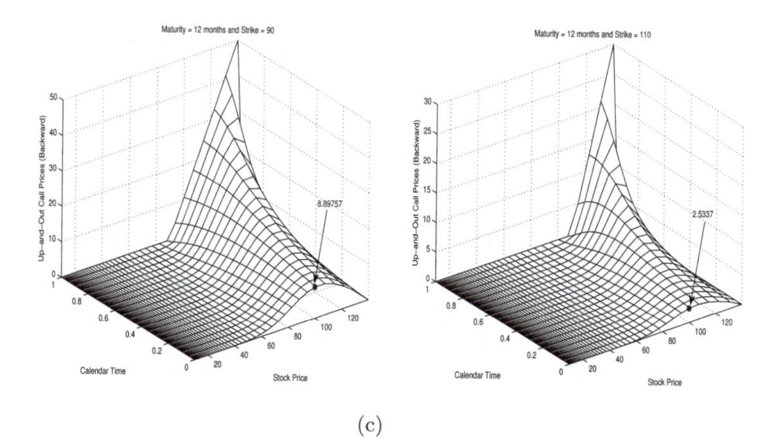


(c)

**FIGURE 5.2**: Up-and-out call prices obtained from using a backward PIDE

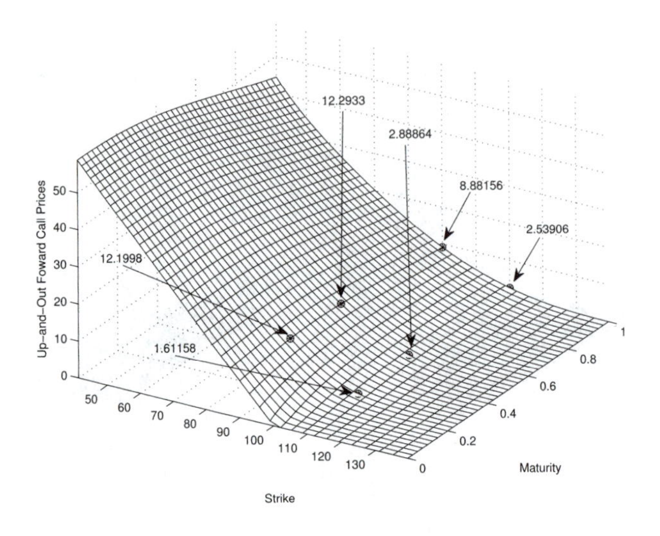


**FIGURE 5.3**: Up-and-out call prices obtained from using a forward PIDE

## 5.5 Calculation of $g_1$ and $g_2$

We wrote our difference equations in terms of $g_1$ and $g_2$ where

$$g_1(\xi) = \int_\xi^\infty \frac{e^{-z}}{z^\alpha} dz \tag{5.41}$$

$$g_2(\xi) = \int_\xi^\infty \frac{e^{-z}}{z^{\alpha+1}} dz \tag{5.42}$$

for $0 \le \alpha < 1$. In the calculation of $g_1$ for $\alpha = 0$

$$g_1(x) = \int_x^\infty \frac{\exp(-t)}{t^\alpha} dt = \int_x^\infty \exp(-t) dt = e^{-x}$$

For $0 < \alpha < 1$ we obtain

$$g_1(x) = \int_x^\infty \frac{\exp(-t)}{t^\alpha} dt = \int_x^\infty e^{-t} t^{-\alpha} dt = \int_x^\infty e^{-t} t^{\beta-1} dt$$

where $\beta = 1 - \alpha$. We observe that the last integral is an *upper incomplete gamma function* for $\beta > 0$. Thus,

$$g_1(x) = \begin{cases} e^{-x} & \alpha = 0 \\ \text{uigf}(x, 1-\alpha) & 0 < \alpha < 1 \end{cases}$$

Note that for special case $x = 0$ we get

$$g_1(0) = \begin{cases} 1 & \alpha = 0 \\ \Gamma(1-\alpha) & 0 < \alpha < 1 \end{cases}$$

In the calculation of $g_2$ for $\alpha = 0$

$$g_2(x) = \int_x^\infty \frac{\exp(-t)}{t^{1+\alpha}} dt = \int_x^\infty \frac{\exp(-t)}{t} dt = \text{expint}(x)$$

For $0 < \alpha < 1$ integration by parts yields

$$
\begin{aligned}
g_2(x) &= \int_x^{\infty} \frac{\exp(-t)}{t^{1+\alpha}} dt \\
&= \frac{\exp(-x)x^{-\alpha}}{\alpha} - \frac{1}{\alpha}\int_x^{\infty} e^{-t}t^{-\alpha}dt \\
&= \frac{\exp(-x)x^{-\alpha}}{\alpha} - \frac{1}{\alpha}\int_x^{\infty} e^{-t}t^{\beta-1}dt
\end{aligned}
$$

where as before $\beta = 1 - \alpha$. Note that as before the last integral is the *upper incomplete gamma function* for $\beta > 0$. Thus,

$$
g_2(x) = \begin{cases} \text{expint}(x) & \alpha = 0 \\ \frac{\exp(-x)x^{-\alpha}}{\alpha} - \frac{1}{\alpha}\text{uigf}(x, 1-\alpha) & 0 < \alpha < 1 \end{cases}
$$

The gamma function is

$$
\Gamma(\beta) = \int_0^{\infty} e^{-t}t^{\beta-1}dt
$$

---

## Problems

1. Assume we are pricing an American call under the CGMY model ($0 < Y < 1$) with strike $K$ at grid point $(x_i, \tau_j)$ where $x_i = x_{\min} + i\Delta x$ for $i = 0, \ldots, N$ and $\Delta x = (x_{\max} - x_{\min})/N$. For the region $y \in (x_N - x_i, \infty)$ we have

$$
\int_{x_N - x_i}^{\infty} (w(x_i + y, \tau_j) - w_{i,j})\, k(y)dy = \int_{x_N - x_i}^{\infty} (w(x_i + y, \tau_j) - w_{i,j})\, \frac{e^{-\lambda_p y}}{\nu y^{1+Y}} dy
$$

   where $w_{i,j}$ is the premium at $(x_i, \tau_j)$. Knowing that $x_i + y$ is outside the grid for this integral, calculate the above integral. Your answer should be expressed as a function of known variables $\nu$, $\lambda_p$, $Y$, $w_{i,j}$ and $g_2(.)$ with

$$
g_2(x) = \int_x^{\infty} \frac{e^{-t}}{t^{1+Y}} dt
$$

2. What happens to option premiums under the CGMY model as we increase $Y$ while keeping other parameters fixed? Why?

3. Price European options under CGMY using the explicit-implicit covered in this chapter to solve the PIDE for the parameter set used to tabulate Table 2.12 and compare your results.

## Case Studies

1. Let $w(x, \tau)$ be the value of a derivative security that satisfies the following PIDE

$$\frac{\partial w}{\partial \tau}(x,\tau) - (r-q)\frac{\partial w}{\partial x}(x,\tau) + rw(x,\tau)$$
$$-\int_{-\infty}^{\infty}\left[w(x+y,\tau) - w(x,\tau) - \frac{\partial w}{\partial x}(x,\tau)(e^y-1)\right]k(y)dy$$

where $k(y)$ is given by

$$\begin{aligned} k(y) &= \frac{e^{-\lambda_p y}}{\nu y^{1+Y}}\mathbf{1}_{y>0} + \frac{e^{-\lambda_n |y|}}{\nu |y|^{1+Y}}\mathbf{1}_{y<0} \\ \lambda_p &= \left(\frac{\theta^2}{\sigma^4} + \frac{2}{\sigma^2\nu}\right)^{\frac{1}{2}} - \frac{\theta}{\sigma^2} \\ \lambda_n &= \left(\frac{\theta^2}{\sigma^4} + \frac{2}{\sigma^2\nu}\right)^{\frac{1}{2}} + \frac{\theta}{\sigma^2} \end{aligned}$$

. and $x = \ln(S)$ and $\tau = T - t$.

For the put option premium, this PIDE must be solved subject to the initial condition

$$w(x,0) = (K - e^x)^+ \tag{5.43}$$

and boundary conditions

$$\begin{aligned} \frac{\partial^2 w}{\partial x^2}(-\infty,\tau) - \frac{\partial w}{\partial x}(-\infty,\tau) &= 0 \ \forall \tau \\ \frac{\partial^2 w}{\partial x^2}(\infty,\tau) - \frac{\partial w}{\partial x}(\infty,\tau) &= 0 \ \forall \tau \end{aligned}$$

Use the explicit-implicit finite difference scheme covered in this chapter to solve the PIDE.

To make sure the implementation is correct, calculate European put option premium in this framework for the following parameters: spot price, $S_0 = \$1300$, strike price $K = 1300$, risk-free interest rate $r = 0.25\%$; dividend rate $q = 1.5\%$, maturity $T = 0.5$ year, $\sigma = 30\%$, $\nu = 0.40$, $\theta = -0.30$, and $Y = 0.0$ and compare it with the results of a transform technique (FFT, fractional FFT, or the COS method) for the variance gamma model.

Then, calculate American put option premiums for the following parameters: spot price, $S_0 = \$1300$, strike price $K = \{1100, 1200, 1300\}$, risk-free interest rate $r = 0.25\%$, dividend rate $q = 1.5\%$; Maturity, $T = 0.5$ year, $\sigma = 30\%$, $\nu = 0.40$, $\theta = -0.30$, and $Y = \{0.0, 0.5, 0.99\}$ by:

(a) Applying Bermudan approach at each time-step.

(b) Applying Brennan–Schwartz algorithm.

(c) Solving the following modified PIDE (synthetic dividend process)

$$
\begin{aligned}
&\frac{\partial w}{\partial \tau}(x,\tau) - (r-q)\frac{\partial w}{\partial x}(x,\tau) + rw(x,\tau) \\
&- \int_{-\infty}^{\infty}\left[w(x+y,\tau) - w(x,\tau) - \frac{\partial w}{\partial x}(x,\tau)(e^y - 1)\right]k(y)dy \\
-\mathbf{1}_{x<x(\tau)}&\left\{rK - qe^x - \int_{x(\tau)-x}^{\infty}\left[w(x+y,\tau) - (K - e^{x+y})\right]k(y)dy\right\} = 0
\end{aligned}
$$

For solving the tridiagonal stiffness matrix in the Brennan-Schwartz algorithm, try both possible approaches: (i) by first making upper diagonal zero and then solving and correcting each element or (ii) by first making lower diagonal zero and then solving and correcting each element. In the case of synthetic dividend, use the following two approaches: (i) use the previous exercise boundary for the current time to solve the matrix equation, (ii) use the previous one to find the current one and use that to re-solve the matrix equation to find a new one.

For each parameter set, plot the critical stock price curves and compare premiums and conclude.

2. Redo Case Study 1 using the fast Fourier transform to numerically solve the integral in the PIDE as suggested in [9] as opposed to the finite difference scheme covered in this chapter.

3. *A fortiori*, the up-and-out call value function $u(S,t)$ solves a backward boundary value problem (BVP), consisting of the backward partial-integro differential equation

$$
\begin{aligned}
&\int_{-\infty}^{\infty}\left[u(S_t e^y, t) - u(S_t,t) - \frac{\partial u}{\partial S}(S_t,t)S_t(e^y - 1)\right]k(y)dy \\
&\quad + \frac{\partial u}{\partial t}(S_t,t) + (r(t) - q(t))S_t\frac{\partial u}{\partial S}(S_t,t) - ru(S_t,t) \quad = \quad 0
\end{aligned}
$$

where $k(y)$ is the same as the one given in Case Study 1. This PIDE is subject to the following boundary conditions:

$$
\begin{aligned}
u(S,T_0) &= (S - K_0)^+, & S &\in [0,H] \\
\lim_{S\downarrow 0} u(S,t) &= 0, & t &\in [0,T_0] \\
\lim_{S\uparrow H} u(S,t) &= 0, & t &\in [0,T_0]
\end{aligned}
$$

Solve the PIDE numerically for the following set of parameters: spot price $S_0 = \$100$, strike price $K = 110$, up barrier $H = 125$, risk-free interest rate $r = 1.5\%$, dividend rate $q = 1.25\%$, maturity $T = 1$ year, and the model parameters $\sigma = 25\%$, $\nu = 0.25$, $\theta = -0.15$, and $Y = 0.6$ using

- Uniform mesh points on $S$
- Uniform mesh points on $\xi$ where

$$
\xi = \frac{\sinh^{-1}((S-B)/\alpha) - c_2}{c_1 - c_2}
$$

where

$$\begin{aligned} c_1 &= \sinh^{-1}\left(\frac{S_{\max} - B}{\alpha}\right) \\ c_2 &= \sinh^{-1}\left(\frac{S_{\min} - B}{\alpha}\right) \\ \alpha &= \frac{S_{\max} - S_{\min}}{20} \end{aligned}$$

Compare approximated values (premium and delta).

4. For $K \in (0, H), T \in [0, \bar{T}]$, and $H > S_0$ the forward PIDE for up-and-out calls (Carr–Hirsa) is the following:

$$\int_{-\infty}^{\infty} [u(Ke^{-y}, T) - u(K, T) - \frac{\partial u(K,T)}{\partial K} K(e^{-y} - 1)] e^y k(y) dy$$
$$+ \frac{\sigma^2(K,T)K^2}{2} \frac{\partial^2(K,T)}{\partial K^2} - [r(T) - q(T)]K \frac{\partial u(K,T)}{\partial K}(K,T) - q(T)U(K,T)$$
$$= \left[\int_0^{\infty} u_k(He^{-y}, T)\nu(x,T) dy + \frac{\sigma^2(H,T)K^2}{2} u_{kkk}(H,T)\right](K - H)$$
$$+ \frac{\partial u(K,T)}{\partial T} + \int_0^{\infty} u(He^{-y}, T) e^y k(y) dx$$

with initial condition

$$u(K, 0) = (S_0 - K)^+, \quad K \in [0, H)$$

and for $H > S_0$. For boundary conditions, we use the following:

$$\lim_{K \downarrow 0} \frac{\partial^2}{\partial K^2} u(K,T) = 0, \quad T \in [0, \bar{T}]$$
$$\lim_{K \uparrow H} \frac{\partial^2}{\partial K^2} u(K,T) = 0, \quad T \in [0, \bar{T}]$$

Under the CGMY framework $k(y)$ is

$$\begin{aligned} k(y) &= \frac{e^{-\lambda_p y}}{\nu y^{1+Y}} \mathbb{1}_{y>0} + \frac{e^{-\lambda_n |y|}}{\nu |y|^{1+Y}} \mathbb{1}_{y<0} \\ \lambda_p &= \left(\frac{\theta^2}{\sigma^4} + \frac{2}{\sigma^2 \nu}\right)^{\frac{1}{2}} - \frac{\theta}{\sigma^2} \\ \lambda_n &= \left(\frac{\theta^2}{\sigma^4} + \frac{2}{\sigma^2 \nu}\right)^{\frac{1}{2}} + \frac{\theta}{\sigma^2} \end{aligned}$$

For the following variables: up barrier, $B = 125$, spot $S_0 = 100$, local volatility surface $\sigma(K,T) = 0$, risk-free rate $r = 1.7\%$, and dividend rate $q = 1.75\%$ and model parameters $\sigma = 25\%$, $\nu = 0.25$, $\theta = -0.15$, and $Y = 0.6$ solve both backward and forward PIDEs to fill in the following table:

| Maturity | | $T_1 = 0.25$ | | $T_2 = 0.5$ | | $T_3 = 1.0$ | |
|---|---|---|---|---|---|---|---|
| Barrier | Strike | **Bwd** | *Fwd* | **Bwd** | *Fwd* | **Bwd** | *Fwd* |
| 125 | 110 | | | | | | |
| | 120 | | | | | | |

# Chapter 6

## Simulation Methods for Derivatives Pricing

For derivatives pricing, in a Markovian framework, the Feynman–Kac formula establishes a link between partial differential equations (PDEs) and stochastic processes. As we have seen finite difference schemes can be used to numerically solve PDEs/PIDEs. As the number of dimensions in Markovian frameworks grows, finite difference schemes become unworkable and unrealistic. Monte Carlo methods may be related to finite difference schemes for solving PDEs/PIDEs via the Feynman–Kac characterization. But unlike finite difference schemes that are limited to Markovian frameworks, Monte Carlo simulation is not and does not suffer from dimensionality issues.

In this chapter we will introduce simulation or Monte Carlo methods for use in derivatives pricing applications. Fundamentally, simulation methods are very simple. The basic algorithm for derivatives pricing using simulation methods can be described very succinctly as follows:

<u>Derivative Valuation via Simulation</u>

for $j = 1, \ldots, \mathrm{N}$

Generate $X_j$

Compute $V_j = \mathrm{C}f(X_j)$

endfor

$C \approx \frac{1}{N} \sum_{j=1}^{N} V_j$

where $X_j$ is a vector of random variables affecting the contract price, whose distributions are determined by the model chosen, $f(X)$ is our pricing function for the contract based on the simulated random variables, and C is the appropriate constant for the measure used in pricing. This works because in pricing derivatives we calculate the following expression $V_T = \mathrm{C}\,\mathbb{E}\left[f(X_T)\right]$ under the corresponding measure and thus we can simply approximate this expectation with a simple average, as long as the market variables were generated to be consistent with the corresponding measure.

Thus the only requirements for using Monte Carlo methods are that we have some set of underlying random variables whose distribution is determined by our model, and we have a pricing function for the contract based on the outcome of these random variables. These lax requirements allow simulation-based pricing to be applied to an extremely wide variety of derivatives pricing problems. This makes simulation-based pricing the most widely applicable of the pricing methods described in this book; as long as we have some probabilistic model of the variables determining the contract value and a pricing function, we can value a contract using these methods. Because of this flexibility, simulation methods are often the only way to price certain kinds of derivative products, typically those with complex path-dependent payoffs.

While the flexibility of simulation methods makes these methods extremely powerful and widely applicable, this comes with a price. Typically simulation methods are the most expensive of all the available pricing methodologies; thus they are often used as a method of last resort, when the complexity of the contract makes simulation methods the only available means of pricing a derivative.

Simulation methods are powerful not just because of their lax requirements; they solve the problem of the curse of dimensionality which causes problems in other pricing methods. We saw in Chapter 2 that FFT methods were $O(n \ln n)$ and COS methods were $O(n)$; however, these methods are only applicable to single underlier assets with specific requirements on the model and/or payoff function. They can solve only one-dimensional problems. In Chapter 4 we saw that PDE methods are typically $O(n^2)$ for single underlier problems and while they can be scaled to handle multiple underlier cases, their complexity is $O(n^{d+1})$ where $d$ is the number of underliers, and the extra dimension is the time dimension. Practically, this means that, while we can handle two underlier cases with PDE/PIDE methods which have complexity $O(n^3)$, anything more than this is infeasible and we would have to turn to simulation methods. This additional level of complexity with each additional dimension is known as the curse of dimensionality in numerical integration, of which most derivatives pricing problems can be considered a subset. Simulation methods are one of the main ways that larger dimensional problems are made tractable.

Derivatives pricing via simulation solves the curse of dimensionality through the law of large numbers and the central limit theorem. The law of large numbers states that

$$\frac{1}{N}\sum_{j=1}^{N} X_j \to \mu \text{ as } N \to \infty \tag{6.1}$$

This allows us to use the simple average of our simulated values as an estimate of the expected value of the derivative contract's payoff and guarantees with enough samples will converge to the correct value of the contract, as long as risk-neutral pricing holds. The central limit theorem states that given a series of identically independent random variables $X$ with mean $\mu$ and standard deviation $\sigma$, as well as the simple average $\overline{X}_N = \frac{1}{N}\sum_{j=1}^{N} X_j$, we have

$$\frac{\overline{X}_N - \mu}{\sigma/\sqrt{N}} \to \mathcal{N}(0,1) \text{ as } N \to \infty \tag{6.2}$$

$$\overline{X}_N - \mu \to \mathcal{N}(0, \sigma^2/N) \tag{6.3}$$

This allows us to measure the accuracy of our estimate or the number of samples we need to achieve a certain degree of accuracy. This also applies in a multidimensional setting, leading to an order of convergence related to $O(N^{1/2})$ regardless of the dimension of the problem. Thus we can sidestep the curse of dimensionality associated with other methods which require evaluation of the pricing function in a large fraction of the multidimensional space. This is especially important for path-dependent contracts, as the dimensionality is not just governed by the price at the maturity of the contract, but the price at every time between the pricing date and maturity. This leads to a theoretically infinite but practically very high, dimensional problem and thus simulation methods may be the only applicable methods.

This chapter describes the most critical elements of simulation-based pricing methods. However, the topic is a vast one and entire books have been written on the subject. We will not attempt cover every aspect of simulation methods; we refer readers to [122] for a complete accounting of simulation methods and their use in finance. For simulation methods and recent advances in derivatives pricing we refer readers to [37], [43], and [41]. For valuing American options by simulation we refer readers to [40], [42], [45], [44], [47], [172], and [46] which cover pricing on single and multi assets and in one- and multi-dimensional cases.

This chapter will be divided into four sections: the first section will deal with the generation of random numbers in general, the second section will deal with generating of random processes using their stochastic differential equations, the third one will give some examples of generating sample paths under a couple of different models, and the last one will review various variance reduction techniques we can use to speed up the calculation of derivatives prices when using simulation methods.

**Models:**
All models which can be described by a set of random variables with a known distribution

**Option Types:**
Any option for which a pricing function exists

**Pros**

1. Allows for pricing under almost any model
2. Allows for pricing of contracts with possibly any payoff
3. Allows for efficient pricing of contracts with many underliers

**Cons**

1. Typically the most expensive pricing methods

---

## 6.1 Random Number Generation

Any simulation method begins with the generation of random numbers. In this section we will review the generation of random numbers, starting with uniform distribution and moving on to methods used to generate random samples from other distributions. Random number generation is a field unto itself, and we will not give a comprehensive review of it; however, we will provide a review of the most essential elements. For a more detailed review of random number generation, see [122].

### 6.1.1 Standard Uniform Distribution

Pseudo random number generation begins with the generation of standard uniform random numbers. There are many different algorithms which can be used to generate standard uniform random numbers, including linear congruential generators, combined generators, and others. They are typically evaluated in terms of their period length, reproducibility, speed, portability and degree and randomness typically described by a battery of statistical tests.

Throughout simulation in this book, the random number generator of L'Ecuyer with Bays–Durham shuffle and added safeguards is used. It returns a uniform random deviate between 0.0 and 1.0 (exclusive of the endpoint values). This module is taken from *Numerical Recipes in C* by Press, Teukolsky, Vetterling, and Flannery [189], page 282. For our purposes, we assume the availability of a sequence $U_1$, $U_2$, ... of independent random variables each satisfying

$$p(U_i \leq u) = \begin{cases} 0, & u < 0 \\ u, & 0 \leq u \leq 1 \\ 1, & u > 1 \end{cases}$$

A standard uniform distribution is called $Unif(0,1)$, which we will abbreviate it to $\mathcal{U}(0,1)$ and each uniformly distributed sample is called $U$ which is obviously between 0 and 1. Unless specified, $U$ is always sampled from $\mathcal{U}(0,1)$, a standard uniform distribution. A function to generate standard univariate random numbers should be available in every programming language and mathematical analysis package.

---

## 6.2 Samples from Various Distributions

Most simulations require sampling random variables or random vectors from probability distributions that are not uniform. There are two general and widely used techniques which can be used to generate samples from other random variables given standard uniform samples:

- Inverse transform method
- Acceptance–rejection method

### 6.2.1 Inverse Transform Method

The inverse transform method allows us to use standard uniform random samples to generate samples from any distribution for which we have a cumulative distribution function $F$. From Chapter 2 we know that we can recover the cumulative distribution function from the characteristic function; we can also apply this method to any distribution for which we have a characteristic function. A random variable which has CDF $F$ must satisfy the property that $P(X < x) = F(x)$ for all $x$. The inverse transform method allows us to generate random variables by applying the following relation:

$$x = F^{-1}(U) \text{ where } U \sim \mathcal{U}(0,1) \tag{6.4}$$

where $F^{-1}$ is the inverse of $F$ and as before $\mathcal{U}(0,1)$ denotes the standard uniform distribution on $[0,1]$. To verify that the inverse transform generates samples from $F$, we check the distribution of $x$ this procedure produces.

$$\begin{aligned} p(X \leq x) &= p(F^{-1}(U) \leq x) && (6.5) \\ &= p(U \leq F(x)) && (6.6) \\ &= F(x) && (6.7) \end{aligned}$$

The inverse of a function is well-defined if it is strictly increasing, and while most cumulative distribution functions are, if the one we are using is not we set the following rule:

$$F^{-1}(u) = \inf\{x : F(x) \leq u\} \tag{6.8}$$

Below we provide some examples of the inverse transform method as applied to some different distributions.

**Example 8** *Uniform distribution $U(a,b)$*

A uniform distribution on $(a,b)$ has the following probability density function:

$$f(x) = \begin{cases} 0 & , \quad x < a \\ \frac{1}{b-a} & , \quad a \leq x < b \\ 0 & , \quad x \geq b \end{cases}$$

and its cumulative distribution is

$$F(x) = \begin{cases} 0 & , \quad x < a \\ \frac{x-a}{b-a} & , \quad a \leq x < b \\ 1 & , \quad x \geq b \end{cases}$$

and its inverse is

$$X = F^{-1}(U) = (b-a)U + a$$

Therefore, having $U \sim \mathcal{U}(0,1)$, we can generate the uniform distribution as follows: $X = (b-a)U + a \sim \mathcal{U}(a,b)$.

**Example 9** *Bernoulli distribution*

The simplest model for binary outcomes is Bernoulli distribution and is used extensively in practice. A random variable $x \in (0,1)$ with parameter $p$ has a Bernoulli distribution if $P(x=0) = 1-p$ and $P(x=1) = p$, i.e., its probability density function is given by

$$f(x) = \begin{cases} 1-p & x = 0 \\ p & x = 1 \end{cases}$$

or equivalently $f(x) = p^x(1-p)^{1-x}$ and its cumulative distribution is

$$F(x) = \begin{cases} 1-p & x = 0 \\ 1 & x = 1 \end{cases}$$

The mean and variance of a Bernoulli random variable are $\mathbb{E}(x) = p$ and $var(x) = p(1-p)$, respectively. Therefore, having $U \sim \mathcal{U}(0,1)$, we can sample from a Bernoulli as follows:

if $U < 1-p$ set $x = 0$

otherwise set $x = 1$

**Example 10** *Exponential distribution*

The exponential distribution with mean $\theta$ has distribution

$$F(x) = 1 - \exp(-x/\theta), \quad x \geq 0 \tag{6.9}$$

This is the distribution of the times between jumps of a Poisson process with rate $\frac{1}{\theta}$. Inverting the exponential distribution gives us the following:

$$X = -\theta \log(1-U) \tag{6.10}$$

This can also be implemented as

$$X = -\theta \log U \tag{6.11}$$

because $U$ and $1-U$ have the same distribution. Therefore, having $U \sim \mathcal{U}(0,1)$, $X = -\theta \log U \sim \exp(\theta)$.

**Example 11** *Poisson distribution*

A random variable $x \in \mathbb{N}^+$ has a Poisson distribution if its probability distribution function is given by

$$f(x) = e^{-\lambda}\frac{\lambda^x}{x!} \tag{6.12}$$

The mean and variance of a Poisson random variable are $\mathbb{E}(x) = \lambda$ and $var(x) = \lambda$, respectively. Sampling from a Poisson is straightforward, given samples from an exponential distribution.

**Example 12** *Arcsine law distribution*

Arcsine law distribution is the distribution of the time that a standard Brownian motion attains its maximum (minimum) over the interval $[0, 1]$. The distribution is as follows:

$$F(x) = \frac{2}{\pi}\arcsin(\sqrt{x}) \quad 0 \leq x \leq 1 \tag{6.13}$$

The inverse transform method for sampling from this distribution is

$$X = \sin^2\left(\frac{U\pi}{2}\right), \quad U \sim \mathcal{U}(0,1) \tag{6.14}$$

$$= \frac{1}{2} - \frac{1}{2}\cos(\pi U) \tag{6.15}$$

In all of the above examples, finding the inverse of the cumulative distribution function analytically was relatively easy. If the inverse of the cumulative distribution function is not analytically available one may find the inverse either (a) numerically; for instance, having the characteristic function analytically, we can evaluate the CDF numerically and form a lookup table or (b) find a very efficient analytical approximation of the CDF which an instance of this is for the standard normal distribution as will be demonstrated later in this Chapter. However, the process of numerically evaluating the CDF to generate a random sample can be computationally expensive. In most cases, if we do not have an analytical form of the inverse function, we would not attempt the inverse transform method and would often use the acceptance–rejection method instead.

### 6.2.2 Acceptance–Rejection Method

The acceptance–rejection method is one of the most widely applicable methods for generating random variables. It involves generating random samples from a more convenient distribution and then rejecting some in order to generate samples from another desired distribution. It is typically used when the form of the target distribution makes it difficult to sample from it directly. Suppose that we want to generate samples from a distribution with density $f$ defined on support $\Upsilon$. Let $g$ be a density on $\Upsilon$ from which we know how to easily generate samples and which has the property

$$f(x) < cg(x) \text{ for all } x \in \Upsilon \tag{6.16}$$

where $c > 1$ is an appropriate bound on $\frac{f(x)}{g(x)}$. To apply the acceptance–rejection method, we generate a sample $X$ from $g$ and accept the sample with the probability $\frac{f(X)}{cg(X)}$. This can easily be implemented by sampling $U$ from $\mathcal{U}(0,1)$ and accepting $X$ if $U \leq \frac{f(X)}{cg(X)}$. If $X$ is

rejected, a new candidate is sampled from $g$ and the acceptance procedure is applied again. The process repeats until the acceptance test is passed; the accepted value is returned as a sample from $f$.
Acceptance–Rejection Method

1. generate $X$ from distribution $g$
2. generate $U$ from distribution $\mathcal{U}(0,1)$
3. if $U \leq \frac{f(X)}{cg(X)}$

   accept and return X

   else

   reject it and go to Step 1

Thus each sample generated via the acceptance–rejection method requires at least one sample from the distribution $g$ and one uniform random sample, along with the amortized cost of each rejection.

In order to prove the validity of this algorithm, we need to prove that if $U$ is sampled from $\mathcal{U}(0,1)$ and $X$ is sampled from distribution $g$ that

$$P\left(X \leq x | U \leq \frac{f(X)}{cg(X)}\right) = F(x) \tag{6.17}$$

as this probability described our sampling procedure.

To prove this, we first need to calculate the probability that $U \leq \frac{f(X)}{cg(X)}$, that is,

$$p = P\left(U \leq \frac{f(X)}{cg(X)}\right)$$

By first conditioning on $X = x$ we get

$$\begin{aligned} P\left(U \leq \frac{f(X)}{cg(X)} | X = x\right) &= P\left(U \leq \frac{f(x)}{cg(x)}\right) \\ &= \frac{f(x)}{cg(x)} \end{aligned}$$

That is because $U \sim \mathcal{U}(0,1)$ and by construction the ratio $\frac{f(X)}{cg(X)}$ is bounded between zero and one. To calculate unconditional probability we integrate $x$ out and obtain

$$\begin{aligned} p &= \int_{-\infty}^{\infty} P\left(U \leq \frac{f(X)}{cg(X)} | X = x\right) g(x)dx \\ &= \int_{-\infty}^{\infty} \frac{f(x)}{cg(x)} g(x)dx \\ &= \frac{1}{c}\int_{-\infty}^{\infty} f(x)dx \\ &= \frac{1}{c} \end{aligned}$$

We can look at $p$ as probability of success. Then the number of times that we generate

$X$ and $U$ in steps 1 and 2 has a geometric probability with probability of success $p$ and therefore

$$P(N = n) = (1 - p)^{n-1} p$$

That means on average the number of iterations required to successfully accept a sample generated by steps 1 and 2 is

$$\begin{aligned} \mathbb{E}(N) &= \frac{1}{p} \\ &= c \end{aligned}$$

Thus we hope to find a $g$ such that $c$ is as small as possible. We know from conditional probability that

$$P\left(X \leq x | U \leq \frac{f(X)}{cg(X)}\right) = \frac{P\left(U \leq \frac{f(X)}{cg(X)} | X \leq x\right) P(X \leq x)}{P\left(U \leq \frac{f(X)}{cg(X)}\right)}$$

Except for the term $P\left(U \leq \frac{f(X)}{cg(X)} | X \leq x\right)$, all other terms are known.

$$\begin{aligned} P(X < x) &= G(x) \\ P\left(U < \frac{f(X)}{cg(X)}\right) &= \frac{1}{c} \end{aligned}$$

And further we can prove that

$$\begin{aligned} P\left(U < \frac{f(X)}{cg(X)} | X < x\right) &= \frac{P\left(U < \frac{f(X)}{cg(X)}, X < x\right)}{P(X < x)} \\ &= \frac{\int_{-\infty}^{x} P\left(U < \frac{f(X)}{cg(X)} | X = u\right) g(u) du}{G(x)} \\ &= \frac{\int_{-\infty}^{x} \frac{f(u)}{cg(u)} g(u) du}{G(x)} \\ &= \frac{\frac{1}{c} \int_{-\infty}^{x} f(u) du}{G(x)} \\ &= \frac{\frac{1}{c} F(x)}{G(x)} \end{aligned}$$

Thus we have

$$\begin{aligned} P\left(X \leq x | U \leq \frac{f(X)}{cg(X)}\right) &= \frac{\frac{\frac{1}{c} F(x)}{G(x)} G(x)}{\frac{1}{c}} \\ &= F(x) \end{aligned}$$

And thus we have proven the validity of this method.

In the following section we provide some examples of how to generate different distributions via the acceptance–rejection method.

#### 6.2.2.1 Standard Normal Distribution via Acceptance–Rejection

This example is borrowed from the write-up in [201] on acceptance–rejection techniques. This method is not the most efficient way of generating normal random variables but it is a way of generating standard normal via the acceptance–rejection method. We wish to generate samples from the standard normal distribution $Z \sim \mathcal{N}(0,1)$. If we can generate samples from the absolute value, $|Z|$, then by symmetry we can obtain $Z$ by independently generating a random variable $S$ which will determine the sign of the sample as positive or negative with probability $\frac{1}{2}$ and setting $Z = S|Z|$. That means we must generate a $U \sim \mathcal{U}(0,1)$ random sample and set $Z = |Z|$ if $U < \frac{1}{2}$ and set $Z = -|Z|$ if $U > \frac{1}{2}$. $|Z|$ is one-sided, non-negative, and obviously has the following probability distribution function:

$$f(x) = \frac{2}{\sqrt{2\pi}} e^{-\frac{x^2}{2}} \tag{6.18}$$

A natural choice is $g(x) = e^{-x}$, $x \geq 0$, the exponential density with mean 1 as a density to sample from and we can easily sample from it. To use the acceptance–rejection method, we need to find $c > 1$ such that $f(x) < cg(x)$ for all $x \geq 0$. In order to do this we define

$$h(x) = \frac{f(x)}{g(x)} = \sqrt{2/\pi} e^{x - \frac{x^2}{2}} \tag{6.19}$$

By simply calculating the maximum of $h(x)$, we find the value of $x$ that maximizes the exponent $x - \frac{x^2}{2}$. It is easy to see that the maximum occurs at $x = 1$. Therefore $c = \sqrt{2e/\pi}$ and so

$$\frac{f(x)}{cg(x)} = e^{-(x-1)^2/2} \tag{6.20}$$

Thus the algorithm for generating $Z$ by the acceptance–rejection method is as follows:

1. generate $X$ from distribution $g$; that is, generate $U \sim \mathcal{U}(0,1)$ and set $X = -\ln(U)$
2. generate $U \sim \mathcal{U}(0,1)$
3. if $U < e^{-(X-1)^2/2}$, set $|Z| = X$, otherwise start from Step 1
4. generate $U \sim \mathcal{U}(0,1)$, set $Z = |Z|$ if $U < 0.5$, otherwise set $Z = -|Z|$.

But $U < e^{-(X-1)^2/2}$ if and only if $(X-1)^2/2 < -\ln U$ and since $-\ln U$ is exponential with mean 1, we can simplify the above algorithm as follows:

1. generate two independent $X_1$ and $X_2$ from distribution $g$; that is, generate $U_1$ and $U_2$ from $\mathcal{U}(0,1)$ and set $X_1 = -\ln(U_1)$ and $X_2 = -\ln(U_2)$
2. if $X_2 > (X_1 - 1)^2/2$, set $|Z| = X_1$, otherwise start from Step 1
3. generate $U \sim \mathcal{U}(0,1)$, set $Z = |Z|$ if $U < 0.5$, otherwise set $Z = -|Z|$.

As stated in [201], an interesting observation is the memoryless property of the exponential distribution, the amount by which $X_2$ exceeds $(X_1 - 1)^2/2$ in step 2 when $X_1$ is accepted, that is, $X \doteq X_2 - (X_1 - 1)^2/2$, is exponential from an exponential distribution with mean 1 and is independent of $X_1$. Therefore we get back an independent exponential for free which could then be used as one of the two exponentials that are needed in step 1, if we were to want to start generating yet another independent $\mathcal{N}(0,1)$ random variable. Thus, after repeated use of this algorithm, the expected number of uniforms required to generate one $Z$ is $(2c+1) - 1 = 2c \approx 2.64$.

#### 6.2.2.2 Poisson Distribution via Acceptance–Rejection

As stated earlier a random variable $x \in \mathbb{N}^+$ has a Poisson distribution if its probability distribution function is given by

$$f(x) = e^{-\lambda}\frac{\lambda^x}{x!} \tag{6.21}$$

where $x$ can be interpreted as the number of arrivals in a unit time. The inter arrival time $x_1, x_2, ...$ are exponentially distributed with a mean of $\frac{1}{\lambda}$, that is $\lambda$ arrivals in a unit time. If there are $n$ arrivals in a unit time, sum of the arrival times of the past $n$ observations has to be less than or equal to one, but if one more arrival time is added, it is greater then one (unit time). From Example 10 we know that times between jumps can be generated by $\frac{-1}{\lambda}\log U_i$ where $U_i \sim \mathcal{U}(0,1)$

$$\sum_{i=1}^{n}\frac{-1}{\lambda}\log U_i \leq 1 < \sum_{i=1}^{n+1}\frac{-1}{\lambda}\log U_i$$

multiplying both sides by $-\lambda$ and using the logarithmic property to get

$$\log\prod_{i=1}^{n}U_i \geq -\lambda > \log\prod_{i=1}^{n+1}U_i$$

or equivalently

$$\prod_{i=1}^{n}U_i \geq e^{-\lambda} > \prod_{i=1}^{n+1}U_i$$

Now, we can apply the acceptance–rejection method to generate a sample from the Poisson distribution.

1. Set $n = 0$, $p = 1$
2. Generate a random number $U_{n+1}$ and replace $p$ by $p \times U_{n+1}$.
3. If $p < e^{-\lambda}$, then accept $x = n$ which means there are $n$ arrivals at this unit time.
4. else, reject the current $n$, increase it by one, return to Step 2.

It begs a question how many random numbers on average are required to generate a Poisson variate. If $x = n$, then $n+1$ random numbers are required. That is because of the $n+1$ random numbers product. For large value of $\lambda$ ($\lambda > 15$) the acceptance–rejection technique becomes too expensive. For large value of $\lambda$ we can use normal distribution to approximate Poisson distribution [4]. When $\lambda$ is large

$$Z = \frac{x-\lambda}{\sqrt{\lambda}}$$

is approximately normally distributed with mean zero and variance one, thus

$$x = \lceil \lambda + \sqrt{\lambda}Z - 0.5 \rceil$$

is used to generate a Poisson random variable. Here $\lceil x \rceil$ rounds $x$ to the nearest integer greater than or equal to $x$. That is why the quantity inside is adjusted by 0.5.

#### 6.2.2.3 Gamma Distribution via Acceptance–Rejection

Suppose $x$ is a gamma random variable gamma($\alpha$, $\frac{1}{\beta}$) where $\alpha$ is the shape parameter and $\beta$ is the scale parameter. Thus this random variable has the distribution function

$$f(x) = \frac{1}{\Gamma(\alpha)}\beta^{\alpha}x^{\alpha-1}e^{-\beta x}$$

The gamma distribution can be simulated using acceptance–rejection by using the exponential density $g(x) = \lambda e^{-\lambda x}$ in which $\frac{1}{\lambda}$ is chosen as the mean of the gamma distribution, and it can be shown that this value is optimal.

First, we will explain why the mean of the gamma distribution is the optimal value for $\lambda$. Define $h(x) = \frac{f(x)}{g(x)}$. We want to determine an upper bound $c$ for $h(x)$ such that $h(x) < c$ for all $x > 0$ and we know that

$$h(x) = \frac{1}{\lambda\Gamma(\alpha)}\beta^{\alpha}x^{\alpha-1}e^{(\lambda-\beta)x} \tag{6.22}$$

We observe that $\lambda < \beta$ must if true if $h(x)$ must be bounded. We find the maximum of $h(x)$ by taking the first derivative and set it equal to zero, which yields $x^* = \frac{\alpha-1}{\beta-\lambda}$. We observe that for $\alpha$ smaller than one, $x^\star$ will be outside the domain. Actually for $\alpha < 1$ $h(x)$ is strictly decreasing and approaches infinity at zero. Therefore, if we restrict $\alpha > 1$ and we have a maximum $x^* > 0$ and $h(x^\star)$

$$h(x^*) = \frac{1}{\lambda\Gamma(\alpha)}\beta^{\alpha}(\alpha-1)^{\alpha-1}\left(\frac{1}{\beta-\lambda}\right)^{\alpha-1}e^{-(\alpha-1)} \tag{6.23}$$

We would like to find $\lambda$ that minimizes $h(x^*)$ which is equivalent to minimizing $\frac{1}{\lambda}\left(\frac{1}{\beta-\lambda}\right)^{\alpha-1}$. We take its derivatives and set it equal to zero and we obtain $\lambda^* = \frac{\beta}{\alpha}$. Note that since $\alpha > 1$ we still have $\lambda < \beta$.

#### 6.2.2.4 Beta Distribution via Acceptance–Rejection

The beta distribution has the following probability density function on $[0, 1]$:

$$f(x) = \frac{1}{B(\alpha,\beta)}x^{\alpha-1}(1-x)^{\beta-1} \tag{6.24}$$

where $\alpha$, $\beta > 0$ are shape parameters and $B(\alpha, \beta))$ is a normalization constant to ensure that the distribution function integrates to one

$$B(\alpha,\beta) = \int_0^1 x^{\alpha-1}(1-x)^{\beta_1} = \frac{\Gamma(\alpha)\Gamma(\beta)}{\Gamma(\alpha+\beta)} \tag{6.25}$$

Note that for $\alpha = \beta = 1$ we get standard uniform distribution. If both parameters are smaller than 1 we get a $U$-shaped distribution. If one parameter is equal to 1 and the other is greater than 1 we get a strictly convex distribution except for being equal to 2 that would be a straight line. In the case where both shape parameters are greater or equal to 1 we get a unimodal distribution.

For the case where both shape parameters are greater than or equal to 1, $f(x)$ has a unimodal distribution and attains its maximum at $\frac{\alpha-1}{\alpha+\beta-2}$. Therefore, if we choose $g$ to be standard uniform density, $g(x) = 1$ for $0 \leq x \leq 1$, then $f(x) \leq cg(x)$ for all $0 \leq x \leq 1$ where $c$ is the value of $f(x)$ evaluated at $\frac{\alpha-1}{\alpha+\beta-2}$.

$$c = \frac{1}{B(\alpha,\beta)}\left(\frac{\alpha-1}{\alpha+\beta-2}\right)^{\alpha-1}\left(\frac{\beta-1}{\alpha+\beta-2}\right)^{\beta-1} \tag{6.26}$$

Then we have

$$\frac{f(X)}{cg(X)} = \frac{f(X)}{c} \tag{6.27}$$

$$= \frac{X^{\alpha-1}(1-X)^{\beta-1}}{\left(\frac{\alpha-1}{\alpha+\beta-2}\right)^{\alpha-1}\left(\frac{\beta-1}{\alpha+\beta-2}\right)^{\beta-1}} \tag{6.28}$$

Thus the acceptance–rejection method for generating beta random variables would be as follows:

1. generate $U_1$ and $U_2$ from $\mathcal{U}(0,1)$
2. if $U_2 \leq \frac{U_1^{\alpha-1}(1-U_1)^{\beta-1}}{\left(\frac{\alpha-1}{\alpha+\beta-2}\right)^{\alpha-1}\left(\frac{\beta-1}{\alpha+\beta-2}\right)^{\beta-1}}$ set $X = U_1$, otherwise start from Step 1

Note that in this case, and in fact in any beta example using $g(x) = 1$, we do not need to know or compute the value of $B(\alpha, \beta)$ at all; it cancels out in the ratio $\frac{f(x)}{cg(x)}$.

While we have demonstrated the fact that we can generate a sample from a beta distribution via acceptance–rejection, there are other ways to generate these samples as well. For example, one can use the basic fact that if $X_1$ is a gamma random variable with the shape parameter $n+1$, and independently $X_2$ is a gamma random variable with shape parameter $m+1$, and both have the same scale parameter, then $X = \frac{X_1}{X_1+X_2}$ is a beta distribution with density $f(x) = bx^n(1-x)^m$.

Thus it suffices to have an efficient algorithm for generating the gamma distribution. In general, when $n$ and $m$ are integers, the gamma becomes Erlang (represented by sums of independently identically distributed exponentials); for example, if $X_1$ and $X_2$ are independent and identically distributed exponentials, then $X = \frac{X_1}{X_1+X_2}$ is uniform on $(0, 1)$.

### 6.2.3 Univariate Standard Normal Random Variables

In the previous sections we have described methods for generating univariate random samples and then using these to generate samples from other distributions via generalized methods. However, standard normal distribution is the most widely used distribution in simulation, and thus efficient generation of standard normal random samples is of paramount importance to many applications. To give a few examples, Brownian motions are closely linked to standard normal variables and in simulating Brownian motion we need to sample from a standard normal distribution. Also, any semi-martingale is a time change Brownian motion [215] and so simulating it will again require standard normal random samples. We already covered a method of generating samples from a standard normal distribution via the acceptance–rejection method. However, because of its central importance to simulation, specialized methods have been developed to generate the samples.

#### 6.2.3.1 Rational Approximation

The rational approximation routine uses rational approximation for lower tail quantile for standard normal distribution function. This routine returns an approximation of the inverse cumulative standard normal distribution function. That is, given $p$, where $0 < p < 1$, it returns an approximation to the $z$ satisfying $\Phi^{-1}(p)$, or equivalently $P(Z \leq z) = p$ where $Z$ is a random variable from the standard normal distribution. Algorithms for doing this use a minimax approximation by rational functions [86]. Here we follow the algorithm by Peter John Acklam [3] and the result has a relative error whose absolute value is less than $1.15e-9$.

Define the low and high regions for $p$ to be

$$\begin{aligned} p_l &= 0.02425 \\ p_h &= 1.0 - p_l \end{aligned}$$

(a) rational approximation for the lower region, that is, if $p < p_l$:

$$q = \sqrt{-2\log(p)}$$
$$z = \frac{(((((c_1q+c_2)q+c_3)q+c_4)q+c_5)q+c_6)}{((((d_1q+d_2)q+d_3)q+d_4)q+1)}$$

(b) rational approximation for the central region, that is, if $p_l \leq p \leq p_h$:

$$q = p - 0.5$$
$$r = q^2$$
$$z = \frac{(((((a_1r+a_2)r+a_3)r+a_4)r+a_5)r+a_6)q}{(((((b_1r+b_2)r+b_3)r+b_4)r+b_5)r+1)}$$

(c) rational approximation for the upper region, that is, if $p > p_h$:

$$q = \sqrt{-2\log(1-p)}$$
$$z = -\frac{(((((c_1q+c_2)q+c_3)q+c_4)q+c_5)q+c_6)}{((((d_1q+d_2)q+d_3)q+d_4)q+1)}$$

where the $a$ vector coefficients are

$$\begin{aligned} a_1 &= -39.69683028665376 \\ a_2 &= 220.9460984245205 \\ a_3 &= -275.9285104469687 \\ a_4 &= 138.3577518672690 \\ a_5 &= -30.66479806614716 \\ a_6 &= 2.506628277459239 \end{aligned}$$

$b$ vector coefficients are

$$\begin{aligned} b_1 &= -54.47609879822406 \\ b_2 &= 161.5858368580409 \\ b_3 &= -155.6989798598866 \\ b_4 &= 66.80131188771972 \\ b_5 &= -13.28068155288572 \end{aligned}$$

$c$ vector coefficients are

$$\begin{aligned} c_1 &= -0.007784894002430293 \\ c_2 &= -0.3223964580411365 \\ c_3 &= -2.400758277161838 \\ c_4 &= -2.549732539343734 \\ c_5 &= 4.374664141464968 \\ c_6 &= 2.938163982698783 \end{aligned}$$

$d$ vector coefficients are

$$\begin{aligned} d_1 &= 0.007784695709041462 \\ d_2 &= 0.3224671290700398 \\ d_3 &= 2.445134137142996 \\ d_4 &= 3.754408661907416 \end{aligned}$$

As mentioned, the relative error of the approximation has an absolute value less than $1.15e-9$. To get full machine precision we can perform one iteration of Halley's rational method.

$$\begin{aligned} e &= \frac{1}{2}\text{erfc}(-\frac{z}{\sqrt{2}}) - p \\ u &= e\sqrt{2\pi}\exp(\frac{z^2}{2}) \\ z &= z - \frac{u}{1+\frac{uz}{2}} \end{aligned}$$

where $\text{erfc}(x)$ is the complementary error function [68] and the last two equalities are from Halley's rational formula [217].

#### 6.2.3.2 Box–Muller Method

Let $X$ and $Y$ be independent and identically distributed (i.i.d.) standard normal random variables $\mathcal{N}(0,1)$ with joint PDF $f(x,y)$.

$$\begin{aligned} f(x,y) &= \frac{1}{\sqrt{2\pi}}e^{-x^2/2}\frac{1}{\sqrt{2\pi}}e^{-y^2/2} \\ &= \frac{1}{2\pi}e^{-\frac{x^2+y^2}{2}} \end{aligned}$$

with

$$\int_{-\infty}^{\infty}\int_{-\infty}^{\infty}\frac{1}{2\pi}e^{-\frac{x^2+y^2}{2}}dxdy = 1$$

We perform the following change of variables:

$$\begin{aligned} R^2 &= X^2 + Y^2 \\ \theta &= \arctan(\frac{Y}{X}) \end{aligned}$$

Under the new coordinate we can see that

$$\begin{aligned} X &= R\cos(\theta) \\ Y &= R\sin(\theta) \end{aligned}$$

and using the fact that $dxdy = RdRd\theta$ we obtain

$$\int_0^{2\pi}\int_0^{\infty}\frac{1}{2\pi}e^{-\frac{R^2}{2}}RdRd\theta = 1$$

We perform one more change of variable by setting $r = R^2$; under this change of variable we have $dr = 2RdR$ and $X$ and $Y$ can be written as coordinate

$$\begin{aligned} X &= \sqrt{r}\cos(\theta) \\ Y &= \sqrt{r}\sin(\theta) \end{aligned}$$

and finally we get

$$\int_0^{2\pi}\int_0^{\infty}\frac{1}{2\pi}e^{-\frac{r}{2}}\frac{1}{2}drd\theta = 1$$

Thus we can write the joint PDF as

$$\begin{aligned} f_{d,\Theta}(r,\theta) &= \frac{1}{2}\frac{1}{2\pi}e^{-r/2} \\ &= \frac{1}{2\pi}\frac{1}{2}e^{-r/2} \\ &= f_{\Theta}(\theta)f_d(r) \end{aligned}$$

where $f_d(r)$ is the PDF of exponential distribution with mean 2, $f_\Theta(\theta)$ is the PDF of uniform distribution on $[0, 2\pi]$, $\mathcal{U}(0, 2\pi)$, with $r = R^2 = x^2 + y^2$, $\theta = \arctan(\frac{y}{x})$. Therefore we can generate two independent normal variables as follows:

- generate $U_1$ and $U_2$ i.i.d. $\mathcal{U}(0,1)$
- $X = \sqrt{-2\ln U_1}\cos(2\pi U_2)$ and $Y = \sqrt{-2\ln U_1}\sin(2\pi U_2)$

As shown in earlier examples, $-2\ln U_1$ is a sample from an exponential with mean 2 and $2\pi U_2$ is a sample from $\mathcal{U}(0, 2\pi)$. Note that at each draw/sample, Box–Muller actually generates a pair of independent standard normal random variables. While the Box–Muller method is effective, it can be computationally expensive, as the evaluation of sin and cos are generally computationally expensive operations.

#### 6.2.3.3 Marsaglia's Polar Method

An improved version of the Box–Muller is Marsaglia's polar method, in which we can circumvent the computationally expensive evaluations of trigonometric functions. Consider $v_1$ and $v_2$ be two independent $\mathcal{U}(-1,1)$,[1] we can show that for $v_1$, $v_2$ with $v_1^2 + v_2^2 < 1$ the following transformation

$$\begin{pmatrix} S \\ \theta \end{pmatrix} = \begin{pmatrix} v_1^2 + v_2^2 \\ \frac{1}{2\pi}\arctan\left(\frac{v_2}{v_1}\right) \end{pmatrix}$$

generates two uniformly distributed random variables $S$ and $\theta$ on $[0,1]$. That means $S \sim \mathcal{U}(-1,1)$. In addition, $v_1$ and $v_2$ are obviously uniformly distributed in the square $[-1,1] \times [-1,1]$. Thus the angle of the vector $(v_1, v_2)$ is also uniformly distributed and can be used to calculate $\cos(\theta)$ and $\sin(\theta)$ using

$$\begin{aligned} \sin\theta &= \frac{v_2}{(v_1^2+v_2^2)^{1/2}} \\ \cos\theta &= \frac{v_1}{(v_1^2+v_2^2)^{1/2}} \end{aligned}$$

Here we use acceptance–rejection to sample points uniformly in the unit disc and then transform these points to normal by polar coordinate. $v_1$ and $v_2$ are i.i.d. $\mathcal{U}(-1,1)$ accepting

[1]Note that if $U_i \sim \mathcal{U}(0,1)$ then $V_i = 2U_i - 1 \sim \mathcal{U}(-1,1)$, which is how we can generate these samples.

**TABLE 6.1**: Elapsed time for sampling 100,000,000 standard normal random variables

| method | elapsed time (milliseconds) |
|---|---|
| acceptance–rejection | 26,095.10 |
| rational approximation | 5,458.20 |
| Box–Muller | 15,077.30 |
| Marsaglia polar | 7,785.83 |

only those pairs inside the unit circle produces points uniformly distributed over a disc of radius one. Conditional on acceptance, $S$ is uniformly distributed between $[0, 1]$. Dividing $v_1$ and $v_2$ by $\sqrt{S}$ projects it from unit disc to the unit circle on which it is uniformly distributed. Moreover $\frac{v_1}{\sqrt{S}}$ and $\frac{v_1}{\sqrt{S}}$ are independent of $S$ conditional on $S \leq 1$. Therefore we can generate two independent normal variables as follows:

1. generate $v_1$ and $v_2$ i.i.d. $\mathcal{U}(-1, 1)$
2. Set $S = v_1^2 + v_2^2$
3. If $S > 1$, then start over

   Otherwise

$$
\begin{aligned}
x &= v_1\sqrt{\frac{-2\ln S}{S}} \\
y &= v_2\sqrt{\frac{-2\ln S}{S}}
\end{aligned}
$$

In this algorithm there are rejections controlled by the condition $S \leq 1$. The probability of $S$ accepted, the area of the unit circle inside the square $[0, 1]^2$ to the area of the square, is

$$P(S < 1) \approx 0.785$$

Therefore, 21% of uniform samples $V_1$ and $V_2$ are rejected, for which $S > 1$. Despite this the algorithm is still more efficient than the one using trigonometric function calls. This is the most common method of generating samples from $\mathcal{N}(0, 1)$.

Table 6.1 shows the elapsed time in milliseconds for generating 100,000,000 samples from a standard normal distribution for the four methods: (a) sampling by acceptance–rejection, (b) Box–Muller, (c) Marsaglia's polar method, and (d) rational approximation. As we see, rational approximation and Marsaglia polar methods are faster than the other two.

### 6.2.4 Multivariate Normal Random Variables

Multivariate normal random variables play a very important role in financial applications, and in this section we will review how they can be generated. One-dimensional standard normal random variables, $\mathcal{N}(0, 1)$, have the following probability and cumulative distribution functions:

$$
\begin{aligned}
\phi(x) &= \frac{1}{\sqrt{2\pi}}e^{-\frac{1}{2}x^2} && (6.29) \\
\Phi(x) &= \frac{1}{\sqrt{2\pi}}\int_{-\infty}^{x} e^{-\frac{1}{2}u^2}du && (6.30) \\
& && (6.31)
\end{aligned}
$$

If $z \sim \mathcal{N}(0,1)$ then $x = \mu + \sigma z$ is a normal distribution with mean $\mu$ and standard deviation $\sigma$, namely, $\mathcal{N}(\mu, \sigma^2)$. A $d$-dimensional normal distribution is characterized by a $d$-vector $\mu$ and $d \times d$ covariance matrix $\Sigma$, where $\Sigma$ is symmetric and positive semi-definite,

$$\Sigma = \begin{pmatrix} \Sigma_{11} & \cdots & \Sigma_{1d} \\ \vdots & \ddots & \vdots \\ \Sigma_{1d} & \cdots & \Sigma_{dd} \end{pmatrix} \tag{6.32}$$

and can be decomposed and written as

$$\Sigma = \sigma \Lambda \sigma = \begin{pmatrix} \sigma_1 & & \\ & \ddots & \\ & & \sigma_d \end{pmatrix} \begin{pmatrix} 1 & \cdots & \rho_{1d} \\ \vdots & \ddots & \vdots \\ \rho_{1d} & \cdots & 1 \end{pmatrix} \begin{pmatrix} \sigma_1 & & \\ & \ddots & \\ & & \sigma_d \end{pmatrix} \tag{6.33}$$

where $\Lambda$ is the correlation matrix and $\sigma_i$ the standard deviation of $i$-th dimension. A $d$-dimensional normal distribution has the following probability distribution function:

$$\phi_{\mu,\Sigma}(x) = \frac{1}{(2\pi)^{d/2}|\Sigma|^{1/2}} e^{-\frac{1}{2}(x-\mu)^\top \Sigma^{-1}(x-\mu)} \tag{6.34}$$

$$\tag{6.35}$$

We know that if $z \sim \mathcal{N}(0, I)$ and $x = \mu + Az$ then $x \sim \mathcal{N}(\mu, AA^\top)$. Thus, the problem of sampling $x$ from the multivariate normal $\mathcal{N}(\mu, \Sigma)$ reduces to finding a matrix $A$ such that $AA^\top = \Sigma$. Matrix $A$ is not unique, among all such $A$s, a lower triangular one is particularly convenient because it reduces the calculation of $\mu + Az$ to the following:

$$\begin{aligned} x_1 &= \mu_1 + a_{11}z_1 \\ x_2 &= \mu_2 + a_{21}z_1 + a_{22}z_2 \\ &\vdots \\ x_n &= \mu_d + a_{d1}z_1 + a_{d2}z_2 + \cdots + a_{dd}z_d \end{aligned} \tag{6.36}$$

The lower triangle $A$ which satisfies $AA^\top = \Sigma$ can be found via Cholesky factorization.

### 6.2.5 Cholesky Factorization

For $d \times d$ covariance matrix $\Sigma$ we need to solve

$$\begin{aligned} \Sigma &= AA^\top \\ &= \begin{pmatrix} a_{11} & & & \\ a_{21} & a_{22} & & \\ \vdots & & \ddots & \\ a_{d1} & a_{d2} & \cdots & a_{dd} \end{pmatrix} \begin{pmatrix} a_{11} & a_{21} & \cdots & a_{d1} \\ & a_{22} & \cdots & a_{d2} \\ & & \ddots & \vdots \\ & & & a_{dd} \end{pmatrix} \end{aligned}$$

Simply by multiplying it, we can see

$$\begin{aligned} a_{11}^2 &= \sigma_{11} \\ a_{21}a_{11} &= \sigma_{21} \\ &\vdots \\ a_{d1}a_{11} &= \sigma_{d1} \\ a_{21}^2 + a_{22}^2 &= \sigma_{22} \\ &\vdots \\ a_{d1}^2 + \cdots + a_{dd}^2 &= \sigma_{dd} \end{aligned}$$

More compactly

$$\sigma_{ii} = \sum_{l=1}^{i} a_{il}^2 \qquad i = 1, \ldots, d$$

$$\sigma_{ij} = \sum_{l=1}^{j} a_{il} a_{jl} \qquad j \le i$$

and we get

$$a_{ii} = \sigma_{ii} - \sum_{l=1}^{i-1} a_{il}^2 \qquad i = 1, \ldots, d$$

$$a_{ij} = \left( \sigma_{ij} - \sum_{l=1}^{j-1} a_{il} a_{jl} \right) / a_{jj} \qquad j < i$$

Thus the algorithm for generating the Cholesky decomposition is as follows:
Pseudo-Code for Cholesky Decomposition

Start with a $d \times d$ zero matrix

for $j = 1, \ldots, d$

for $i = j, \ldots, d$

$x_i = \sigma_{ij}$

for $k = 1, \ldots, j-1$

$x_i = x_i - a_{jk} a_{ik}$

end for

$a_{ij} = x_i / \sqrt{x_j}$

end for

end for

#### 6.2.5.1 Simulating Multivariate Distributions with Specific Correlations

As we will discuss later in Chapter 7 on calibration techniques, assuming one Brownian motion (one factor) captures the behavior of the entire term structure is not realistic. For that reason, we typically use three or four to confine its evolution, one factor for very short maturity (3-month to 2-year), a second one f or short (2-year to 5-year), a third one for medium (5-year to 10-year), and the fourth and the last for very long maturities (15-year to 30-year). In order to simulate such a four factor model, we have to generate correlated standard normal random variables. Assume we have generated 40,000 i.i.d. standard normal random variables and resize them into four vectors of length 10,000; we expect to get an identity matrix as its correlation. Table 6.2 shows the correlation matrix that is pretty close to an identity matrix.

For factor correlation, we use a correlation of ten years of 3-month LIBOR rates, 5-year, 10-year, 30-year swap rates, which is illustrated in Table 6.3. We use Cholesky factorization

**TABLE 6.2**: Correlation of standard normal random vectors

| $\rho$ | $Z_1$ | $Z_2$ | $Z_3$ | $Z_4$ |
|---|---|---|---|---|
| $Z_1$ | 1 | -0.0034 | -0.0216 | -0.0083 |
| $Z_2$ | -0.0034 | 1 | 0.0047 | 0.01 |
| $Z_3$ | -0.0216 | 0.0047 | 1 | 0.0219 |
| $Z_4$ | -0.0083 | 0.01 | 0.0219 | 1 |

**TABLE 6.3**: Historical correlation of LIBOR and swap rates

| $\rho$ | 3m LIBOR | 5-year swap | 10-year swap | 30-year swap |
|---|---|---|---|---|
| 3m LIBOR | 1 | 0.1638 | 0.0817 | 0.0814 |
| 5y swap | 0.1638 | 1 | 0.7118 | 0.8595 |
| 10y swap | 0.0817 | 0.7118 | 1 | 0.6816 |
| 30y swap | 0.0814 | 0.8595 | 0.6816 | 1 |

to obtain

$$A = \begin{pmatrix} 1 & 0 & 0 & 0 \\ 0.1638 & 0.9865 & 0 & 0 \\ 0.0817 & 0.7080 & 0.7015 & 0 \\ 0.0814 & 0.8595 & 0.0965 & 0.4983 \end{pmatrix}$$

then use (6.36) to obtain

$$\begin{aligned} x_1 &= z_1 \\ x_2 &= 0.1638z_1 + 0.9865z_2 \\ x_3 &= 0.0817z_1 + 0.7080z_3 + 0.7015z_3 \\ x_n &= 0.0814z_1 + 0.8595z_2 + 0.0965z_3 + 0.4983z_4 \end{aligned}$$

Having $X$s we can now compute the correlation of them. Their correlations are displayed in Table 6.4. Comparing correlations between the historical correlation matrix in Table 6.3 and correlations in Table 6.4 we see that they are pretty close as expected.

**TABLE 6.4**: Correlation of standard normal random vectors after Cholesky factorization

| $\rho$ | $X_1$ | $X_2$ | $X_3$ | $X_4$ |
|---|---|---|---|---|
| $X_1$ | 1 | 0.1624 | 0.0646 | 0.0724 |
| $X_2$ | 0.1624 | 1 | 0.7087 | 0.8558 |
| $X_3$ | 0.0646 | 0.7087 | 1 | 0.6855 |
| $X_4$ | 0.0724 | 0.8558 | 0.6855 | 1 |

## 6.3 Models of Dependence

Here we discuss three models of dependence. Our discussion closely follows the work in [163]. The models under consideration are (a) full rank Gaussian copula model, (b) correlating Gaussian components in a variance gamma representation, and (c) linear mixtures of independent Lévy processes.

### 6.3.1 Full Rank Gaussian Copula Model

From one viewpoint there is no use to compute correlations of non-Gaussian variates as the result does not lead us to any ability of writing down the joint probability law. We simply have correlation estimates. On the other hand, if data are transformed to standard normal variates before correlation is computed, then computed correlations may be used to write down the joint multivariate normal law of the transformed Gaussian variates with original data being a non-linear transform of what just mentioned.

Let $X = (x_1, x_2, \ldots, x_n)$ be a vector of dimension $N$. Marginal distribution for each $x_i$ is given by

$$P(x_i \leq x) = F_i(x) \tag{6.37}$$

We may transform the marginal to standard normal variates by

$$Z_i = \Phi^{-1}(F_i(x_i)) \tag{6.38}$$

where $\Phi$ is the CDF of the standard normal variable. By construction $Z_i \sim \mathcal{N}(0,1)$ and we can recover $x_i$ by

$$x_i = F_i^{-1}(\Phi(Z_i)) \tag{6.39}$$

We suppose vector $Z = (Z_1, Z_2, \ldots, Z_N)$ is standard multivariate normal with the correlation matrix $C$.

### 6.3.2 Correlating Gaussian Components in a Variance Gamma Representation

Suppose that the marginal distributions are centered variance gamma with

$$X_i = \theta_i(g_i - 1) + \sigma_i\sqrt{g_i}Z_i \tag{6.40}$$

where $g_i$ and $Z_i$ are gamma and standard normal random variables.

We now further assume $Z$ is multivariate normal with the correlation matrix $C$. The joint probability density and the characteristic function are not available in closed form as we have to integrate over a large number of independent gamma densities but they emerge as products of square roots that do not separate out in either the density or characteristic function. However, the joint law is easily simulated from a multivariate normal simulation together with drawings from gamma densities.

### 6.3.3 Linear Mixtures of Independent Lévy Processes

Assume that

$$X = AY \tag{6.41}$$

where $A$ is a mixing matrix and $Y$ is independent. Given the characteristic function

$$\phi_j(u) = \mathbb{E}(e^{iuY_i}) \tag{6.42}$$

the joint characteristic function of $X$ may be easily derived as

$$\phi_X(u) = \prod_{j=1}^{N} \phi_i((A^\top u)_j) \tag{6.43}$$

---

## 6.4 Brownian Bridge

To simulate a standard Wiener process at time $t$ we use the fact that

$$W_t - W_0 \sim \sqrt{t}z \quad \text{where } z \sim \mathcal{N}(0,1) \tag{6.44}$$

and knowing that $W_0 = 0$ it simply becomes

$$W_t \sim \sqrt{t}z \tag{6.45}$$

Suppose we have simulated $W_{t_1}$ and $W_{t_2}$. It is now desired to fill in points in the interval $[t_1, t_2]$, that is, to interpolate between the already generated points $W_{t_1}$ and $W_{t_2}$. To do this we use a Brownian bridge that is required to go through the values $W_{t_1}$ and $W_{t_2}$.

A Brownian bridge $x$ is a process that at time $t_1$ has the value $a$ and at time $t_2$ has the value $b$. Between $t_1$ and $t_2$, $x$ behaves like a Brownian motion. A Brownian bridge satisfies the following stochastic differential equation:

$$dx_t = \frac{b - x_t}{t_2 - t}dt + dB_t, \quad x_{t_1} = a \tag{6.46}$$

where $B_t$ is a standard Brownian motion.

The linear SDE (6.46) can be solved explicitly and the solution is

$$x_t = a\frac{t_2 - t}{t_2 - t_1} + b\frac{t - t_1}{t_2 - t_1} + (t_2 - t)\int_{t_1}^{t} \frac{dB_u}{t_2 - u} \tag{6.47}$$

Knowing that the conditional distribution of $x_t$ is normal, it can be shown that its mean and variance are

$$\mathbb{E}_{t_1}(x_t) = a + (b - a)\frac{t - t_1}{t_2 - t_1} \tag{6.48}$$

$$Var_{t_1}(x_t) = \frac{(t_2 - t)(t - t_1)}{t_2 - t_1} \tag{6.49}$$

As an example, let us assume we want to construct a Brownian bridge between 0 and $T$ to calculate its values at $t_j$ for $j = 1, \ldots, m-1$ with $0 < t_1 < t_2 < \cdots < t_{m-1} < t_m = T$. We first generate $W_T$ at $T = t_m$, then use a Brownian bridge to get the entire path at $\{t_1, t_2, t_3, \ldots, t_{m-1}\}$. Using the value of $W_T$, and $W_{t_0} = W_0 = 0$, it generates $W_{t_1}$. It generates $W_{t_2}$ using $W_{t_1}$ and $W_T$, and it generates $W_{t_3}$ using $W_{t_2}$ and $W_T$. The construction proceeds until we reach $t_{m-1}$. Thus, the discretely sampled Brownian path is generated by determining its values at $T, t_1, t_2, t_3, \ldots, t_{m-1}$ according to

$$\begin{aligned}
W_T &= \sqrt{T} z_1 \\
W_{t_1} &= W_{t_0} + (W_T - W_{t_0})\frac{t_1 - t_0}{T - t_0} + \sqrt{\frac{(T - t_1)(t_1 - t_0)}{T - t_0}} z_2 \\
W_{t_2} &= W_{t_1} + (W_T - W_{t_1})\frac{t_2 - t_1}{T - t_1} + \sqrt{\frac{(T - t_2)(t_2 - t_1)}{T - t_1}} z_3 \\
&\vdots \\
W_{t_{m-1}} &= W_{t_{m-2}} + (W_T - W_{t_{m-2}})\frac{t_{m-1} - t_{m-2}}{T - t_{m-2}} + \sqrt{\frac{(T - t_{m-1})(t_{m-1} - t_{m-2})}{T - t_{m-2}}} z_m
\end{aligned} \tag{6.50}$$

where $z_i$ for $i = 1, \ldots, m$ are independent and identically distributed standard normal random variables $\mathcal{N}(0, 1)$.

---

## 6.5 Monte Carlo Integration

Consider the following integral

$$\int_{I_s} f(x)dx$$

where $I_s$ is the $s$-dimensional unit cube, $I_s = [0,1] \times \cdots \times [0,1]$. Assume we want to numerically evaluate the integral for $s = 20$. For just 10 discretization points on each dimension we will have $10^{20}$ number of grid points. The difficulty in computing the integral due to the high dimensionality is referred to as the curse of dimensionality. To overcome this hurdle, we employ Monte Carlo integration. Instead of trying to compute the integral via some numerical integration scheme, we sample the set $x_1, \ldots, x_N$, uniformly distributed $N$ vectors and evaluate the function $f$ at a set of points $x_1, \ldots, x_N$ and evaluate the following sum as an approximation for the integral:

$$\theta_N = \frac{1}{N} \sum_{i=1}^{N} f(x_i) \widehat{I_s}$$

where $\widehat{I_s}$ is the volume[2] of the integration domain $I_s$. Therefore the summation is an approximation for the integral

$$\int_{I_s} f(x)dx \approx \theta_N$$

[2] In this example $I_s = [0,1] \times \cdots \times [0,1]$ so it has volume 1.

From the law of large numbers we have

$$\begin{aligned}\lim_{N\uparrow\infty}\theta_N &= \lim_{N\uparrow\infty}\frac{1}{N}\sum_{i=1}^{N}f(x_i)\widehat{I}_s \\ &= \widehat{I}_s\mathbb{E}(f(x)) \\ &= \widehat{I}_s\int_{I_s}f(x)\frac{1}{\widehat{I}_s}dx \\ &= \int_{I_s}f(x)dx\end{aligned}$$

where we use the fact that the probability of $x_i$, that is uniform over $I_s$, is $\frac{1}{\widehat{I}_s}$. Now we are interested in how fast $\theta_N$ approaches the integral as $N$ approaches infinity. To quantify the convergence rate we define the error term $\delta_N$ as

$$\begin{aligned}\delta_N &\equiv \int_{I_s}f(x)dx-\theta_N \\ &= \int_{I_s}f(x)dx-\frac{1}{N}\sum_{i=1}^{N}f(x_i)\widehat{I}_s \\ &= \frac{1}{N}\sum_{i=1}^{N}\left(\int_{I_s}f(x)dx-f(x_i)\widehat{I}_s\right) \\ &= \frac{1}{N}\sum_{i=1}^{N}\left(\int_{I_s}f(x)\frac{1}{\widehat{I}_s}dx-f(x_i)\right)\widehat{I}_s \\ &= \frac{\widehat{I}_s}{N}\sum_{i=1}^{N}\left(\int_{I_s}f(x)\frac{1}{\widehat{I}_s}dx-f(x_i)\right)\end{aligned}$$

Denote

$$\Delta f(x_i)=\int_{I_s}f(x)\frac{1}{\widehat{I}_s}dx-f(x_i)$$

and we can see that

$$\delta_N \equiv \frac{\widehat{I}_s}{N}\sum_{i=1}^{N}\Delta f(x_i)$$

It is easy to show that $\Delta f(x_i)$ has zero mean that is

$$\mathbb{E}(\Delta f(x_i))=0$$

and also considering that for $i\neq j$, $x_i$ and $x_j$ are independent, then $\Delta f(x_i)$ and $\Delta f(x_j)$ are uncorrelated.

$$\mathbb{E}(\Delta f(x_i)\Delta f(x_i))=\mathbb{E}(\Delta f(x_i))\mathbb{E}(\Delta f(x_j))=0$$

Having these properties we can analyze the variance of $\delta_N$, that is,

$$\begin{aligned}\mathrm{Var}(\delta_N) &= \mathbb{E}(\delta_N^2)-(\mathbb{E}(\delta_N))^2 \\ &= \frac{\widehat{I}_s^2}{N^2}\sum_{i=1}^{N}\mathbb{E}(\Delta f(x_i)^2)^2+0 \\ &= \frac{\widehat{I}_s^2}{N}\left\{\int_{I_s}f^2(x)\frac{1}{\widehat{I}_s}dx-\left(\int_{I_s}f(x)\frac{1}{\widehat{I}_s}dx\right)^2\right\}\end{aligned}$$

Define

$$\sigma^2(f) = \int_{I_s} f^2(x)\frac{1}{\widehat{I_s}}dx - \left(\int_{I_s} f(x)\frac{1}{\widehat{I_s}}dx\right)^2$$

Then we get the following for the variance of the error term

$$\text{Var}(\delta_N) = \frac{\widehat{I_s^2}}{N}\sigma^2(f)$$

Observations are that the error in Monte Carlo integration can be reduced by increasing $N$ the number of sampling points. The rate of convergence is

$$\sqrt{\text{Var}(\delta_N)} \sim \frac{1}{\sqrt{N}}$$

From the variance we can see how Monte Carlo integration resolves the curse of dimensionality. As an example, for the integral of $s = 20$, if we sample $N = 10^6$ points, then the Monte Carlo error is approximately $O\left(\frac{1}{\sqrt{N}}\right) \sim \frac{1}{\sqrt{10^6}} = 10^{-3}$. It is important to notice that the magnitude in error term $\delta_N$ is independent of the dimension $s$. However, knowing that the error convergence rate is $O\left(\frac{1}{\sqrt{N}}\right)$ makes the convergence pretty slow. In order to improve this slow convergence, we should employ some kind of variance reduction method, which will be discussed later.

Knowing the variance, we can set up an error bound as an indicator of how accurate the Monte Carlo result is. This is not a strict bound. Despite this, by convention, a Monte Carlo result is often written as

$$\int_{I_s} f(x)dx \approx \frac{\widehat{I_s}}{N}\sum_{i=1}^{N} f(x_i) \pm \frac{\widehat{I_s^2}}{N}\sigma^2(f)$$

to indicate the sense of accuracy.

Note that if $f(x)$ is constant, $f(x) \equiv c$, then

$$\begin{aligned}
\sigma^2(f) &= \int_{I_s} f^2(x)\frac{1}{\widehat{I_s}}dx - \left(\int_{I_s} f(x)\frac{1}{\widehat{I_s}}dx\right)^2 \\
&= \int_{I_s} c^2\frac{1}{\widehat{I_s}}dx - \left(\int_{I_s} c\frac{1}{\widehat{I_s}}dx\right)^2 \\
&= c^2\int_{I_s} \frac{1}{\widehat{I_s}}dx - c^2\left(\int_{I_s} \frac{1}{\widehat{I_s}}dx\right)^2 \\
&= c^2 - c^2 \\
&= 0
\end{aligned}$$

that is, what we expect to get if an integrand is a constant and obviously there is no need to do the integral. The way to interpret $\sigma^2(f)$ is how much the function $f(x)$ deviates from a constant. If we can find a transformation that changes the coordinate $x$ to a new coordinate $y$ in which the transformed function is rather flat, then we can reduce the variance of Monte Carlo integration. This is the essence of variance reduction. Another observation is that $\sigma^2(f)$ involves two integrals, which can be estimated approximately through the Monte Carlo method by viewing $\frac{1}{\widehat{I_s}}$ inside the integrals as the uniform probability density:

$$\sigma^2(f) \approx \frac{1}{N}\sum_{i=1}^{N} f^2(x_i) - \left(\frac{1}{N}\sum_{i=1}^{N} f(x_i)\right)^2$$

where $x_i$ are sampled uniformly from the domain $I_s$.

### 6.5.1 Quasi-Monte Carlo Methods

Quasi-Monte Carlo methods can be regarded as a deterministic equivalent of classical Monte Carlo. They are used to evaluate multi-dimensional integrals with no closed-form solution. Consider

$$\int_{I_s} f(x)dx$$

over the $s$-dimensional unit cube, $I_s = [0,1] \times \cdots \times [0,1]$. As explained earlier in classical Monte Carlo integration we select set points $x_1, ..., x_N$, that is, a sequence of pseudo random numbers, and approximate the integral by

$$\theta_N = \frac{1}{N} \sum_{i=1}^{N} f(x_i) \widehat{I}_s$$

In quasi-Monte Carlo methods we select points deterministically. Specifically, quasi-Monte Carlo methods produce a deterministic sequence of points that provides the best possible spread in $I_s$. These deterministic sequences are referred to as low-discrepancy sequences (e.g., Niederreiter [182], Fang and Wang [113]). There exist a variety of different low-discrepancy sequences. Quasi-random number generators produce highly uniform samples of the unit hypercube. They are designed to minimize the discrepancy between the distribution of generated points and a distribution with equal proportions of points in each sub-cube of a uniform partition of the hypercube [151]. As a result, quasi-random number generators systematically fill the holes in any initial segment of the generated quasi-random sequence. Examples include the Halton sequence [127], the Sobol sequence [205], the Faure sequence [114], and the Niederreiter sequence [182]. In Figure 6.1 uniform random variables where uniform sampled from a unit square are plotted. An example of a two-dimensional low-discrepancy sequence from a Halton set is plotted in Figure 6.2 where it is clear from the graph that unlike the uniform random variables there is nothing random about these points. Unlike pseudo-random sequences, quasi-random numbers are too uniform to pass tradi-

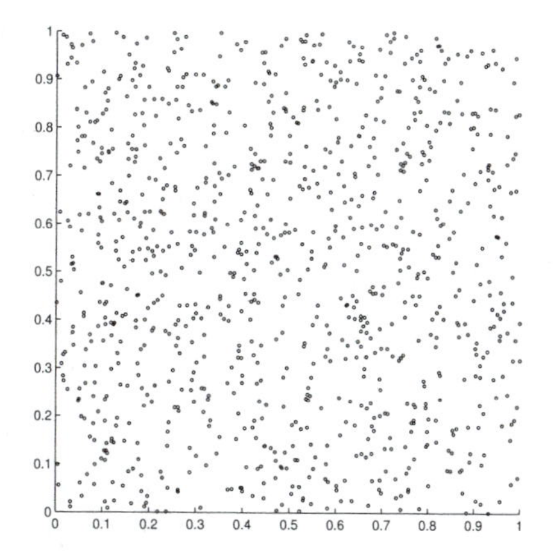


**FIGURE 6.1**: Plot of uniform random variables in unit square

tional tests of randomness (Kolmogorov–Smirnov test). In general, quasi-random sequences fail many statistical tests for randomness. Approximating true randomness, however, is not their goal. Quasi-random sequences seek to fill space uniformly, and to do so in such a way that initial segments approximate this behavior up to a specified density. However, due to their deterministic nature, statistical methods do not apply to quasi-Monte Carlo methods. Moreover, determining the error of quasi-Monte Carlo method analytically can be extremely difficult [51].

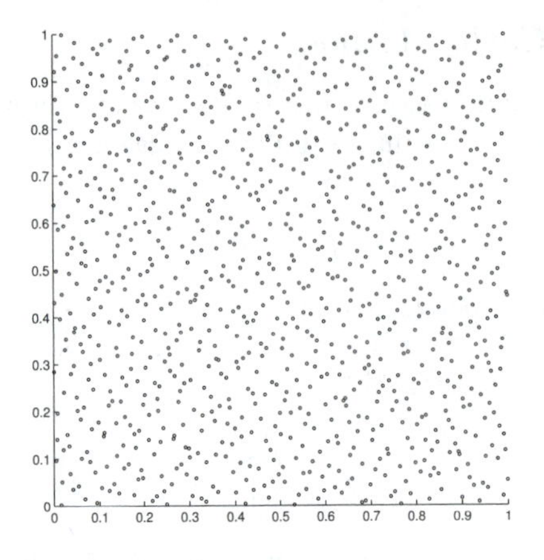

**FIGURE 6.2**: Plot of low discrepancy points (Halton set) sampled from a unit square

### 6.5.2 Latin Hypercube Sampling Methods

Latin hypercube sequences, though not quasi-random in the sense of minimizing discrepancy, nevertheless these sequences produce sparse uniform samples useful in experimental designs. Latin hypercube sampling is a statistical method for generating a distribution of plausible collections of parameter values from a multidimensional distribution. The sampling method is often applied in uncertainty analysis. The technique was introduced by McKay, Beckman, and Conover [176]. It was further elaborated by Iman, Helton, and Campbell [145], [144]. A comparison of Latin hypercube sampling with other techniques is given in Iman and Helton [143].

In the context of statistical sampling, a square grid containing sample positions is a Latin square if and only if there is only one sample in each row and each column. A Latin hypercube is the generalization of this concept to an arbitrary number of dimensions, whereby each sample is the only one in each axis-aligned hyperplane containing it.

---

## 6.6 Numerical Integration of Stochastic Differential Equations

In the last section, we reviewed methods for generating random variables. However, most models of asset prices or rates model the evolution of those prices or rates over time. As we saw in Chapter 4, many models can be expressed in terms of a stochastic differential equation (SDE). In this section we review how to simulate the evolution of a random process using its SDE. If we can solve the SDE directly then there is no need to simulate the SDE across a set of discrete points to arrive at a simulation of the process at some terminal time; we can just use the solution of the SDE to simulate samples at any point in time directly. However, it is rarely the case that the solution of an SDE is independent of its path and therefore in most cases we need to discretize the SDE and then simulate over each time interval in order to simulate paths of the process across time.

Consider the following generic one-dimensional SDE:

$$\begin{aligned} dX_t &= \mu(X_t,t)dt + \sigma(X_t,t)dW_t \qquad t_0 \leq t \leq T \\ X(t_0) &= X_0 \end{aligned}$$

where $W = \{W_t, 0 \leq t \leq T\}$ is a one-dimensional standard Wiener process, $\mu$ and $\sigma$ are the drift and the diffusion coefficient respectively. The assumption is that $\mu$ and $\sigma$ are defined and measurable.

The Itô-Taylor expansion for this SDE is

$$\begin{aligned} X_t &= X_{t_0} + \mu(X_{t_0}) \int_{t_0}^{t} ds + \sigma(X_{t_0}) \int_{t_0}^{t} dW(s) \\ &+ \frac{1}{2}\sigma(X_{t_0})\sigma'(X_{t_0}) \left( [W(t) - W(t_0)]^2 - (t - t_0) \right) + R \end{aligned} \tag{6.51}$$

where $R$ is the remainder. Once we have the Itô-Taylor expansion, we can construct numerical integration schemes for the SDE. A SDE defines the evolution of a random process over continuous time period, at infinitely many points in time. Thus we cannot simulate an SDE exactly subject to practical computational constraints. Thus, when we simulate an SDE we generate samples of the discretized version of SDE at a finite number of points

$$\hat{X}_{\Delta t}, \hat{X}_{2\Delta t}, \ldots, \hat{X}_{m\Delta t} \tag{6.52}$$

where $m$ is the number of time steps and $\Delta t$ is the time step assuming equidistant subintervals, $\Delta t = \frac{T-0}{m}$. To write it more formally

$$\hat{X}_{t_1}, \hat{X}_{t_2}, \ldots, \hat{X}_{t_j}, \ldots, \hat{X}_{t_m}$$

where $t_j = t_0 + j\Delta t = j\Delta t$ for $j = 1, \ldots, m$. As $\Delta t$ approaches zero, our discretized path will converge toward the theoretical continuous-time path. For the interval $[t_j, t_{i+1}]$ , by choosing

$$\begin{aligned} t_0 &= t_j, \\ t &= t_{j+1}, \\ \Delta t &= t_{j+1} - t_j, \\ \Delta W_j &= W(t_{j+1}) - W(t_j), \end{aligned}$$

we get the following expression for (6.51)

$$X_{t_{j+1}} = X_{t_j} + \mu(X_{t_j})\Delta t + \sigma(X_{t_j})\Delta W_j + \frac{1}{2}\sigma(X_{t_j})\sigma'(X_{t_j}) \left( (\Delta W_j)^2 - \Delta t \right) + R \tag{6.53}$$

There are various schemes for simulating SDEs of this form, and the most common ones are

Euler scheme

Milstein scheme

Runge–Kutta scheme

we will review each of these in turn.

### 6.6.1 Euler Scheme

The Euler scheme is the simplest discretization scheme available for discretizing SDEs. Keeping the first three terms in Equation (6.53) gives us the explicit Euler method

$$\begin{aligned} \hat{X}_{t_{j+1}} &= \hat{X}_{t_j} + \mu(\hat{X}_{t_j}, t_j)\Delta t + \sigma(\hat{X}_{t_j}, t_j)\Delta W_j \\ &= \hat{X}_{t_j} + \mu(\hat{X}_{t_j}, t_j)\Delta t + \sigma(\hat{X}_{t_j}, t_j)\sqrt{\Delta t} Z_j \end{aligned}$$

where $Z_j$ are i.i.d. $\mathcal{N}(0, 1)$. This approximation expands the drift term to $O(\Delta t)$ but only expands the diffusion term to $O(\sqrt{\Delta t})$, omitting the second term involving the diffusion implied by Itô's lemma.

### 6.6.2 Milstein Scheme

The Milstein scheme improves upon the Euler discretization by adding a second diffusion term, expanding the diffusion term to $O(\Delta t)$. Milstein scheme is obtained by simply keeping all terms of $O(\Delta t)$ in Equation (6.53), that is

$$\begin{aligned}\hat{X}_{t_{j+1}} &= \hat{X}_{t_j}+\mu(X_{t_j},t_j)\Delta t+\sigma(X_{t_j},t_j)\Delta W_j+\frac{1}{2}\sigma(X_{t_j},t_j)\sigma'(X_{t_j},t_j)[(\Delta W_j)^2-\Delta t] \\ &= \hat{X}_{t_j}+\mu(\hat{X}_{t_j},t_j)\Delta t+\sigma(\hat{X}_{t_j},t_j)\sqrt{\Delta t}Z_j+\frac{1}{2}\sigma(X_{t_j},t_j)\sigma'(X_{t_j},t_j)\Delta t(Z_j^2-1) \quad (6.54)\end{aligned}$$

where $\sigma'(x,t)=\frac{\partial}{\partial x}(\sigma(x,t))$. So while the Milstein scheme has a higher order in discretization, it requires knowing the first derivative of the the volatility function.

### 6.6.3 Runge–Kutta Scheme

While the Milstein scheme improves on the accuracy of the Euler scheme, it requires both knowledge of the first derivative of the volatility function, which may not be available at all or may be expensive to compute. The Runge–Kutta scheme allows us to avoid using the first derivative of the volatility function, by using the Runge–Kutta approximation, while still keeping the same order of accuracy.

Start from Taylor expansion of $\sigma(X_i+\Delta X_i)$

$$\sigma(X_i+\Delta X_i)=\sigma(X_i)+\sigma'(X_i)\Delta X+O((\Delta X)^2)$$

For $\Delta X_i=\mu(X_i)\Delta t+\sigma(X_i)\Delta W_i$ we have

$$\begin{aligned}\sigma(X_i+\Delta X_i)-\sigma(X_i) &= \sigma'(X_i)[\mu(X_i)\Delta t+\sigma(X_i)\Delta W_i]+O((\Delta X)^2) \\ &= \sigma'(X_i)\sigma(X_i)\Delta W_i+O(\Delta t) \quad (6.55)\end{aligned}$$

as $(\Delta X)^2\sim O(\Delta t)$. Another look at the term $\sigma(X_i+\Delta X_i)$ after substituting for $\Delta X_i$ by adding and subtracting the term $\sigma(X_i)\sqrt{\Delta t}$ to get

$$\begin{aligned}\sigma(X_i+\Delta X_i) &= \sigma(X_i+\mu(X_i)\Delta t+\sigma(X_i)\Delta W_i) \\ &= \sigma(X_i+\mu(X_i)\Delta t+\sigma(X_i)\sqrt{\Delta t}+\sigma(X_i)(\Delta W_i-\sqrt{\Delta t})) \quad (6.56)\end{aligned}$$

now using another Taylor expansion to get

$$\begin{aligned}\sigma(X_i+\Delta X_i) &= \sigma(X_i+\mu(X_i)\Delta t+\sigma(X_i)\sqrt{\Delta t}+\sigma(X_i)(\Delta W_i-\sqrt{\Delta t})) \\ &= \sigma(X_i+\mu(X_i)\Delta t+\sigma(X_i)\sqrt{\Delta t}) \\ &+ \sigma'(X_i+\mu(X_i)\Delta t+\sigma(X_i)\sqrt{\Delta t})\sigma(X_i)(\Delta W_i-\sqrt{\Delta t})+O((\Delta W_i-\sqrt{\Delta t})^2) \\ &= \sigma(X_i+\mu(X_i)\Delta t+\sigma(X_i)\sqrt{\Delta t}) \\ &+ \sigma'(X_i+\mu(X_i)\Delta t+\sigma(X_i)\sqrt{\Delta t})\sigma(X_i)(\Delta W_i-\sqrt{\Delta t})+O(\Delta t) \quad (6.57)\end{aligned}$$

Yet another Taylor expansion for the term $\sigma'(X_i+\mu(X_i)\Delta t+\sigma(X_i)\sqrt{\Delta t})$ to get

$$\begin{aligned}\sigma'(X_i+\mu(X_i)\Delta t+\sigma(X_i)\sqrt{\Delta t}) &= \sigma'(X_i)+(\mu(X_i)\Delta t+\sigma(X_i)\sqrt{\Delta t})\sigma''(X_i)+\ldots \\ &= \sigma'(X_i)+O(\sqrt{\Delta t}) \quad (6.58)\end{aligned}$$

Substituting (6.58) into (6.57) we get

$$\sigma(X_i+\Delta X_i)=\sigma(X_i+\mu(X_i)\Delta t+\sigma(X_i)\sqrt{\Delta t})+\sigma'(X_i)\sigma(X_i)(\Delta W_i-\sqrt{\Delta t})+O(\Delta t) \quad (6.59)$$

Now substituting (6.59) into (6.55) and canceling the common term $\sigma'(X_i)\sigma(X_i)\Delta W_i$ we get

$$\sigma(X_i + \mu(X_i)\Delta t + \sigma(X_i)\sqrt{\Delta t}) - \sigma(\Delta X_i) = \sigma'(X_i)\sigma(X_i)\sqrt{\Delta t} + O(\Delta t)$$

which implies

$$\sigma'(X_i)\sigma(X_i) = \frac{1}{\sqrt{\Delta t}}[\sigma(X_i + \mu(X_i)\Delta t + \sigma(X_i)\sqrt{\Delta t}) - \sigma(X_i)] + O(\sqrt{\Delta t})$$

To obtain Runge–Kutta scheme, we substitute (6.60) into the Milstein scheme (6.54). Thus, we have the following Runge–Kutta scheme

$$\begin{aligned}
\widehat{X}_i &= X_i + \mu(X_i)\Delta t + \sigma(X_i)\sqrt{\Delta t} \\
X_{i+1} &= X_i + \mu(X_i)\Delta t + \sigma(X_i)\Delta W_i \\
&+ \frac{1}{2\sqrt{\Delta t}}\left[\sigma(\widehat{X}_i) - \sigma(X_i)\right]((\Delta W_i)^2 - \Delta t)
\end{aligned}$$

A higher order Runge–Kutta scheme can be constructed in a similar way. For an overview of methods of Runge–Kutta type for SDEs look at [50], for derivations of new classes of stochastic Runge–Kutta schemes look at [1], [211] and [79]. Note that both the Milstein and Runge–Kutta schemes reduce to the Euler scheme if $\sigma(X_i)$ is constant.

## 6.7 Simulating SDEs under Different Models

In this section we will give a few practical examples of simulation via SDEs for a number of different models.

### 6.7.1 Geometric Brownian Motion

As we saw in Chapter 1, geometric Brownian motion (GBM) is governed by the following SDE:

$$dX_t = \mu X_t dt + \sigma X_t dW_t \tag{6.60}$$

However, we explicitly know the solution to this SDE via Itô's lemma

$$X_T = X_0 \exp\left\{(\mu - \frac{\sigma^2}{2})T + \sigma W_T\right\} \tag{6.61}$$

Thus for geometric Brownian motion, the distribution of $X_T$ is known and we can simulate $X_T$ directly, so there is no need for discretization. We already know that $W_T \sim \mathcal{N}(0,T) \sim \sqrt{T}\mathcal{N}(0,1)$, which implies

$$X_T = X_0 \exp\left\{(\mu - \frac{\sigma^2}{2})T + \sigma\sqrt{T}Z\right\} \tag{6.62}$$

where $Z \sim \mathcal{N}(0,1)$. Unfortunately this is not true for most models.

### 6.7.2 Ornstein–Uhlenbeck Process

As we saw in Chapter 1, the Ornstein–Uhlenbeck (OU) process is governed by the following SDE:

$$dX_t = \kappa(\theta - X_t)dt + \sigma dW_t \tag{6.63}$$

And again, we know the solution to this SDE via Itô's lemma

$$X_T = e^{-\kappa T}X_0 + \theta(1 - e^{-\kappa T}) + \sigma e^{\kappa T}\int_0^T e^{\kappa s}dW_s \tag{6.64}$$

Unlike the previous example, however, $X_T$ now depends on the entire path of the Brownian motion. However, for this example the distribution of $X_T$ is also known and we can simulate $X_T$ directly without having to discretize the SDE.

### 6.7.3 CIR Process

As we saw in Chapter 1, the CIR process is governed by the following SDE:

$$dX_t = \kappa(\theta(t) - X_t)dt + \sigma\sqrt{X_t}dW_t \tag{6.65}$$

There is no explicit solution to the SDE. We do not necessarily need an explicit solution to determine the distribution of $X_T$. In case of $\theta(t) = \theta$ we know that $X_T$ is distributed as a non-central chi-squared from which we can easily simulate. Unfortunately, once we move to a CIR with a time varying $\theta(t)$, the distribution of $X_T$ is not available, and one method of simulating $X_T$ indirectly is discretizing and simulating the SDE. This situation, where we do not know the distribution of $X_T$, is typical. Therefore, it is often necessary to simulate an SDE.

For illustrative purposes, in Figure 6.3 we plot two simulated paths of Vasicek versus CIR. In the simulation process we use the same normal random numbers for both for a fair comparison. For both cases we use the following parameters: $\sigma = 0.2$, $\kappa = 0.5$, $\theta = 0.05$, and $r_0 = 0.05$.

### 6.7.4 Heston Stochastic Volatility Model

The Heston stochastic volatility model is defined by the coupled two-dimensional SDE as covered in Chapter 1

$$\begin{aligned} dS_t &= rS_t dt + \sqrt{v_t}S_t dW_S(t), \\ dv_t &= \kappa(\theta - v_t)dt + \sigma\sqrt{v_t}dW_v(t), \end{aligned}$$

where the two Brownian components $W_S(t)$ and $W_v(t)$ are correlated with rate $\rho$. Conditional on time $s$ a Euler discretization of the variance process for $t > s$ reads

$$v(t) = v(s) - \kappa(\theta - v(s))\Delta t + \sigma\sqrt{v(s)}\sqrt{\Delta t}z_v$$

where $\Delta t = s - t$ with $z_v \sim \mathcal{N}(0,1)$. One can show the above scheme can go negative with positive probability and that is the main issue with this scheme [213], [10]. There are several fixes in the literature on this issue. Lord et al. [173] unify several Euler schemes in the following framework:

$$v(t) = f_1(v(s)) - \kappa(\theta - f_2(v(s)))\Delta t + \sigma\sqrt{f_3(v(s))}\sqrt{\Delta t}z_v$$

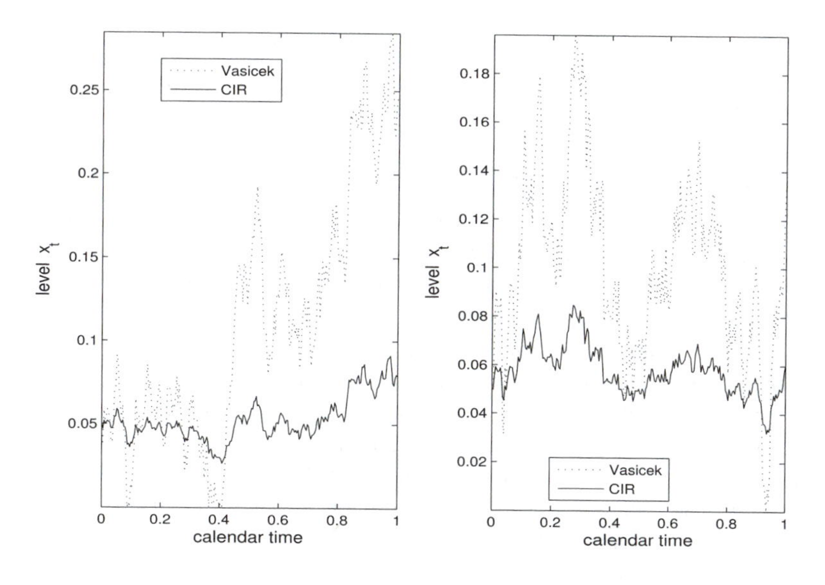


**FIGURE 6.3**: Simulated paths of Vasicek versus CIR

where all should satisfy $f_i(x) = x$ for $x \geq 0$ and $f_3(x) \geq 0$ for all $x$. The one that seems to work the best is produced by full truncation scheme that chooses $f_1(x) = x$ , $f_2(x) = f_3(x) = x^+$ where $x^+ = \max(x, 0)$. The resulting scheme is

$$v(t) = v(s) - \kappa(\theta - v(s)^+)\Delta t + \sigma\sqrt{v(s)^+}\sqrt{\Delta t}z_v \tag{6.66}$$

Now having a discretization scheme for the variance process, we now need to specify the simulation schemes of the asset price process. The most straightforward choices would be to either directly apply an Euler discretization scheme to the asset process or to simulate the price process from its exact (conditional) distribution. Direct discretization yields the following Euler scheme

$$S(t) = S(s) + rS(t)\Delta t + S(s)\sqrt{v(t)^+}\sqrt{\Delta t}z_S$$

Alternatively the exact solution by applying Itôs lemma yields

$$S(t) = S(s)\exp\left[\int_s^t (r - \frac{1}{2}v(u))du + \int_s^t \sqrt{v(u)}dW_S(u)\right]$$

By taking the log and utilizing Euler discretization we obtain the following scheme

$$\log(S(t)) = \log(S(s)) + [r - \frac{1}{2}v(s)]\Delta t + \sqrt{v(s)^+}\sqrt{\Delta t}z_S \tag{6.67}$$

where $z_S = \rho z_v + \sqrt{1-\rho^2}z$ with $z \sim \mathcal{N}(0, 1)$.

#### 6.7.4.1 Full Truncation Algorithm

Following the above argument, the full truncation scheme for the Heston can be summarized by the following algorithm:

Generate a random sample $z_v$ from a standard normal distribution.

Having $v(s)$, compute $v(t)$ from Equation (6.66).

Generate a random sample $z$ from the standard normal distribution and set $z_S = \rho z_v + \sqrt{1-\rho^2} z$.

Having $\log S(s)$, compute $\log S(t)$ using Equation (6.67).

In [213], the authors give an overview of the existing schemes for simulation of the Heston model. They found it rather remarkable that Euler full truncation scheme [173] outperforms by far many more complex schemes like the *almost exact simulation method* of Smith [204] and the Kahl and Jäckel scheme [158] in terms of computational efficiency. It is pointed that even though the Euler full truncation method is simple and straightforward to implement, it produces biased estimates for coarse time intervals. In their setting, they found at least one has to use 32 time steps per year to obtain reasonable small biases. This is of no surprise considering that the Euler scheme uses no analytical properties of the non-central chi-squared distribution of the variance process. The drift interpolation scheme of Broadie and Kaya [48] and the scheme proposed by Andersen in [10] do not have the biases and do not suffer from computational inefficiency.

### 6.7.5 Variance Gamma Process

Formally the VG process $X(t;\sigma,\nu,\theta)$ is obtained by evaluating Brownian motion with drift $\theta$ and volatility $\sigma$ at a random time given by a gamma process $\gamma(t;1,\nu)$ with mean rate unity and variance rate $\nu$ as

$$X(t;\sigma,\nu,\theta) = \theta\gamma(t;1,\nu) + \sigma W(\gamma(t;1,\nu))$$

with the characteristic function

$$\phi(u) = \mathbb{E}(e^{iuX_t}) = \left(\frac{1}{1 - iu\theta\nu + \sigma^2 u^2 \nu/2}\right)^{t/\nu}$$

We suppose the stock price process is given by the geometric VG law with parameters $\sigma$, $\nu$, $\theta$ and the log price at time $t$ is given by

$$\ln S_t = \ln S_0 + (r - q + \omega)t + X(t;\sigma,\nu,\theta)$$

where

$$\begin{aligned} \omega &= -\frac{1}{t}\log(\phi(-i)) \\ &= \frac{1}{\nu}\ln(1 - \theta\nu - \sigma^2\nu/2) \end{aligned}$$

is the usual Jensen's inequality correction ensuring that the mean rate of return on the asset is risk neutrally $(r-q)$.

In addition to the volatility of the normal distribution $\sigma$, there are parameters that control for (i) kurtosis, $\nu$ (long tailedness, a symmetric increase in the left and right tail probabilities relative to the normal for the return distribution) and (ii) skewness, $\theta$, that allows for the asymmetry of the left and right tails of the return density. An additional attractive feature of the model is that it nests the lognormal density and the Black–Scholes formula as a parametric special case ($\nu = 0$ and $\theta = 0$).[3] Also, there is a closed-form formula

[3] In the case of $\nu = 0$ and $\theta = 0$, we have

$$\lim_{\nu\downarrow 0 \text{ and } \theta\downarrow 0} \omega = \frac{1}{\nu}\ln(1 - \theta\nu - \sigma^2\nu/2) = -\frac{1}{2}\sigma^2$$

for pricing European options when the underlying asset follows the log VG process in terms of the special functions of mathematics as explained in [175].

Assume $N$ equidistant time intervals of length $h$ where $h = T/N$. To sample for a time interval of length $h$, we sample from a gamma distribution with mean $h$ and variance $\nu h$. A gamma process with the shape parameter $\alpha$ and the scale parameter $\beta$ has the following probability distribution function:

$$f(x, \alpha, \beta) = \frac{1}{\Gamma(\alpha)\beta^{\alpha}} x^{\alpha-1} e^{-\frac{x}{\beta}}$$

where its mean and variance are

$$\begin{aligned} \mu &= \alpha\beta \\ \sigma^2 &= \alpha\beta^2 \end{aligned}$$

In the case of the variance gamma process we have

$$\begin{aligned} \alpha\beta &= h \\ \alpha\beta^2 &= \nu h \end{aligned}$$

which implies

$$\begin{aligned} \alpha &= \frac{h}{\nu} \\ \beta &= \nu \end{aligned}$$

Therefore a sample for the VG process, $X(h; \sigma, \nu, \theta)$, is

$$\theta g(h/\nu, \nu) + \sigma\sqrt{g(h/\nu, \nu)}z$$

where $z \sim \mathcal{N}(0, 1)$ and $g(h/\nu, \nu) \sim \text{gamrand}(h/\nu, \nu)$.

The following is an algorithm for simulating a VG process.

for $i = 1, \ldots, N$

$$\begin{aligned} &z \sim \mathcal{N}(0, 1) \\ &g \sim \text{gamrand}(h/\nu, \nu) \\ &X_i = \theta g + \sigma\sqrt{g}z \end{aligned}$$

end

For the log of stock price we have

for $i = 1, \ldots, N$

$$\log S_i = \log S_{i-1} + (r - q)h + \omega h + X_i$$

end

Also

$$X(t; \sigma, 0, 0) = \sigma W(\gamma(t; 1, 0)) = \sigma W_t$$

which yields

$$\ln S_t = \ln S_0 + (r - q - \frac{1}{2}\sigma^2)t + \sigma W_t$$

which is exactly the evolution for geometric Brownian motion (Black–Scholes process).

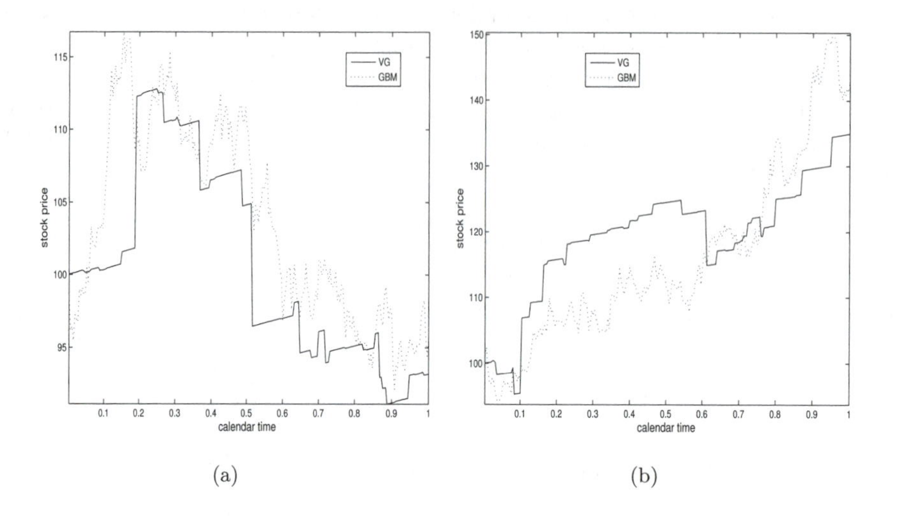


**FIGURE 6.4**: VG simulated paths versus GBM simulated paths

where $\omega = \frac{1}{\nu}\log(1 - \theta\nu - \sigma^2\nu/2)$

In Figures 6.4(a) and 6.4(b) we plot two simulated paths for VG stock price evolution versus stock price GBM. In the simulation process we use the same normal random numbers for both for a fair comparison. For VG we use the following parameters: $\sigma = 0.2$, $\nu = 0.2$, $\theta = -0.1$, and for Black–Scholes $\sigma = 0.2$. For risk-free interest rate and dividend rate $r = .01$ and $q = 0.02$, respectively.

### 6.7.6 Variance Gamma with Stochastic Arrival (VGSA) Process

As quoted in [54] the basic intuition underlying the approach to stochastic volatility arises from the Brownian scaling property. This property relates changes in scale to changes in time and thus random changes in volatility can alternatively be captured by random changes in time. This rate of time change must be mean reverting if the random time changes are to persist. The classic example of a mean-reverting positive process is the square root process of Cox, Ingersoll, and Ross. To obtain variance gamma with stochastic arrival (VGSA), as explained in [54], we take the VG process which is a homogeneous Lévy process and build in stochastic volatility by evaluating it at a continuous time change given by the integral of a Cox, Ingersoll, and Ross [82] (CIR) process. The mean reversion of the CIR process introduces the clustering phenomena often referred to as volatility persistence. This enables us to calibrate to market price surfaces that go across strike and maturity simultaneously. This process is tractable in the analytical expressions for its characteristic function.

Formally we define the CIR process $y(t)$ as the solution to the stochastic differential equation

$$dy_t = \kappa(\eta - y_t)dt + \lambda\sqrt{y_t}dW_t$$

where $W_t$ is a Brownian motion, $\eta$ is the long-term rate of time change, $\kappa$ is the rate of mean reversion, and $\lambda$ is the volatility of the time change.

The process $y(t)$ is the instantaneous rate of time change and so the time change is given by $Y(t)$ where

$$Y(t) = \int_0^t y(u)du$$

The characteristic function for $Y(t)$ is given by

$$\begin{aligned} \mathbb{E}(e^{iuY(t)}) &= \phi(u, t, y(0), \kappa, \eta, \lambda) \\ &= A(t, u)e^{B(t,u)y(0)} \end{aligned} \tag{6.68}$$

where

$$\begin{aligned} A(t, u) &= \frac{\exp\left(\frac{\kappa^2 \eta t}{\lambda^2}\right)}{\left(\cosh(\gamma t/2) + \frac{\kappa}{\gamma}\sinh(\gamma t/2)\right)^{2\kappa\eta/\lambda^2}} \\ B(t, u) &= \frac{2iu}{\kappa + \gamma\coth(\gamma t/2)} \end{aligned}$$

with

$$\gamma = \sqrt{\kappa^2 - 2\lambda^2 iu}$$

The stochastic volatility Lévy process, termed the VGSA process, is defined by

$$\begin{aligned} Z(t) &= X_{VG}(Y(t); \sigma, \nu, \theta) \\ &= \theta\gamma(Y(t); 1, \nu) + \sigma W(\gamma(Y(t); 1, \nu)) \end{aligned} \tag{6.69}$$

Thus $\sigma$, $\nu$, $\theta$, $\kappa$, $\eta$, and $\lambda$ are the six parameters defining the process. Its characteristic function is given by

$$\mathbb{E}(e^{iuZ_{VGSA}(t)}) = \phi(-i\Psi_{VG}(u), t, \frac{1}{\nu}, \kappa, \eta, \lambda)$$

where $\phi$ is the characteristic function of $Y(t)$ given in (6.68)and $\Psi_{VG}$ is the log characteristic function of the variance gamma process at unit time, namely,

$$\Psi_{VG}(u) = -\frac{1}{\nu}\log\left(1 - iu\theta\nu + \sigma^2\nu u^2/2\right)$$

We define the stock process at time $t$ by the random variable

$$S(t) = S(0)\frac{e^{(r-q)t+Z(t)}}{\mathbb{E}[e^{Z(t)}]} \tag{6.70}$$

We note that

$$\mathbb{E}[e^{Z(t)}] = \phi(-i\Psi_{VG}(-i), t, \frac{1}{\nu}, \kappa, \eta, \lambda) \tag{6.71}$$

which is equivalent to $e^{-\omega t}$ in the VG case. Therefore the characteristic function of the log of the stock price at time $t$ is given by

$$\mathbb{E}[e^{iu\log S_t}] = \exp(iu(\log S_0 + (r-q)t)) \times \frac{\phi(-i\Psi_{VG}(u), t, \frac{1}{\nu}, \kappa, \eta, \lambda)}{\phi(-i\Psi_{VG}(-i), t, \frac{1}{\nu}, \kappa, \eta, \lambda)^{iu}}$$

With a closed form for the VGSA characteristic function for the log price, one can employ various transform techniques to price European call and put options [60], [21], and [111]. The resulting model may be used to estimate parameter values consistent with market option prices for vanilla options across the entire strike and maturity spectrum.

To simulate the VGSA process, as before, we assume $N$ equidistant time intervals of length $h$ where $h = T/N$. We wish to simulate the VGSA process over each subinterval $h$ at time $t+h$. From Equation (6.69) we can write

$$\begin{aligned}\Delta Z_t &= Z(t) - Z(t-h) \\ &= \theta\gamma(Y(t);1,\nu) + \sigma W(\gamma(Y(t);1,\nu)) \\ &- (\theta\gamma(Y(t-h);1,\nu) + \sigma W(\gamma(Y(t-h);1,\nu))) \\ &= \theta(\gamma(Y(t);1,\nu) - \gamma(Y(t-h);1,\nu)) \\ &+ \sigma\sqrt{\gamma(Y(t);1,\nu) - \gamma(Y(t-h);1,\nu)}z\end{aligned}$$

where $z$ is $\mathcal{N}(0,1)$. The gamma process $\gamma(Y(t);1,\nu)$ with mean $Y(t)$ and variance $Y(t)\nu$ has the following shape and scale parameters

$$\begin{aligned}\alpha &= \frac{Y(t)}{\nu} \\ \beta &= \nu\end{aligned}$$

That implies

$$\begin{aligned}\gamma(Y(t);1,\nu) - \gamma(Y(t-h);1,\nu)) &= \text{Gamma}\left(\frac{Y(t)}{\nu},\nu\right) - \text{Gamma}\left(\frac{Y(t-h)}{\nu},\nu\right) \\ &= \text{Gamma}\left(\frac{Y(t)-Y(t-h)}{\nu},\nu\right)\end{aligned}$$

by the summation property of gamma processes. Hence

$$\Delta Z_t = \theta\text{Gamma}\left(\frac{Y(t)-Y(t-h)}{\nu},\nu\right) + \sigma\sqrt{\text{Gamma}\left(\frac{Y(t)-Y(t-h)}{\nu},\nu\right)}\, z \qquad (6.72)$$

where $Y(t) - Y(t-h) = \int_{t-h}^{t} y(u)du$. For simulation, we first start by discretization of the CIR process. Milstein discretization of the CIR process gives

$$y_j = y_{j-1} + \kappa(\eta - y_{j-1})h + \lambda\sqrt{y_{j-1}}\sqrt{h}z + \frac{\lambda^2}{4}h(z^2-1)$$

where $y_j$ is an approximation to $y(t_j)$ with $t_j = jh$ for $j = 0,\ldots,N$ and $z \sim \mathcal{N}(0,1)$. For the interval $(t_{j-1}, t_j)$ the new clock is given by the integral

$$\int_{t_{j-1}}^{t_j} y(u)du$$

and by the trapezoidal rule we get

$$\int_{t_{j-1}}^{t_j} y(u)du = \frac{h}{2}(y_{j-1} + y_j)$$

Therefore to simulate a VGSA process we do the following:

for $j = 1, \ldots, N$

$$z \sim \mathcal{N}(0,1)$$
$$y_j = y_{j-1} + \kappa(\eta - y_{j-1})h + \lambda\sqrt{y_{j-1}}\sqrt{h}z + \frac{\lambda^2}{4}h(z^2-1)$$
$$\hat{t}_j = \frac{h}{2}(y_j + y_{j-1})$$
$$g = \text{gamrand}(\frac{\hat{t}_j}{\nu}, \nu)$$
$$z \sim \mathcal{N}(0,1)$$
$$X_j = \theta g + \sigma\sqrt{g}z$$

end

where $X_j$ is an approximation to $\Delta Z_{t_j}$. An alternative is to use the scaling property of gamma processes and do simulation as follows:

for $j = 1, \ldots, N$

$$z \sim \mathcal{N}(0,1)$$
$$y_j = y_{j-1} + \kappa(\eta - y_{j-1})h + \lambda\sqrt{y_{j-1}}\sqrt{h}z + \frac{\lambda^2}{4}h(z^2-1)$$
$$\widehat{t}_j = \frac{h}{2}(y_j + y_{j-1})$$
$$\widehat{\sigma} = \sigma\sqrt{\widehat{t}_j}$$
$$\widehat{\nu} = \nu/\widehat{t}_j$$
$$\hat{\theta} = \theta\widehat{t}_j$$
$$g = \text{gamrand}(\frac{1}{\hat{\nu}}, \hat{\nu})$$
$$z \sim \mathcal{N}(0,1)$$
$$X_j = \widehat{\theta}g + \widehat{\sigma}\sqrt{g}z$$

end

For the logarithmic of stock price, from Equation (6.70) we have

$$\begin{aligned}
\log S_t &= \log S_0 + (r-q)t + Z(t) - \log(\mathbb{E}(e^{Z(t)})\\
\log S_{t-h} &= \log S_0 + (r-q)(t-h) + Z(t-h) - \log(\mathbb{E}(e^{Z(t-h)})
\end{aligned}$$

Subtracting the two equations yields

$$\begin{aligned}
\log S_t &= \log S_{t-h} + (r-q)h + Z(t) - Z(t-h) + \log(\mathbb{E}(e^{Z(t-h)}) - \log(\mathbb{E}(e^{Z(t)})\\
&= \log S_{t-h} + (r-q)h + \Delta Z_t + \log(\mathbb{E}(e^{Z(t-h)}) - \log(\mathbb{E}(e^{Z(t)})\\
&= \log S_{t-h} + (r-q)h + \Delta Z_t + \Delta\omega_t
\end{aligned}$$

Thus the simulation of the logarithmic of stock price under VGSA becomes

for $j = 1, \ldots, N$

$$\widehat{\Delta\omega}_j = \log(\phi(-i\Psi_{VG}(-i), (j-1)h, \tfrac{1}{\nu}, \kappa, \eta, \lambda)) - \log(\phi(-i\Psi_{VG}(-i), jh, \tfrac{1}{\nu}, \kappa, \eta, \lambda))$$
$$\log S_j = \log S_{j-1} + (r-q)h + \widehat{\Delta\omega}_j + X_j$$

end

## 6.8 Output/Simulation Analysis

Once we have decided on a discretization scheme, we need a general algorithm for derivatives pricing. As we saw in Section 1.3, most derivatives pricing problems can be expressed as an expectation of the form $\theta = \mathbb{E}(f(X_T))$ where $X_t$ is the $t$-time value of the underlying security and $X_T$ is the $T$-time simulated price under our model, assuming $T$ is the maturity or a time of importance to us, and where the expectation is taken under the correct measure. To estimate this expectation, we can use the following algorithm:
Valuation via SDE Simulation

for $j = 1, \ldots,$ N

$\widehat{X} = X_0$

$t = 0$

for $i = 1, \ldots, M$

$Z \sim \mathcal{N}(0,1)$

$\hat{X} = \text{SDEStep}(\widehat{X}, t)$

$t = t + h$

end

$f_j = f(\widehat{X})$

end

$\widehat{\theta}_N = (f_1 + \cdots + f_N)/N$

$\widehat{\sigma}_N^2 = \sum_{j=1}^N (f_j - \widehat{\theta}_N)^2/(N-1)$

where $\text{SDEStep}(\hat{X}, t)$ is the discretization function based on the scheme we have chosen. We can use the central limit theorem to calculate the $100(1-\alpha)\%$ confidence interval, which is $\left[\widehat{\theta}_N - z_{1-\frac{\alpha}{2}} \frac{\widehat{\sigma}_N}{\sqrt{N}} \quad \widehat{\theta}_N + z_{1-\frac{\alpha}{2}} \frac{\widehat{\sigma}_N}{\sqrt{N}}\right]$ where $z_{1-\frac{\alpha}{2}}$ is a $z$ such that $P(-z \leq Z \leq z) = 1 - \alpha$. That implies

$$\begin{aligned} \Phi(z) &= P(Z \leq z) = 1 - \frac{\alpha}{2} \\ z &= \Phi^{-1}(\Phi(z)) = \Phi^{-1}(1 - \frac{\alpha}{2}) \end{aligned}$$

We see that we can make the confidence interval tighter and thus our estimates more accurate by either (a) reducing the variance, $\hat{\sigma}_N^2$, and/or (b) increasing the number of simulation paths, $N$. In the next section, we will be focusing on variance reduction techniques and provide examples on performance of each method.

## 6.9 Variance Reduction Techniques

As we mentioned in the beginning of this chapter, while simulation methods for derivatives pricing are the most flexible, the cost of the flexibility is typically the highest computational cost. And as we mentioned in the beginning of the last section, there are two ways to improve the accuracy of our estimates, either computing additional samples, or we can reduce the variance of the estimator itself. Increasing the number of samples, $N$, can be costly, especially if the pricing function or the simulation is computationally intensive. Thus we would prefer to focus on reducing the variance of the estimator assuming $N$ is fixed. Variance reduction techniques are a set of techniques which are designed to reduce the variance of our estimator and thus reduce the time we need to calculate our estimates. Typically variance reduction methods allow us to use information about the pricing problem or the model to adjust the simulation, allowing our estimator to take advantage of this information so we apply the randomness of our sampling in the most efficient way possible. As such, variance reduction techniques typically have to be tailored to the problem itself, and not all techniques are applicable to all pricing problems. There might be cases that no technique would reduce the variance. The six principal methods used for variance reduction in financial applications are:

Control Variate Method

Antithetic Variates

Conditional Monte Carlo

Importance Sampling

Stratified Sampling

Common Random Number

This section will review these methods of variance reduction; however, a full accounting of these methods is beyond the scope of this book. We refer readers to [122] for more information about variance reduction techniques. Throughout this section, we adopt notations used by Martin Haugh, my colleague and friend at Columbia University, in his simulation course.

### 6.9.1 Control Variate Method

The control variate method is based upon using a random variable which is associated with the quantity we have to estimate, but which has a known or easily computable expected value, to adjust our estimator and reduce its variance. If we can find such a random variable, which is well correlated with the quantity we wish to estimate but whose mean is known, we can use our explicit knowledge about this high correlation to reduce our estimator variance.

Suppose we want to estimate

$$\theta = \mathbb{E}(h(X)) \tag{6.73}$$

Assume that $Z$ is also an output of the simulation and moreover assume that $\mathbb{E}(Z)$ is known and $Z$ is correlated with the random variable whose expectation we wish to find. Then we can construct an unbiased estimator of $\theta$ as follows:

$\hat{\theta} = Y$ The usual estimator

$\hat{\theta}_c = Y + c(Z - \mathbb{E}(Z))$ for some number $c$

It is clear that

$$\mathbb{E}(\hat{\theta}_c) = \theta \tag{6.74}$$

so the new estimator will still converge to the correct value by the law of large numbers. The question we should ask is can we find a $c$ such that $\hat{\theta}_c$ has a lower variance than $\hat{\theta}$. Let us look at the variance of $\hat{\theta}_c$.

$$Var(\hat{\theta}_c) = Var(\hat{\theta}) + c^2 Var(Z) + 2cCov(Y, Z) \tag{6.75}$$

We choose $c$ such that $Var(\hat{\theta}_c)$ is minimized and that is

$$c^* = -\frac{Cov(Y, Z)}{Var(Z)} \tag{6.76}$$

and substituting $c^*$ in $Var(\hat{\theta}_c)$ we obtain

$$Var(\hat{\theta}_{c^*}) = Var(\hat{\theta}) - \frac{Cov(Y, Z)^2}{Var(Z)} \tag{6.77}$$

So as long as $Cov(Y, Z)$ is non-zero we can reduce the variance. In this case, $Z$ is called a control variate for $Y$.

**Example 13** *Pricing an Asian option using a control variate*

An arithmetic geometric Asian option has the following payoff at maturity $T$:

$$\max\left(\frac{\sum_{i=1}^m S_{t_i}}{m} - K, 0\right)$$

where $K$ is the strike price and $S_{t_i}$ is the level of the underlying asset (stock) at $i$-th monitoring time $t_i$. We assume $m$ total monitoring times $0 < t_1 < t_2 < \cdots < t_{m-1} < t_m = T$. To price it via simulation we do it as follows. Assuming $S_t$ follows geometric Brownian motion,

$$\text{compute } S_{t_1} = S_{t_0} \exp\left((r-q-\tfrac{\sigma^2}{2})(t_1-t_0)+\sigma\sqrt{t_1-t_0}z_1\right) \text{ where } z_1 \sim \mathcal{N}(0,1)$$

$$\text{compute } S_{t_2} = S_{t_1} \exp\left((r-q-\tfrac{\sigma^2}{2})(t_2-t_1)+\sigma\sqrt{t_2-t_1}z_2\right) \text{ where } z_2 \sim \mathcal{N}(0,1)$$

$$\vdots$$

$$\text{compute } S_{t_m} = S_{t_{m-1}} \exp\left((r-q-\tfrac{\sigma^2}{2})(t_m-t_{m-1})+\sigma\sqrt{t_m-t_{m-1}}z_m\right) \text{ where } z_m \sim \mathcal{N}(0,1)$$

or in short

$$S_{t_i} = S_{t_0} \exp\left((r-q-\frac{\sigma^2}{2})(t_i-t_0)+\sigma\sum_{j=1}^{i}\sqrt{t_j-t_{j-1}}z_j\right) \tag{6.78}$$

To reduce variance one might use either geometric Brownian motion or the geometric Asian option as the control variate. The closed-form solution for the geometric Asian option is given by

$$C = S_0 e^{(b-r)T}\Phi(d_1) - Ke^{-rT}\Phi(d_2)$$

**TABLE 6.5**: Asian premiums with and without control variates

| | without control variate | | with European option as control variate | | | with geometric Asian option as control variate | | |
|---|---|---|---|---|---|---|---|---|
| $M$ | price | stdev | $c^\star$ | price | stdev | $c^\star$ | price | stdev |
| 1000 | 18.5613 | 21.3603 | -0.4470 | 17.6039 | 11.5645 | -1.0796 | 17.9862 | 1.7200 |
| 10000 | 16.9393 | 20.1972 | -0.4665 | 17.0014 | 10.4591 | -1.0710 | 17.0311 | 1.1893 |
| 50000 | 17.0900 | 20.5631 | -0.4671 | 17.0515 | 10.5713 | -1.0735 | 17.1246 | 1.1891 |
| 100000 | 16.9462 | 20.3033 | -0.4675 | 17.0572 | 10.4341 | -1.0726 | 17.0419 | 1.1742 |
| 500000 | 17.0314 | 20.4656 | -0.4657 | 17.0139 | 10.5054 | -1.0740 | 17.0923 | 1.2005 |

where $\Phi$ is the cumulative distribution function of a standard normal random variable and

$$d_1 = \frac{\ln \frac{S_0}{K} + (b + \frac{1}{2}\sigma_A)T}{\sigma_A T}$$

$$d_2 = d_1 - \sigma_A T$$

$$\sigma_A = \frac{\sigma}{\sqrt{3}}$$

$$b = r - q$$

Table 6.5 shows numerical results for pricing an Asian option via simulation with and without control variates. The parameter set for this example is $S_0 = 100$, $K = 90$, $\sigma = 0.3$, $r = 5\%$, number of monitoring times $N = 20$, maturity $T = 2$. In Table 6.5 we illustrate the optimal value $c^\star$ for each control variate as well as its standard deviation. As will see using geometric Asian as control variate has much smaller standard deviation than using European option as control variate as expected.

### 6.9.2 Antithetic Variates Method

The antithetic variates method is based on using correlations between pairs of samples to reduce the variance of our desired estimator.

As before, let us suppose we want to estimate

$$\theta = \mathbb{E}(h(X)) = \mathbb{E}(Y) \tag{6.79}$$

By using knowledge about the correlation of the underlying samples of $X$ we can induce anti-correlation in the resulting samples of the contract payoff $Y$ which will reduce the variance of our estimator. Suppose we have generated two samples, $Y_1$ and $Y_2$. Then an unbiased estimator of $\theta$ is

$$\hat{\theta} = \frac{Y_1 + Y_2}{2} \tag{6.80}$$

and

$$Var(\hat{\theta}) = \frac{Var(Y_1) + Var(Y_2) + 2Cov(Y_1, Y_2)}{4} \tag{6.81}$$

If $Y_1$ and $Y_2$ are i.i.d. then

$$Var(\hat{\theta}) = Var(Y)/2 \tag{6.82}$$

However, we can reduce $Var(\hat{\theta})$ if we have $Cov(Y_1, Y_2) < 0$. Typically this is done by explicitly inducing negative covariance in the samples $Y$. This can be done fairly easily in most cases if the simulation methods we use are based on uniform sampling and inverse transform methods. Since $U$ and $1 - U$ have the same distribution, but are perfectly negatively correlated when they are used as a basis for the inverse transform method, as the inverse distribution function is monotone this negative correlation is preserved. This can also be easily done if our simulation is based on a symmetric distribution like the normal distribution; we can simply invert the signs of the samples to generate two perfectly anti-correlated samples. While the pricing function $h(X)$ may reduce the effects of the anti-correlation, typically it preserves some of them and reduces the variance.

**Example 14** *Pricing an Asian option using antithetic variates*

We already covered the steps in simulating an arithmetic geometric Asian option in Example 13. In this example, we aim to reduce variance using antithetic variate. Applying an antithetic variate to reduce variance in Example 13 we have

$$\text{compute } S^+_{t_1} = S_{t_0} \exp\left((r-q-\tfrac{\sigma^2}{2})(t_1-t_0)+\sigma\sqrt{t_1-t_0}z_1\right)$$

$$\text{compute } S^+_{t_2} = S^+_{t_1} \exp\left((r-q-\tfrac{\sigma^2}{2})(t_2-t_1)+\sigma\sqrt{t_2-t_1}z_2\right)$$

$$\vdots$$

$$\text{compute } S^+_{t_m} = S^+_{t_{m-1}} \exp\left((r-q-\tfrac{\sigma^2}{2})(t_m-t_{m-1})+\sigma\sqrt{t_m-t_{m-1}}z_m\right)$$

and

$$\text{compute } S^-_{t_1} = S_{t_0} \exp\left((r-q-\tfrac{\sigma^2}{2})(t_1-t_0)-\sigma\sqrt{t_1-t_0}z_1\right)$$

$$\text{compute } S^-_{t_2} = S^-_{t_1} \exp\left((r-q-\tfrac{\sigma^2}{2})(t_2-t_1)-\sigma\sqrt{t_2-t_1}z_2\right)$$

$$\vdots$$

$$\text{compute } S^-_{t_m} = S^-_{t_{m-1}} \exp\left((r-q-\tfrac{\sigma^2}{2})(t_m-t_{m-1})-\sigma\sqrt{t_m-t_{m-1}}z_m\right)$$

For each simulation path we have

$$\frac{\max\left(\frac{\sum_{i=1}^m S^+_{t_i}}{m} - K, 0\right) + \max\left(\frac{\sum_{i=1}^m S^-_{t_i}}{m} - K, 0\right)}{2}$$

and the premium would be the average of these quantities. Table 6.6 compares the results for the Asian call option without and with antithetic variates.

### 6.9.3 Conditional Monte Carlo Methods

Conditional Monte Carlo methods are based on conditioning the expectation which determines our contract value on as much information as possible which is either known explicitly or which can be calculated easily, in order to reduce the variance of the resulting expectation.

Suppose we wish to estimate

$$\theta = \mathbb{E}(h(X)) = \mathbb{E}(Y) \tag{6.83}$$

**TABLE 6.6**: Asian option premiums using the antithetic variates method

| | without antithetic variates | | using antithetic variates | |
|---|---|---|---|---|
| $M$ | price | stdev | price | stdev |
| 1000 | 18.5613 | 21.3603 | 17.3759 | 9.2149 |
| 10000 | 16.9393 | 20.1972 | 17.2101 | 9.0221 |
| 50000 | 17.0900 | 20.5631 | 17.0499 | 8.7409 |
| 100000 | 16.9462 | 20.3033 | 17.0066 | 8.8537 |
| 500000 | 17.0314 | 20.4656 | 17.0386 | 8.8476 |

where $X = (X_1, \ldots, X_m)$. Computing $\theta$ analytically is equivalent to solving an $m$-dim integral. However, it may be possible to evaluate part of the integral analytically. If so, we will use simulation to solve the other part and that way reduce the variance.

Before explaining the method, we give a quick review of conditional expectation and conditional variance. Suppose $X$ and $Z$ are random vectors and $Y = h(X)$. Suppose

$$V = \mathbb{E}(Y|Z) \tag{6.84}$$

Then $V$ itself is a random vector that depends on $Z$. We may write

$$V = g(Z) \tag{6.85}$$

We also know that

$$\mathbb{E}(V) = \mathbb{E}(\mathbb{E}(Y|Z)) = \mathbb{E}(Y) \tag{6.86}$$

so that if we are trying to estimate

$$\theta = \mathbb{E}(h(X)) = \mathbb{E}(Y) \tag{6.87}$$

one possibility would be to simulate $V$ instead of simulating $Y$. It is still necessary to compute $Var(Y)$ and $Var(\mathbb{E}(Y|Z))$. Recall that the conditional variance formula is

$$Var(Y) = \mathbb{E}(Var(Y|Z)) + Var(\mathbb{E}[Y|Z])) \tag{6.88}$$

Therefore

$$Var(Y) \geq Var(\mathbb{E}[Y|Z]) = Var(V) \tag{6.89}$$

So we can conclude that $V$ is a better estimator of $\theta$ than $Y$ since we can eliminate the variance contributed by $Var(\mathbb{E}[Y|Z]))$. To be able to reduce variance via conditional Monte Carlo we must have another variable $Z$ that satisfies the following:

$Z$ can be easily simulated

$V = g(Z) = \mathbb{E}(Y|Z)$ can be computed exactly

#### 6.9.3.1 Algorithm for Conditional Monte Carlo Simulation

We can implement conditional Monte Carlo simulation algorithm as follows:

For $j = 1, \ldots, \mathrm{N}$

generate $Z_j$

compute $g(Z_j) = \mathbb{E}(Y|Z_j)$

set $V_j = g(Z_j)$

end

$\hat{\theta}_{N,CM} = \hat{V}_N = \sum_{j=1}^{N} V_j/N$

$\hat{\sigma}^2_{N,CM} = \sum_{j=1}^{N}(V_j - \hat{V}_N)^2/(N-1)$

and the $100(1-\alpha)\%$ confidence interval is $\left[\hat{\theta}_{N,CM} - z_{1-\frac{\alpha}{2}}\frac{\hat{\sigma}_{N,CM}}{\sqrt{N}} \quad \hat{\theta}_N + z_{1-\frac{\alpha}{2}}\frac{\hat{\sigma}_{N,CM}}{\sqrt{N}}\right]$.

**Example 15** *Pricing delayed/forward start options using conditional Monte Carlo*

In delayed/forward start options, the strike is set at some future time as a factor of the level of stock at that time. The time and the factor are specified in the contract in advance. To price the contract using simulation we do as follows:

for $j = 1, \ldots,$ N

generate $Z_1 \sim \mathcal{N}(0,1)$ and compute $S_{t_1} = S_0 \exp((r-q-\frac{\sigma^2}{2})t_1 + \sigma\sqrt{t_1}Z_1)$

set $K = \lambda S_{t_1}$

generate $Z_2 \sim \mathcal{N}(0,1)$ and compute $S_T = S_{t_1} \exp((r-q-\frac{\sigma^2}{2})(T-t_1) + \sigma\sqrt{T-t_1}Z_2)$

discounted payoff $f_j = e^{-rT}\max(S_T - K, 0)$

end

$\hat{\theta} = \frac{1}{N}\sum_{j=1}^{N} f_j$

In the above simulation, for each simulated path we have to generate two normal random numbers. Using conditional Monte Carlo we can reduce that to one. By conditioning on the level of stock at $t_1$ we can price the option from $t_1$ to $T$ using the Black–Scholes formula, hence

for $j = 1, \ldots,$ N

generate $Z_1 \sim \mathcal{N}(0,1)$ and compute $S_{t_1} = S_0 \exp((r-q-\frac{\sigma^2}{2})t_1 + \sigma\sqrt{t_1}Z_1)$

set $K = \lambda S_{t_1}$

$f_j = e^{-rt_1} \times \text{BlackScholes}(S_{t_1}, K, \sigma, r, q, T - t_1)$

end for

$\hat{\theta}_{CM} = \frac{1}{N}\sum_{j=1}^{N} f_j$

The closed-form solution for delayed/forward start European options with spot $S_{t_1}$ and strike price $\lambda S_{t_1}$ is calculated as follows:

$$e^{-r(t_1-t_0)}\int_0^\infty \text{BlackScholes}(S_{t_1}, \lambda S_{t_1}, \sigma, r, q, T - t_1) f(S_{t_1}|S_{t_0}) dS_{t_1}$$

**TABLE 6.7**: Delayed start European options (closed-form solution is 31.5482)

| | without conditional Monte Carlo | | with conditional Monte Carlo | |
|---|---|---|---|---|
| $M$ | premium | stdev | premium | stdev |
| 1000 | 32.5218 | 53.5932 | 31.6853 | 9.8011 |
| 10000 | 31.0281 | 49.7165 | 31.6426 | 9.7275 |
| 50000 | 31.8873 | 51.8664 | 31.5728 | 9.6227 |
| 100000 | 31.3153 | 50.8236 | 31.5115 | 9.6616 |
| 500000 | 31.5470 | 51.3883 | 31.5587 | 9.6868 |

Or equivalently (due to linear homogeneity)

$$e^{-r(t_1-t_0)}\text{BlackScholes}(1,\lambda,\sigma,r,q,T-t_1)\int_0^\infty S_{t_1} f(S_{t_1}|S_{t_0})dS_{t_1}$$

where $f(S_{t_1}|S_{t_0})$ is the conditional lognormal density. By integrating out $S_{t_1}$, we can show that the price is

$$S_0 e^{-r(t_1-t_0)}\text{BlackScholes}(1,\lambda,\sigma,r,q,T-t_1)$$

Table 6.7 shows the results for the parameter set $S_0 = 100$, $\sigma = 0.3$, $r = 5\%$, maturity $T = 4$ years, $t_1 = \frac{T}{4}$, and $\lambda = 0.9$. Cliquet options are a series of forward start options where the strike of each forward is set at the expiry of the previous forward start options. Cliquet options are widely traded and they are embedded in many structured products. The valuation of cliquet options are hard and it is claimed that some dealers are mispricing them [186].

### 6.9.4 Importance Sampling Methods

Importance sampling is one of the most powerful methods of variance reduction available. The method is based on using knowledge of the pricing problem to focus our sampling on critical areas of interest, values of the underlying variable which yield important results, which in turn reduces the variance of our estimator. One obvious example of where this can be applied is out-of-the-money options, where payoffs only occur at very large (or small) values of the underlier, and so sampling in these regions makes our estimator less variable.

Suppose we are interested in computing

$$\theta = \mathbb{E}_f(h(X)) \tag{6.90}$$

where $X$ had a probability distribution function $f$. Let $g$ be another probability distribution function with the property that $g(x) \neq 0$ whenever $f(x) \neq 0$. Then

$$\begin{aligned}
\theta &= \mathbb{E}_f(h(X)) = \int h(x)f(x)dx && (6.91)\\
&= \int h(x)\frac{f(x)}{g(x)}g(x)dx && (6.92)\\
&= \mathbb{E}_g\left(h(X)\frac{f(X)}{g(X)}\right) && (6.93)\\
&= \mathbb{E}_g(h^*(X)) && (6.94)
\end{aligned}$$

where $\frac{f(X)}{g(X)}$ is known as the likelihood ratio and should be easily computable. This has very important implications for estimating $\theta$. In our original estimation algorithm, we generate $N$ samples of $X$ from $f(.)$ and set

$$\hat{\theta}_N = \frac{1}{N}\sum_{j=1}^{N} h(X_j) \tag{6.95}$$

In the alternative estimation algorithm, however, we generate $N$ values from $g(.)$ and set

$$\hat{\theta}_{N,IS} = \frac{1}{N}\sum_{j=1}^{N} h(X_j)\frac{f(X_j)}{g(X_j)} \tag{6.96}$$

$\hat{\theta}_{N,IS}$ is then an importance sampling estimator of $\theta$.

#### 6.9.4.1 Variance Reduction via Importance Sampling

Variance reduction is achieved via importance sampling by knowing that both of the following expectations are equivalent:

$$\begin{aligned} \theta &= \mathbb{E}_f(h(X)) & (6.97)\\ &= \mathbb{E}_g(h^*(X)) & (6.98) \end{aligned}$$

and this gives rise to two estimators

$h(X)$ where $X \sim f(.)$

$h^*(X)$ where $X \sim g(.)$

The variance of the importance sampling estimator is given by

$$\begin{aligned} Var_g(h^*(X)) &= \int h^*(x)^2 g(x)dx - \theta^2 \\ &= \int \frac{h^2(x)f^2(x)}{g^2(x)} g(x)dx - \theta^2 \\ &= \int \frac{h^2(x)f(x)}{g(x)} f(x)dx - \theta^2 \end{aligned}$$

Also we know that

$$Var_f(h(X)) = \int h^2(x)f(x)dx - \theta^2$$

So

$$Var_f(h(X)) - Var_g(h^*(X)) = \int h^2(x)\left(1 - \frac{f(x)}{g(x)}\right) f(x)dx \tag{6.99}$$

In order to achieve a variance reduction the integral should be positive. For this to happen we should have

$\frac{f(x)}{g(x)} > 1$ where $h(x)f(x)$ is small

$\frac{f(x)}{g(x)} < 1$ where $h(x)f(x)$ is large

Let us say that there is a region, $\mathfrak{A}$, where $h(x)f(x)$ is large. Then by the above argument we would like to choose $g$ so that $f(x)/g(x)$ is small whenever $x$ is in $\mathfrak{A}$, which means we would like a density $g$ that puts more weight on $\mathfrak{A}$.

Note that when $h$ involves a rare event so that $h(x) = 0$ over most of the state space, it can then be particularly valuable to choose $g$ so that we often sample from that part of the state space where $h(x) \neq 0$. Also note that if we choose $g$ to be

$$g(x) = h(x)f(x)/\theta \tag{6.100}$$

then we have $Var_g(h^\star(X)) = \theta^2 - \theta^2 = 0$. We know that in reality $\theta$ is not known and actually it is the quantity that we are trying to calculate. However, it tells us the closer $g$ is to the shape of $h(x)f(x)$ the lesser the variance.

**Example 16** *Estimating a rare normal event using importance sampling*

Consider the problem of estimating

$$\theta = P(X > 8)$$

where $X \sim \mathcal{N}(0,1)$. If one tries to estimate $\theta$ via simulation without doing importance sampling, we will often get zero as this event is extremely rare. Now we try to estimate $\theta$ by doing importance sampling with a new random variable $Y \sim \mathcal{N}(\mu, 1)$ with some appropriate choice for $\mu$.

$$\begin{aligned}
\theta &= P(X > 8) \\
&= \mathbb{E}_f(\mathbb{1}_{X>8}) \\
&= \int \mathbb{1}_{z>8} \frac{1}{\sqrt{2\pi}} e^{-z^2/2} dz \\
&= \int \mathbb{1}_{z>8} \frac{\frac{1}{\sqrt{2\pi}} e^{-z^2/2}}{\frac{1}{\sqrt{2\pi}} e^{-(z-\mu)^2/2}} \frac{1}{\sqrt{2\pi}} e^{-(z-\mu)^2/2} dz \\
&= \int \mathbb{1}_{z>8} e^{-\mu z + \mu^2/2} \frac{1}{\sqrt{2\pi}} e^{-(z-\mu)^2/2} dz \\
&= \mathbb{E}_g(\mathbb{1}_{X>8} e^{-\mu X + \mu^2/2})
\end{aligned}$$

where $g(.)$ is the probability distribution function of $\mathcal{N}(\mu, 1)$. It is clear that $g(.)$ attains its maximum at $x = \mu$. Therefore an optimal choice for $\mu$ could be

$$\begin{aligned}
\mu &= \arg\max_x h(x)f(x) \\
&= \arg\max_x \mathbb{1}_{x>8} \frac{1}{\sqrt{2\pi}} e^{-x^2/2} \\
&= \arg\max_{x\geq 8} e^{-x^2/2} \\
&= 8
\end{aligned}$$

Another very useful application of importance sampling is for deep out-of-the-money options.

### 6.9.5 Stratified Sampling Methods

Stratified sampling is a probability sampling technique which involves the division of a population into subgroups known as strata, then randomly selects the final subjects proportionally from the different strata. Suppose for some random variable $Y$ we want to estimate

$\theta = \mathbb{E}(Y)$. Assuming $X$ is another random variable such that (a) for any arbitrary interval $\delta \subset \mathbb{R}$ we can easily calculate $P(X \in \delta)$, (b) it is easy to generate $Y$ given $X$ in $\delta$ that is $(Y|X \in \delta)$. Forming $m$ non-overlapping divisions $\delta_i$ with $\bigcup_{i=1}^m \delta_i = \mathbb{R}$ we have $\sum_{i=1}^m p_i = 1$ where $p_i = P(X \in \delta_i)$.

Let $\theta_i = \mathbb{E}(Y|X \in \delta_i)$ and $\sigma_i^2 = \text{Var}(Y|X \in \delta_i)$ and define yet another random variable $I$ by setting it equal to $i$ if $X \in \delta_i$ and let $Y^{(i)} = (Y|I = i) = (Y|X \in \delta_i)$.

Now, we can write

$$\begin{aligned}
\theta &= \mathbb{E}(Y) \\
&= \mathbb{E}(\mathbb{E}(Y|I)) \\
&= p_1\mathbb{E}(Y|I=1) + p_2\mathbb{E}(Y|I=2) + \cdots + p_m\mathbb{E}(Y|I=m) \\
&= p_1\theta_1 + p_2\theta_2 + \cdots + p_m\theta_m
\end{aligned}$$

If we use a total of $n$ samples to estimate $\theta$ and $n_i$ samples to estimate $\theta_i$, where $n = n_1 + n_2 + \cdots + n_m$, then an estimate of $\theta$ is given by

$$\widehat{\theta} = p_1\widehat{\theta}_1 + p_2\widehat{\theta}_2 + \cdots + p_m\widehat{\theta}_m$$

If for each $i$, $\widehat{\theta}_i$ is an unbiased estimate of $\theta_i$, then clearly $\widehat{\theta}_n$ would be an unbiased estimate of $\theta$ as well. The very first choice for $n_i$ might be $n_i = np_i$. For this choice of $n_i$ we obtain

$$\begin{aligned}
\text{Var}(\widehat{\theta}_n) &= \text{Var}(p_1\widehat{\theta}_1 + \cdots + p_m\widehat{\theta}_m) \\
&= p_1^2\frac{\sigma_1^2}{n_1} + \cdots + p_m^2\frac{\sigma_m^2}{n_m} \\
&= p_1\frac{\sigma_1^2}{n} + \cdots + p_m\frac{\sigma_m^2}{n} \\
&= \frac{1}{n}\sum_{i=1}^m p_i\sigma_i^2
\end{aligned}$$

where $\frac{\sigma_i^2}{n_i} = \text{Var}(\widehat{\theta}_i)$. Recall that the estimator of $\theta$ without any variance reduction has a variance of $\sigma^2/n$. We want to show $\sum_{i=1}^m p_i\sigma_i^2 < \sigma^2$. From the definition of conditional variance we have

$$\begin{aligned}
\sigma^2 &= \text{Var}(Y) \\
&= \mathbb{E}(\text{Var}(Y|I)) + \text{Var}(\mathbb{E}(Y|I)) \\
&\geq \mathbb{E}(\text{Var}(Y|I)) \\
&= \sum_{i=1}^m p_i\sigma_i^2
\end{aligned}$$

Thus the above stratification reduces the variance. One might ask, is it possible to do better. The optimal stratification is to choose $n_i$ to minimize $\text{Var}(\widehat{\theta}_n)$, that is,

$$\begin{aligned}
&\min_{n_i} \sum_{i=1}^m p_i^2\frac{\sigma_i^2}{n_i} \\
&\text{subject to } \sum_{i=1}^m n_i = n
\end{aligned}$$

Applying a Lagrange multiplier, the optimal solution to this constrained problem is

$$n_i^\star = \frac{p_i\sigma_i}{\sum_{i=1}^m p_i\sigma_i} n \tag{6.101}$$

Substitute $n_i^\star$ into the variance equation and we get the following term for minimized variance:

$$\mathrm{Var}(\widehat{\theta}_n) = \frac{1}{n}\left(\sum_{i=1}^{m} p_i \sigma_i\right)^2$$

#### 6.9.5.1 Findings and Observations

Equation (6.101) indicates sampling more from regions with high $p$ and/or high $\sigma$, which makes sense. A drawback with optimal $n_i^\star$ is that we need to know values of $\sigma_i$ in advance in order to compute $n_i^\star$ and obviously they are not known. As before we can set a pilot program to find an estimate of $\sigma_i$. Using those estimated $\sigma_i$ to calculate $n_i^\star$ and run the main program.

It is clear to see if the population density varies greatly within a division, stratified sampling ensures that estimates can be made with equal accuracy in different parts of that division, and that comparisons of sub-regions can be made with equal statistical power. Randomizing stratification can be used to improve the final result.

#### 6.9.5.2 Algorithm for Stratified Sampling Methods

We can implement the stratified sampling method algorithm as follows:

Choose $m$ and set $\delta_i$ and $n_i$ for $i = 1, \ldots, m$

Calculate $p_i$ for $i = 1, \ldots, m$

Set $\hat{\theta}_{n,st} = 0$ and $\hat{\sigma}^2_{n,st} = 0$

for $i = 1, \ldots, m$

set $s_i = 0$ and $u_i = 0$

for $j = 1, \ldots, n_i$

generate $X \in \delta_i$ and given $X$ generate $Y_j^{(i)}$

$s_i = s_i + Y_j^{(i)}$

$u_i = u_i + (Y_j^{(i)})^2$

end for

calculate mean which is $\hat{\theta}_i = \frac{s_i}{n_i}$

calculate variance which is $\hat{\sigma}_i^2 = \frac{u_i - s_i^2/n_i}{n_i - 1}$

$\hat{\theta}_{n,st} = \hat{\theta}_{n,st} + p_i \hat{\theta}_i$

$\hat{\sigma}^2_{n,st} = \hat{\sigma}^2_{n,st} + \hat{\sigma}_i^2 p_i^2 / n_i$

end for

and the $100(1-\alpha)\%$ confidence interval is $\left[\hat{\theta}_{n,st} - z_{1-\frac{\alpha}{2}} \hat{\sigma}_{n,st} \quad \hat{\theta}_{n,st} + z_{1-\frac{\alpha}{2}} \hat{\sigma}_{n,st}\right]$.

**Example 17** *Pricing a European put option via stratified sampling*

**TABLE 6.8**: Non-overlap subintervals for $X \sim \mathcal{N}(0,1)$ and their probabilities

| $\delta_i$ | $p_i$ | $n_i = np_i$ | $n_i^\star$ |
|---|---|---|---|
| $(-\infty,-2)$ | $\Phi(-2)-\Phi(-\infty)=0.0228$ | 2275 | 7021 |
| $[-2,-1)$ | $\Phi(-1)-\Phi(-2)=0.1359$ | 13591 | 41281 |
| $[-1,1)$ | $\Phi(1)-\Phi(-1)=0.6827$ | 68269 | 51698 |
| $[1,2)$ | $\Phi(2)-\Phi(1)=0.1359$ | 13591 | 0 |
| $[2,\infty)$ | $\Phi(\infty)-\Phi(2)=0.0228$ | 2275 | 0 |

**TABLE 6.9**: European put option premium using a stratified sampling method (Black–Scholes premium 26.2172)

| | without using stratified sampling | | with suboptimal stratified sampling | | with optimal stratified sampling | |
|---|---|---|---|---|---|---|
| $n$ | premium | stdev | premium | stdev | premium | stdev |
| 100000 | 26.4776 | 0.1935 | 26.1030 | 0.0648 | 26.2024 | 0.0480 |

Assume a spot price of 1260 for S&P 500. We want to price an out-of-the-money European put for a strike of 1100 and a maturity of 3 months, $T = 1/4$. Assume volatility of 35%, risk-free rate of interest 0.25% and dividend rate 1%. Under Black–Scholes we get 26.2172 for its premium. Now we employ a stratified sampling method with regular and optimal $p_i$ and compare results. Set $X = Z \sim \mathcal{N}(0,1)$. Assume five non-overlapping sub-intervals for $X$ as shown in Table 6.8. To guarantee $Z \sim \mathcal{N}(0,1) \in \delta_i$, we draw $\widehat{U} \sim \mathcal{U}(\Phi(l_{\delta_i}), \Phi(u_{\delta_i}))$[4] where $l_{\delta_i}$ and $u_{\delta_i}$ are lower and upper bounds of the subinterval $\delta_i$ and $\Phi$ is the CDF of the standard normal random variable. Having $\widehat{U}$, we calculate $Z = \Phi^{-1}(\widehat{U})$ and plug it into

$$S_{T,j}^{(i)} = S_0 e^{(r-q-\frac{\sigma^2}{2})T+\sigma\sqrt{T}Z_j^{(i)}}$$

to calculate $S_{T,j}^{(i)}$ and subsequently calculate $Y_j^{(i)} = e^{-rT}\max(K - S_{T,j}^{(i)}, 0)$. Table 6.9 illustrates results for simulation with no variance reduction, simulation with a sub-optimal stratified sampling method, and simulation with an optimal stratified sampling method. For $n_i^\star$ we first use a pilot program (500 samples for each $i$) to find $\sigma_i$ for each $i$. As we see the optimal stratified sampling method yields the best variance reduction but for this case it seems to be pretty marginal.

**Example 18** *Pricing an Asian option via stratified sampling methods*

A natural choice is to set $X = \sum_{i=1}^m y_i$ where $y_i \sim \mathcal{N}(0, T/m)$. We know that if $y_i \sim \mathcal{N}(0, T/m)$, then $X = \sum_{i=1}^m y_i \sim \mathcal{N}(0,T)$. It is pretty straightforward to calculate $p_i$. Assume five non-overlapping sub-intervals for $X$ as shown in Table 6.10. To guarantee that $X \in \delta_i$ we use the idea of a Brownian bridge. Instead of calculating $S_{t_i}$ according to the following formula

$$S_{t_i} = S_{t_0} \exp\left((r-q-\frac{\sigma^2}{2})(t_i-t_0)+\sigma\sum_{j=1}^{i}\sqrt{t_j-t_{j-1}}z_j\right) \tag{6.102}$$

[4] If $U \sim \mathcal{U}(0,1)$, then $\hat{U} = a + (b-a)U \sim \mathcal{U}(a,b)$.

**TABLE 6.10**: Non-overlap subintervals for $X \sim \mathcal{N}(0,T)$ and their probabilities

| $\delta_i$ | $p_i$ | $n_i = np_i$ | $n_i^{call}$ | $n_i^{put}$ |
|---|---|---|---|---|
| $(-\infty,-0.8)$ | 0.1289 | 25790 | 6231 | 92357 |
| $[-0.8,-0.4)$ | 0.1569 | 31371 | 18155 | 62617 |
| $[-0.4,0.4)$ | 0.4284 | 85679 | 95362 | 45025 |
| $[0.4,0.8)$ | 0.1569 | 31371 | 35933 | 0 |
| $[0.8,\infty)$ | 0.1289 | 25790 | 44320 | 0 |

**TABLE 6.11**: Asian option premium using a stratified sampling method

| | without using stratified sampling | | with suboptimal stratified sampling | | with optimal stratified sampling | |
|---|---|---|---|---|---|---|
| $M$ | Premium | stdev | Premium | stdev | Premium | stdev |
| 50000 | 20.5127 | 51.0703 | 20.8177 | 14.1565 | 20.8630 | 13.9744 |
| 100000 | 20.7692 | 51.1957 | 21.0222 | 14.0383 | 21.1066 | 14.0082 |
| 200000 | 21.0376 | 51.7771 | 20.9079 | 14.0220 | 20.9522 | 13.9823 |

we do as follows:

$$S_{t_i} = S_{t_0} \exp\left( (r-q-\frac{\sigma^2}{2})(t_i-t_0)+\sigma W_{t_i} \right) \tag{6.103}$$

where $W_{t_i}$ are exactly generated according to Equation (6.51).

For an Asian put option, we consider a strike price of \$1100, maturity of half a year, monitoring interval of one month (6 monitoring times), volatility of 35%, risk-free rate of interest 0.25%, and dividend rate 1%. Table 6.11 illustrates simulation results with no variance reduction, simulation with a sub-optimal stratified sampling method, and simulation with an optimal stratified sampling method.

### 6.9.6 Common Random Numbers

To calculate Greeks, we typically use the finite difference approximation

$$\begin{aligned} \Delta &= \frac{\partial C}{\partial S} \\ &= \frac{C(S+\Delta S) - C(S-\Delta S)}{\Delta S} \end{aligned}$$

In calculating $C(S+\Delta S)$ and $C(S-\Delta S)$ via Monte Carlo simulation, we use the same random variable at each time step and for each path starting from $S_0+\Delta S$ and $S_0 - \Delta S$ respectively as opposed to different random numbers. That would reduce variance in calculating Greeks. The following example illustrates the effect of common random numbers in calculating Greeks.

**Example 19** *Calculating Greeks in Black–Scholes using common random numbers*

As an example, we use common random numbers to calculate $\Delta$ in a Black–Scholes case. Table 6.12 shows the results for $\Delta$ with and without common random numbers

**TABLE 6.12**: Black–Scholes $\Delta = 0.7479$

| | without using common number | | with using common number | |
|---|---|---|---|---|
| $M$ | value | stdev | value | stdev |
| 1000 | 0.5578 | 5.6484 | 0.755 | 0.5896 |
| 10000 | 0.7603 | 6.0394 | 0.7498 | 0.5832 |
| 50000 | 0.7001 | 5.9939 | 0.7444 | 0.5877 |
| 100000 | 0.7270 | 5.9491 | 0.7458 | 0.5873 |

---

## Problems

1. (a) By rejection from a Gaussian distribution,

$$g(x) = \frac{1}{\sqrt{2\pi}\sigma} \exp\left[-\frac{1}{2}\left(\frac{x}{\sigma}\right)^2\right]$$

we can sample from the following density:

$$f(x) = \frac{1}{A}e^{-x^4/4}$$

where $A = \int_{-\infty}^{+\infty} e^{-x^4/4}dx$. Find an optimal value of $\sigma$ such that the rejection method becomes efficient.

(b) Explain how to sample a random variable from a standard double exponential distribution using the *Inverse Transform Method.* The probability density function of the standard double exponential is

$$f(x) = \frac{1}{2}e^{-|x|}$$

(c) Consider the problem of estimating

$$\theta = P(Z > 8)$$

where $Z \sim \mathcal{N}(0, 1)$.

(a) Estimate $\theta$ via simulation without doing importance sampling.

(b) Estimate $\theta$ by doing importance sampling with a new random variable $Y \sim \mathcal{N}(\mu, 1)$ with some appropriate choice for $\mu$.

2. Assume that the stock price follows the following process:

$$dS_t = rS_t dt + \sigma S_t dW_t$$

Now, consider simulation of a European put with the following parameters: $S_0 = 100$, $K = 110$, $r = 4.75\%$, $\sigma = 20\%$, and maturity $T = 0.5$.

(a) Use the Euler method with the time step size $\Delta t = 0.0005$ to generate $10,000$ realizations of $S_T$ and compute the value of the discounted payoff

$$V_0^{(i)} = e^{-rT}(K - S_T^{(i)})^+$$

Estimate the mean and the variance of $V_0$. Compare the value of the put obtained using Monte Carlo simulation with the Black–Scholes formula.

(b) Repeat the above analysis using the antithetic variates. What can you conclude?

(c) Repeat the analysis in (b) using the Milstein method. Is there any significant improvement? If yes, why? If not, why not?

3. Assume that the stock price follows the following process:

$$dS_t = rS_t dt + \sigma(S_t, t)S_t dW_t$$

For the following parameters: spot price $S_0 = 100$, strike price $K = 110$, risk-free rate $r = 4.75\%$, local volatility surface $\sigma(S, t) = 0.5e^{-t}(100/S)^{0.3}$, and maturity $T = 1.0$, price a European put option via

(a) finite differences

(b) Monte Carlo simulation

(c) Monte Carlo simulation with a control variate

and compare.

4. In the Heston stochastic volatility model, stock price follows the process:

$$\begin{aligned} dS_t &= rS_t dt + \sqrt{v_t} S_t dW_t^{(1)} \\ dv_t &= \kappa(\theta - v_t)dt + \sigma\sqrt{v_t} dW_t^{(2)} \end{aligned}$$

where the two Brownian components $W_t^{(1)}$ and $W_t^{(2)}$ are correlated with rate $\rho$. The parameters $\kappa$, $\theta$, and $\sigma$ have certain physical meanings: $\kappa$ is the mean reversion speed, $\theta$ is the long run variance, and $\sigma$ is the volatility of the volatility.

The characteristic function for the log of the stock price process is given by

$$\begin{aligned} \Phi(u) &= \mathbb{E}(e^{iu \ln S_t}) \\ &= \frac{\exp\{\frac{\kappa\theta t(\kappa - i\rho\sigma u)}{\sigma^2} + iutr + iu \ln S_0\}}{(\cosh \frac{\gamma t}{2} + \frac{\kappa - i\rho\sigma u}{\gamma} \sinh \frac{\gamma t}{2})^{\frac{2\kappa\theta}{\sigma^2}}} \exp\left\{-\frac{(u^2 + iu)v_0}{\gamma \coth \frac{\gamma t}{2} + \kappa - i\rho\sigma u}\right\} \end{aligned}$$

where $\gamma = \sqrt{\sigma^2(u^2 + iu) + (\kappa - i\rho\sigma u)^2}$, and $S_0$ and $v_0$ are the initial values for the price process and the volatility process, respectively. Apply the following methods to price a European call:

- Fast Fourier transform technique
- Monte Carlo simulation (with and without variance reduction)

and compare you results for the following parameters: spot price, $S_0 = \$100$, strike price $K = 90$, maturity $T = 1$ year, risk-free interest rate $r = 5.25\%$, volatility of volatility $\sigma = 30\%$, $\kappa = 1$, $\theta = 0.08$, $\rho = -0.8$, and $\nu_0 = 0.04$.

5. In this problem we aim to investigate biases in the Euler full truncation scheme discussed in Section 6.7.4. Assume that the stock price follows Heston stochastic volatility

$$\begin{aligned} dS_t &= (r - q)S_t dt + \sqrt{v_t} S_t dW_t^1, \\ dv_t &= \kappa(\theta - v_t)dt + \lambda\sqrt{v_t} dW_t^2 \end{aligned}$$

Consider pricing a European put option via simulation for the following parameters: $S_0 = 100$, $K = 110$, $r = 1.5\%$, $\sigma = 30\%$, $\kappa = 1$, $\theta = 0.08$, $\rho = -0.8$, $\nu_0 = 0.04$ and maturity $T = 1$.

Use the Euler full truncation scheme to price the option for the time step sizes $\Delta t = \frac{1}{12}, \frac{1}{50}, \frac{1}{250}$ to generate 100,000 realizations of $S_T$ and compute the value of the discounted payoff. Compare the value of the put obtained using Monte Carlo simulation with those of fractional fast Fourier transform and conclude.

6. Let $b(t;\theta,\sigma) \equiv \theta t + \sigma W(t)$ be a Brownian motion with constant drift rate $\theta$ and volatility $\sigma$, where $W(t)$ is a standard Brownian motion. Denote by $\gamma(t;\nu)$ the gamma process with independent gamma increments of mean $h$ and variance $\nu h$ over non-overlapping intervals of length $h$.

   The three parameter VG process, $X(t;\sigma,\theta,\nu)$, is defined by

   $$X(t;\sigma,\theta,\nu) = b(\gamma(t;\nu),\theta,\sigma)$$

   We see that the process $X(t)$ is a Brownian motion with drift evaluated at a gamma time change. The characteristic function for the time $t$ level of the VG process is

   $$\phi_{X(t)}(u) = \mathbb{E}(e^{iuX(t)}) = \left(\frac{1}{1 - iu\theta\nu + \sigma^2 u^2 \nu/2}\right)^{\frac{t}{\nu}} \tag{6.104}$$

   The VG dynamics of the stock price mirrors that of geometric Brownian motion for a stock paying a continuous dividend yield of $q$ in an economy with a constant continuously compounded interest rate of $r$. The risk-neutral drift rate for the stock price is $r - q$ and the forward stock price is modeled as the exponential of a VG process normalized by its expectation. Let $S(t)$ be the stock price at time $t$. The VG risk-neutral process for the stock price is given by

   $$S(t) = S(0)e^{(r-q)t + X(t) + \omega t} \tag{6.105}$$

   where the normalization factor $e^{\omega t}$ ensures that $\mathbb{E}_0[S(t)] = S(0)e^{(r-q)t}$. It follows from the characteristic function evaluated at $-i$ that

   $$\omega = \frac{1}{\nu}\ln(1 - \sigma^2\nu/2 - \theta\nu)$$

   By the definition of risk neutrality, the price of a European put option with strike $K$ and maturity $T$ is

   $$p(S(0);K,t) = e^{-rT}\mathbb{E}_0((K - S(T))^+).$$

   For the following parameters: spot price $S_0 = \$100$, strike price $K = 105$, maturity $T = 1$ year, risk-free interest rate $r = 4.75\%$, continuous dividend rate $q = 1.25\%$, $\sigma = 25\%$, $\nu = 0.50$, $\theta = -0.3$, price a European put option via

   (a) FFT technique

   (b) simulation

   and compare.

7. Assume that the stock price follows the following process:

   $$dS_t = rS_t dt + \sigma S_t dW_t$$

   Assume that we would like to calculate the price of a European put option that has the following payoff:

   $$h(X) = \begin{cases} (K_1 - S_T)^+ & : \text{ if } S_{T/2} < H \\ (K_2 - S_T)^+ & : \text{ otherwise} \end{cases}$$

where $X = (S_{T/2}, S_T)$. The price of the European put option can be written as

$$P(S_0, t = 0; K_1, K_2, T) = e^{-rT}\mathbb{E}\left(\mathbb{1}_{S_{T/2}<H}\{(K_1 - S_T)^+\} + \mathbb{1}_{S_{T/2}>=H}\{(K_2 - S_T)^+\}\right)$$

Now, explain how one would estimate the price of this option via simulation using just one normal random variable per path.

8. Assume we can easily sample from $\mathcal{U}(0,1)$. Also assume that we can easily calculate both $\Phi(x)$, the cumulative distribution function of $\mathcal{N}(0,1)$

$$\Phi(x) = \int_{-\infty}^{x} \frac{1}{\sqrt{2\pi}} e^{-\frac{1}{2}x^2} dx$$

and its inverse $\Phi^{-1}(x)$.

Show how to sample from the following truncated normal distribution:

$$g(x) = \frac{1}{\sqrt{2\pi}A\sigma} \exp\left[-\frac{1}{2}\left(\frac{x-\mu}{\sigma}\right)^2\right]$$

where $A = \int_a^b \frac{1}{\sqrt{2\pi}\sigma} \exp\left[-\frac{1}{2}\left(\frac{x-\mu}{\sigma}\right)^2\right] dx$.

9. Consider estimating the following integral

$$\theta = \int_{-\infty}^{0} e^{-x^2} dx$$

(a) Describe the *standard* Monte Carlo integration method for estimating $\theta$. Hint: First convert it to a definite integral by change of a variable.

(b) Describe another Monte Carlo simulation method to estimate $\theta$ that does not require a change of variable in the integration. Hint: You might consider writing

$$\theta = \int_{-\infty}^{0} e^{-x^2} \frac{h(x)}{h(x)} dx$$

for an appropriate density function $h(x)$. You do not necessarily have to specify a particular $h(.)$, but you should state what properties $h(.)$ should possess and how you would use it to estimate $\theta$.

10. Suppose the following is the probability density function of the random variable $X$.

$$f(x) = \begin{cases} 0 & : \quad x < -3 \\ \kappa(x+3) & : \quad -3 \le x \le 0 \\ \kappa(3-x) & : \quad 0 \le x \le 3 \\ 0 & : \quad x > 3 \end{cases}$$

(a) What is the value of $\kappa$?

(b) Utilize the *inverse transform method* to generate a sample of $X$ given a uniform random variable $U \sim \mathcal{U}(0,1)$.

(c) Utilize the *acceptance–rejection* algorithm to generate a sample of $X$ given a uniform random number $U \sim \mathcal{U}(0,1)$. How many uniform random variables on average will be required to generate one sample of $X$?

# Part II

# Calibration and Estimation

# Chapter 7

## Model Calibration

The first six chapters of this book have dealt with different pricing methods for a number of different derivatives under a variety of models. To recap, in the case of having the characteristic function of the underlying process in an analytical/semi-analytical form, we can utilize one of the transform techniques to price European options or some weakly path-dependent derivatives, as explained in Chapter 2. In case of not having the closed form for the characteristic function of the process or interested in pricing path-dependent options, one can apply numerical methods to solve a PDE/PIDE to price most derivative contracts as long as the process is Markov as covered in Chapters 4 and 5. In case the process is Markov with high dimensional framework or non-Markov or perhaps the payoff structure is quite complex, then we are forced to use Monte Carlo methods as covered in Chapter 6.

In Table 7.1, we have summarized pricing methods that can be applied for valuing derivatives under processes covered sofar in a matrix of solution methods available indexed by the model, computational method and payoff type. In that table, vanilla implies European-type options, weak means weak path dependency, derivatives like one-touch barriers, and exotics mean derivatives with complex payoff and/or strong path dependency like American options. The check mark symbol indicates the method can be applied overall. The cross mark symbol points out the method cannot be applied in most cases or if it does would be fairly complicated to be used.

However, all of these pricing models require a set of model parameters in order to fully define the dynamics of each model. Additionally, none of these models is applicable to real world derivative markets unless the model is made congruent with some set of actual market prices. The process of adjusting these model parameters such that the model prices are compatible with market prices is called *calibration.* This essential step in derivative pricing can be used to price exotic derivatives utilizing the prices of their more liquid counterparts. Lastly, they can also spot arbitrage opportunities among liquidly traded derivatives, among other applications.

Pricing routines for derivative contracts take as input two mutually exclusive sets of parameters: (a) Contractual/Market parameters and (b) Model parameters. Contractual/Market parameters reflect attributes of the derivative which are specified in the contract, such as maturity, strike price and the like, which are obviously model free. Model parameters, associate with the choice of the model for evolution of the underlying process; are subjective and absolutely model dependent.

Until now we have assumed that model parameters for all the models considered were known. In order to make a model relevant to real markets and applicable for pricing, risk management, or trading, we must perform calibration, which is the process of determining a parameter set such that model prices and market prices match very closely for a given set of liquidly traded instruments. These liquid instruments are called benchmark or calibration instruments and their calibrated prices are typically recorded together in a market snapshot. The so-called *calibration* procedure delivers the optimal parameter set for the model based on these calibration instruments. For many applications, this set represents the majority if not the entirety, of the set of liquidly traded derivatives in a given market. Furthermore,

**TABLE 7.1**: Pricing schemes for various different models/processes

| | model | Pricing Model/Method | | | | | | | | |
|---|---|---|---|---|---|---|---|---|---|---|
| | | Transform Techniques | | | PDEs/PIDEs | | | MC simulation | | |
| | | vanilla | weak | exotic | vanilla | weak | exotic | vanilla | weak | exotic |
| Eq/Fx/Cmdty/Credit | GBM | $\checkmark$ | $\checkmark$ | $\times$ | $\checkmark$ | $\checkmark$ | $\checkmark$ | $\checkmark$ | $\checkmark$ | $\checkmark$ |
| | LV | $\times$ | $\times$ | $\times$ | $\checkmark$ | $\checkmark$ | $\checkmark$ | $\checkmark$ | $\checkmark$ | $\checkmark$ |
| | CEV | $\times$ | $\times$ | $\times$ | $\times$ | $\times$ | $\times$ | $\checkmark$ | $\checkmark$ | $\checkmark$ |
| | Heston SV | $\checkmark$ | $\checkmark$ | $\times$ | $\checkmark$ | $\checkmark$ | $\checkmark$ | $\checkmark$ | $\checkmark$ | $\checkmark$ |
| | SLV | $\times$ | $\times$ | $\times$ | $\checkmark$ | $\checkmark$ | $\checkmark$ | $\checkmark$ | $\checkmark$ | $\checkmark$ |
| | VG/NIG | $\checkmark$ | $\checkmark$ | $\times$ | $\checkmark$ | $\checkmark$ | $\checkmark$ | $\checkmark$ | $\checkmark$ | $\checkmark$ |
| | CGMY | $\checkmark$ | $\checkmark$ | $\times$ | $\checkmark$ | $\checkmark$ | $\checkmark$ | $\times$ | $\times$ | $\times$ |
| | VGSA | $\checkmark$ | $\checkmark$ | $\times$ | $\checkmark$ | $\checkmark$ | $\checkmark$ | $\checkmark$ | $\checkmark$ | $\checkmark$ |
| | CGMYSA | $\checkmark$ | $\checkmark$ | $\times$ | $\checkmark$ | $\checkmark$ | $\checkmark$ | $\times$ | $\times$ | $\times$ |
| 1-factor | NIGSA | $\checkmark$ | $\checkmark$ | $\times$ | $\checkmark$ | $\checkmark$ | $\checkmark$ | $\checkmark$ | $\checkmark$ | $\checkmark$ |
| | OU/Vasicek | $\checkmark$ | $\checkmark$ | $\times$ | $\checkmark$ | $\checkmark$ | $\checkmark$ | $\checkmark$ | $\checkmark$ | $\checkmark$ |
| | CIR | $\checkmark$ | $\checkmark$ | $\times$ | $\checkmark$ | $\checkmark$ | $\checkmark$ | $\checkmark$ | $\checkmark$ | $\checkmark$ |
| | Hull-White | $\checkmark$ | $\checkmark$ | $\times$ | $\checkmark$ | $\checkmark$ | $\checkmark$ | $\checkmark$ | $\checkmark$ | $\checkmark$ |
| | Ho-Lee | $\checkmark$ | $\checkmark$ | $\times$ | $\checkmark$ | $\checkmark$ | $\checkmark$ | $\checkmark$ | $\checkmark$ | $\checkmark$ |
| n-factor | Vasicek | $\checkmark$ | $\checkmark$ | $\times$ | $\checkmark$ | $\checkmark$ | $\checkmark$ | $\checkmark$ | $\checkmark$ | $\checkmark$ |
| | CIR | $\checkmark$ | $\checkmark$ | $\times$ | $\checkmark$ | $\checkmark$ | $\checkmark$ | $\checkmark$ | $\checkmark$ | $\checkmark$ |
| | ATSM | $\checkmark$ | $\checkmark$ | $\times$ | $\times$ | $\times$ | $\times$ | $\checkmark$ | $\checkmark$ | $\checkmark$ |
| | HJM | $\times$ | $\times$ | $\times$ | $\times$ | $\times$ | $\times$ | $\checkmark$ | $\checkmark$ | $\checkmark$ |
| | LMM | $\times$ | $\times$ | $\times$ | $\times$ | $\times$ | $\times$ | $\checkmark$ | $\checkmark$ | $\checkmark$ |

calibration is typically done very frequently (e.g. 2-3 times per day), in order to keep model derived prices close to their real world equivalents.

If using a model in which the set of its parameters is larger than the set of prices for the calibration instruments, then the solution obviously would not be unique. This type of problem is over-parameterized. This is often the case in markets with small numbers of derivatives and complex dynamics. In cases of over-parametrization, no set of model parameters can be found which forces the model prices to exactly match the market prices of the calibration instruments; so in practice, an approximate solution is determined by solving an appropriately constructed optimization problem. Conversely, there could be a case that the model class is too narrow to reproduce a full set of prices for the set of calibration instruments; then the solution does not exist. Moreover, this type of problem is under-parameterized. The best case for it is option pricing in the Black–Scholes framework where we have only one free parameter, volatility, yet many liquidly traded options. In the case of under-parametrization the model, is typically calibrated in such a way that smaller subsets of calibration instruments have their market prices matched with the model prices under different model parameters. The most obvious example is the classic Black–Scholes volatility surface, where a calibrated volatility exists for every liquidly traded option. Hence, the model calibration is typically an ill-posed problem [74].

There are a number of different ways one can choose the objective function for a calibration routine depending on the desired results and their uses. We will assume that $F(\Theta, \Lambda_i)$ is a pricing model[1] which is derived from a stochastic model with uncalibrated model parameters $\Theta$ and the known contractual/market parameters $\Lambda_i$ for instrument $i$. We will use

[1] As covered in earlier chapters, model prices are calculated analytically, semi-analytically via Fourier/transform methods or numerically by solving a partial (integro) differential equation or via Monte Carlo simulation.

the shorthand $C_i^\Theta$ for the model price of the $i$-th instrument using parameter set $\Theta$. In most generic cases, we can write the optimization problem as

$$\inf_{\Theta \in \mathbb{O}} \sum_{i=1}^{I} H(C_i^\Theta - C_i) \tag{7.1}$$

where $\mathbb{O}$ is the space for all possible parameter sets and $H$ is an objective function applied to the discrepancy $C_i^\Theta - C_i$, between market and model prices. Once we have formulated this calibration problem, an optimization algorithm is then applied to compute a solution and determine the calibrated model parameters.

The primary concern when performing calibration is the stability of the reconstructed model prices as a function of the input market prices. In case of under-parameterized or fully parameterized calibration problem, the market prices for a full set or subset of the calibration instruments can often be reproduced exactly if the model is sufficiently robust. However, if the resulting calibrated model parameters fluctuate wildly for every calibration, the usefulness of the model in terms of providing consistent pricing of exotics, detecting arbitrage, or gauging risk is dubious as the model is more than likely will have little predictive value. Furthermore, the model will likely not have correctly approximated the underlying evolution of market prices. In the case of an over-parameterized model, this problem can become even more acute as the calibration problem cannot be solved exactly. So as not only the derived model prices have uncertainty related to the inability of the model to reproduce the current market prices, but they also have uncertainty related to the stability of the underlying parameters [74].

The rest of this chapter will deal with the various aspects of calibration routines, used for approximating the current market prices of the calibration instruments. We will first discuss different methods for formulating the optimization problem for under-parameterized calibrations. Explaining a few different formulations and their possible uses. Next, we will present a number of different models, promoting their common uses. Furthermore, we will explore real world examples under the guise of calibration of the models and discuss the results. Finally, we will move on to a discussion of model risk, which relates to the validity of one's calibrated model and we will review its acceptable use for different applications.

---

## 7.1 Calibration Formulation

The purposes of optimal formulation of the optimization problem which drives our calibration is largely dependent on two attributes: planned application of the resulting calibrated prices and sensitivities. When utilizing the calibrated model for the pricing of exotic derivatives, the most important aspect of the calibration is its ability to correctly reproduce the prices and hedge ratios for the calibrated instruments, as well as exotic derivatives and making sure that the results are stable in time. Exact pricing in risk management may be less important than the stability of calibrated results over longer periods of time. This is due to the fact, that calibration is performed less often in this setting. For arbitrage/mispricing detection purposes, it could be advantageous to adjust the calibration to exclude or intentionally underweight the model results for instruments believed to be mispriced in order to identify arbitrage opportunities. Because there exist various goals for different calibration routines, we will review a number of different methods for formulating the calibration problem.

### 7.1.1 General Formulation

General formulation for the objective function $H$ could be as follows:

$$H(C_i^\Theta - C_i) = w_i|C_i^\Theta - C_i|^p$$

or

$$H(C_i^\Theta - C_i) = w_i|\frac{C_i^\Theta - C_i}{C_i}|^p$$

or

$$H(C_i^\Theta - C_i) = w_i|\ln C_i^\Theta - \ln C_i|^p$$

where $p \geq 1$ and $w_i$ is a positive weight often chosen inversely proportional to the squared of bid-offer spread of $C_i$. The results may vary with the choice of the objective function.

### 7.1.2 Weighted Least-Squares Formulation

The most common formulation of the calibration problem is the least-squares formulation:

$$\inf_{\Theta \in \mathbb{O}} \sum_{i=1}^{I} w_i(C_i^\Theta - C_i)^2 \tag{7.2}$$

Least-squares solutions to model fitting problems are popular in a number of disciplines and derivatives, pricing is no exception; typically, this is the most popular formulation of the calibration. Furthermore, many other formulations will assume market prices are equally valid and as such, will set all the weights in this least-squares formulation to one. Somewhat more accurate formulations will account for the fact that most derivative prices are defined only up to a bid-offer spread. A model may generate prices compatible with the market but may not exactly fit the mid-market prices for any given $\Theta \in \mathbb{O}$. Therefore, we may reformulate calibration as a least-squares problem where $C_i$ is the mid-market quote for the $i$-th instrument and $w_i$ its corresponding weight which is positive and often chosen inversely proportional to the square of the bid-offer spread of $C_i$. Thereby placing smaller weight on prices with larger spreads to express the additional uncertainty in those prices. Additionally, if the calibration routine is going to be used for arbitrage detection, the formulation may be made to exclude prices for certain calibration instruments which are believed to be mispriced or the weights for these instruments may be set very low. Then the generated model prices may be used to gauge relative mispricing for various different subsets of the calibrated instruments.

### 7.1.3 Regularized Calibration Formulations

In most cases, model price $C_i^\Theta$ depends continuously on $\Theta$ and the parameter space $\mathbb{O}$ is a subset of a finite dimensional space and in this case the least-squares formulation, it always grants a solution. However, because the calibration problem is often under-parameterized, the solution may not be unique. As the case maybe for the existence of several local minima, any of which could be taken as the true global optimal solution when using: (a) different optimization routines, (b) different starting point, (c) different objective function, or (d) different restriction on the parameter set. Thus, even if the number of observed prices is much higher than the number of model parameters, this does not imply unique identifiability of parameters.

Regularization methods can be used to overcome this problem. A common method is to have a convex penalty term $R$, called the regularization term, added to the pricing error and solve the auxiliary problem. This regularization technique should be used to solve an ill-posed problem or to prevent overfitting. It introduces an additional requirement, such as restriction for smoothness or bounds on the vector space norm.

An example of regularization methods for calibration formulation is presented by Marco Avellaneda in [19]. This paper presents a minimum entropy algorithm to fit the U.S. LIBOR curve and shows the corresponding sensitivities of fixed income securities to the input price. The author also proposes several regularizations at the level of forward-curve building.

Another example of regularization is done by Herbert Egger and Heinz W. Engl [108] where they utilize a Tikhonov regularization method for calibration problems by applying the Tikhonov regularization to the inverse problem of option

pricing. They focus on the stability of Tikhonov regularization and study the convergence rate. This technique adds a regularization term to the minimum object, and therefore avoids the non-unique solution phenomenon. In [65], the authors discuss a convex regularization framework for local volatility calibration in derivative markets.

## 7.2 Calibration of a Single Underlier Model

In this section we will examine the calibration procedure for a number of different models, specifically on the evolution of a single underlier. Although, many derivatives may be traded on this underlier, it will include most equity, commodity, and foreign exchange (FX) markets where derivatives are usually dependent on only a single underlying asset or rate. Thus, when employing the common uses of these models, we will utilize common techniques for calibration under these models, present calibration results from real world data sets, and discuss the results.

### 7.2.1 Black–Scholes Model

The Black–Scholes model, which assumes a geometric Brownian motion for the underlying asset, is the simplest stochastic model that can be used for evolution of an underlying process (stock price, exchange rate, commodity price, etc.) that guarantees that the price stays positive. As discussed previously, under the risk-neutral measure, the underlying price, $S_t$, satisfies the following stochastic differential equation.

$$dS_t = (r - q)S_t dt + \sigma S_t dW_t$$

where $r$, $q$ and $\sigma$ are a continuous interest rate, a continuous dividend rate and the instantaneous volatility, respectively. The exact solution to the SDE is given by

$$S_t = S_0 e^{(r-q-\sigma^2/2)t+\sigma W_t}$$

The Black–Scholes PDE, gives us the price of derivative securities depending on the terminal and boundary conditions we apply to it. However, the only non-observable model parameter in this model, is the volatility of the underlying asset. In that parameter set, where we need to determine during calibration, it is simply $\Theta = \{\sigma\}$, the implied volatility of the traded option. Thus, this model is under-parameterized, as are most markets options are traded at a variety of different strikes and maturities. Attempting to fit all market prices simultaneously would be unrealistic because we would be assuming a constant volatility for all traded options with different strike prices and maturities. It is well documented that at-the-money option's volatility is lower than out-of-the-money options volatility [95]. Therefore, when calibrating to at-the-money options, we would underestimate out-of-the-money prices. In contrast, in calibrating to out-of-the-money options, we would overestimate at-the-money premiums. This is called, the volatility smile phenomenon ([95],[194]). This is a well-known issue with the Black–Scholes model. However, because of the simplicity of the Black–Scholes model, its analytical tractability, and the general robustness of generated hedging ratios, the market continues to embrace it. While electing to quote option prices not in terms of premiums, instead in terms of the so-called *implied volatility.*

The calibration procedure under the Black–Scholes model, involves solving for the implied volatility. That is, given a single European option premium, find for the volatility that makes the Black–Scholes price the same as the market price.

As long as a call premium is between $(S_0 - K)^+$ and $S_0$ and a put premium is between $(K - S_0)^+$ and $K$ by applying the bisection method or Newton-Raphson method one should be able to find the implied volatility with no issues. Therefore, generating the discrete points on an implied volatility surface consists of simply doing this single option calibration for all quoted option prices.

### 7.2.2 Local Volatility Model

While the Black–Scholes model is the simplest formulation for derivative pricing and is still utilized for many other simpler derivative contracts. Here the need for a volatility surface, which implies different underlying parameters for every quoted option is needed and the model's inability to correctly model the evolution of the underlying asset. Limiting the usefulness in pricing and hedging more exotic derivative contracts and thus extensions were developed.

To overcome the shortcomings of the Black–Scholes model, Derman and Kani [95] proposed a local volatility model which parameterized underlier volatility in terms of the current underlier price and the calendar time. Extending the parameterization of the standard Black–Scholes model to include both underlier price and time which will allow us to simultaneously calibrate to many or all liquid vanilla option contracts. In addition, as the same time, allowing this model to be more realistically applied to exotic derivatives. The local volatility model has the following SDE:

$$dS_t = (r(t) - q(t))S_t dt + \sigma(S_t, t)S_t dW_t$$

Here the assumption is a deterministic term structure for both interest rate and dividend rate. Prices of options under this model, will satisfy the so-called generalized Black–Scholes PDE, which gives the pricing of derivatives securities, depending on the terminal condition and boundary conditions we apply to it.

$$\frac{\partial V}{\partial t} + \frac{1}{2}\sigma^2(S,t)S^2\frac{\partial^2 V}{\partial S^2} + (r(t) - q(t))S\frac{\partial V}{\partial S} = r(t)V(S,t)$$

The popularity of the local volatility model is due to its simplicity. Arriving at solutions and derivative prices under the local volatility model requires, only a few simple modifications of the Black–Scholes model. This along with its additional flexibility in consistently pricing a full set of options, is why most traders and firms actively utilize this model. Marking and pricing of derivatives is simple and calculating hedge ratios is straightforward. Moreover, as explained in [136], local volatility models re-engineer semi-martingale models for vanilla options. However, for path-dependent options, local volatility and semi-martingales could behave very differently [137].

The set of calibration parameters under the local model is $\Theta = \{\sigma(S_t, t)\}$ and the set of contract parameters and deterministic market parameters is $\Lambda = \{S_0, K, T, r(t), q(t)\}$.

Thus, the objective of the calibration procedure for local volatility models is to find the local volatility surface $\sigma(S, t)$. So that model prices will closely match market prices for options at available strikes and maturities. One way to accomplish this is to formulate the calibration problem as follows:

$$\arg\min_{\Theta} \frac{1}{M} \frac{1}{N} \sum_{i=1}^{N} \sum_{j=1}^{M} \|V_{ij} - \widehat{V}_{ij}\|$$

where $V_{ij}$ and $\widehat{V}_{ij}$ are market and model prices for strike $K_i$ and maturity $T_j$, respectively. Note that for each set of strike and maturity we assume the same volatility surface $\sigma(S, t)$, yet we have to solve the generalized Black–Scholes PDE for each set of maturity and strike separately, in order to generate the market prices for any and all available options. For a fixed strike price the option payoff is the same, but it is paid at a different time (maturity), which would result in a different set of option prices. For the fixed maturity, the option has a different payoff depending on the strike price that would again result in a different set of option prices.

#### 7.2.2.1 Forward Partial Differential Equations for European Options

The pricing and hedging derivatives in a manner consistent with the volatility smile has been a major research area for over a decade and the development of the local volatility model was a significant step in improving performance in this area. However, as noted in the last section, the basic construction of the calibration problem for local volatility surfaces, when utilizing the generalized Black–Scholes PDE. It requires us to solve the PDE, again for every calibrated derivative price and for every iteration of the optimization routine. This involves a great many PDE solutions and if possible, we would like to be able to solve for every option price in the strike and maturity grid simultaneously. This could reduce our computation time for the calibration procedure significantly.

A breakthrough occurred in the mid-nineties with the recognition that in certain models, European option values satisfied forward evolution equations, in which the independent variables are the options strike and maturity. Specifically, Bruno Dupire [104] showed that under deterministic carrying costs and a diffusion process for the underlying price, no arbitrage implies that European option prices satisfy a certain partial differential equation (PDE), now called the Dupire equation. If we assume that one can observe European option prices at all strikes and maturities, then this forward PDE can be used to explicitly determine the underlying's instantaneous volatility as a function of the underlying's price and time.

Once this volatility function is known, the value function for European, American and many exotic options can be determined by a wide array of standard methods. Because this value function relates to theoretical prices of these instruments to the underlying's price and time, it can also be used to determine many hedge parameters (Greeks) of interest as well. In addition to their usefulness in determining the volatility function, forward equations also serve another useful purpose. Once the volatility function is known, either by an explicit specification or by a prior calibration, the forward PDE, can be solved via finite differences to efficiently value a collection of European options of different strikes and maturities, all written on the same underlying asset. Assuming a known local volatility surface, this will allow us to solve for every option price in strike and maturity space simultaneously on the same grid and as pointed out in [16], all the Greeks of interest satisfy the same forward PDE and hence can also be efficiently determined in the same way. This will allow us to solve the implied volatility calibration problem, in a much more computationally efficient manner.

As stated in Section 4.6, the Dupire PDE gives European call prices for all strikes and maturities.

$$-\frac{\partial C}{\partial T}+\frac{1}{2}\sigma^2(K,T)K^2\frac{\partial^2 C}{\partial K^2}-(r(T)-q(T))\,K\frac{\partial C}{\partial K}=q(T)C$$

Assuming market quotes of option prices $C(K_i,T_j)$, we can then calculate the local volatility surface from market prices explicitly using the following inversion formula:

$$\sigma(K,T)=\left(\frac{\frac{\partial C}{\partial T}+(r(T)-q(T))\,K\frac{\partial C}{\partial K}+q(T)C}{\frac{1}{2}K^2\frac{\partial^2 C}{\partial K^2}}\right)^{1/2} \tag{7.3}$$

To apply the local volatility inversion formula from the Dupire equation, we must work under the assumption that we have a very smooth surface for the call price premiums in terms of strike and maturity, $C(K,T)$. In addition, we must also be able to calculate the calendar spread, $\frac{\partial C}{\partial T}$, butterfly spread, $\frac{\partial C}{\partial K}$ and second partial derivative with respect to strike price, $\frac{\partial^2 C}{\partial K^2}$, at arbitrary points on the surface of call price premiums.

This is the key difficulty in implementing this method; option prices are only available in the market on a finite grid of strike prices and maturities, where interpolation schemes must be invoked to infer prices for the intermediate strike prices and maturities. Here interpolations used, may or may not be consistent with the requirements of the absence of at least static arbitrage across the strike price and maturity spectrum. Even when this is accomplished, the interpolation schemes can introduce non-differentiability at various levels, leading to local volatility functions that are erratic and inspire little confidence. The task of properly interpolating the surface of option prices consistent with observed market prices, is essentially the task of formulating and estimating a market with a consistent option pricing model. As illustrated in [74], the local volatility can be very sensitive to small changes in inputs. In the coming section on calibration of local volatility surfaces, we will explore a number of different methods proposed to construct a smooth, stable, and consistent local volatility surfaces.

#### 7.2.2.2 Construction of the Local Volatility Surface

The formulated calibration which will in theory solve for a fully specified local volatility surface, in practice will not have enough market quoted option prices to generate a fully specified and smooth surface for option premiums, which makes constructing a local volatility surface very challenging. This stems from the fact that we typically solve the option pricing PDE, on a mesh, which is much finer than the grid of available option premiums arranged by strike price and maturity. Therefore making this calibration problem is over-parameterized. Thus, it is usually the case that we must interpolate/extrapolate from market prices to get option prices at any strike price and maturity and then apply some smoothing technique to get a smooth surface.

Naturally, the first attempt at this solution is bi-cubic spline interpolation. This could be done on either the call prices or implied volatilities to calculate call prices or implied volatilities for a range of strike prices and maturities and then substitute interpolated/extrapolated values into the calibration routine to construct the local volatility surface. However, the resulting local volatility surface is generally a very non-smooth and non-differentiable surface, as reported in [137].

Constructing a smooth local volatility surface from a limited set of option prices has been a topic of interest for some years now and there have been a number of papers written on the topic. One approach is, to use a suitable functional form for the local volatility surface that reflects the shape of the implied volatility surface in a given market, usually having a smile or smirk for shorter maturities and decaying with time and flattening for longer maturities. This functional form, might take four to six parameters, and we can redefine this to be the set of model parameters that we will solve for in our calibration routine.

In [2], the authors propose several algorithms to find the volatility surface. This paper does not deal with local volatility models, but with the Black–Scholes model and how to fit the Dupire equation utilizing plain vanilla European option prices. Even though the principles are applicable to local volatility surface calibration as well. The authors reach the conclusion that using mesh adaptation/multi-level strategies can reduce the computing time significantly.

The work in [71] follows the same train of thought. This paper discusses the well recognized fact that index option markets typically exhibit a volatility smile and therefore demonstrate that the constant volatility of Black–Scholes option pricing formula is not realistic. The very existence of different implied volatilities for different options, demonstrates that the constant volatility model does not adequately describe the underlying price dynamics. The authors demonstrate that under a one-factor continuous diffusion model, the constant volatility method, with different implied volatilities applied to options of different strike prices and maturities, is able to price the vanilla options accurately and the hedge ratios that are computed by using these implied volatilities, can be erroneous.

The authors propose a method for computing a smooth local volatility function assuming that the underlying asset follows a one-factor continuous diffusion model. In the paper, they emphasize that accurately approximating the local volatility function under a one-factor model, is crucial in hedging even simple European and in pricing exotic options. To approximate a local volatility function, they use a spline functional approach, wherein the local volatility function is represented, by a spline whose values at chosen knots are determined by solving a constrained nonlinear optimization problem. This optimization formulation, is amenable to various option evaluation methods and in the paper they use as an example of a partial differential equation implementation.

Using a synthetic European call option example, they illustrate the ability of the proposed method to reconstruct the unknown local volatility function. Also demonstrated is the spline volatility function, which yields smaller average absolute hedging error than the implied/constant volatility method due to more accurate hedge parameters. In addition, market European call option data on the S&P 500 stock index, was used to compute the local volatility function under their spline function method it shows that in both the S&P 500 index option and futures option markets, the average hedging error using the volatility function approach, is always smaller than that of the implied/constant volatility error. Moreover, it is smaller for a sufficiently long hedge horizon.

In a followup paper [69], they illustrate the spline volatility function proposed in [71]. It yields smaller than average, the absolute hedging error is compared with the implied/constant volatility method for more accurate hedge parameters. When comparing the hedge performance in the S&P 500 index option, as well as the futures option markets, which observe similar results in the delta from the implied/constant volatility method, is typically greater than that of the local deterministic volatility function approach. In both the S&P 500 index option and futures option markets, the observation is that the average hedging error using the volatility function approach is consistently smaller than that of the implied/constant volatility error. Moreover, the average absolute hedging error utilizing the volatility function, is smaller than that of the implied/constant volatility method for a sufficiently long hedge horizon. For approximately 17 days for the S&P 500 index options and 6 days for the S&P 500 futures options.

Another method for calibrating a pricing model to a set of market quoted option prices, is presented in [20]. In this paper the authors describe an algorithm which yields an arbitrage free diffusion process, that minimizes the relative entropy distance to a prior diffusion and then solve a constrained (min-max) optimal control problem utilizing a finite-difference scheme for a Bellman parabolic equation, combined with a gradient based optimization routine. The number of unknowns in the optimization step is made equal to the number of market quoted option prices that need to be matched. That number is independent of the mesh size used for the scheme. This results in an efficient nonparametric calibration method, that can match an arbitrary number of option prices, to any desired degree of accuracy. The algorithm can be used to interpolate implied volatilities of traded options, in both the strike and maturity dimensions. Furthermore it can also be used to price exotic derivatives. The stability and qualitative properties of the computed volatility surface are reviewed, including the effect of the Bayesian prior on the shape of the surface and on the implied volatility smile/skew.

Another work which deals with stable calibrations of a full set of market traded options is [13], in which the authors illustrate how to construct an unconditionally stable finite difference lattice consistent with the equity option volatility smile. In particular, their work shows how to extend the method of forward induction on Arrow–Debreu securities to generate local instantaneous volatilities in implicit and semi-implicit (Crank–Nicolson) lattices. The technique developed in the paper provides an accurate fit to the entire volatility smile and offers convergence properties and high flexibility of asset- and time-space partitioning. In contrast to standard algorithms based on binomial trees, this approach is well suited to price options with discontinuous payouts (e.g., knock-out and barrier options) and does not suffer from problems arising from negative branching probabilities. However, the previous two approaches presented in [20] and [13], suffer from the drawback that the constructed calibration surface is non-smooth.

Another attempt at stable local volatility surface construction is presented in [85], which describes a method that uses trinomial trees and Tikhonov regularization to calibrate the local volatility. The author demonstrates an implementation of this method on the inverse problem of calibrating a local volatility function from observed vanilla option prices in a generalized Black–Scholes model and claims the methodology is numerical stability [65].

A more generic algorithm for estimating parameters of option pricing models from a set of observed option prices is presented in [128], wherein the authors propose a probabilistic approach. The approach is based on a stochastic optimization algorithm which generates a random sample from the set of global minima of the in-sample pricing error and allows for the existence of multiple global minima. Starting from an independently and identically distributed population of candidate solutions, drawn from a prior distribution of the set of model parameters, the population of parameters is updated through cycles of independent random moves followed by *selection* according to pricing performance. The authors examine conditions under which an evolving population converges to a sample of calibrated models. The heterogeneity of the obtained sample can be used to quantify the degree of ill-posedness of the inverse problem. It provides a natural example of a coherent measure of risk, which is compatible with observed prices of vanilla options and takes into account the model uncertainty, resulting from incomplete identification of the model. They go on to describe a fully specified algorithm in the case of a diffusion model, with the goal of retrieving the unknown local volatility surface from a finite set of option prices and illustrate its performance on simulated and empirical data sets of index options.

In [137], the authors propose inferring a local volatility surface, from the calibrated parameters by simply calculating call prices for a range of strike prices and maturities for the model under consideration and substituting those premiums in Equation (7.3) to obtain the local volatility surface. Any stochastic volatility model, such as the Heston stochastic volatility model or the variance gamma with stochastic arrival (VGSA) model, can be successfully calibrated to market European option prices across all strikes and maturities. The obvious result being those calibrated models can deliver a smooth surface for call/put prices in both dimensions. For the VGSA model, they have shown the scheme generates a pretty smooth local volatility surface. The obtained local volatility surface is fully dependent on the model originally used to interpolate European option prices in strike and maturity space. Obviously it is not going to be unique. Figures 7.3 and 7.9 illustrate the constructed local volatility surface for the S&P 500 stock index as of December 10, 2000 using this approach for the VGSA and Heston stochastic volatility models, respectively.

We can see in these figures, volatility surfaces look pretty similar. The drawback to this method, is that the use of stochastic volatility models tend to compensate for elevated deep out-of-the-money option prices, by pushing up volatility to a level that might be unrealistic, as reported in [136].

Setting aside computational problems in construction of the local volatility surface , there are more significant semantic issues, as cited in [94]. Looking at the future local volatilities in these models consistent with todays implied volatilities, are not reassuring. Local volatility models have a scale that depends specifically on future index levels and time. Far in the future, the local volatilities are roughly flat, predicting a future smile that is much flatter than current smiles, an uncomfortable and unrealistic forecast that contradicts the nature of the skew. If these models forecast unrealistic future volatilities, then the question begs, how much one can trust prices and hedge ratios from these models? For all these reasons it is compelling to look at models of a different nature.

### 7.2.3 Constant Elasticity of Variance (CEV) Model

The shortcomings of the Black–Scholes model in terms of calibration, as compared to a full set of market prices for options, have been well known for many years. As discussed in the last section, local volatility models provide enough scale of freedom to fit an arbitrary number of market option prices. Also noted, is that construction of a smooth volatility surface, is a very difficult task and as such, some sought more parsimonious models for the underlying asset.

The existence of the volatility smile, prompted the idea that the Black–Scholes model with constant volatility was not sufficient and even before the full local volatility model was developed, models which included volatility, with asset price dependency were used. The *constant elasticity of variance* (CEV) process [81] developed by Cox, assumes that the asset price follows this process

$$dS_t = (r - q)S_t dt + \delta S_t^{\beta+1} dW_t$$

for $t > 0$, $S_0 > 0$. The two model parameters are $\delta$ and $\beta$, where the latter can be interpreted as the *elasticity of the local volatility function* and the former is a scale parameter that may be used to calibrate the initial instantaneous volatility. Though not as flexible as the local volatility model, this model does have more flexibility in matching the stock return distribution implied by vanilla options at a given maturity date than the Black–Scholes model. In addition, for the CEV process with $\beta < 0$ and $r - q > 0$, the price of the plain vanilla put is also given by a quasi closed-form solution. This complex functional form is derived in detail in [89] and [90]. The price of the plain vanilla call is obtained using put-call parity.

Below is an example of the CEV model calibration. Table 7.2 shows CEV parameters obtained from calibration of out-of-the-money call and put vanilla option values for the S&P 500 of October 19, 2000.

We note the relative constancy of the parameter $\sigma$ across maturity. The parameter $\beta$ declines with maturity and it seems that this parameter is attempting to cope with both

**TABLE 7.2**: CEV parameters obtained from calibration of the S&P 500 on October 19, 2000

| Time to maturity | $\sigma$ | $\beta$ | $r$ | $q$ | Spot |
|---|---|---|---|---|---|
| 0.07934 | 0.2162 | -2.1100 | 0.0663 | 0.0125 | 1389.459 |
| 0.15585 | 0.2239 | -4.2195 | 0.0663 | 0.0128 | 1389.869 |
| 0.40504 | 0.2202 | -2.6892 | 0.0667 | 0.0119 | 1389.459 |
| 0.65424 | 0.2208 | -2.1729 | 0.0660 | 0.0117 | 1389.708 |
| 0.92273 | 0.2257 | -1.9863 | 0.0654 | 0.0116 | 1390.906 |

changes in implied skewness and kurtosis. Though it is not clear, which is the dominating influence. We observe from Figure 7.1 CEV does pretty poorly in calibration to out-of-the-money options with short maturities; however, for longer maturities the fit seems to be good.

### 7.2.4 Heston Stochastic Volatility Model

While the CEV model is an improvement upon the Black–Scholes model in terms of flexibility for calibration, in that it allows for volatility to vary with the underlier price. The volatility still has no time component. This is somewhat at odds, with the observation in many markets of a term structure of implied volatilities under the Black–Scholes model, indicating an effective change in volatility over time. The Heston stochastic volatility model, presented in [134], models volatility itself as stochastic, allowing for both asset level and time effects to be expressed in this model. Under the Heston model the asset price has the following form:

$$\begin{aligned}
dS_t &= (r-q)S_t dt + \sqrt{v_t} S_t dW_t^1 \\
dv_t &= \kappa(\eta - v_t)dt + \lambda\sqrt{v_t} dW_t^2 \\
dW_t^1 dW_t^2 &= \rho dt
\end{aligned}$$

This allows volatility to be stochastic but models it as mean-reverting,. Reflecting the observation that volatility generally does not diffuse to extreme levels and remains somewhat range bound. The parameter $\eta$, represents the long term mean volatility, while $\kappa$ represents the rate of mean reversion. The variable $\lambda$ represents the volatility of variance and $\rho$, represents the correlation between the two driving Brownian motions. As follows, the parameters to be calibrated are $\Theta = \{\kappa, \eta, \lambda, \rho, v_0\}$. For the square-root process the variance stays positive and if $2\kappa\eta > \lambda^2$, then it never reaches zero. An observation is that the drift term of the variance process is asymptotically stable if $\kappa > 0$ and equilibrium point is $v_t = \eta$. As demonstrated in Chapter 2, the characteristic function of the log of the underlying process is available under this model and so we can price European and some weakly path-dependent options via transform methods. Other derivative prices are available via numerical solution of the two-dimensional PDE, as discussed in Chapter 4.

Minimizing the objective function is clearly a nonlinear programming problem, with the nonlinear constrain $2\kappa\eta > \lambda^2$. This condition ensures that the volatility process cannot reach zero. Unfortunately the objective function is far from being convex and it turns out that usually there exist many local minima. As a consequence, we should try various starting points to make sure we get the optimal parameter set. We might use a penalty function, $R(\Theta, \Theta_0)$, to make the calibration some additional stability as suggested in [179].

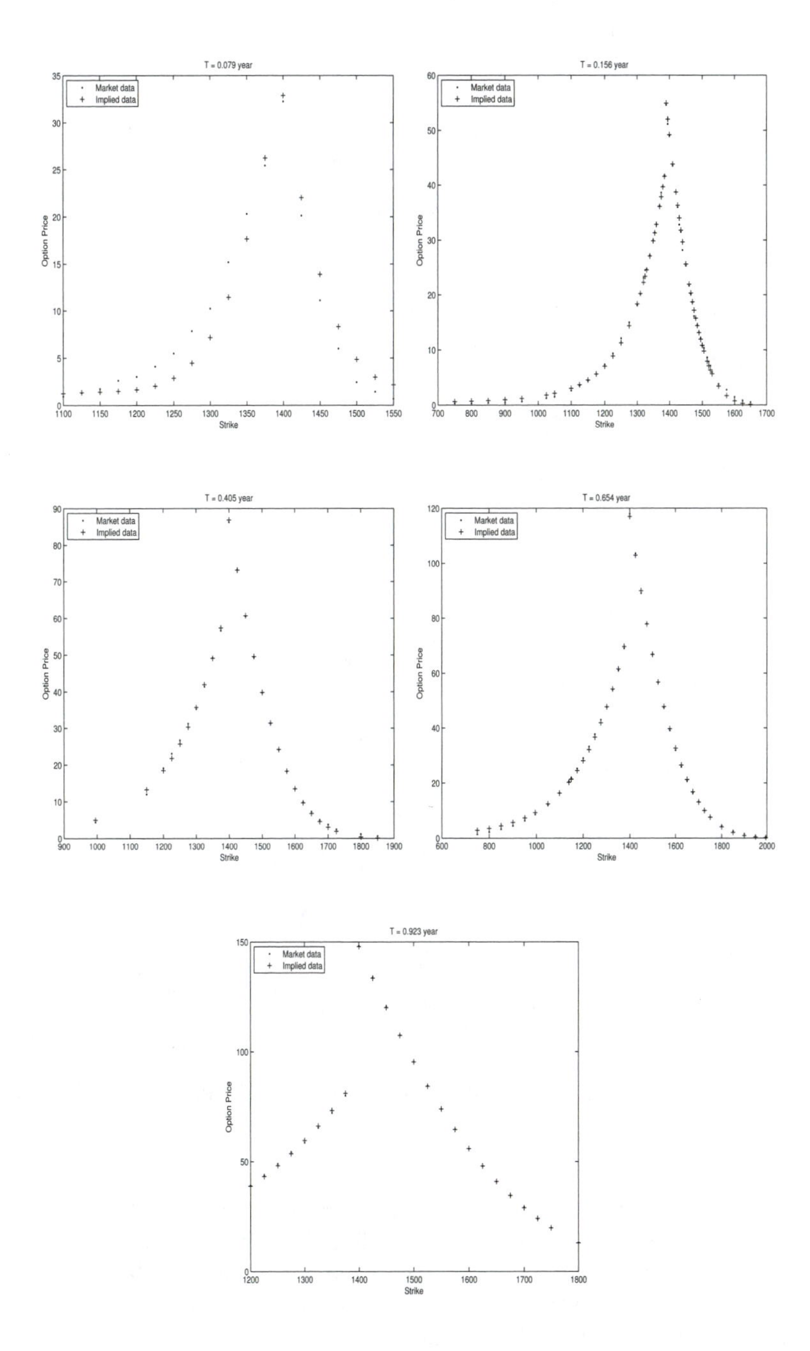


**FIGURE 7.1**: CEV vs. market premiums for the S&P 500 on October 19, 2000 for different maturities

For example we might set $R(\Theta, \Theta_0) = \|\Theta - \Theta_0\|^2$ the distance from the original starting parameter set.

We provide an example of the Heston stochastic volatility model parameters generated from calibration of S&P 500 prices on October 19, 2000.

$$
\begin{aligned}
\lambda &= 1.0143 \\
\kappa &= 4.9549 \\
\eta &= 0.0562 \\
\rho &= -0.6552 \\
v_0 &= 0.057
\end{aligned}
$$

In Figure 7.2 we display Heston stochastic volatility premiums vs. market premiums for S&P

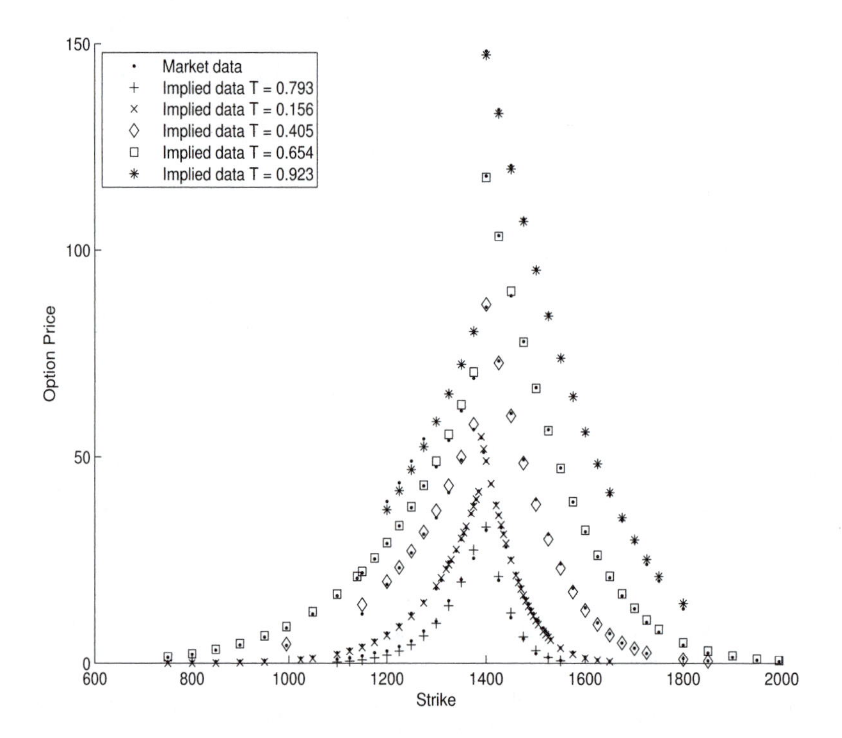


**FIGURE 7.2**: Heston vs. market premiums for the S&P 500 on October 19, 2000 for various maturities

500 option premiums on October 19, 2000 across all maturities. The fit across all maturities seems to be good. Having parameters from calibration, we can calculate premiums for any strike and any maturity and construct the vanilla call surface. Having the call surface utilizing Equation (7.3) to find the local volatility surface implied from the call surface. The local volatility surface is illustrated in Figure 7.3.

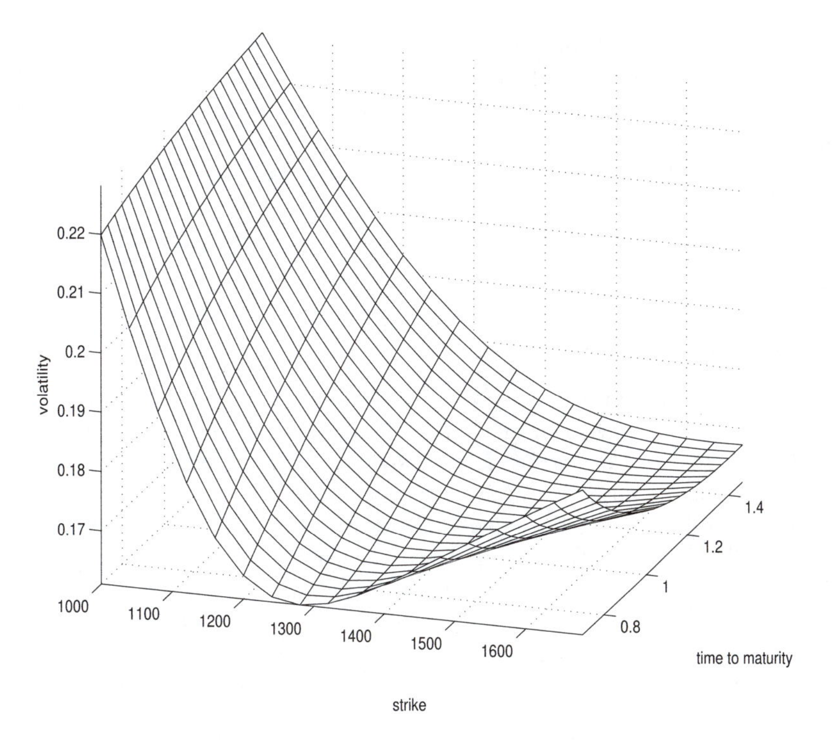


**FIGURE 7.3**: Local volatility surface obtained from the call price surface of the Heston stochastic volatility model

### 7.2.5 Mixing Model — Stochastic Local Volatility (SLV) Model

As described in Section 1.2.4, the stochastic local volatility (SLV) model is a mixture of the local volatility model and the stochastic volatility model given by [209]

$$\begin{aligned} dS_t &= (r-q)S_t dt + L(S_t,t)V_t S_t dW_t^1 \\ dV_t &= \kappa(\eta - V_t)dt + \lambda V_t dW_t^2 \\ dW_t^1 dW_t^2 &= \rho dt \end{aligned}$$

Two parameter sets need to be calibrated. One is stochastic parameters and the other one is the local volatility component $L(S_t,t)$, called leverage surface. As stated in [171], the calibration procedure is an ill-posed and unstable inverse problem. This fact is shown by the classical formula that connects the local and implied volatilities in the presence of the term structure.

Given a set of stochastic parameters, we can calibrate the leverage surface to match the vanilla market. Given different sets of stochastic parameters we can fit the vanilla market quite well by re-calibrating the leverage surface $L$. However, this will correspond to a different dynamics.

First, we calculate local volatility surface $\sigma(S_t,t)$ using the Dupire formula. The marginal distribution of the stochastic local volatility model is the same as the marginal distribution of the local volatility model [126]. Using integral form of the conditional expectation and Kolmogorov PDE, by the convergence of $L$, we obtain $L(S,t)$ numerically. We solve the PDE using a finite difference method [209].

For stochastic parameters, we find $\kappa, \theta, \lambda$ and $\rho$ to match vanilla market as closely as possible in pure stochastic model. Then we choose a mixing fraction that represents the percentage of stochastic model in the stochastic local volatility model. We can calibrate the mixing fraction parameter by letting the model match key barrier option prices or matching the historical dynamics of the volatility surface. We calibrate leverage surface to match the vanilla market more often than stochastic parameters. In practice, stochastic parameters are not supposed to be recalibrated frequently.

We refer readers to [209] and [191] for more detail on calibration of the stochastic local volatility model.

### 7.2.6 Variance Gamma Model

The presence of sometimes very large volatility smiles for option prices, especially for short durations, can lead to very large changes in modeled volatilities under diffusion models. This will allow for non-constant volatility, such as the local volatility, CEV, and Heston models. This issue has led many researchers to concentrate their efforts on modeling underlying asset prices with jump models, which allow for discrete jumps in asset prices which may more readily explain the volatility smile. One such model, is the variance gamma model which was described in Chapter 1. As explained, it is a three-parameter model with the calibration parameters being $\Theta = \{\sigma, \theta, \nu\}$, volatility, skewness, and kurtosis respectively.

Results indicate that for a fixed maturity the model does an adequate job fitting across various strike prices. This would suggest that for each maturity, we should do a separate calibration and as a result would have a separate set of parameters. In calibration, we typically observe that the volatility parameter of the variance gamma model reduces from shorter to longer maturities, however, overall volatility stays in a tight range. Kurtosis increases for longer maturity; skewness, on the other hand, reduces. For equity options we typically observe negative skew.

Here we present a couple of calibration cases for the VG model. In the first case, we look into the limiting behavior of the variance gamma model. We show that if out-of-the-money call and put prices from a Black–Scholes model with a constant volatility are provided as market prices, the VG model can detect premiums are coming from Black–Scholes with a constant volatility and that indicates that VG can recover a pure diffusion model under the special case of $\nu = 0$ and $\theta = 0$.

Parameters used for this example are spot price \$100, volatility $\sigma = 0.40$, maturity $T = 1.0$ year, risk-free rate $r = 5\%$, dividend rate $q = 0.0\%$, strike prices ranging from $80, 85, 90, \ldots, 125$. Initial parameters used for VG model are $\sigma = 0.10$, $\nu = 0.10$, and $\theta = 0.10$. Obtained parameters from calibration using simplex optimization are $\sigma = 0.400304$, $\nu = 0.041651$, $\theta = 0.037529$. In Figure 7.4 we plot VG calibrated prices versus Black–Scholes prices. Obviously, they match very closely. For illustrative purposes on how prices depart from Black–Scholes, we do the following scenarios. In the first graph we keep $\theta$ fixed and start increasing $\nu$ and in the second graph $\nu$ is fixed while $\theta$ changes. As shown in Figure 7.5(a), for out-of-the-money puts premiums decrease as $\nu$ increases. For out-of-the-money calls it is exactly the reverse, premiums increase as $\nu$ increases. As shown in 7.5(b), in general premiums increase as we depart from zero skewness for both out-of-the-money calls and puts.

In the second calibration case, we obtain variance gamma parameters from calibration to S&P 500 options. Table 7.3 displays the variance gamma parameters obtained from calibration of S&P 500 out-of-the-money European option prices (calls/puts) on October 19, 2000, across various different maturities. We note negative values for $\theta$ that decline with maturity. This is a reflection of negative skewness induced by risk aversion that declines in the implied annualized risk-neutral density. The parameter $\nu$ is rising with maturity and reflects an increase in excess kurtosis for the implied annualized risk-neutral density. In this

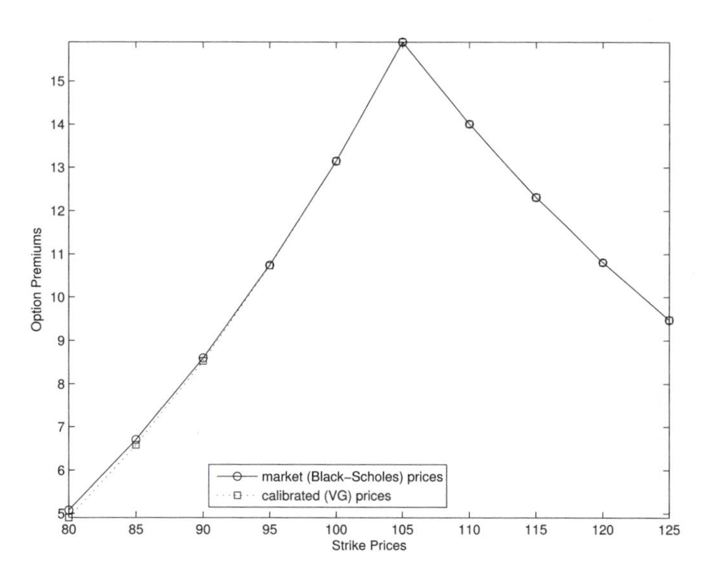


**FIGURE 7.4**: VG calibrated to Black–Scholes

**TABLE 7.3**: VG parameters obtained from calibration of the S&P 500 on October 19, 2000

| Time to maturity | $\sigma$ | $\nu$ | $\theta$ | $r$ | $q$ | Spot |
|---|---|---|---|---|---|---|
| 0.07934 | 0.2085 | 0.0735 | -0.4986 | 0.0663 | 0.0125 | 1389.459 |
| 0.15585 | 0.2100 | 0.1267 | -0.3599 | 0.0663 | 0.0128 | 1389.869 |
| 0.40504 | 0.1925 | 0.2509 | -0.2820 | 0.0667 | 0.0119 | 1389.459 |
| 0.65424 | 0.1902 | 0.4352 | -0.2283 | 0.0660 | 0.0117 | 1389.708 |
| 0.92273 | 0.1939 | 0.6088 | -0.1991 | 0.0654 | 0.0116 | 1390.906 |

case, the volatility parameter is fairly constant across maturity. We observe from Figure 7.6, unlike the CEV model, the variance gamma model does pretty well for short maturities as well as longer maturities.

### 7.2.7 CGMY Model

As shown in Chapter 2, the characteristic function of the log of the underlying process is available under CGMY, and so we can price European and some exotic options via transform methods. Other derivative prices are available via numerical solution of the PIDE discussed in Chapter 5. CGMY parameters are obtained from calibration
to S&P 500 prices on October 19, 2000. We observe from Figure 7.7 CGMY like the variance gamma model does well for short maturities as well as longer maturities.

### 7.2.8 Variance Gamma with Stochastic Arrival Model

As described in Chapter 1, the variance gamma with stochastic arrival model is an extension of the variance gamma model which allows for a stochastic volatility model through stochastic arrival of jump times. The parameter set for the model is $\Theta = \{\sigma, \nu, \theta\, \kappa, \nu, \lambda\}$.

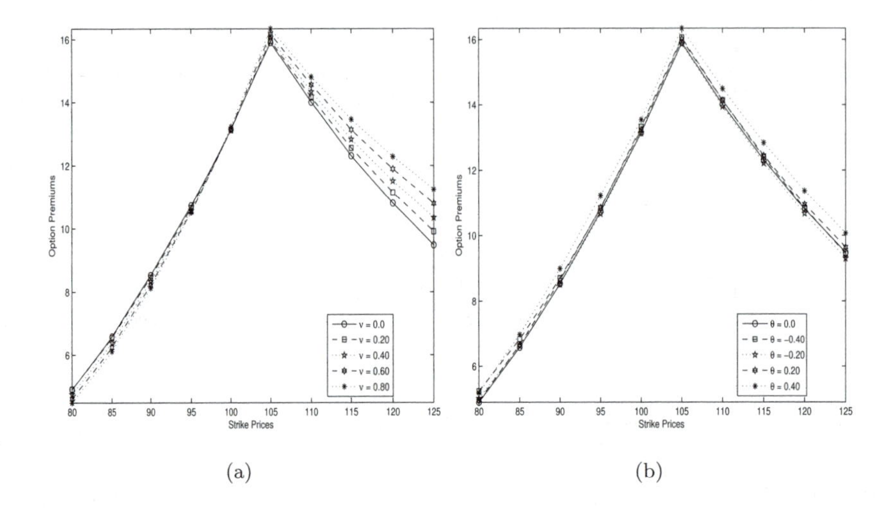


**FIGURE 7.5**: VG with various $\nu$ and $\theta$

**TABLE 7.4**: CGMY parameters obtained from calibration of the S&P 500 on October 19, 2000

| Time to maturity | $C$ | $G$ | $M$ | $Y$ | $r$ | $q$ | Spot |
|---|---|---|---|---|---|---|---|
| 0.07934 | 0.6083 | 7.6829 | 39.9301 | 0.8571 | 0.0663 | 0.0125 | 1389.459 |
| 0.15585 | 0.6688 | 6.6770 | 27.1778 | 0.7266 | 0.0663 | 0.0128 | 1389.869 |
| 0.40504 | 0.2077 | 3.5402 | 27.3654 | 0.9573 | 0.0667 | 0.0119 | 1389.459 |
| 0.65424 | 0.2125 | 3.0505 | 24.5504 | 0.8695 | 0.0660 | 0.0117 | 1389.708 |
| 0.92273 | 0.1344 | 2.0388 | 25.9019 | 0.9684 | 0.0654 | 0.0116 | 1390.906 |

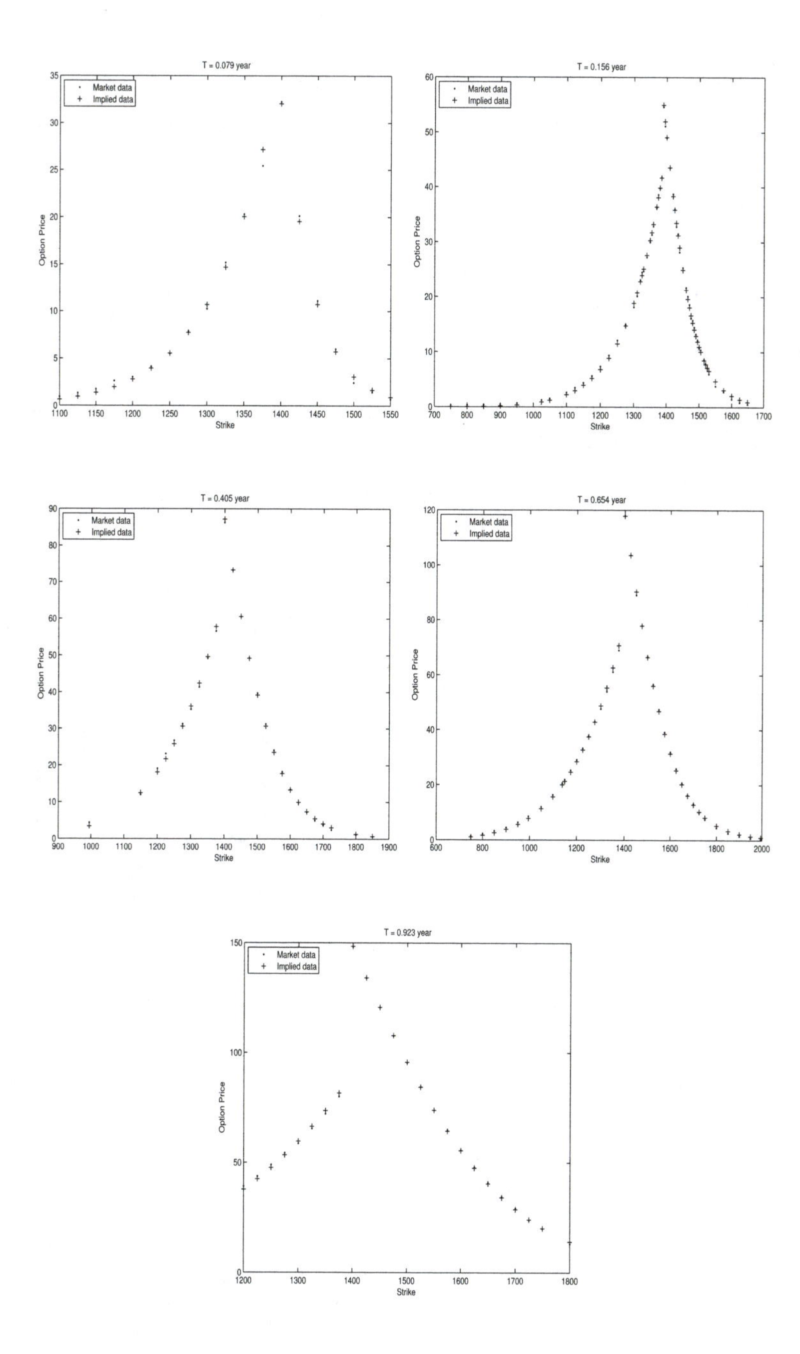


**FIGURE 7.6**: VG vs. market premiums for the S&P 500 on October 19, 2000 for different maturities

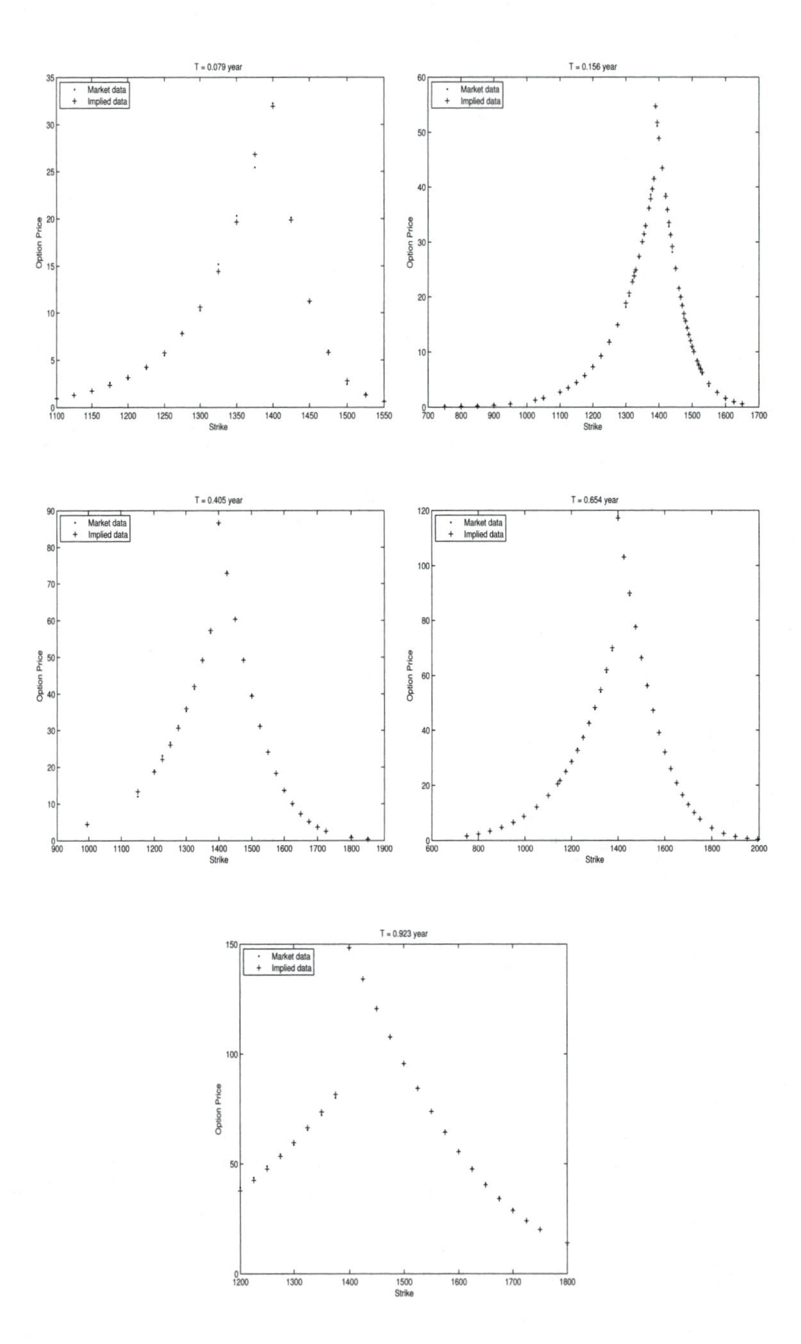


**FIGURE 7.7**: CGMY vs. market premiums for the S&P 500 on October 19, 2000 for different maturities

Unlike VG, the model can be steadily calibrated across both maturity and strike simultaneously.

To illustrate, we provide an example of the VGSA parameters generated from calibration of S&P 500 prices on December 13, 2000.

$$\begin{aligned}
\sigma &= 0.1022 \\
\nu &= 0.1819 \\
\theta &= -0.0761 \\
\kappa &= 8.1143 \\
\eta &= 2.8060 \\
\lambda &= 10.3646
\end{aligned}$$

We note that $\sigma$, $\nu$ and $\theta$ are very close to the shorter maturity of VG parameters. As in VG, $\theta < 0$ and this is a reflection of negative skewness. The left tail of the Lévy density has a slower rate of exponential decay and hence we have higher prices for equal percentage down moves compared with up moves. The rate of mean reversion reflects a half life of 3.82 months and this is consistent with a reasonable level of volatility persistence. Long term levels of activity are a third of their current levels; hence market prices for hedging moves in the future are lower when compared with the costs of hedging near term market moves. Note that the volatility of volatility is substantial in comparison to the level of long term volatility.

In Figure 7.8 we display VGSA premiums versus market premiums for the S&P 500 on October 19, 2000 across all maturities. The fit across all maturities seems to be good. Using parameters from calibration, we can calculate the premiums for any strike and maturity and construct the vanilla call surface. Having the call surface using Equation (7.3) to find the local volatility surface implied from the call surface. The local volatility surface is illustrated in Figure 7.9.

At the first glance local volatility surfaces from Heston and VGSA look pretty similar but considering that for very short maturity Heston was underpricing we expect to get higher volatilities for shorter maturities. To better visualize this we also plot both surfaces against each other. That would give us a better sense of comparison on how much they differ as shown in Figure 7.10. We can see that the local volatility levels implied from Heston is a bit larger than the local volatility levels implied from VGSA.

### 7.2.9 Lévy Models

Calibration of Lévy models across a full set of options, including options of varying maturity and strike, remains a somewhat unstable problem when using a least-squares based formulation. In [76], the authors present a non-parametric method for calibrating jump diffusion and more generally exponential Lévy models, to a finite set of observed option prices. They demonstrate that the usual formulations of the inverse problem via non-linear least squares are ill-posed and propose a regularization method based on relative entropy. To this end, they reformulate the calibration problem into a problem of finding a risk-neutral exponential Lévy model that reproduces the observed option prices and has the smallest possible relative entropy with respect to a chosen prior model. This approach allows us to reconcile the idea of calibration by relative entropy minimization with the notion of risk-neutral valuation in a continuous time model. They demonstrate a numerical implementation of the method using a gradient-based optimization algorithm and show that the entropy penalty resolves the numerical instability of the calibration problem.

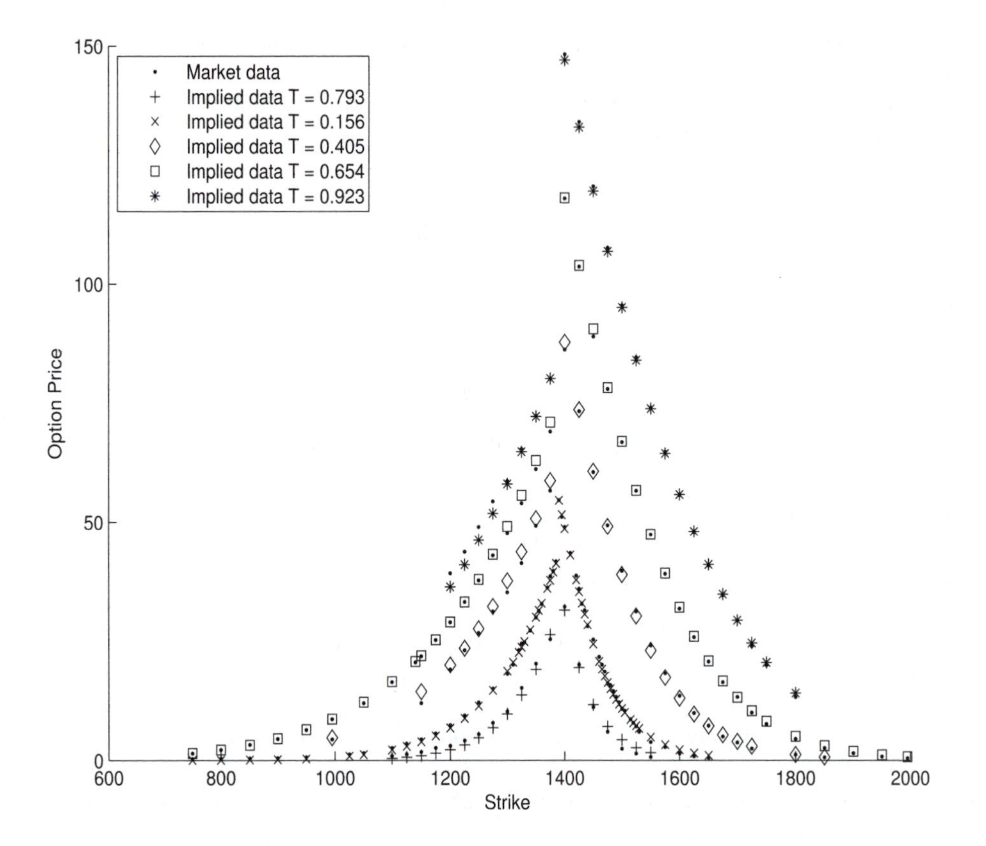


**FIGURE 7.8**: VGSA vs. market premiums for the S&P 500 on October 19, 2000 for various maturities

## 7.3 Interest Rate Models

The models we have discussed thus far deal with derivatives whose value depends on a single underlying asset as in single name equity, foreign exchange and commodities and the calibration routines we have discussed are designed to closely match the prices of a full set of derivatives which are traded in the market. However, these models are not applicable to interest rate derivatives, whose value can depend on the evolution of the entire yield curve. Models designed to capture the evolution of the yield curve, are often called term structure models and have been a topic of research for decades and in this section we will discuss an increasingly complex series of these models.

Term structure models are particularly onerous to develop and calibrate for a number of reasons. Although the first models for yield curves contained only one factor, models which capture the full set of possibilities in terms of the shape of the yield curve require modeling multiple correlated underliers. In addition, the shape of the yield curve is usually restricted by real world arbitrage conditions, disallowing negative forward rates for instance, which dictate the possible future states of the yield curve.

The primary goal of calibration in term structure models and usually the most difficult, is the process of determining the singular volatility (in the case of a single factor model) or correlation matrix (in the case of single factor models) that are used in the term structure

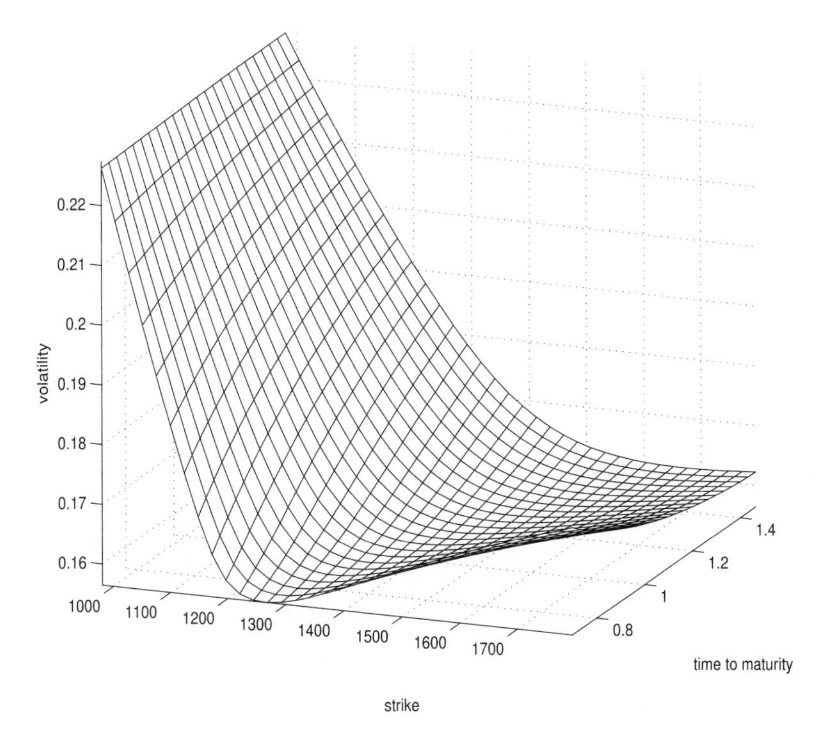


**FIGURE 7.9**: Local volatility surface obtained from the call price surface of VGSA

model. Additionally, parameters relating to the mean or means of the relevant factors, including parameters relating to the long term mean, mean reversion, or its term structure, will need to be calibrated as well, depending on the model used.

There are various different highly liquid instruments that we can use as calibration instruments for interest rate models. They include LIBOR rates, swap rates, zero-coupon bonds and options on bonds, just to name a few. Most of these different rates can be expressed as a simple function of zero coupon bond prices[2], $P(t,T)$, which are of fundamental interest in fixed income pricing. For instance, LIBOR rates are simply compounded interest rates, which relate to the zero-coupon bond prices as follows:

$$L(t,T) = \frac{100}{T-t}\left(\frac{1}{P(t,T)} - 1\right) \tag{7.4}$$

where the maturities are compounded based on the ACT/360 day count convention, starting two business days forward. This is exactly the same as Equation (7.24) where derivation of it is done in Section 7.9. Swap rates relate to the zero-coupon prices by

$$s(t,T) \quad = \quad 100\frac{1-P(t,T)}{\Delta\sum_{j=1}^{N} P(t,T_j)} \tag{7.5}$$

where $T_j$ for $j = 1,\ldots,n$ are reference dates with $T_n = T$ denoting the swap term or the maturity of the swap and $\Delta$ denotes the length of the period $[T_{j-1}, T_j]$ in the corresponding day count convention. The tenor of $s(t,T)$ is the time $T-t$. For U.S. swap contracts, the

[2] $P(t,T)$ is the value of zero-coupon bond at time $t$, the amount willing to pay at $t$ to receive at one dollar at maturity time $T$. Obviously $P(T,T) = 1$.

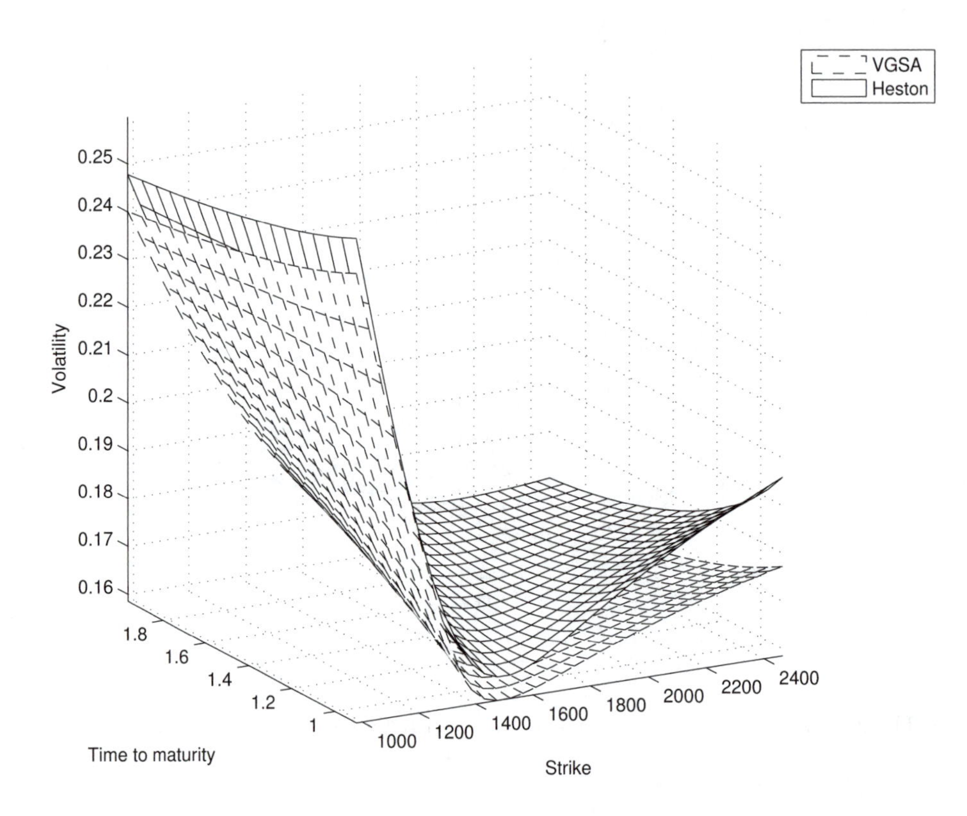


**FIGURE 7.10**: Local volatility surface implied from Heston vs. local volatility surface implied from VGSA

number of payments is every six months, $\Delta = \frac{1}{2}$, a half-year tenor. In addition to these instruments, markets for interest rate caps and floors, as well as swaptions, is very liquid and remains a valuable source of market price data which can be used to help calibrate volatility and correlations.

The number of markets one can use to calibrate a chosen interest rate model depends largely on the degrees of freedom provided by the model. For instance, the simplest short rate models will not even be able to reproduce the current term structure of interest rates perfectly. When these models are augmented with time-varying parameters they can typically be made to perfectly calibrate to the current yield curve, but usually cannot be simultaneously and perfectly calibrated against the cap, floor, and swaption markets. For risk management or marking purposes, we typically want to use a single framework for pricing and calculating hedge ratios and as a consequence would be of interest to be able to calibrate the whole set of market price simultaneously. However, it is market making and trading that would not be the goal.

The majority of these models require the current yield curve to be specified to determine the initial state. However, constructing the current yield curve from bond prices, LIBOR rates, swap rates, or interest rate futures is a non-trivial problem in and of itself. It is made more difficult by the fact that the yield curve can go out twenty or thirty years and there are not enough traded instruments to represent every tenor, but we often need to construct a smooth curve based on a few instruments going out thirty years. The procedure used to extract today's yield curve from the market quoted rates and prices is called the *cooking process or construction of the yield curve.* We will briefly cover this procedure in Section 7.7 at the end of this chapter.

### 7.3.1 Short Rate Models

The first attempts to model the evolution of interest rates followed the same pattern as models in markets with a single underlying asset. We model the evolution of the term structure of interest rates in terms of the instantaneous short rate $r_s$ at time $s$. This is the instantaneous continuously compounded interest rate, so discounting the exponential of integral of this rate from the current time to any future time and taking the expectation gives the zero-coupon bond price for that maturity, that is

$$P(t,T) = \mathbb{E}_t^{\mathbb{Q}}\left(e^{-\int_t^T r_s ds}\right)$$

Modeling interest rates in terms of a single factor short rate leads to a simple transition from models for other assets to interest rates models. However, because we have only a single underlying factor, these models typically imply perfect correlation between all points on the yield curve, and thus we lose the ability to model changes in the shape of the yield curve. This can be a critical disadvantage when attempting to price or hedge interest rate dependent instruments whose value is highly dependent on the shape of the yield curve and the correlation between different points in the term structure of rates. That being said, short rate models still remain popular because of their parsimoniousness and ease of implementation, especially in situations where the level of rates, and not the shape of the term structure, is of primary importance. Another reason for its popularity is the fact that by construction it yields a functional form for $p(t,T)$ which a smooth curve.

#### 7.3.1.1 Vasicek Model

One of the first short rate models developed was the Vasicek model. Like the Black–Scholes model, the Vasicek model is the simplest model for the evolution of interest rates. The Vasicek model assumes that the instantaneous short rate, $r_t$, follows the following stochastic differential equation:

$$dr_t = \kappa(\theta - r_t)dt + \sigma dW_t$$

This is very similar to the Bachelier model except in this case, the short rate is given a mean reversion component, with $\theta$ being the long term mean of the short rate and $\kappa$ being the mean reversion rate. This was included in even the earliest interest rate models in recognition of the fact that interest rates rarely see the type of diversions from their long term mean that is seen in other market variables; in the long term they tend to remain somewhat range bound. In addition, the mean reversion can help to mitigate one of the disadvantages of the Vasicek model, which allows for negative interest rates, which are a rarity in real markets[3]. By including mean reversion, the model can limit the number of scenarios in which negative rates can occur. The exact solution to the Vasicek SDE given by

$$r_t = e^{\kappa t}r_0 + \theta(1 - e^{-\kappa t}) + \sigma e^{-\kappa t}\int_0^t e^{\kappa s}dW_s$$

While this solution is useful, the most important quantity when pricing fixed income instruments is not the actual short rate, but the zero-coupon bond prices. Under Vasicek,

[3]This is possible when interest rate is smaller than inflation rate.

**TABLE 7.5**: LIBOR rates

| LIBOR rates | | |
|---|---|---|
| maturity (months) | Oct. 29, 2008 rate (%) | Feb. 14, 2011 rate (%) |
| 1 | 3.1175 | 0.2647 |
| 2 | 3.2738 | 0.2890 |
| 3 | 3.4200 | 0.3140 |
| 6 | 3.4275 | 0.4657 |
| 12 | 3.4213 | 0.7975 |

zero-coupon bond prices are given by

$$\begin{aligned} P(t,T) &= \mathbb{E}_t^{\mathbb{Q}}\left(e^{-\int_t^T r_s ds}\right) \\ &= e^{A(t,T)-B(t,T)r_t} \end{aligned}$$

where the loading factors are

$$\begin{aligned} B(t,T) &= \frac{1-e^{-\kappa(T-t)}}{\kappa} \\ A(t,T) &= (\theta - \frac{\sigma^2}{2\kappa^2})[B(t,T)-(T-t)] - \frac{\sigma^2}{4\kappa}B^2(t,T) \end{aligned}$$

Note that the function $P(t,T)$ is time-homogeneous. From these results, if interested in the evolution of zero-coupon bond prices under this model, we can show that zero-coupon bond prices, $P(t,T)$, satisfy the following SDE:

$$dP(t,T) = r_t P(t,T)dt - P(t,T)B(t,T)\sigma dW_t$$

The calibration procedure for the Vasicek model is straightforward. It is a four-parameter model and the set of parameters which need to be calibrated is $\Theta = \{\kappa, \theta, \sigma, r_0\}$. It is important to note that the parsimonious set of parameters associated with this model provides very little in the way of degrees of freedom, which will be essential in model calibration. Because we have only four free parameters, it is impossible to perfectly calibrate the model to the current term structure of interest rates, let alone cap, floor, or swaption implied volatilities. This limits the practical usefulness of the model.

For illustration purposes, we calibrate the Vasicek model to LIBOR rates and swap rates. We will use LIBOR rates with maturities of one, two, three, six, and twelve months and swap rates with maturities of two, three, five, seven, ten, fifteen and thirty years.

Tables 7.5 and 7.6 contain LIBOR and swap rates used for calibration as of October 29, 2008 and February 14, 2011 (using Bloomberg data).

For any guestimate for the parameter set, we calculate zero-coupon bond prices using the closed-form solution. We then use zero-coupon bond prices to compute the LIBOR and swap rates implied by our model utilizing Equations (7.4) and (7.5), respectively. We define our objective function as the sum of the squares of relative errors (SSRE) between LIBOR and swap rates implied by the model and the market LIBOR and swap rates. In other words, if model LIBOR rates are $L_{MODEL}$, market LIBOR rates are $L_{MARKET}$, model swap rates

**TABLE 7.6**: Swap rates

| Swap rates | | |
|---|---|---|
| term (year) | Oct. 29, 2008 rate (%) | Feb. 14, 2011 rate (%) |
| 2 | 2.6967 | 1.0481 |
| 3 | 3.1557 | 1.5577 |
| 5 | 3.8111 | 2.5569 |
| 7 | 4.1497 | 3.1850 |
| 10 | 4.3638 | 3.7225 |
| 15 | 4.3753 | 4.1683 |
| 30 | 4.2772 | 4.4407 |

are $S_{MODEL}$, and market swap rates are $S_{MARKET}$, our objective is to minimize

$$\begin{aligned} SSRE &= \sum_{i=1}^{I} \left( \left( L_{MODEL(i)} - L_{MARKET(i)} \right) / L_{MARKET(i)} \right)^2 \\ &+ \sum_{j=1}^{J} \left( \left( S_{MODEL(j)} - S_{MARKET(j)} \right) / S_{MARKET(j)} \right)^2 \end{aligned}$$

We find the minimum SSRE using a numerical optimizer. Note that throughout this report, results are sensitive to the exact optimization parameters used. We use the simplex method with a tolerance of $1e-4$, and $1,000$ maximum iterations. Using data from October 29, 2008, we obtain the following parameters from calibration: $\kappa = 0.1153$, $\theta = 0.0532$, $\sigma = 0.0028$ and $r_0 = 0.0309$. Using data from February 14, 2011, we obtain the following parameters from calibration: $\kappa = 0.1717$, $\theta = 0.0670$, and $\sigma = 0.0009$. We also obtain $r_t = 0.0020$.

The top panels of Figures 7.11 and 7.12 display zero-coupon bond prices for the Vasicek model using the calibrated parameters. The middle panel displays the market LIBOR rates as well as the LIBOR rates produced by the calibrated Vasicek model. The bottom panel displays the market swap curve, as well as the swap curve produced by the calibrated Vasicek model. The Vasicek model provides a good fit on February 14, 2011, but a poor fit on October 29, 2008. This is likely because a single factor model cannot capture the odd shape of market rates prevalent on that date.

#### 7.3.1.2 Pricing Swaptions with the Vasicek Model

After calibrating to the market LIBOR rates and swap curve, we produce swaption prices through simulation. Using a discretized version of the Vasicek model for the instantaneous short rate, we produce a large number of realized interest rate paths. Each path gives us a realized swaption value, and the price of the swaption is the discounted expected value (e.g., the discounted average across paths).

Table 7.7 contains market swaption prices as of October 29, 2008 and February 14, 2011. Table 7.8 contains the results from pricing swaptions via simulation using our calibrated parameters. In this scenario, all of our swaption prices are zero. This occurs because our simulation results in swaptions which never finish in the money. While our simple model performs reasonably well at fitting market swap and LIBOR rates, it fails to produce reasonable swaption prices.

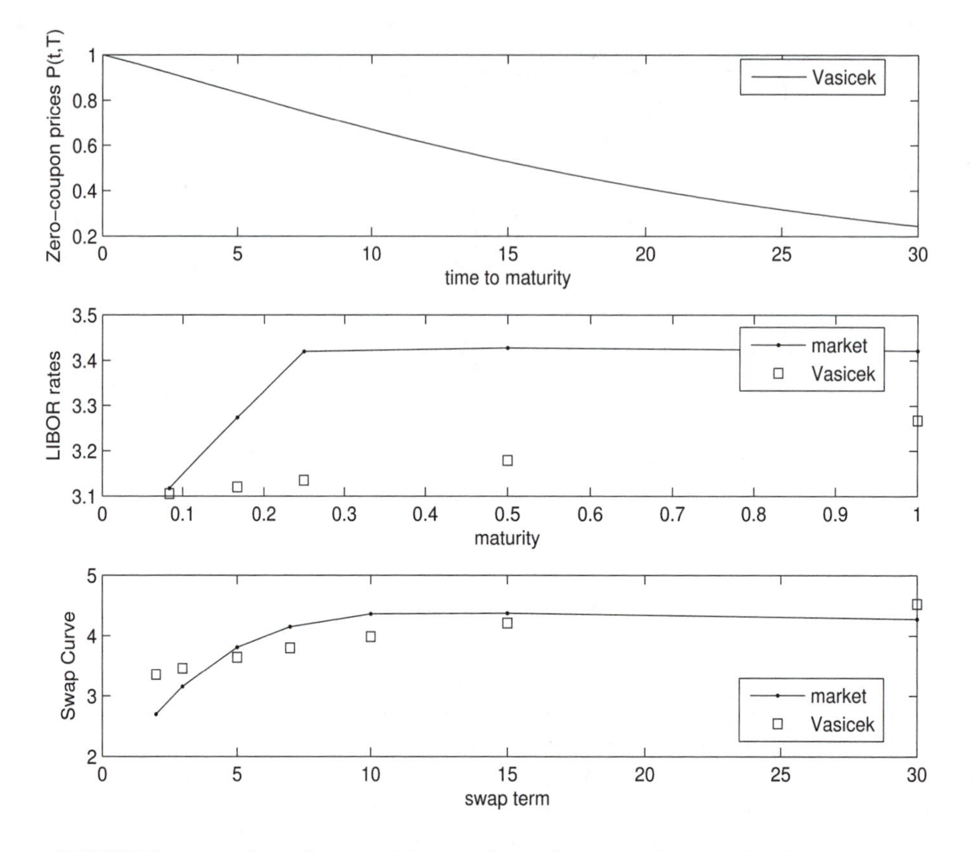


**FIGURE 7.11**: Vasicek model (single factor) vs. market on October 29, 2008

#### 7.3.1.3 Alternative Vasicek Model Calibration

Our first calibration of the Vasicek model focused on choosing model parameters to mimic the behavior of interest rates directly. While we achieved reasonable fit to LIBOR and swap rates, our model failed to produce reasonable swaption prices. In an attempt to produce more reasonable swaption prices, we test an alternative calibration, where we add a relative error term for the fit to selected swaption values to our objective function. Specifically, if LIBOR and swap rates are indicated as above, model swaption rates are $O_{MODEL}$, market swaption rates are $O_{MARKET}$ and we consider $N$ different option maturities on $M$ different swap maturities, our objective is to minimize:

$$
\begin{aligned}
SSRE &= \sum_{i=1}^{I}\left(\left(L_{MODEL(i)} - L_{MARKET(i)}\right)/L_{MARKET(i)}\right)^2 \\
&+ \sum_{j=1}^{J}\left(\left(S_{MODEL(j)} - S_{MARKET(j)}\right)/S_{MARKET(j)}\right)^2 \\
&+ \sum_{m=1}^{M}\sum_{n=1}^{N}\left(\left(O_{MODEL(m,n)} - O_{MARKET(m,n)}\right)/O_{MARKET(m,n)}\right)^2
\end{aligned}
$$

for example, our objective function is extended by including relative error terms for four swaptions corresponding to the four pairs of shortest and longest option and swap maturities.

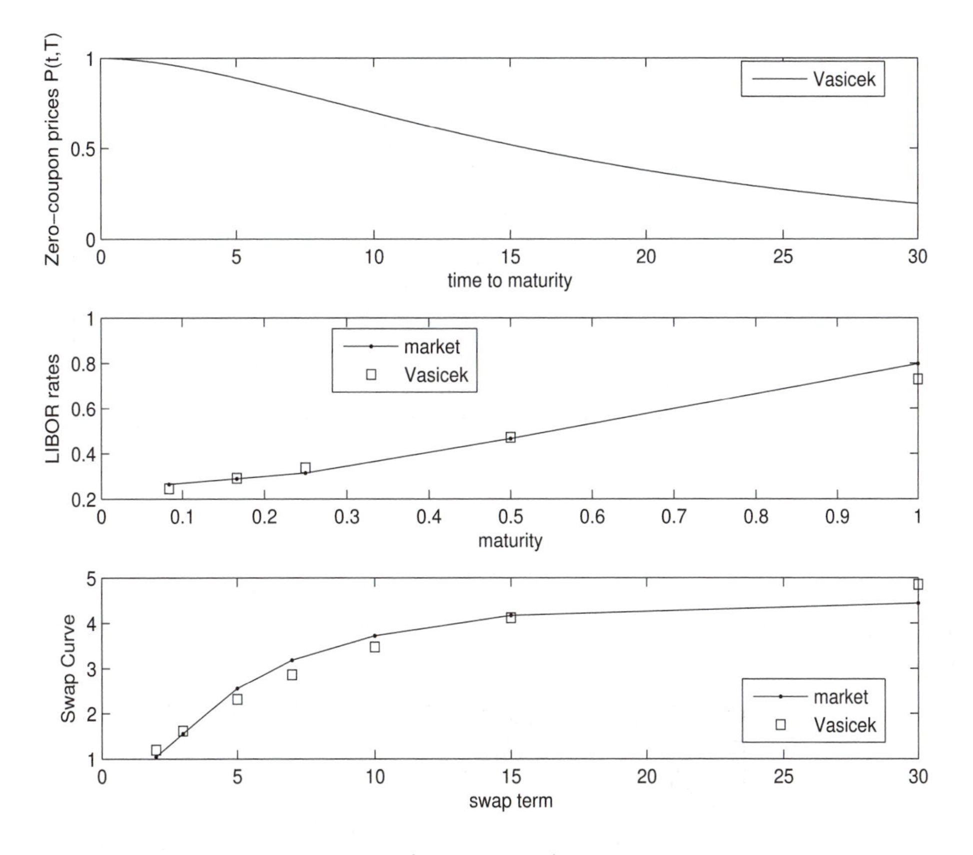


**FIGURE 7.12**: Vasicek model (single factor) vs. market on February 14, 2011

We find the minimum SSRE of the alternative objective using a numerical optimizer. Using data from October 29, 2008, we obtain the following parameters from calibration: $\kappa = 1.7726$, $\theta = 0.0394$, $\sigma = 0.0777$ and $r_0 = 0.0301$. Using data from February 14, 2011, we obtain the following parameters from calibration: $\kappa = 4.0528$, $\theta = 0.0246$, $\sigma = 0.5008$ and $r_0 = -0.0051$. While the fit to the term structure of interest rates is much poorer, here we are primarily interested in how well we price swaptions.

Table 7.9 contains the results from pricing swaptions via simulation using our parameters calibrated under the alternative objective function. A comparison of these values to Table 7.7 shows that while our errors are still large, our swaption prices are at least non-zero, and the shape of the pricing surface is beginning to very roughly correspond to market prices. The sum of squared errors (SSE) between simulated prices and actual prices has decreased, which we see by comparing Tables 7.9 and 7.8. However, we also see that swaption prices depend primary on option maturity, with very little variation in pricing across swap maturities, which is not the case for market prices. This indicates that our simple model may be too limited to reflect market behavior as expected.

#### 7.3.1.4 CIR Model

One of the major drawbacks of the Vasicek model is that the instantaneous short rate can become negative, implying negative interest rates. In order to address this shortcoming, the CIR model was developed. In the CIR process we assume that the instantaneous short

**TABLE 7.7**: Market at-the-money swaption prices (maturity in years)

| Oct. 29, 2008 | | | | Feb. 14, 2011 | | | |
|---|---|---|---|---|---|---|---|
| | Swap Maturity | | | | Swap Maturity | | |
| Option Maturity | 2 | 5 | 10 | Option Maturity | 2 | 5 | 10 |
| 1 | 227.6 | 547.0 | 896.4 | 1 | 174 | 428 | 730 |
| 2 | 286.4 | 663.8 | 1094.8 | 2 | 256 | 591 | 993 |
| 5 | 325.8 | 745.4 | 1251.2 | 5 | 340 | 756 | 1265 |
| 10 | 269.8 | 634.4 | 1090.2 | 10 | 307 | 681 | 1127 |

**TABLE 7.8**: Simulated swaption prices from the Vasicek model (single factor, maturity in years)

| Oct. 29, 2008 | | | | Feb. 14, 2011 | | | |
|---|---|---|---|---|---|---|---|
| | Swap Maturity | | | | Swap Maturity | | |
| Option Maturity | 2 | 5 | 10 | Option Maturity | 2 | 5 | 10 |
| 1 | 0 | 0 | 0 | 1 | 0 | 0 | 0 |
| 2 | 0 | 0 | 0 | 2 | 0 | 0 | 0 |
| 5 | 0 | 0 | 0 | 5 | 0 | 0 | 0 |
| 10 | 0 | 0 | 0 | 10 | 0 | 0 | 0 |
| SSE: 6.8e6 | | | | SSE: 6.3e6 | | | |

**TABLE 7.9**: Simulated swaption prices from the Vasicek model, with the addition of selected swaption price relative error to objective function (single factor, maturity in years)

| Oct. 29, 2008 | | | | Feb. 14, 2011 | | | |
|---|---|---|---|---|---|---|---|
| | Swap Maturity | | | | Swap Maturity | | |
| Option Maturity | 2 | 5 | 10 | Option Maturity | 2 | 5 | 10 |
| 1 | 203.80 | 214.52 | 214.51 | 1 | 488.75 | 488.76 | 488.75 |
| 2 | 223.94 | 235.79 | 235.82 | 2 | 560.13 | 560.16 | 560.16 |
| 5 | 253.97 | 267.43 | 267.46 | 5 | 846.02 | 846.07 | 846.07 |
| 10 | 318.93 | 335.84 | 335.88 | 10 | 1678.44 | 1678.54 | 1678.54 |
| SSE: 3.4e6 | | | | SSE: 4.1e6 | | | |

rate, $r_t$, follows the following stochastic differential equation:

$$dr_t = \kappa(\theta - r_t)dt + \sigma\sqrt{r_t}dW_t$$

We note that this is very similar to the Vasicek model, with the single addition of the $\sqrt{r_t}$ term. This term will go to zero as the short rate approaches zero, effectively eliminating volatility as the short rate declines. The addition of this term will force $r_t$ to remain non-negative in the CIR model, unlike the Vasicek model. However, one large downside is that the addition of the volatility limiting term causes the CIR model, unlike the Vasicek model, to be non-Gaussian. That means that pricing under the model is not nearly as straightforward as Vasicek. However, a closed-form solution does exist for zero-coupon bond prices. That is,

$$\begin{aligned} P(t,T) &= \mathbb{E}_t^{\mathbb{Q}}\left(e^{-\int_t^T r_s ds}\right) \\ &= e^{A(t,T)-B(t,T)r_t} \end{aligned}$$

where

$$\begin{aligned} A(t,T) &= \frac{2\kappa\theta}{\sigma^2}\ln\left(\frac{\exp(\kappa(T-t)/2)}{\cosh(\gamma(T-t)/2)+\frac{\kappa}{\gamma}\sinh(\gamma(T-t)/2)}\right) \\ B(t,T) &= \frac{2}{\kappa+\gamma\coth(\gamma(T-t)/2)} \end{aligned}$$

with

$$\gamma = \sqrt{\kappa^2 + 2\sigma^2}$$

Closed-form solutions do not exist for most interest rate derivatives under this model, so if we are interested in pricing derivatives we generally must use either Monte Carlo simulation or numerical solutions of the PDE. While in the CIR model, we do have to deal with the issue of negative rates, its similarly small parameter set means it too cannot be calibrated to the current term structure of interest rates.

The calibration procedure for the CIR model is analogous to the calibration procedure for the Vasicek model. We have three parameters to be calibrated: $\Theta = \{\kappa, \theta, \sigma, r_0\}$. While the CIR model eliminates the negative interest rate problem, we still have only four free parameters, so we cannot expect to match the current term structure of interest rates exactly, nor price derivatives accurately.

We now repeat our earlier calibration exercise, using the CIR model to model LIBOR and swap rates. We make use of the same market data as before, given in Tables 7.5 and 7.6 (LIBOR and swap rates as of October 29, 2008 and February 14, 2011). We start with an initial guess of our parameters and use the closed-form solution for zero-coupon bond prices to find the zero-coupon bond prices given by the model and our choice of parameters. We then use the zero-coupon bond prices to find the LIBOR and swap rates implied by our model. Our objective function is defined as before: we seek to minimize the sum of the squares of relative errors (SSRE) between model LIBOR and swap rates and market LIBOR and swap rates. We find the minimum SSRE, using a numerical optimizer. Using data from October 29, 2008, we obtain the following parameters from calibration: $\kappa = 0.0912$, $\theta = 0.0589$, and $\sigma = 0.0430$. We also obtain $r_t = 0.0309$. Using data from February 14, 2011, we obtain the following parameters from calibration: $\kappa = 0.0044$, $\theta = 2.2662$, and $\sigma = 0.2364$. We also obtain $r_t = 0.0021$.

The top panels of Figures 7.13 and 7.14 display zero-coupon bond prices for the CIR model using the calibrated parameters. The middle panel displays the market LIBOR rates

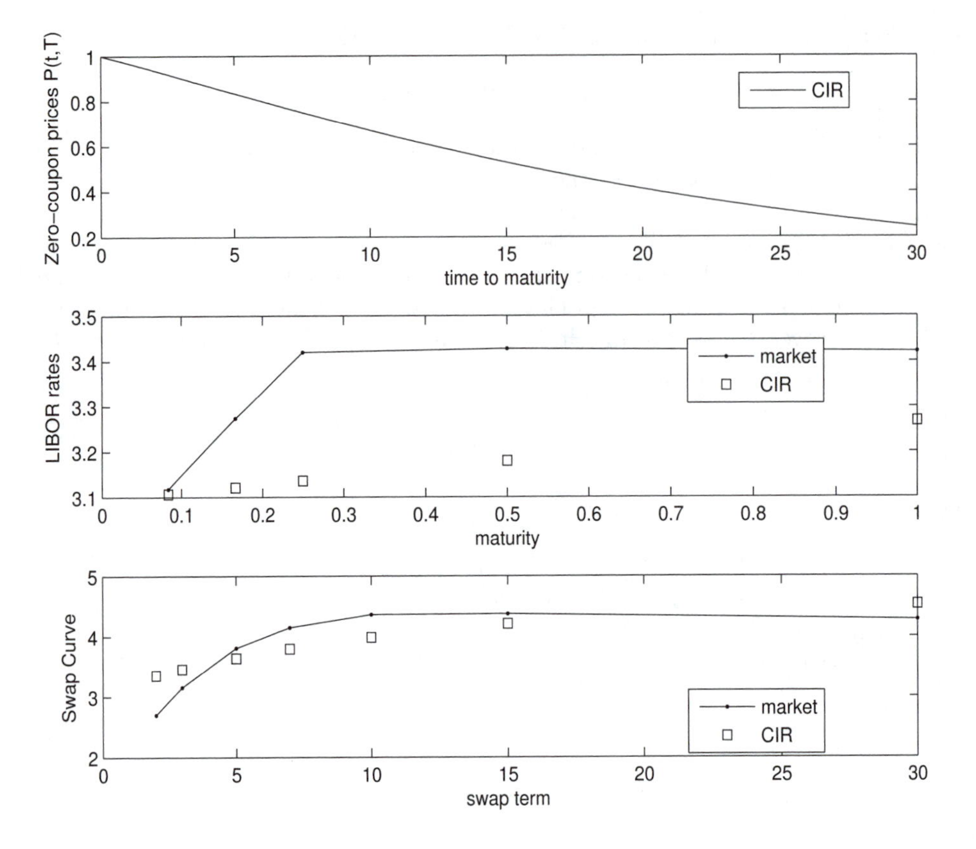


**FIGURE 7.13**: CIR model vs. market on October 29, 2008

as well as the LIBOR rates produced by the calibrated CIR model. The bottom panel displays the market swap curve, as well as the swap curve produced by the calibrated CIR model. Similarly to the Vasicek model, the CIR model provides a good fit on February 14, 2011, but a poor fit on October 29, 2008. Again this is likely because a single factor model cannot capture the shape of market rates prevalent on that date.

#### 7.3.1.5 Pricing Swaptions with the CIR Model

After calibrating to the market LIBOR rates and swap curve, we produce swaption prices through simulation. Using a discretized version of the CIR model for the instantaneous short rate, we produce a large number of realized interest rate paths. Each path gives us a realized swaption value, and the price of the swaption is the discounted expected value (e.g., the discounted average across paths).

Table 7.7 contains the market swaption prices. Table 7.10 contains the results from pricing swaptions via simulation using our calibrated parameters. We see that the CIR one-factor model produces somewhat reasonable swaption prices for October 29, 2008. For February 14, 2011, prices at short option maturities are reasonable, but quickly grow unreasonably large at longer option maturities.

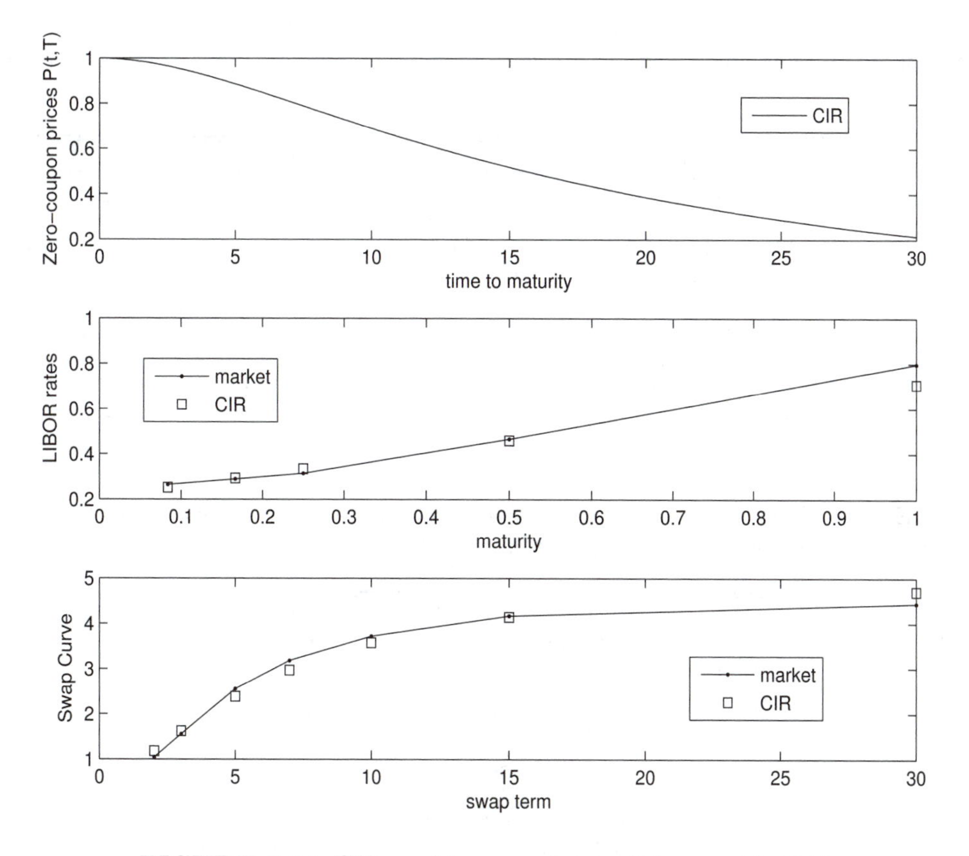


**FIGURE 7.14**: CIR model vs. market on February 14, 2011

#### 7.3.1.6 Alternative CIR Model Calibration

Our first calibration of the CIR model focused on choosing model parameters to mimic the behavior of interest rates directly. While some of our swaption prices were reasonable, others were much too large and we would like to see if we can improve upon this.

In an attempt to produce swaption prices closer to those observed in the market, we again test an alternative calibration, where we add a relative error term for the fit to selected swaption values to our objective function. Our objective function is extended by including relative error terms for four swaptions corresponding to the four pairs of shortest and longest option and swap maturities.

We find the minimum SSRE of the alternative objective using a numerical optimizer. Using data from October 29, 2008, we obtain the following parameters from calibration: $\kappa = 1.4573$, $\theta = 0.0442$, and $\sigma = 0.3359$. We also obtain $r_t = 0.0185$. Using data from February 14, 2011, we obtain the following parameters from calibration: $\kappa = 1.5718$, $\theta = 0.0210$ and $\sigma = 0.4990$. We also obtain $r_t = 0.0003$. While the fit to the term structure of interest rates is much poorer, here we are primarily interested in how well we price swaptions.

Table 7.11 contains the results from pricing swaptions via simulation using our parameters calibrated under the alternative objective function. A comparison of these values to Table 7.7 shows that our errors are still large. Comparing Tables 7.11 and 7.10, we see that the SSE has decreased for February 14, 2011, while it has increased for October 29, 2008. If we look at the relative errors for October 29, 2008, for just the swaptions that were used in our objective function, we see that the SSRE for those swaption prices did decrease.

**TABLE 7.10**: Simulated swaption prices from the CIR model (single factor, maturity in years)

| Oct. 29, 2008 | | | | Feb. 14, 2011 | | | |
|---|---|---|---|---|---|---|---|
| | Swap Maturity | | | | Swap Maturity | | |
| Option Maturity | 2 | 5 | 10 | Option Maturity | 2 | 5 | 10 |
| 1 | 79.1 | 198.6 | 315.6 | 1 | 207.0 | 505.2 | 684.4 |
| 2 | 119.2 | 297.2 | 468.0 | 2 | 422.7 | 980.6 | 1280.4 |
| 5 | 214.2 | 527.4 | 819.7 | 5 | 1560.5 | 3145.3 | 3832.7 |
| 10 | 380.7 | 925.2 | 1419.7 | 10 | 28566.5 | 39743.2 | 42396.1 |
| SSE: 1.5e6 | | | | SSE: 4.0e9 | | | |

**TABLE 7.11**: Simulated swaption prices from the CIR model, with the addition of selected swaption price relative error to objective function (single factor, maturity in years)

| Oct. 29, 2008 | | | | Feb. 14, 2011 | | | |
|---|---|---|---|---|---|---|---|
| | Swap Maturity | | | | Swap Maturity | | |
| Option Maturity | 2 | 5 | 10 | Option Maturity | 2 | 5 | 10 |
| 1 | 178.0 | 199.6 | 201.1 | 1 | 161.9 | 176.8 | 177.7 |
| 2 | 215.7 | 238.9 | 239.4 | 2 | 209.5 | 226.6 | 226.8 |
| 5 | 251.2 | 277.4 | 277.6 | 5 | 234.8 | 253.6 | 253.7 |
| 10 | 322.5 | 356.0 | 356.3 | 10 | 282.8 | 305.4 | 305.6 |
| SSE: 3.3e6 | | | | SSE: 3.2e6 | | | |

However, overall the fit is worse. Once could potentially improve upon this result by using an alternative objective function, such as incorporating a greater number of swaption points or using SSE in place of SSRE.

A comparison to Table 7.9 shows that the single factor CIR model does not perform any better than the Vasicek model for swaption pricing. Interestingly, similar to the Vasicek calibration, we see that swaption prices now depend primarily on option maturity, with very little variation in pricing across swap maturities, which is not the case for market prices and was not the case in our initial calibration. It seems that we have sacrificed reflecting the proper shape of market prices in order to reduce overall error. This indicates that our simple model may be too limited to reflect market behavior — a model with more parameters should be able to do both.

#### 7.3.1.7 Ho–Lee Model

The CIR model added an additional term to the Vasicek model in order to prevent negative interest rates; however, both have the same mean-reverting dynamics for the short rate. As discussed, the small number of parameters which can be calibrated in these models typically prohibits calibration of the model to the current term structure of interest rates. To allow for more degrees of freedom, which will allow us to calibrate to the current yield curve, Ho and Lee introduced a term structure of means of the short rate. By allowing a term structure of parameters of arbitrary degree into the model, we explicitly allow for perfect calibration of the model to the current term structure of interest rates.

The Ho–Lee model assumes that the instantaneous interest rate follows the following SDE:

$$dr_t = \theta(t)dt + \sigma dW_t$$

where $\theta(t)$ is a deterministic function and $W_t$ is a $\mathbb{Q}$-Brownian motion.

For the Ho–Lee model, we give the full derivation of the zero-coupon bond price. The same procedure can be applied for the other models. Like the previously discussed models, calibrating the Ho–Lee model is dependent on being able to value zero coupon bonds, which can be used to price a number of the most liquidly traded interest rate derivatives. Deriving the price of zero-coupon bonds in this model is somewhat less straightforward than in the previously discussed models, but can be done as follows: First we let $X_T = \int_0^T W_t dt$, which has a Gaussian distribution and Gaussian increments as it is the integral (sum) of Brownian motions. Therefore we can find an analytical expression for $\mathbb{E}^{\mathbb{Q}}[\exp(-X_T)]$. To do that we need to find the mean and variance of $X_T$.

$$\begin{aligned}
\mathbb{E}(X_T) &= \mathbb{E}\left(\int_0^T W_t dt\right) = \int_0^T \mathbb{E}(W_t)dt = 0 \\
\mathbb{E}(X_T^2) &= \mathbb{E}\left(\int_0^T \int_0^T W_t W_s dt ds\right)
\end{aligned}$$

Applying Fubini's theorem to interchange expectation and the integral and using the fact that $\mathbb{E}(W_t W_s) = t \wedge s$ we obtain

$$\mathbb{E}(X_T^2) = \int_0^T \int_0^T \mathbb{E}(W_t W_s) ds dt = \int_0^T \int_0^T (t \wedge s) ds dt = \int_0^T \int_0^t s ds dt + \int_0^T \int_t^T t ds dt = \frac{T^3}{3}$$

So $X_T \sim \mathcal{N}(0, \frac{T^3}{3})$. From the properties of the moment generating function for normal distribution $\mathcal{N}(\mu, \sigma^2)$ we know that if $X$ is normal then $\mathbb{E}(\exp(\theta X)) = \exp(\theta\mu + \frac{\theta^2\sigma^2}{2})$. Hence $\mathbb{E}(\exp(-X_T)) = \exp(\frac{T^3}{6})$. Now by integrating the SDE we can see that

$$r(u) = r(t) + \int_t^u \theta(s)ds + \sigma(W_u - W_t)$$

for all $u$ in $[t, T]$. Therefore for zero-coupon bond prices, $P(t, T)$ can be expressed as

$$\begin{aligned}
P(t,T) &= \mathbb{E}_t^{\mathbb{Q}}\left(e^{-\int_t^T r(u)du}\right) \\
&= e^{-r(t)(T-t)-\int_t^T \int_t^u \theta(s)dsdu} \mathbb{E}_t^{\mathbb{Q}}\left(e^{-\sigma \int_t^T (W_u - W_t)du}\right) \\
&= e^{-r(t)(T-t)-\int_t^T \int_t^u \theta(s)dsdu + \frac{\sigma^2 (T-t)^3}{6}}
\end{aligned}$$

From this equation we can see that $P(t, T)$ is an affine function and can be written as

$$P(t,T) = e^{A(t,T)-B(t,T)r_t} \tag{7.6}$$

where $B(t, T) = T - t$. Using these results we can see that the time-zero, $t = 0$, zero-coupon bond price $P(0, T)$ observed in the market for maturity $T$ is $P(0,T) = \exp(-\int_0^T f(0,t)dt)$ or equivalently $f(0,T) = -\frac{\partial \log P(0,T)}{\partial T}$. From this equation we have

$$P(0,T) = e^{-r(0)T-\int_0^T \int_0^u \theta(s)dsdu + \frac{\sigma^2 T^3}{6}}$$

Taking the logarithm and equating we have

$$\int_0^T f(0,t)dt = r(0)T + \int_0^T \int_0^u \theta(s)dsdu - \frac{\sigma^2 T^3}{6}$$

Differentiating twice with respect to $T$ and putting maturity to $t$ we will get

$$\theta(T) = \frac{\partial}{\partial T} f(0,T) + \sigma^2 T.$$

Hence if we are given the initial instantaneous forward rate curve $f(0,T)$ we can calculate the term structure of interest rate means, $\theta(T)$.

Not only can we price zero-coupon bonds and derive the term structure of means analytically in this model, but we can also price a European option on a zero-coupon bond analytically as well. Define $F(x) = (x-K)^+$, the payoff of a call option on $x$, $F(x) = (K-x)^+$, the payoff of a put option on $x$. We can express the time $t$ price $V_t$ of a European option on a zero-coupon bond that expires at time $T$ with payoff $V_T = F(P(T,U))$ with the expression[4]

$$\begin{aligned} V_t &= \mathbb{E}_t^{\mathbb{Q}}[\frac{B_t}{B_T} V_T] \\ &= \mathbb{E}_t^{\mathbb{Q}}[\frac{B_t}{B_T} F(P(T,U))] \\ &= P(t,T)\mathbb{E}_t^{Q}[\frac{B_t P(T,T)}{P(t,T)B_T} F(P(T,U))] \\ &= P(t,T)E_t^{\mathbb{P}^T}[F(Z(T,U))] \end{aligned}$$

where $B_t = e^{\int_0^t r_s ds}$ for $t < T < U$ is the money-market account. Here we have expressed the expectation under the forward measure $\mathbb{P}^T$, where $P(t,T)$ is the numeraire and $\frac{B_t P(T,T)}{P(t,T)B_T}$ is the Radon-Nikodym derivative. For detail see Section 1.3.3. This expression is valid because $P(T,T) = 1$.

To find an analytical expression for the expectation we proceed as follows. First, applying Itô's lemma to (7.6) to see that

$$dP(t,T) = P(t,T)(r_t dt - \sigma B(t,T)dW_t)$$

Then, define $Z(t,U) = \frac{P(t,U)}{P(t,T)}$ and apply Itô's lemma again to show that

$$dZ(t,U) = Z(t,U)S(t,T)[S(t,T) - S(t,U)]dt + Z(t,U)[S(t,T) - S(t,U)]dW_t \tag{7.7}$$

where $S(t,T) = \sigma B(t,T)$. Now, under the definition of the forward measure $\mathbb{P}^T$, $Z(t,U)$ is a $\mathbb{P}^T$-martingale. By Girsanov's theorem we have

$$W_t^{\mathbb{P}^T} = \int_0^t S(u,T)du + W_t$$

which is a $\mathbb{P}^T$ standard Brownian motion. Thus we can now write (7.7) as

$$dZ(t,U) = Z(t,U)[S(t,T) - S(t,U)]dW_t^{\mathbb{P}^T} \tag{7.8}$$

From this we can conclude that the $\mathbb{P}^T$-distribution of $Z(T,U)$ is log-normal, and more specifically we have

$$\ln Z(T,U) \sim \mathcal{N}(\ln(P(t,U)/P(t,T)) - \frac{1}{2}\sigma^2(U-T)^2(T-t), \sigma^2(U-T)^2(T-t))$$

[4]To be clear, $T$ is the maturity of the option and $U$ is the bond maturity.

Hence we get a Black–Scholes like formula for the price of the call option, that is,

$$C_t = P(t,T)\mathbb{E}_t^{\mathbb{P}^T}[(Z(T,U)-K)^+] = P(t,U)\Phi(m) - KP(t,U)\Phi(m-n)$$

where

$$\begin{aligned} m &= \frac{1}{n}\ln\left(\frac{Z(t,U)}{K}\right) + \frac{1}{2}n \\ n^2 &= \sigma^2(U-T)^2(T-t) \end{aligned}$$

The parameter set of the Ho–Lee model for calibration is $\Theta = \{\theta(t), \sigma, r_0\}$.

#### 7.3.1.8 Hull–White (Extended Vasicek) Model

While the Ho–Lee model has the ability to handle a term structure of mean short rates and thus can be perfectly calibrated to the current yield curve, it does not necessarily model the mean reverting behavior of interest rates that is frequently seen empirically. To allow the modeling of both this mean reverting behavior and maintain the freedom to specify a term structure of means, Hull and White extended the Vasicek model to allow for a deterministic term structure of long term means along with mean reverting behavior. In this model the short rate follows the following SDE:

$$dr_t = \kappa(\theta(t) - r_t)dt + \sigma dW_t$$

where $\kappa$ is fixed as in Vasicek but $\theta(t)$ is a deterministic function as in the Ho–Lee model and $W_t$ is a $\mathbb{Q}$-Brownian motion. The parameter set we need to determine for calibration is $\Theta = \{\kappa, \theta(t), \sigma, r_0\}$. For pricing zero-coupon bonds and options we typically assume some functional or parametric form for $\theta(t)$ and calculate zero-coupon bonds and options via simulation.

### 7.3.2 Multi-Factor Short Rate Models

The previously discussed short rate models can also be extended to a multi-factor setting, wherein we assume that the instantaneous short rate has the following form:

$$r_t = \sum_{i=1}^{n} X_t^i$$

where $n$ is the number of factors and $X_t^i$ is the $i^{th}$ factor.

Moving to the multi-factor setting has the advantage of giving us more degrees of freedom in terms of calibration. We should be able to improve our fit to the term structure as well as improve our pricing of the many liquidly traded instruments which depend on the term structure of interest rates. For example, maintaining a term structure as part of the dynamics of the short rate but adding multiple factors allows us to simultaneously calibrate perfectly the current yield curve while still leaving enough free parameters to more accurately approximate the volatility structure of rates implied by the prices of caps, floors, and swaptions. The amount of flexibility we have in calibrating to the option market will depend on the number of factors we allow in the model.

However, moving to a multi-factor model also increases the complexity of our calibration. In particular, for unrestricted multi-factor models, we have no closed-form solution for zero-coupon bond prices. We can still determine zero-coupon bond prices, but are forced to rely on simulation. This means that if we wish to calibrate a set of parameters, every iteration of the calibration must run multiple simulations. For any reasonable degree of pricing accuracy, the number of simulations required for each step is prohibitively high to conduct an accurate calibration in well vectorized code.

We could perhaps calibrate an unrestricted multi-factor model using well written C++ or a similar high performance language. However, we are interested only in providing easily reproducible examples for illustrative purposes, so instead we restrict ourselves to models with closed-form solutions for zero-coupon bond prices.

#### 7.3.2.1 Multi-Factor Vasicek Model

Extending the Vasicek model to a multi-factor setting is very straightforward, assuming the short rate follows as

$$r_t = \sum_{i=1}^{n} X_t^i$$

where

$$dX_t = A(\theta - X_t)dt + \Sigma dW_t$$

with $n$ being the number of factors, $A$ a lower $n \times n$ lower diagonal matrix, $\theta$ an $n \times 1$ vector, $\Sigma$ an $n \times n$ positive-definite volatility matrix, and $W_t$ an $n$-dimensional Brownian motion. The exact solution to this SDE is

$$X_t = e^{-At}X_0 + \int_0^t e^{-A(t-s)}\theta ds + \int_0^t e^{-A(t-s)}\Sigma dW_s$$

The parameter set that is being calibrated is $\Theta = \{A, \Sigma, \theta\}$.

#### 7.3.2.2 Multi-Factor CIR Model

The CIR model can also be extended to a multi-factor setting and in a very similar way to the Vasicek model. The short rate again is merely a sum of factors:

$$r_t = \sum_{i=1}^{n} x_t^{(i)}$$

and in this case the factors simply follow a CIR type process

$$dx_t^{(i)} = \kappa_i(\theta_i - x_t^{(i)})dt + \sigma_i\sqrt{x_t^{(i)}}dW_t^{(i)}$$

If we assume that $W_t^{(i)}$ are mutually independent then we have

$$\begin{aligned}
P(t,T) &= \mathbb{E}_t^{\mathbb{Q}}\left(e^{-\int_t^T r_s ds}\right) \\
&= \mathbb{E}_t^{\mathbb{Q}}\left(e^{-\int_t^T \sum_{i=1}^n x_s^{(i)} ds}\right) \\
&= \Pi_{i=1}^n \mathbb{E}_t^{\mathbb{Q}}\left(e^{-\int_t^T x_s^{(i)} ds}\right) \\
&= \Pi_{i=1}^n e^{A_i(t,T) - B_i(t,T)x_t^{(i)}}
\end{aligned}$$

where as before

$$A_i(t,T) = \frac{2\kappa_i\theta_i}{\sigma_i^2}\ln\left(\frac{\exp(\kappa_i(T-t)/2)}{\cosh(\gamma_i(T-t)/2)+\frac{\kappa_i}{\gamma_i}\sinh(\gamma_i(T-t)/2)}\right)$$

$$B_i(t,T) = \frac{2}{\kappa_i+\gamma_i\coth(\gamma_i(T-t)/2)}$$

with

$$\gamma_i = \sqrt{\kappa_i^2+2\sigma_i^2}$$

In case $\rho_{ij} = \text{corr}(W_t^{(i)}, W_t^{(j)})$ is non-zero then we should develop some analytical/semi-analytical way of pricing; if it is not available use simulation for pricing.

#### 7.3.2.3 CIR Two-Factor Model Calibration

We demonstrate a calibration for the two-factor case. The calibration procedure for the CIR two-factor model is very similar to the calibration procedure for the single factor CIR model. We now have three parameters each of two factors to be calibrated: $\Theta = \{\kappa, \theta, \sigma\}$.

We repeat our earlier calibration exercise, using the CIR two-factor model to model LIBOR and swap rates. We make use of the same market data as before, given in Tables 7.5 and 7.6 (LIBOR and swap rates as of October 29, 2008 and February 14, 2011).

We begin by calibrating a single factor CIR model using the same procedure as before. After finding the single factor calibrated values $\kappa_1$, $\theta_1$, and $\sigma_1$, we use the results as the initial guess for our two-factor model calibration. For example, our initial guesses are $\kappa_2 = [\kappa_1, \kappa_1]$, $\theta_2 = [\theta_1, \theta_1]$, and $\sigma_2 = [\sigma_1, \sigma_1]$. Given these parameters, we use the closed-form solution for the zero-coupon bond prices given by the model and our choice of parameters. We then use the zero-coupon bond prices to find the LIBOR and swap rates implied by our model in the same manner as before. Our objective function is also defined as before: we seek to minimize the sum of the squares of relative errors (SSRE) between model LIBOR and swap rates and market LIBOR and swap rates.

We find the minimum SSRE, using Nelder–Mead simplex method as an optimizer. Using data from October 29, 2008, we obtain the following parameters from calibration: $\kappa = [0.1226, -0.0149]$, $\theta = [0.0520, 0.0635]$, and $\sigma = [0.1153, 0.3464]$. We also obtain $r_t = [0.1256, -0.0914]$. Using data from February 14, 2011, we obtain the following parameters from calibration: $\kappa = [-0.0303, 0.0635]$, $\theta = [0.0990, 0.2073]$, and $\sigma = [0.0157, 0.0662]$. We also obtain $r_t = [0.0024, -0.0003]$.

The top panels of Figures 7.15 and 7.16 display zero-coupon bond prices for the CIR models using the calibrated parameters. The middle panel displays the market LIBOR rates as well as the LIBOR rates produced by the calibrated CIR models. The bottom panel displays the market swap curve, as well as the swap curve produced by the calibrated CIR models. In all panels, results for both one- and two-factor models are shown. While it is difficult to see visually, the two-factor model does provide a slightly better fit to the term structure of rates, and our objective function returns noticeably lower values. However, overall, our fit for the term structure on October 29, 2008 is still poor. Likely this is due to the *hump* in the term structure of LIBOR rates on this date. A model with additional factors would be necessary to capture this behavior.

#### 7.3.2.4 Pricing Swaptions with the CIR Two-Factor Model

After calibrating to the market LIBOR rates and swap curve, we produce swaption prices through simulation. As before, we use a discretized version of the CIR model for

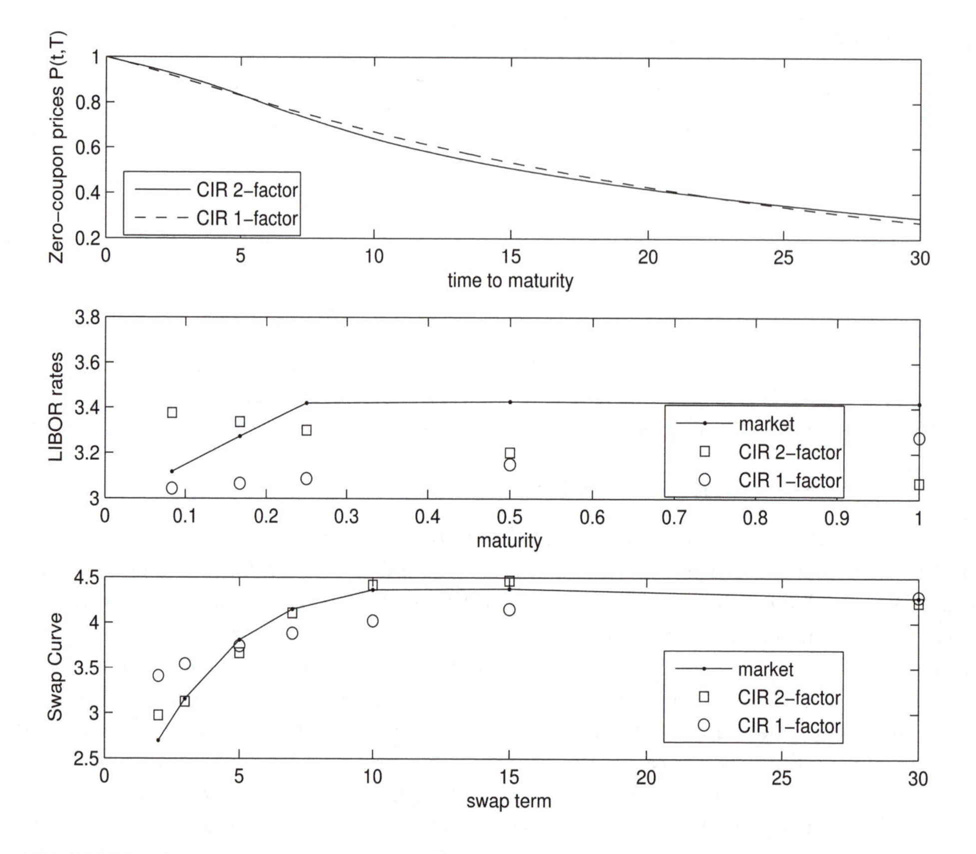


**FIGURE 7.15**: CIR one- and two-factor models vs. market on October 29, 2008

the instantaneous short rate (but now with 2 factors), and we produce a large number of realized interest rate paths. Each path gives us a realized swaption value, and the price of the swaption is the discounted expected value (e.g., the discounted average across paths).

Table 7.7 contains the market swaption prices. Table 7.12 contains the results from pricing swaptions via simulation using our calibrated parameters. Interestingly, we see that the CIR two-factor model produces swaption prices closer to the market on February 14, 2011. However, the lower SSE is primarily because the swaption prices for longer option maturities are more reasonable. The negative swaption values for short option maturities are clearly unreasonable. On October 29, 2008, the swaption prices from the two-factor model are significantly worse, with an SSE two orders of magnitude larger. While this is surprising, our objective function did not include fit to swaption prices at this stage, so it may indicate that swaption prices were not consistent with the term structure of interest rates on that date.

#### 7.3.2.5 Alternative CIR Two-Factor Model Calibration

As before, our first calibration of the CIR two-factor model focused on choosing model parameters to mimic the behavior of interest rates directly. While our fit to the term structure improved relative to the CIR one-factor model, our swaption prices were very poor, and we would like to see if we can improve upon this.

We again test our alternative calibration, where we add a relative error term for the fit to selected swaption values to our objective function. Our objective function is extended by

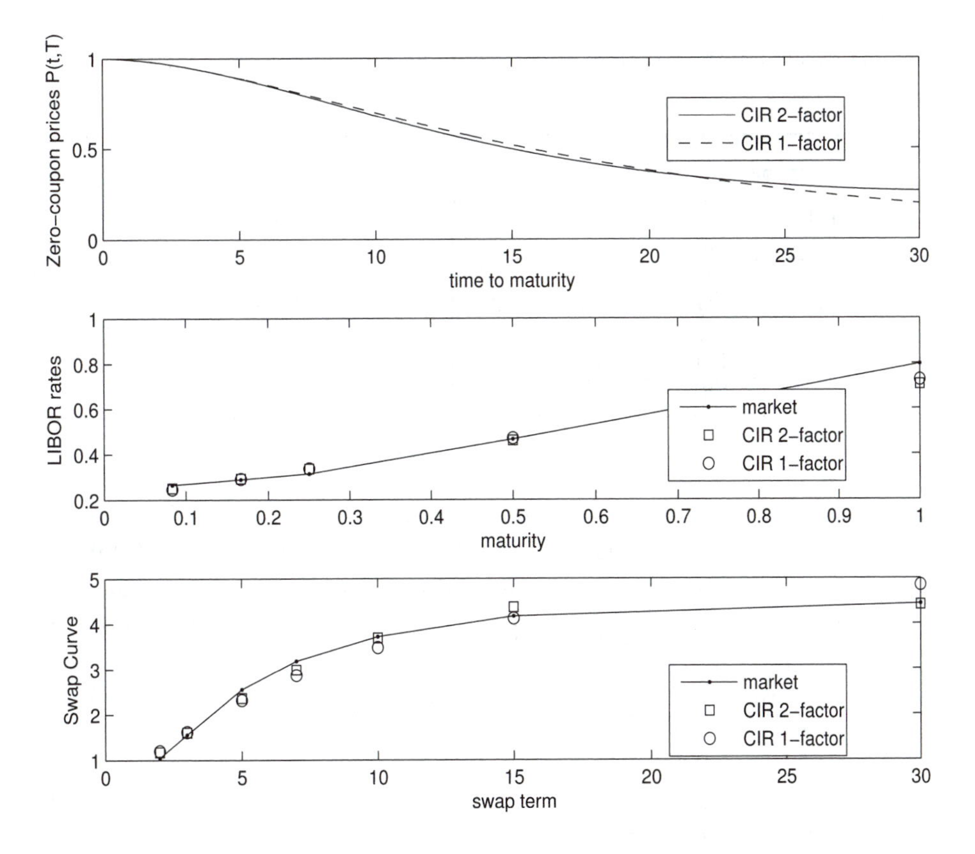


**FIGURE 7.16**: CIR one- and two-factor models vs. market on February 14, 2011

including relative error terms for four swaptions corresponding to the four pairs of shortest and longest option and swap maturities.

We find the minimum SSRE using simplex method. Using data from October 29, 2008, we obtain the following parameters from calibration: $\kappa = [0.0725, 0.2384]$, $\theta = [0.0030, 0.0290]$, and $\sigma = [0.0447, 0.1035]$. We also obtain $r_t = [0.0118, 0.0104]$. Using data from February 14, 2011, we obtain the following parameters from calibration: $\kappa = [0.0386, 0.1120]$, $\theta = [0.0779, 0.0304]$, and $\sigma = [0.0612, 0.0516]$. We also obtain $r_t = [0.0009, 0.0027]$. The fit to the term structure of interest rates is much poorer, but here we are primarily interested in how well we price swaptions.

Table 7.13 contains the results from pricing swaptions via simulation using our parameters calibrated under the alternative objective function. A comparison of these values to Table 7.7 shows that our errors are still larger than we would like, but that our simulated prices are finally beginning to take shape in a realistic manner. Comparing Tables 7.13 (two-factor model with swaption pricing in the objective function) and 7.12 (two-factor model with basic objective function), we see that the SSE has decreased greatly for both dates. Even more importantly, comparing Tables 7.13 (two-factor model with swaption pricing in the objective function) and 7.11 (single factor model with swaption pricing in the objective function), we see that the SSE has decreased noticeably for both dates. This means that using the same objective function, the CIR two-factor model substantially outperforms the CIR single factor model. In fact, for both dates these are the most accurate prices (lowest SSE) we have produced so far. In addition, we finally see that our swaption prices are

**TABLE 7.12**: Simulated swaption prices from the CIR model (two-factor, maturity in years)

| Oct. 29, 2008 | | | | Feb. 14, 2011 | | | |
|---|---|---|---|---|---|---|---|
| | Swap Maturity | | | | Swap Maturity | | |
| Option Maturity | 2 | 5 | 10 | Option Maturity | 2 | 5 | 10 |
| 1 | 2461.9 | 4632.7 | 5318.9 | 1 | -39.4 | -68.3 | -10.3 |
| 2 | 2361.3 | 4360.7 | 5132.6 | 2 | 64.5 | 234.1 | 565.1 |
| 5 | 2345.1 | 4661.5 | 5976.5 | 5 | 471.2 | 1352.8 | 2576.9 |
| 10 | 4444.8 | 9015.4 | 11725.1 | 10 | 2045.6 | 5325.8 | 9144.0 |
| SSE: 3.2e8 | | | | SSE: 9.2e7 | | | |

**TABLE 7.13**: Simulated swaption prices from the CIR model, with the addition of selected swaption price relative error to objective function (two-factor, maturity in years)

| Oct. 29, 2008 | | | | Feb. 14, 2011 | | | |
|---|---|---|---|---|---|---|---|
| | Swap Maturity | | | | Swap Maturity | | |
| Option Maturity | 2 | 5 | 10 | Option Maturity | 2 | 5 | 10 |
| 1 | 109.9 | 245.3 | 350.3 | 1 | 34.6 | 91.7 | 156.4 |
| 2 | 161.2 | 355.6 | 501.8 | 2 | 67.7 | 177.3 | 295.5 |
| 5 | 275.4 | 596.2 | 825.4 | 5 | 172.6 | 441.7 | 713.8 |
| 10 | 441.7 | 944.4 | 1286.2 | 10 | 412.3 | 1027.7 | 1614.7 |
| SSE: 1.2e6 | | | | SSE: 1.9e6 | | | |

dependent on both option and swap maturities (proper shape), without incredibly large errors.

#### 7.3.2.6 Findings

In summary, we have seen that

- A single factor model can give a good fit to the term structure of interest rates in limited circumstances. For some market dates, a single factor model cannot adequately capture the shape of the term structure.
- In terms of calibration, the CIR model is not likely to outperform the Vasicek model, but does have the advantage that we will not encounter negative interest rates in simulations.
- Single factor models provide an extremely poor fit to derivative instruments which depend on the term structure of interest rates. This holds even when we include the errors in pricing of the derivatives in our objective function.
- Increasing the number of factors in our model improves our fit to the term structure of interest rates, and provides more flexibility in fitting the term structure for unusual market data. However, even two factors can be insufficient to capture unusual term structure shapes.

- Increasing the number of factors in our model also allows us to provide much better pricing for derivative instruments, provided we include the pricing errors of those instruments in our objective function. If we do not include the pricing errors of the derivative instruments in our objective and only calibrate to the term structure, we may see improved term structure fit but larger derivative pricing errors. This may indicate that market derivative pricing is inconsistent with the market term structure of interest rates, or that our models are still too simplistic to capture market behavior.

### 7.3.3 Affine Term Structure Models

We can expand on the previous models by allowing for a slightly more flexible representation of the short rate. If we add a deterministic scalar component and weight the sum of factors, we arrive at a short rate model wherein the short rate is an affine function of the underlying factors. Under this class of models, the short rate can be expressed as

$$r_t = a_r + b_r^\top x_t$$

where $a_r$ is a scalar and $b_r$ is an $n \times 1$ vector with $n$ being the number of the factors. We assume the vector of underlying factors $x_t$ follows the matrix $OU$ equation driven by Brownian noise

$$dx_t = (-b_\gamma - Bx_t)dt + \Sigma dW_t$$

In general, matrix $B$ may be full, with all eigenvalues having a positive real part.

If we are interested in using a stochastic interest rate model with stochastic volatility, we can incorporate stochastic volatility by making the square root equation for variance $v_t$ stochastic and defining its SDE as

$$dv_t = \kappa(\theta - v_t)dt + \lambda\sqrt{v_t}dZ_t$$

The variables $W_t$ and $Z_t$ represent $n \times 1$ and $1 \times 1$ dimensional univariate Brownian processes, respectively. The correlation between $W_t$ and $Z_t$ is denoted by the vector $\rho = d < W_t, Z_t >$. We now reformulate the equation for $x_t$ as

$$dx_t = (-b_\gamma - Bx_t)dt + \Sigma\sqrt{v_t}dW_t$$

We can have a discrete time model in nature; thus there is no need for discretization. In [138], the authors propose a discrete time stochastic volatility model in an affine framework

$$x_{t+1} = x_t + (-b_\gamma - Bx_t)\Delta t + \Sigma\sqrt{v_{t+1}\Delta t}z_{t+1}$$

where $b_\gamma$ is an $n \times 1$ vector, $B$ is a lower-diagonal $n \times n$ matrix, $\Sigma$ is an $n \times n$ volatility matrix, and $z_t$, $t = 1, \ldots, T$ is a standard Gaussian sequence.

The dynamics for the variance sequence can be derived explicitly and are given by a double gamma model. The gamma variate $Gamma(\gamma, c)$ has shape parameter $\gamma$ and scale parameter $c$ with density $f(x)$, where

$$f(x) = \frac{c^\gamma x^{\gamma-1}e^{-cx}}{\Gamma(\gamma)}, \quad x > 0$$

Denote by $u \sim Gamma(\gamma, c)$ an independent draw of such a variate. The sequence $(v_t, t \geq 0)$ satisfies

$$\begin{aligned} v_{t+1} &\sim Gamma\left(\lambda v_t + y_{t+1}, d\right) \\ y_{t+1} &\sim Gamma\left(\gamma, c\right) \end{aligned}$$

Under both continuous-time and discrete-time framework caps, floors, and swaptions can be priced semi-analytically. In [141], the authors derive the characteristic function of the log of the future forward rate under both models. Having the characteristic functions, they employ Fourier techniques to price caps. In [132] and [138] they derive the characteristic function of the log swap rate under swap measure for continuous-time and discrete-time models respectively and as previously done employ Fourier techniques (FFT) to price swaptions. To be able to price both caps and swaptions analytically is quite unique in the short rate models and indeed any interest rate models and facilitates the calibration procedure.

Once parameters are calibrated from the cap, floor, and swaption prices, we can price other instruments via Monte Carlo simulation. The underlying factors can be simulated using a single Euler expansion.

### 7.3.4 Forward Rate (HJM) Models

All of the models considered thus far model only the evolution of the instantaneous short rate models. Given this rather simple parameterization of the term structure of interest rates, researchers have been able to produce a surprising array of models which successfully calibrate to a wide variety of markets. While the simplest models cannot even calibrate to the current yield curve, more complex models can calibrate to the current term structure and additionally to cap, floor, and swaption markets with a large degree of success, incorporating both the current rates and volatility structure of rates.

However, the quantity being modeled is still a single unobservable and instantaneous rate, which causes difficulties in the calibration process. For instance, simultaneously calibrating perfectly to the current yield curve while capturing the covariance structure of forward rates is difficult. In addition, forward rates play a much more central role in market traded fixed income and interest rate derivative instruments and thus are much more readily observable. As such it becomes more natural to model the instantaneous forward rate. Let $f(t,s)$ denote the instantaneous forward rate at calendar time $t$ for the forward period $[s, s+dt]$. If the instantaneous forward rates are known, then calculating zero-coupon bond prices is straightforward.

$$P(t,T) = e^{-\int_t^T f(t,u)du}$$

Thus modeling the evolution of forward rates can be both a more natural and a similarly tractable approach to modeling the term structure of interest rates. The framework of forward rate models has a theoretically infinite number of underlying stochastic factors, since at any given point in time there are infinitely many distinct instantaneous short rates, one for every maturity. In the most abstract sense, the evolution of the term structure depends on the theoretically infinite dimensional volatility structure of these rates. Thus this framework ends up being extremely general; indeed most significant short rate models can be seen as special cases of it, and in practical implementations we require a more specific model which reduces the model to a finite dimension.

Modeling the evolution of instantaneous forward rates was pioneered by Heath, Jarrow, and Morton, and models of this type are known as HJM models. HJM models thus describe the evolution of forward rates in the following form:

$$df(t,s) = \mu(t,s)dt + \sigma(t,s)^\top dW_t$$

where $W_t$ is $d$-dimensional Brownian motion under risk-neutral measure $\mathbb{Q}$. However, we do not have as much freedom in specifying the model parameters as in short rate models. No-arbitrage constraints dictate the following restriction on drift [131]:

$$\mu(t,s) = \sigma(t,s)^\top \int_t^s \sigma(t,u)du$$

and so the evolution of forward rates is described by

$$df(t,s) = \left(\sigma(t,s)^\top \int_t^s \sigma(t,u)du\right) dt + \sigma(t,s)^\top dW_t$$

Note that for all $s > t$ the same set of shocks (Brownian motions) being applied. The assumption of the existence of an equivalent martingale measure $\mathbb{Q}$ is a non-trivial one. For a given choice of the diffusion process $\sigma(t,s)$ we have to check to make sure that it exists[63]. The set of model parameters which need to be calculated in HJM includes the initial term structure of instantaneous forward rates and the volatility structure of the rates, $\Theta = \{f(0,t), \sigma(t,s)\}$. However, today's instantaneous forward curve $f(0,t)$ is not directly observable, like the instantaneous short rate. It must be derived from the term structure of zero-coupon bond prices.

$$f(0,t) = -\frac{\partial \log P(0,t)}{\partial t}$$

However, in practice zero-coupon bond prices only exist for a finite number of maturities. This means that deriving $f(0,t)$ for all $t$ requires some modeling as well, especially because the derived zero coupon bond curve has to be smooth, considering that we need to differentiate it. In practice market rates such as swap rates and futures are also used along with zero-coupon bond prices to help construct the yield curve. This process is explained in detail in Section 7.7.

To specify $\sigma(t,s)$ we have to choose $\sigma(t,s)$ so that the model prices and market prices of the calibration instruments closely match. Given the very large number of free variables[5] in $\sigma(t,s)$, depending on how many forward rates are modeled, this is typically achieved by assuming a functional form for $\sigma(t,s)$. While this seems like a natural solution, picking the right functional form and calibrating it stably has proven very challenging to achieve.

One model which assumes a functional form of $\sigma(t,s)$ which has gained wide exposure in the interest rate derivative community is the linear diffusion Heath–Jarrow–Morton model [131] specified by the SDE

$$df(t,s) = \mu(t,s)dt + V(t,s)\min(f(t,s),\lambda)\rho(t,s)dW_t \tag{7.9}$$

where $V$ is a deterministic function called the volatility matrix ($V : [0,T] \times [0,T] \to \mathbb{R}$), $\rho$ is the factor structure ($\rho : \mathbb{R} \to \mathbb{R}^d$), and $\lambda > 0$ the rate cutoff. It can be shown that this model exists and generates non-negative forward rates [131]. Under this model, caplets and swaptions are approximated very accurately with the Black formula, which implies calibration of the volatility matrix $V$ to Eurodollar futures options and caps is straightforward and the calibration to swaptions is more manageable. However, this model does not price in-the-money and out-of-the-money options correctly in comparison with at-the-money options. This is due to the fact that the distribution generated by this model has too much weight for states in which rates are large. Researchers have considered other diffusion processes for the forward rates. The square-root diffusion HJM model [63] is specified by the SDE

$$df(t,s) = \mu(t,s)dt + V(t,s)\sqrt{f(t,s)}\rho(t,s)dW_t \tag{7.10}$$

It can be shown that this model exists and generates non-negative forward rates [63]. The normal HJM model is given by

$$df(t,s) = \mu(t,s)dt + V(t,s)\rho(t,s)dW_t \tag{7.11}$$

[5]Note that for $t < s$ we have $\sigma(t,s) = 0$ because the forward rate ceases to exist.

This model exists, but generates negative forward rates, and thus is not arbitrage-free in an economy with cash. This model is useful to have because of its analytical tractability, and because the volatility smiles generated by this model are sometimes a good local approximation to the volatility surface observed in the market.

#### 7.3.4.1 Discrete-Time Version of HJM

While the HJM framework is very general and encompasses many different models as special cases, most classes of HJM models are non-Markovian. This implies that one cannot apply PDE-based techniques associated with the Feynman–Kac formula for pricing derivatives securities. This is a significant drawback and it means that much more computationally expensive Monte Carlo simulation techniques must be used to value derivatives under this model.

In order to simulate $f(t,s)$ for $s > t$ it is necessary to discretize both time, $t$, and maturity, $s$. There are many ways to do this, with perhaps the most obvious method being to simulate the continuous-time model using an Euler scheme on $0 = t_0 < t_1 < \cdots < t_m = T$. Since we need to discretize both time and maturity, however, it would be very computationally expensive to simulate the Euler scheme for small values of the time step $h$ where $h = t_{i+1} - t_i$. Instead we might choose to simulate the Euler scheme for larger values of $h$. However, the quality of the approximation then deteriorates and so it might be preferable to directly develop discrete-time (but still continuous-state) arbitrage-free HJM models instead. This approach is used commonly in practice and we will describe one approach.

We will assume that the same partition, $0 = t_0 < t_1 < \cdots < t_m = T$, is used to discretize both time, $t$, and maturity, $s$, for $0 < t < s < T$. Let $\hat{f}(t_i, t_j)$ denote the forward rate at time $t_i$ for borrowing or lending between $t_j$ and $t_{j+1}$. Then we can write the $t_i$ zero-coupon bond price for maturity $t_j$ as

$$P(t_i, t_j) = \exp\left(-\sum_{k=i}^{j-1} \hat{f}(t_i, t_k) h_k\right)$$

In the multi-factor case we assume a model of the form

$$\hat{f}(t_i, t_j) = \hat{f}(t_{i-1}, t_j) + \mu(t_{i-1}, t_j) h_{i-1} + \sum_{l=1}^{d} \hat{\sigma}_l(t_{i-1}, t_j) \sqrt{h_{i-1}} Z_{i,l}, \quad j = i, \ldots, m$$

where $Z_i = (Z_{i,1}, \ldots, Z_{i,d})$ for $i = 1, \ldots, m$ are independent $\mathcal{N}(0, I)$ random vectors.

Applying the martingale condition to compute the form of the drift function for this discrete version, we can show that multi-factor HJM models have the following drift restriction in discrete time:

$$\mu(t_{i-1}, t_j) h_j = \sum_{l=1}^{d} \left[ \frac{1}{2} \left( \sum_{k=i}^{j} \hat{\sigma}_l(t_{i-1}, t_k) h_k \right)^2 - \frac{1}{2} \left( \sum_{k=i}^{j-1} \hat{\sigma}_l(t_{i-1}, t_k) h_k \right)^2 \right]$$

The proof of it is left as an exercise for the reader. From equations (7.9), (7.10), and (7.11) we have

$$\hat{\sigma}_l(t_i, t_k) = g(f(t_i, t_j)) V(t_i, t_j) \rho(t_j - t_i)$$

where $g$ depends on the model used (e.g., for the square root model $g(x) = \sqrt{x}$). We can also compute the form of the drift function for use with various different numeraires. Now that we have written HJM in discrete time, pricing derivative securities via simulation under this model becomes straightforward once we specify the discrete time volatility matrix $V(t_i, t_j)$ and the discrete time factor structure $\rho(t_j - t_i)$. This is the daunting task of calibration of HJM.

#### 7.3.4.2 Factor Structure Selection

There are a number of issues which must be taken into account when selecting the factor structure $\rho$ and a number of factors which are used to model the interest rates. To begin with, analysis of historical interest rate data shows that correlations between interest rates are reasonably stable over time (as compared to, say, interest rate levels or volatilities). This leads to the choice of the factors as time homogeneous functions, e.g., functions of relative forward time, $s > t$. Given this assumption, there are some obvious methods to apply, namely, principal component analysis and explicit specification of the factors via a set of basis functions.

Principal component analysis of swap rates for the major currencies has shown that roughly 95% of the variance of the changes to the curve can be explained using few factors (roughly three factors). These factors are interpreted as an overall level of interest rates, a curve steepening, and a curve bowing. It is unclear, however, that the above analysis can be extended to forward rates, since the forward rates themselves are generally not observable in the market and must be derived from some sort of model (e.g., a yield curve spline). Indeed, it has been argued that because par rates are essentially integrals of forward rates, one cannot tell how many factors are driving the forward curve by only observing the curve because much of the high frequency behavior is integrated out. Therefore, we can also choose the number of factors driving the forward curve based on other criteria. For example, if we wish to accurately reproduce a prespecified matrix of correlations between $N$ forward rates, a priori we would expect to have to use roughly $N^2$ factors. For large $N$, of course, this introduces far too many sources of uncertainty. A reasonable compromise is thus made by determining the smallest number of factors needed to reproduce correlations to within some desired accuracy.

The implementation also allows one to use a set of prespecified basis functions and weights for the factors. These functions and weights are chosen to reproduce the correlations of various spot and forward rates to a value and an admissible tolerance.

### 7.3.5 LIBOR Market Models

The HJM framework and subsequent models represented a significant breakthrough in the modeling of the term structure of interest rates. They describe the dynamics of the entire term structure of interest rates, along with their arbitrage constraints, and thus from this framework we are able to construct a finite dimensional model with as many free parameters as are necessary for calibration. However, this comes at a price as calibration of the volatility structure remains very difficult in practice.

In addition, neither short rate or HJM models can reproduce the pricing formulas commonly used in the most liquid interest rate derivative markets, that of caps and swaptions. Just as the market prices of most non-interest rate options are quoted in implied volatilities, where the assumed model is the Black–Scholes model, prices for caps and swaptions are quoted in implied volatilities where the assumed model is Black's model. Black's model in these cases assumes that either the LIBOR forward rate or swap rate is log normal with zero drift.

The LIBOR market models developed by Brace, Gatarek and Musiela[38] were the first to be able to reproduce Black's model in a more rigorous framework, with the LIBOR forward model pricing caps using the market standard Black's formula for caps and the swap market model pricing swaptions using the standard Black's formula for swaptions. While these two models are mutually incompatible, the ability to price options in one major market analytically is a large advantage. In addition, accurate analytical approximations exist for pricing swaptions in the LIBOR market model.

In addition, the HJM framework still models the evolution of instantaneous rates, which are unobservable in the market. As such, determining the initial instantaneous forward rate curve $f(0,t)$ can be very difficult and error prone, especially given the fact that is the derivative of observed quantities. The term structure market models were developed as a response to this problem. Instead of modeling the evolution of the instantaneous forward rates, LIBOR market models directly model the evolution of market quoted non-instantaneous LIBOR forward or swap rates. This makes calibration to current market prices extremely straightforward since those rates are in fact directly observable in the market.

In the LIBOR market models, a set of $n$ LIBOR rates is modeled as a diffusion process $L_i(t)$ for $i = 1, \ldots, n$ as

$$dL_i(t) = \mu_i(t)L_i(t)dt + L_i(t)\sigma_i(t)^\top dW_t$$

Under this model we can analytically price caps/floors, and using a lognormal approximation we can price swaption [52]. The calibration problem is to find a symmetric semi-definite matrix $\sigma$ such that the model prices closely match market prices. This problem can be recast as a semi-definite programming problem [88].

---

## 7.4 Credit Derivative Models

In this book, we do not cover credit derivative models and pricing. Credit derivative is a vast topic and entire books have been written about it. The intention in this short section is merely to state similar techniques and processes covered in this book are applied and used in pricing credit derivatives. Our citations by no means cover what is in the literature as it is an immense field.

For instance, similar formulas as shown in Equation (7.3) can be obtained in credit derivative pricing models, for inverting portfolio default rates from collateralized debt obligation (CDO) tranche spreads [75] and pure jump models with a state-dependent jump intensity local Lévy model [55]. The work in [75] presents a method that recovers the default intensity of a portfolio from CDO spreads. The proposed method consists of two parts: a non-parametric method to recover expected tranche notional from CDO spreads, and an inversion formula to compute the local intensity functions from the expected tranche notional. The authors in [7] present a generic one-factor Lévy model for pricing synthetic CDOs. Laurent and Gregory [169] discuss new approaches to the pricing of basket credit derivatives and CDOs. They show that loss distribution, required in the valuation of CDOs, can be obtained by fast Fourier transform which is semi-analytical.

## 7.5 Model Risk

In the preceding sections we have discussed how to formulate model calibration problems which allow us to calibrate models such that they replicate market price as closely as possible. We have discussed a number of different models and how they perform differently in a realistic setting. However, the very need for a calibration step and the need to recalibrate our models to market prices proves implicitly that our models are just that, models of the real world, an attempt to explain enormously complex economic processes with simple stochastic evolution of a few market variables. Inevitably, we will not be able to fully capture the true complexity of the processes which drive the formation of market prices, and so there will always remain some amount of model risk associated with our attempts.

What is model risk? Here we examine the notion of model risk by reviewing academic literature on model risk and its assessment..

Emanuel Derman in his work [93] gives an overview of model risk. He defines seven types of model risk which we quote here:

1. Inapplicability of modeling. As defined in [93], this is the most fundamental of risk, the risk that modeling is just not applicable. The true underlying dynamics of the problem may be so complex that a simplified model may not be appropriate.

2. Incorrect model. Ultimately all models are incorrect on some level since they represent an attempt to simplify a complex problem into its most important features and mathematically model those features. However, the model chosen must be able to reproduce the critical features of the underlying problem at least approximately in order to be applicable to a given domain and in doing so help us better understand the dynamics of the system under consideration. If the model fails to do so, then we most likely have chosen the wrong model for the problem.

3. Correct model, incorrect solution. Even if modeling is appropriate, as we have seen in the preceding chapters and sections, computing solutions for these models can be very difficult and calibrating them to market prices even harder. Thus there is always the risk that even the right model can be solved in such a way as to yield an incorrect solution.

4. Correct model, inappropriate use. If we assume that we have chosen a correct model and computed a correct solution under that model, there is still the risk that the model results will be used inappropriately. This has often been a problem in the modern history of mathematical finance where those who utilize models and their results fail to understand their assumptions and limitations.

5. Badly approximate solution. Even if our choice of model is correct and we can compute a solution, the accuracy of the solution may not be appropriate for the use to which it is applied. Most of the pricing algorithms we have presented in this book have a tradeoff between the accuracy of the solution computed and the computation time of the algorithm in question and when using models we must be sure that they are computed with an accuracy that is acceptable. This may vary widely by use case; pricing performed for portfolio risk management purposes may require considerably less accuracy than hedge ratios computed and used on a highly leveraged trading book. Thus it is important to ensure the right level of accuracy dictated by the intended use.

6. Incorrect implementation. Assuming all of the other modeling choices have been correct, simple programming errors can cause costly and embarrassing errors in the routines which actually do the modeling in a practical setting. One must strive to thoroughly test the actual implementation of models in a real world setting.

7. Unstable data, non-stationary solutions. The ability to acquire and process accurate market data, though it is rarely discussed in the literature, is one of the most critical steps in the practical implementation of any model. Without accurate data no calibration can take place and any modeling attempts would be fruitless. Thus we must be able to obtain stable and reliable market data. If accurate market data is itself highly unstable, then this may indicate that modeling itself may be inappropriate.

The paper provides some guidelines on how to avoid model risk. The author suggests that models be regarded as interdisciplinary endeavors, that complex models be tested on simple cases first to verify that even with the added complexity, basic results can be reproduced, that models should be tested at their boundary conditions to determine under what conditions a model will fail, and that small discrepancies should not be ignored as they may indicate more serious problems.

One of the most important uses of pricing models is the ability not only to calibrate the model to liquid markets for derivatives, but extrapolate the prices for exotic derivatives under these models. In this function, model risk becomes very important in that many models may have sufficient parametrization to reproduce the prices for regularly quoted derivatives; however, they may differ significantly in their extrapolated prices for exotics. In [137], Hirsa, Courtadon, and Madan assess the effect of model risk on the valuation of barrier options. They calibrate four different models: (a) the local volatility model, (b) the constant elasticity of volatility model , (c) variance gamma model, and (d) variance gamma with stochastic arrival model to the European option market and then use the calibrated models to price path-dependent options. They conclude that even though those calibrated models can reproduce European option prices very closely, for barrier option prices could behave very differently.

The work in [197] is a similar to [137] with some vigorous analysis on the conclusion in [137]. Kyprianou, Schoutens, and Wilmott in [197] show that several advanced equity option models incorporating stochastic volatility can be calibrated very nicely to a realistic implied volatility surface. Specifically, they focus on the Heston stochastic volatility model (with and without jumps in the stock price process), the Barndorf-Nielsen–Shephard model, and Lévy models with stochastic time.

All these models are capable of accurately describing the marginal distribution of stock prices or indices and hence lead to almost identical European vanilla option prices when calibrated as shown in [137]. As such, we can hardly discriminate between the different processes on the basis of their smile-conform pricing characteristics. Therefore we are tempted to apply them to a range of exotics. However, due to the different structure in path behavior between these models, they find that the resulting exotics prices can vary significantly. This is particularly true for derivatives that depend on the realized moments of (daily) log returns. An already traded example of these derivatives is the variance swap. A comparison of these moment derivatives premiums demonstrates an even bigger discrepancy between the aforementioned models.

Note that an almost identical calibration means that at the time points of the maturities of the calibration data set the marginal distribution is fitted accurately to the risk-neutral distribution implied by the market. If we have different models leading to such almost perfect calibrations, all models have at most the same marginal distributions. It should be clear that even if at all time points $0 < t < T$ marginal distributions among different models coincide, this does not imply that exotic prices should also be the same.

The authors of [197] show surprisingly large differences between prices for exotics among these models even when calibrated to the same underlying option prices consistent with the conclusion in [137]. They demonstrate price differences for one-touch barriers of over 200 percent. For lookback call options a price range of more than 15 percent among the models was observed. A similar conclusion was valid for digital barrier premiums. Even for cliquet options, which only depend on the stock realizations over a limited amount of time points, prices vary substantially among the models. Moment derivatives, like variance swaps, amplified the pricing disparities. These results demonstrate a very material amount of model risk involved when using calibrated models to extrapolate the pricing and hedging parameters of exotic derivatives.

In [162] the authors take a different view of model risk. They analyze model risk separately for pricing models and risk measurement models. They define model risk in pricing models to be the risk arising from the use of a model which cannot accurately evaluate market prices or which is not a mainstream model in the market. Alternatively, they define model risk in risk measurement as the risk of not accurately estimating the probability of future losses.

Expanding on these definitions, they define the sources of model risk in pricing models to be:

1. Use of wrong assumptions
2. Errors in estimations of parameters
3. Errors resulting from discretization
4. Errors in market data

There are some overlaps with the work in [93]. They define the sources of model risk in risk measurement models to be:

1. The difference between assumed and actual distributions
2. Errors in the logical framework of the model

The authors in [162] suggest a number of practical steps to control model risks from a quantitative perspective. In the case of pricing models, they suggest the use of multiple alternative models to determine the pricing and hedge ratios of the various assets which utilize the models in question and the establishment of capital reserves to allow for the difference in estimations, as well as position limits which take into account the differences in model estimates. In the case of risk measurement models they again suggest the use of different models, but they also suggest scenario analysis be undertaken for various different extreme historical or imagined stressful scenarios. In addition, they suggest position limits can be established based on information obtained from scenario analysis. From a qualitative perspective, improvement of risk management systems is suggested, including their organization, authorization, and lines of reporting, as well as human resources. Further, examination of models, periodic review of models, and maintenance of proper communications with end users of models are considered practical steps which can improve model risk management from a qualitative perspective.

Not only do the authors [162] provide suggestions of how to improve model risk management, but they also present various cases in which actual model risks have been realized and use market data to empirically analyze the problems of the models. Specifically, they use examples of index swaps, mark cap, and the experiences of Long-Term Capital Management to develop a general description of how model risks arise in real world situations. They also proceed to analyze long-term foreign exchange options, barrier options on stock prices, and a strangle short strategy to identify some salient features of model risk.

Rama Cont discusses model risk in [73] by introducing two methods: one based on a coherent risk measure comparable to market price, another is based on a convex risk measure. The coherent risk measure is a risk measure that has the properties of monotonicity, sub-additivity, homogeneity, and translational invariance. The convex risk measure is a risk measure that has the property of convexity.

## 7.6 Optimization and Optimization Methodology

In the preceding sections we have discussed a number of different models used in a variety of markets, their pros and cons as well as their empirical performance in practical calibration problems. However, these models only help us define the objective function of the calibration routine which allows us to match model and market prices. In all of these discussions we have left out one critical step in the calibration process, the actual optimization routine used to solve the calibration problem and generate our calibrated model parameter set.

Because of the complexity of these pricing functions and the multivariate nature of the parameter sets, the optimization problem presented in the majority of calibrations tends to be a multivariate nonlinear optimization problem, one of the harder optimization problems to solve. Many books have been written solely covering the different methods which can be utilized to solve these types of optimization problems and so a comprehensive treatment of the subject is well beyond the scope of this text. However, we will provide a basic description of some of the most common optimization methodologies which can be used to solve calibration problems and provide additional references for those readers who are interested in a complete treatment.

There are some important attributes of the optimization problem we are trying to solve which will guide our choice of optimization technique. One important attribute is the computational complexity of the pricing algorithm and thus the objective function. A very computationally intensive pricing function will likely result in one favoring algorithms which perform fewer evaluations of the object function, for instance opting for solution techniques which search only for a local optimum solution instead of a global one which will generally require many more evaluations.

Another critically important attribute is the availability and computational complexity of calculating gradients and Hessians for the pricing method and thus the objective function. For many pricing algorithms analytically calculable gradients and Hessians do not exist and must be numerically approximated. This can be an extremely time intensive process, especially given the fact that if our parameter set consists of $n$ scalar variables we typically must evaluate the pricing function $O(n)$ times to derive the gradient numerically and $O(n^2)$ times to derive the Hessian numerically, instead of just evaluating the pricing function once to get the value of the objective function at a single point. For this reason, with computationally intensive pricing functions we may prefer to use gradient free methods.

Finally, another critical decision is whether or not we want to attempt to find the global optimal solution in our parameter space or whether we will be satisfied with a local optimum. Many of the commonly used optimization techniques are designed solely to find a locally optimal solution by searching locally for improvements over the current solution. However, there are a number of augmentations which can be used to improve the chances that a global optimum is found, such as multi-start optimization or simulated annealing. These techniques do come at a significant cost in terms of the number of objective function evaluations, so for very complex pricing functions we may prefer a local optimum. Additionally, some techniques are not based on local searches of the parameter space and thus are more suited to finding globally optimal solutions, for instance genetic algorithms, but this often comes at the cost of a slower convergence rate. Before going over various optimization methodologies it is important to mention the following points:

The likelihood may very well be non-differentiable; therefore we should use a search algorithm that is not based on gradient calculation.

From the point of view, the direction set (e.g., Powell algorithm) is a good choice.

(c) As the number of parameters increases, convergence of the optimization process becomes more difficult as the likelihood function gets pretty flat.

What are advantages and disadvantages of this algorithm.

### 7.6.1 Grid Search

Grid search or brute-force search is a straightforward method for searching for the optimal solution (global minima) which is easy to implement but computationally very costly to run to completion. This method does not require derivatives of the objective function. We typically start with a very coarse grid and narrow it down to a finer grid until the global minima is reached.

Assume the dimension of the parameter set is $n$. On each direction/axis we choose a range by choosing a minimum and a maximum for the parameter at that direction and divide the range into $m_i-1$ equidistant subintervals. These points define our grid in the $i$-th direction. With $m_i$ points for each axis, we create $\prod_{i=1}^{n} m_i$ vectors. For high dimensional search space, this scheme would result in a very large number of vectors (assuming $n = 10$ and $m_i = 7$ implies $7^{10} = 282,475,249$ function evaluations). Having those points, we perform a simple search loop to find a vector from the set of $\prod_{i=1}^{n} m_i$ vectors that minimizes the objective function. We could then form a finer mesh around that point to find yet a better point.

1. Draw axis — draw $m_i$ points starting from $x_{min}^{(i)}$, that is, $x_{j_i}^{(i)} = x_{min}^{(i)} + (j-1)\lambda_i$ for $i = 1, \ldots, n$ and $j_i = 1, \ldots, m_i$, where $\lambda_i$ is the step size in direction $i$. This algorithm might need to be adjusted due to the constraints.

2. For vector $x$, that is,

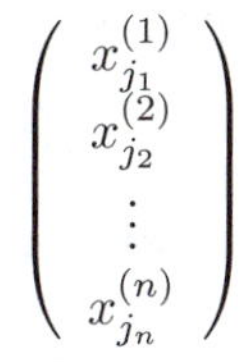

$$\begin{pmatrix} x_{j_1}^{(1)} \\ x_{j_2}^{(2)} \\ \vdots \\ x_{j_n}^{(n)} \end{pmatrix}$$

evaluate the objective function for each $x$ over the loop $j_i = 1, \ldots, m_i$ for all $i$ and find $x$ that minimizes $f$ and call it $x^\star$.

3. Relocate and contract. In the new iteration, we would reduce $\lambda_i$ and perhaps reduce $m_i$ to get a smaller searching space around $x^\star$ to get a yet better $x$. This is repeated until a certain criterion is met.

In general, in optimization the starting point (vector) could be very crucial. An arbitrary starting point could cause a very slow convergence and in most cases we find a local minima as opposed to a global minima. This is even more pronounced for higher dimensional cases. The grid search typically used to find a good starting point and the parameter set found in this routine are passed as a starting point to yet another optimizer.

### 7.6.2 Nelder–Mead Simplex Method

The Nelder–Mead simplex method is a non-linear optimization method which is derived from previous simplex methods used to solve linear optimization methods. A simplex is a polytype which has $N+1$ vertices in $N$ dimensions, meaning a line in one dimension, a triangle in two, a tetrahedron in three, and so on. The original method capitalizes on the insight that the inequality in a linear problem forms a polytype and one can walk along the edges of this polytype to find a solution to the linear programming problem.

However, the constraints in a non-linear problem do not necessarily form a simplex. Instead the Nelder–Mead method starts with an initial simplex in the parameter space. The least optimal vertex on the simplex is calculated and replaced with a new vertex, which can be derived in a number of ways. One common implementation is to reflect across the centroid of the simplex to get the new point, and if this represents an improvement, expand the simplex in this direction further. If this is not an improvement, then we have crossed a local minimum and shrink the simplex around it. This method is usually used for unconstrained problems, but with some tweaks and twists that will be discussed shortly, it can be used to solve constrained problems as well.

The Nelder–Mead algorithm can be laid out as follows:

1. Sort vertex values $f(x)$

$$f(x_1) \leq f(x_2) \leq f(x_3) \leq \cdots \leq f(x_{n+1})$$

2. Calculate the centroid $\overline{x}$ of the $n$ best points

$$\overline{x} = \sum_1^n w_i x_i$$

   where $w_i$ are the weights and a common choice is $w_i = \frac{1}{n}$

3. Reflect – the worst vertex

$$x_r = \overline{x} + \alpha(\overline{x} - x_{n+1})$$

   If $f(x_1) \leq f(x_r) < f(x_n)$, replace $x_{n+1}$ with $x_r$, return to step 1.

4. Expand — if $f(x_r) < f(x_1)$, we have found a new most optimal point

$$x_e = \overline{x} + \gamma(\overline{x} - x_{n+1})$$

   Replace $x_{n+1}$ with the better one between $x_e$ and $x_r$, return to step 1.

5. Contract — if $f(x_r) \geq f(x_n)$,

$$x_c = \overline{x} + \rho(\overline{x} - x_{n+1})$$

   If $f(x_c) \leq f(x_{n+1})$, replace $x_{n+1}$ with the $x_c$, go to step 6.

6. Shrink — set $x_i = x_1 + \sigma(x_i - x_1)$ for $i = 2, \ldots, n+1$, return to step 1.

$\alpha$, $\gamma$, $\rho$ and $\sigma$ are positive scalars, generally set to be 0.5.

### 7.6.3 Genetic Algorithm

A radically different approach to the optimization problem is genetic algorithms (GA), which are optimization methods that use the concept of gene evolutionary biology. In other words, the object value (gene) evolves to a better value (evolved gene) through inheritance, crossover, selection, and mutation processes. Here we introduce one algorithm for function maximization.

- Initialize

  with $n$ starting vectors $x_1, x_2, \ldots, x_n$, evaluate function $f$ at $x_i$ for $i = 1, \ldots, n$ and call them

$$f_i = f(x_i)$$

- Inheritance

  set $\hat{f}_i = (f_i - \min(f_1, \ldots, f_n))^{(1+\log(iter)/100)}$ for $i = 1, \ldots, n$

  calculate $\hat{p}_i = \frac{\sum_{k=1}^{i} \hat{f}_k}{\sum_{k=1}^{n} \hat{f}_k}$ for $i = 1, \ldots, n$

  calculate $N_i = \sum_{k=1}^{n} \mathbb{I}_{(r_k > p_i)} + 1$ for $i = 1, \ldots, n$

  where $r_k \in \mathcal{N}(0,1)$ for $k = 1, \ldots, n$; then reproduce new $x_i$ as $x_i = x_{N_i}$

- Crossover

  $M$ is the number of unique $x$

$$\begin{aligned} \psi &= \frac{M}{n} \\ \eta &= \max(0.2, \min([1\eta - \psi + \psi_0])) \\ \psi_0 &= \psi \end{aligned}$$

  Generate $\frac{n}{2}$ random variable $r_i \in \mathcal{N}(0,1)$ for $i = 1, \ldots, \frac{n}{2}$

$$N_u = \sum_{k=1}^{n/2} \mathbb{I}_{(r_k < \eta)}$$

  If $N > 0$ choose crossover point

$$r_i \in \mathcal{N}(0,1), \quad i = 1, 2, \ldots, N_u \tag{7.12}$$

$$u_i = \text{floor}(r_i(K-1)) + 1 \tag{7.13}$$

  where $K$ is the parameter space dimension for $i = 1, \ldots, N_u$; cross part of the parameter vectors

$$x_i(u_i + 1 : K) \overset{switch}{\longleftrightarrow} x_{i+N_u/2}(u_i + 1 : K) \tag{7.14}$$

- Mutation

  Draw mutated parameter set

$$r_i \in \mathcal{N}(0,1), \quad i = 1, \ldots, N \tag{7.15}$$

  If $r_i < \gamma$, mutation happened

$$x_i = (1 + r_i)x_0 \tag{7.16}$$

  where $\gamma$ is the mutation rate and $x_0$ is the starting parameter value.

Both the simplex method and genetic algorithms do not need derivative or gradient information. They are basically intelligent search algorithms which can minimize non-smooth functions. This is a large advantage for optimization functions whose evaluation is very computationally intensive, typically when the pricing function for some derivative is very difficult to evaluate. Genetic algorithms can even minimize discontinuous functions too, but the conceptual difference between simplex methods and genetic algorithms is that the latter are designed to find the global optimal solution, while the former are designed to find only locally optimal solutions.

### 7.6.4 Davidson, Fletcher, and Powell (DFP) Method

The procedure in the Davidson, Fletcher, and Powell (DFP) method is as follows:

1. Initialize — set a starting $x_0$ and a real positive definite matrix $D_0$. $D_0$ can be an approximation to the inverse Hessian matrix $H^{-1}(x_0)$ where $H(x_0)$ is positive definite.

2. Find $\alpha$. Define

$$y_k(\alpha) = f(x_k) - \alpha D_k \nabla f(x_k), \ \alpha > 0$$

   then find $\alpha$ that minimizes $y_k$ by a search procedure.

3. Update

$$\begin{aligned} x_{k+1} &= x_k - \alpha D_k \nabla f(x_k) \\ D_{k+1} &= D_k + \frac{d_k^\top d_k}{b_k d_k} - \frac{(D_k b_k)^\top (D_k b_k)}{b_k D_k b_k^\top} \end{aligned}$$

   where

$$\begin{aligned} d_k &= x_{k+1} - x_k \\ b_k &= \nabla f(x_{k+1}) - \nabla f(x_k) \end{aligned}$$

### 7.6.5 Powell Method

The Powell method, or the Powell conjugate gradient descent method, is an optimization method that does not require the function to be differentiable or to know its derivatives.

If we assume the dimension of parameter space is $n$ then we initialize this algorithm by providing $n$ linear independent vectors. The algorithm will search for the optimal solution along each vector, in each direction, and then update the vectors by the combining the initial vectors provided. The algorithm is relatively simple compared with other optimization algorithms and also has the advantage of not requiring derivatives, which, as discussed previously, may greatly reduce the computational complexity of the algorithm.

Below we provide a short summary of the steps in the algorithm:

1. Initialize — start with an $n$-dimensional search vector. Successively generate $n-1$ new search vectors, $u_i$, that are normal to each previously generated search vector. Then generate a starting point $x_0$.

2. Minimum of $p_i$ — for $i = 1, \ldots, n$ find $\lambda_i$ that minimizes $f(x_{i-1} + \lambda_i u_i)$, the minimum along search direction $u_i$, then update the current point

$$x_i = x_{i-1} + \lambda_i u_i$$

3. Update search directions — for $i = 1, \ldots, n-1$

$$\begin{aligned} -u_i &= u_{i+1} \\ u_n &= x_n - x_0 \end{aligned}$$

4. Update the current point $x_0$ — find the $\lambda$ that minimizes the $f(x_n + \lambda(x_n - x_0))$, then let

$$x_0 = x_0 + \lambda(x_n - x_0)$$

### 7.6.6 Using Unconstrained Optimization for Linear Constrained Input

If we have an unconstrained optimization method available for a specific problem, and this problem has only a simple range constraint, we can adjust most unconstrained optimization algorithms to incorporate these constraints instead of changing to a more complex constrained method. We can do this by simply mapping the unconstrained space to the constrained space. In many cases unconstrained optimization algorithms are considerably less computationally intensive than constrained ones, and so this may save a lot of time.

We can map an unconstrained parameter $x$ to a new parameter $y$ which is constrained using the following algorithm:

for $x$ being mapped to $[c \;\; +\infty)$ do

$$y = |x| + c$$

for $x$ being mapped to $(-\infty \;\; d]$ do

$$y = -|x| + d$$

for $x$ being mapped to $[c \quad d]$ do

if $c < x < d$

set $y = x$

else

first calculate

$$\begin{aligned} \text{range} &= d - c \\ n &= \text{floor}((x - c)/\text{range}) \end{aligned}$$

if $n$ is even, set

$$y = x - n \times \text{range}$$

else if $n$ is odd, set

$$y = x + (n + 1) \times \text{range} - (x - c)$$

This mapping is a periodic extension. By this mapping method, we transform the constrained problem to an unconstrained problem, where several unconstrained optimization methods are available. This map is a linear map; one could use other maps that are nonlinear.

### 7.6.7 Trust Region Methods for Constrained Problems

Trust region methods focus on the small region around the search point and use approximations for the changes in function value in this small area, which simplifies the search process. In this method, we *trust* the following quadratic model holds in a small region around $x_k$:

$$\psi_k(s) = g_k^\top s + \frac{1}{2} s^\top B_k s$$

where $s = x_{k+1} - x_k$, $g(x) = \nabla f(x)$, $B$ is the approximation to the Hessian matrix $\nabla^2 f(x_k)$. Thus $\psi_k(s)$ is an approximation of the function value at $x_{k+1}$. Once we have this approximation we can solve a subproblem to minimize $\psi_k(s)$ in the small region around the current point and once a proper $s$ is chosen, we move one step forward to $x_{k+1}$.

Here we introduce a constrained optimization algorithm using the trust region method, which was proposed by Coleman and Li [70]. Assume $x$ is constrained as $l \leq x \leq u$ and we define vector $v(x)$ that for each component $1 \leq i \leq m$

- if $g_i < 0, u_i < +\infty$, then $v_i = x_i - u_i$
- if $g_i \geq 0, l_i > -\infty$, then $v_i = x_i - l_i$
- if $g_i < 0, u_i = +\infty$, then $v_i = -1$
- if $g_i \geq 0, l_i = -\infty$, then $v_i = 1$

and we define a diagonal matrix

$$D(x) = diag(|v(x)|^{-\frac{1}{2}})$$

or equivalently

$$\begin{aligned} D(x)_{i,i} &= |v(x)_i|^{-\frac{1}{2}} \\ D(x)_{i,j} &= 0 \qquad (i \neq j) \end{aligned}$$

for $k = 0, 1, \ldots, n$

1. Compute

$$\psi_k(s) = g_k^\top s + \frac{1}{2} s^\top (B_K + C_k) s$$

where $B_k$ is the approximation to the Hessian matrix $\nabla^2 f(x_k)$, $C_k = D_k diag(g_k) J_k^v D_k$ where $J_k^v$ is the Jacobian matrix of $|v(x)|$ when $|v(x)|$ is differentiable and set $J_{k_i}^v = 0$ when $g_{k_i} = 0$.

2. Compute the solution $p_k$ that minimize

$$\psi(s) = g_k^\top s + \frac{1}{2} s^\top B_k s : \|D_k s\| \leq \Delta_k$$

3. Compute $\rho_k$ – first we define

$$\alpha_k^*[d_k] = \theta_k \tau_k^* d_k$$

then

$$
\begin{aligned}
s_k &= \alpha_k^*[p_k] && (7.17)\\
\rho_k^c &= \frac{\psi_k(s_k)}{\psi_k^*[-D_k^{-2}g_k]} && (7.18)\\
\rho_k^f &= \frac{f(x_k+s_k)-f(x_k)+\frac{1}{2}s_k^\top C_k s_k}{\psi_k(s_k)} && (7.19)
\end{aligned}
$$

4. Update $x$ — if $\rho_k^f > \mu$ and $\rho_k^c > \beta$, then $x_{k+1} = x_k + s_k$, otherwise $x_{k+1} = x_k$.
5. Update $D_k$ and $\Delta_k$.

### 7.6.8 Expectation–Maximization (EM) Algorithm

The expectation–maximization (EM) algorithm is a method for finding maximum likelihood or maximum of posteriori estimates of parameters in statistical models, where the model depends on unobserved latent variables. The EM algorithm is an iterative method which alternates between performing an expectation step, which computes the expectation of the log-likelihood evaluated using the current estimate for the latent variables, and a maximization step, which computes parameters maximizing the expected log-likelihood that was found on the expectation step. These estimated parameters are then used to determine the distribution of the latent variables in the next expectation step.

start from $\Theta'$ and a threshold level $\varepsilon$

set $\Theta_0^* = \Theta'$

for $i = 1, \ldots, n$

  $\Theta_n^* = \max \mathbb{E}(\log \mathcal{L}(\Theta \mid \Theta_{n-1}^*))$ (E step)

  if $\|\Theta_n^* - \Theta_{n-1}^*\| < \varepsilon$ (M step)

    break and end the main for loop

  end

end

---

## 7.7 Construction of the Discount Curve

As mentioned in previous sections, many interest rate models assume knowledge of the current yield curve or forward curve. While these models are complex in their own right, cooking the current yield curve from market rate quotes is a fairly complex process in and of itself. This section will briefly cover how the curve can be constructed from LIBOR instruments using real historical market rates. In constructing the yield curve we iteratively derive market implied LIBOR zero-coupon bond yields based on the no-arbitrage derived analytical definitions of different market rates. However, there are a limited number of LIBOR based instruments actively traded in the market, and certainly not enough with maturity dates that cover every date in a date range of thirty years. This means that for most dates on which one would want to get a rate, there is no LIBOR yield instrument that has a maturity corresponding to that date and hence no rate is available. To overcome this problem, mathematical techniques have been developed to determine the approximate LIBOR yield rate on any given date by applying interpolation or smoothing techniques to currently available market data.

### 7.7.1 LIBOR Yield Instruments

In the LIBOR yield curve construction we use the following instruments: LIBOR rates, Eurodollar futures, and swap rates. LIBOR rates cover maturities up to one year, Eurodollar futures cover maturities from three months to five years,[6] and swaps cover maturities from two to thirty years.

In Tables 7.14, 7.15, and 7.16 we provide LIBOR rates, Eurodollar futures, and swap rates, respectively, at market close on January 19, 2007.[7] Just a minute point that LIBOR rates are published at 11:00AM GMT and so the rates would not all be synced. LIBOR

**TABLE 7.14**: LIBOR rates at market close on January 19, 2007

| LIBOR rates | |
|---|---|
| Ticker | Quote |
| US00O/N | 5.28875 |
| US0001W | 5.30375 |
| US0002W | 5.31125 |
| US0001M | 5.32000 |
| US0002M | 5.34438 |
| US0003M | 5.36000 |
| US0004M | 5.37000 |
| US0005M | 5.38000 |
| US0006M | 5.39000 |
| US0007M | 5.39000 |
| US0008M | 5.39250 |
| US0009M | 5.39250 |
| US0010M | 5.39000 |
| US0011M | 5.38875 |
| US0012M | 5.38688 |

rates are quoted as simple interest rates. So one dollar held overnight will earn a cash interest rate of 5.28875% (on an ACT/360 basis), that is,

$$1.0 + (5.28875/100.0)\frac{1}{360.0} \tag{7.20}$$

LIBOR futures are quoted on a price basis (100-yield), but to convert this yield to a forward yield, we need to apply a convexity adjustment to account for futures-forward bias caused by the daily cash settlement of the futures. The first futures contract in Table 7.15 has settlement of 3/21/2007 and final settlement of 06/20/2007 (approximately three months out corresponding to the settlement date of the following Eurodollar contract). The convention for calculating the forward rate based on the contract quote and convexity adjustment

[6]The first two years are more liquid.

[7]In these tables we follow Bloomberg symbols.

**TABLE 7.15**: Eurodollar futures at market close on January 19, 2007

| Eurodollar Futures | | |
|---|---|---|
| Ticker | Settlement | Quote |
| EDH7 | 03/21/07 | 94.640 |
| EDM7 | 06/20/07 | 94.675 |
| EDU7 | 09/19/07 | 94.775 |
| EDZ7 | 12/19/07 | 94.890 |
| EDH8 | 03/19/08 | 94.965 |
| EDM8 | 06/18/08 | 95.005 |
| EDU8 | 09/17/08 | 95.030 |
| EDZ8 | 12/17/08 | 95.040 |
| EDH9 | 03/18/09 | 95.050 |
| EDM9 | 06/17/09 | 95.035 |
| EDU9 | 09/16/09 | 95.015 |
| EDZ9 | 12/16/09 | 94.985 |
| EDH0 | 03/17/10 | 94.970 |
| EDM0 | 06/16/10 | 94.940 |
| EDU0 | 09/15/10 | 94.915 |
| EDZ0 | 12/15/10 | 94.875 |
| EDH1 | 03/17/11 | 94.860 |
| EDM1 | 06/16/11 | 94.840 |
| EDU1 | 09/15/11 | 94.815 |
| EDZ1 | 12/15/11 | 94.785 |

(which is assumed to be zero) is:

$$\begin{aligned} F(t, T_1, T_2) &= 100.0 - \text{quote} - \text{convexity adjustment} \\ &= 100.0 - 94.640 - 0.0 \\ &= 5.32 \end{aligned}$$

where $t = 01/19/2007$, $T_1 = 3/21/2007$ and $T_2 = 6/20/2007$. Swap rates are quoted on a spread to treasury basis. For example, 5-year swap is quoted as a par rate based on a spread to on-the-run five year.

We notice that LIBOR rates are simple interest rates on a spot basis; the LIBOR futures imply yields on a forward basis and the swaps imply yields on a par basis. Since each of these

**TABLE 7.16**: Swap rates at market close on January 19, 2007

| Swap rates | |
|---|---|
| Ticker | Quote |
| USSW2 | 5.2671 |
| USSW3 | 5.1993 |
| USSW4 | 5.1830 |
| USSW5 | 5.1834 |
| USSW7 | 5.2094 |
| USSW10 | 5.2640 |
| USSW30 | 5.3856 |

rates types are not the same they cannot be directly compared unless they are *normalized.* The simplest way to do this is to convert all of the rates to discount factors. A discount factor is a way of expressing interest at different maturities in terms of a discount to a dollar and has the following meaning. If $P(t,T)$ is the discount factor from today $t$ to time $T$ then its meaning is the price one would be willing to pay today to get a dollar back at time $T$. Discount factors are easy to work with because their definition is simple, not involving different payment frequencies, compounding types, or day counts, unlike rates or yields. Note that in our definition discount factors are equivalent to zero-coupon bond prices.

Thus, in order to use LIBOR rates, Eurodollar futures implied forward rates, and swap rates to construct the discount curve, we must first convert each of these corresponding rates to discount factors, which we do in the following sections.

#### 7.7.1.1 Simple Interest Rates to Discount Factors

The discount factor $P(t,T)$ for a cash instrument which matures at $T$ and pays at the simple interest rate (non-compounded) $F(t,T)$ on an ACT/360 basis is given by

$$P(t,T) = \frac{1}{1 + F(t,T) \times (T-t)/360}$$

This is derived by simply reversing the simple interest accrual formula.

#### 7.7.1.2 Forward Rates to Discount Factors

The discount factor at time $S$, $P(t,S)$, can be expressed in terms of the forward rate $F(t;T,S)$ and the discount factor at time $P(t,T)$ (on an ACT/360 basis) using

$$P(t,S) = \frac{P(t,T)}{1 + F(t;T,S) \times (S-T)/360}$$

This expression is used for calculating the discount factor for Eurodollar futures. This formula is derived from the no-arbitrage required equivalence of holding time deposit from today to $T$ which pays rate $F(t,T)$ and then rolling it into time deposit at time $T$ which pays rate $F(t;T,S)$ and holding a time deposit until $S$, to which the discount factor of $P(t,S)$ is applied.

#### 7.7.1.3 Swap Rates to Discount Factors

We know that the swap rate at time $t$ from the swap term $T_n$ is

$$R_{\text{swap}}(t) = \frac{P(t,T_0) - P(t,T_n)}{\Delta \sum_{i=1}^{n} P(t,T_i)} \tag{7.21}$$

where $\Delta = \frac{1}{2}$ as mentioned earlier in the chapter. Solving for $P(t,T_n)$ we get

$$P(t,T_n) = \frac{P(t,T_0) - \Delta R_{\text{swap}}(t) \sum_{i=1}^{n-1} P(t,T_i)}{1 + \Delta R_{\text{swap}}(t)} \tag{7.22}$$

Note that in case of par swap rate we have $t = T_0$, which implies $P(t,T_0) = 1$. It is important to note that solving for a discount factor at time $T_n$ requires that discount factors for the previous payment intervals must be known. This is one of the complexities introduced by par rates, in that solving for the discount factor at some maturity date $T_n$ depends on having the discount factors at all payments prior to $T_n$.

## 7.7.2 Constructing the Yield Curve

Thus far we have covered how to convert market quotes for LIBOR rates, Eurodollar futures, and swap rates into discount factors. Now we will provide an outline of the entire construction process. The LIBOR yield curve construction process is broken down into two distinct parts. The first part of the process involves constructing the short end of the curve using *cash instruments* and *Eurodollar futures*. The short end has maturities going out for a term of about two years. The second part of the process uses the results from the short end plus the swap instruments to cook the remainder of the curve.

Note that for LIBOR rates we do not go beyond three months and switch to Eurodollar futures for maturities out to two years. Beyond two years the market for Eurodollar futures becomes illiquid and we switch to swaps. When the construction process is complete we should be able to calculate a discount factor for every day, starting from the curve date and ending at the maturity date of the longest swap instrument.

### 7.7.2.1 Construction of the Short End of the Curve

Using the conversion formulas described earlier, discount factors are calculated for using cash LIBOR rates and Eurodollar futures prices. There are fifteen LIBOR rates and twenty Eurodollar futures and they cover a maturity range from overnight to five years. In order to capture the most liquidly traded rates and futures, we use LIBOR rates up to the three month rate and Eurodollar futures with maturities of up to two years.

Looking at the provided sample data, we see that we will use six cash instruments and eight Eurodollar futures. The curve date is January 19, 2007 and the last Eurodollar futures (EDZ8) has as its last possible settlement date 3/18/2009. This means there are 790 days between the curve date and the date of the final settlement of the last Eurodollar futures used. Since there are fourteen instruments used for the short end, when we apply the conversion methods we will be able to calculate the discount factors for each of these points, and since the discount factor for the curve date is 1.0, we now have calculated the discount factor for fifteen out of a possible 790 points for the short end of the curve.

Consider the beginning sequence of LIBOR rates for 01/19/2007, as shown in Table 7.14. Let us calculate the discount factor for the first LIBOR rate (overnight). It matures on 01/20/2007 and pays a simple interest rate of 5.28875% from the time interval 01/19/2007 to 01/20/2007. From the provided equation the discount factor at 01/20/2007 or $P(t, t+1)$ (1 day from the curve date) is calculated as

$$\begin{aligned} P(t, t+1) &= \frac{1}{1 + (5.28875/100) \times 1/360} \\ &= 0.999853111857074 \end{aligned}$$

In the same way we can calculate the discount factor for the second LIBOR rate in Table 7.14. The second rate matures on 01/26/2007 and pays a simple interest rate of 5.30375% from the dates 01/19/2007 to 01/26/2007. On an ACT/360 basis there are 7 days between 01/19/2007 and 1/26/2007. So $P(t, t+7)$ (7 days from the curve date) is calculated as

$$\begin{aligned} P(t, t+7) &= \frac{1}{1 + (5.30375/100) \times 7/360} \\ &= 0.998969777730265 \end{aligned}$$

We also calculate the discount factor for the fifth LIBOR rate. The fifth rate matures on 03/20/2007 and pays a simple interest rate of 5.34438% from the dates 01/19/2007 to

03/20/2007. On an ACT/360 basis there are 60 days between 01/19/2007 and 3/20/2007. So $P(t, t+60)$ (60 days from the curve date) is calculated as

$$\begin{aligned} P(t,t+60) &= \frac{1}{1+(5.34438/100)\times 60/360} \\ &= 0.991171339527427 \end{aligned}$$

Table 7.17 illustrates calculated discount factors using LIBOR rates.

**TABLE 7.17**: Discount factors for the first three months of the LIBOR yield curve

| Discount factors | | |
|---|---|---|
| 01/19/07 | $P(t,t)$ | 1.00000000000 |
| 01/20/07 | $P(t,t+1)$ | 0.99985311185 |
| 01/26/07 | $P(t,t+7)$ | 0.99896977773 |
| 02/02/07 | $P(t,t+14)$ | 0.99793877132 |
| 02/18/07 | $P(t,t+30)$ | 0.99558623436 |
| 03/20/07 | $P(t,t+60)$ | 0.99117133953 |
| 04/19/07 | $P(t,t+90)$ | 0.98677718571 |

Now we are going to calculate the discount factor for the first Eurodollar futures contract. The first Eurodollar futures listed (EDH7) has a first possible settlement date of 03/21/2007 and a last possible settlement date of 06/20/2007, which corresponds to the first possible settlement date of the next contract. To calculate the discount factor at 06/20/2007 we can use the expression in the preceding section for computing the discount factor given a forward rate $F(t, T, S)$ and a discount factor at $S$, $P(t, S)$.

With our data: $T = 03/21/2007$, $S = 06/20/2007$, and $P(t, T) = 0.991024303826553$ computed using a not-a-knot cubic spline interpolation.[8] Thus we have $S - T = 91$. First we calculate $F(t, T, S)$. From the data

$$\begin{aligned} F(t,T,S) &= 100.0 - 94.64 \\ &= 5.360 \end{aligned}$$

Using the provided expression to derive a discount factor from a forward rate on an ACT/360 basis, we get

$$\begin{aligned} P(t,S) &= \frac{P(t,T)}{1+F(t,T,S)/100)\times(S-T)/360} \\ &= \frac{0.991024303826553}{1+(5.360/100)\times 91/360} \\ &= 0.977776518420312 \end{aligned}$$

Table 7.18 shows the results of applying the same technique to the rest of the futures contracts combined with earlier results from LIBOR rates. We have now calculated the discount factors for the first two years at fifteen distinct points. To get discount factors for remaining points on the short end of the curve we employ some interpolation or smoothing methodology to calculate the discount factors in the gaps, for those days where there is no instrument from which to derive a discount factor. This will be discussed in more detail in Section 7.7.3.

[8] We will discuss different types of splines in Section 7.7.3.

**TABLE 7.18**: Discount factors for the first two years of the LIBOR yield curve

| Discount factors | | |
|---|---|---|
| 01/19/07 | $P(t,t)$ | 1.00000000000 |
| 01/20/07 | $P(t,t+1)$ | 0.99985311185 |
| 01/26/07 | $P(t,t+7)$ | 0.99896977773 |
| 02/02/07 | $P(t,t+14)$ | 0.99793877132 |
| 02/18/07 | $P(t,t+30)$ | 0.99558623436 |
| 03/20/07 | $P(t,t+60)$ | 0.99117133953 |
| 04/19/07 | $P(t,t+90)$ | 0.98677718571 |
| 06/20/07 | $P(t,t+152)$ | 0.97777651842 |
| 09/19/07 | $P(t,t+243)$ | 0.96479004245 |
| 12/19/07 | $P(t,t+334)$ | 0.95221354974 |
| 03/19/08 | $P(t,t+425)$ | 0.94007070863 |
| 06/18/08 | $P(t,t+516)$ | 0.92825645354 |
| 09/17/08 | $P(t,t+607)$ | 0.91668219497 |
| 12/17/08 | $P(t,t+698)$ | 0.90530875084 |
| 03/18/09 | $P(t,t+789)$ | 0.89409873953 |

#### 7.7.2.2 Construction of the Long End of the Curve

To construct the long end of the yield curve we use the swap instruments and the results for the short end of the curve derived in the last section. The first step is to formulate a swap curve that has a swap rate on a semi-annual basis at each payment date for swap instruments. However, swap rates are not quoted at this level of granularity, so we will have to use smoothing techniques on the market swap data to derive implied swap rates for each of these dates. Once we have a swap rate for each cashflow date for the swap instruments we can then solve for the discount factors at each of these payment dates. After these discount factors are solved for we can apply a smoothing method to get the discount factors for dates which are not swap payment dates. Note that this cooking process allows swap cashflows to occur on non-business days; however, a more precise construction will only allow swap payments to be made on business days. At this step, we skip a single rate bootstrap; by adding it this would be more clear.

Since swaps generally settle in two business days, the swap will start on 01/21/2007 and have its first coupon payment on 07/21/2007. Swap rates on or before 1/21/2009 were solved using discount factors constructed for the short end of the curve. The results of this step are listed in Table 7.19. Using interpolated swap rates in Table 7.19 and provided swap

**TABLE 7.19**: Swap curve from Eurodollar futures

| date | swap yield |
|---|---|
| 7/21/2007 | 5.4830 |
| 1/21/2008 | 5.4382 |
| 7/21/2008 | 5.3401 |

rates in Table 7.16, a more complete swap curve is derived by using a not-a-knot spline interpolator to get a swap yield on a semi-annual basis from 01/19/2007 to 01/21/2037. This will result in a curve that has sixty points. The results of this step are shown in Table 7.20. The swap curve formed by this process is then used to successively solve for

**TABLE 7.20**: Interpolated swap curve

| date | swap yield | date | swap yield | date | swap yield |
|---|---|---|---|---|---|
| 7/21/2007 | 5.4830 | 7/21/2017 | 5.2734 | 7/21/2027 | 5.4320 |
| 1/21/2008 | 5.4382 | 1/21/2018 | 5.2830 | 1/21/2028 | 5.4362 |
| 7/21/2008 | 5.3401 | 7/21/2018 | 5.2924 | 7/21/2028 | 5.4397 |
| 1/21/2009 | 5.2671 | 1/21/2019 | 5.3020 | 1/21/2029 | 5.4427 |
| 7/21/2009 | 5.2238 | 7/21/2019 | 5.3114 | 7/21/2029 | 5.4451 |
| 1/21/2010 | 5.1993 | 1/21/2020 | 5.3208 | 1/21/2030 | 5.4468 |
| 7/21/2010 | 5.1876 | 7/21/2020 | 5.3300 | 7/21/2030 | 5.4478 |
| 1/21/2011 | 5.1830 | 1/21/2021 | 5.3392 | 1/21/2031 | 5.4481 |
| 7/21/2011 | 5.1819 | 7/21/2021 | 5.3481 | 7/21/2031 | 5.4477 |
| 1/21/2012 | 5.1834 | 1/21/2022 | 5.3569 | 1/21/2032 | 5.4465 |
| 7/21/2012 | 5.1874 | 7/21/2022 | 5.3654 | 7/21/2032 | 5.4445 |
| 1/21/2013 | 5.1935 | 1/21/2023 | 5.3737 | 1/21/2033 | 5.4416 |
| 7/21/2013 | 5.2010 | 7/21/2023 | 5.3816 | 7/21/2033 | 5.4380 |
| 1/21/2014 | 5.2094 | 1/21/2024 | 5.3894 | 1/21/2034 | 5.4333 |
| 7/21/2014 | 5.2180 | 7/21/2024 | 5.3967 | 7/21/2034 | 5.4279 |
| 1/21/2015 | 5.2269 | 1/21/2025 | 5.4037 | 1/21/2035 | 5.4214 |
| 7/21/2015 | 5.2359 | 7/21/2025 | 5.4102 | 7/21/2035 | 5.4141 |
| 1/21/2016 | 5.2452 | 1/21/2026 | 5.4164 | 1/21/2036 | 5.4056 |
| 7/21/2016 | 5.2545 | 7/21/2026 | 5.4221 | 7/21/2036 | 5.3962 |
| 1/21/2017 | 5.2640 | 1/21/2027 | 5.4273 | 1/21/2037 | 5.3856 |

the discount factors starting at the first cashflow date beyond the last date derived for the short end of the curve and ending at the maturity date of the longest swap. We leave this step as an exercise at the end of the chapter. This yield curve is still relatively sparse when compared to the thousands of days for which we might want to calculate a yield, and so we must use some interpolation technique to solve for all the remaining discount factors.

### 7.7.3 Polynomial Splines for Constructing Discount Curves

Due to the lack of liquidly traded instruments, spline interpolation plays a crucial role in deriving implied swap rates and implied discount rates for those maturities for which no traded instruments exist.

The very first smoothing technique that usually comes to mind is linear interpolation. In most realistic settings we will not use this approach because it will render very inaccurate results; however, it is useful for illustrative purposes. To give a simple example of smoothing using linear interpolation, we can apply it to the following example. Suppose we had rates for two cash instruments. The first one matures 7 days from today and has a rate of 2% and the other one matures in 28 days and has a rate of 2.05%. Suppose we want to get an estimate of what a cash instrument rate would be that matures in 14 days. Using linear interpolation the answer would be

$$2.0 + \frac{2.05 - 2.0}{28 - 7}(14 - 7) = 2.0175\%$$

The same type of principle is applied to the data we observe on a daily basis, except the mathematics and finance behind it is more complicated. By definition, however, the results of a linear interpolation have a discontinuous first derivative at the knots, and thus we do not get a smooth curve.

The second smoothing technique that comes to mind will be cubic splines. Cubic splines can be used successfully to generate smooth curves, unlike linear interpolation, and ensure a continuous first and second derivative for the spline function across all the knots which make up our known swap rates or discount rates.

### 7.7.3.1 Hermite Spline

A cubic Hermite spline is a third-degree spline with each polynomial of the spline in Hermite form. The Hermite form consists of two control points and two control tangents for each polynomial. For interpolation on a grid with points $x_k$ for $k = 1, \ldots, n$, interpolation is performed on the subinterval $(x_k, x_{k+1})$ at a time (given that tangent values are predetermined). Interpolating $x$ in the interval of $(x_k, x_{k+1})$ can be done with the formula

$$p(t) = p_k g_0(t) + m_k h g_1(t) + p_{k+1} g_2(t) + m_{k+1} h g_3(t)$$

with $h = x_{k+1} - x_k$, $t = \frac{x - x_k}{h}$ and

$$\begin{aligned}
g_0(t) &= 2t^3 - 3t^2 + 1 \\
g_1(t) &= t^3 - 2t^2 + t \\
g_2(t) &= -2t^3 + 3t^2 \\
g_3(t) &= t^3 - t^2
\end{aligned}$$

Here $p_k$ and $m_k$ are the value and tangent at $x_k$ respectively.

A Kochanek-Bartels (KB) spline [107] is a further generalization on how to choose the tangents given the data points $p_{k-1}$, $p_k$, and $p_{k+1}$ with three possible parameters: tension, bias, and a continuity parameter. To interpolate a curve with $n$ cubic Hermite curve segments, for each curve we have a starting point $p_k$ and an ending point $p_{k+1}$ with starting tangent $m_k$ and ending tangent $m_{k+1}$ defined by

$$\begin{aligned}
m_k &= \frac{(1-\tau)(1+b)(1+c)}{2}(p_k - p_{k-1}) + \frac{(1-\tau)(1-b)(1-c)}{2}(p_{k+1} - p_k) \\
m_{k+1} &= \frac{(1-\tau)(1+b)(1-c)}{2}(p_{k+1} - p_k) + \frac{(1-\tau)(1-b)(1+c)}{2}(p_{k+2} - p_{k+1})
\end{aligned}$$

where $\tau$ is the tension, $b$ is the bias, and $c$ is the continuity parameter. When all three parameters are set to zero, the KB spline becomes the Catmull–Rom spline [64], in which tangents are defined as

$$\begin{aligned}
m_k &= \frac{p_{k+1} - p_{k-1}}{2} \\
m_{k+1} &= \frac{p_{k+2} - p_k}{2}
\end{aligned}$$

Changing $\tau$ changes the length of the tangent at the control point, a smaller tangent leading to a tightening and a larger tangent leading to slackening. When $c$ is zero, the curve has a continuous tangent vector at the control point. As $|c|$ increases (up to 1), the resulting curve would have a *corner* at the control point, the direction of which depends on the sign of $c$. Finally, when $b$ equals zero, the left and right one-sided tangents are equally weighted. For $b$ near $-1$, the outgoing tangent dominates the direction of the curve through the control point (undershooting). For $b$ near 1, the incoming tangent dominates (overshooting).

#### 7.7.3.2 Natural Cubic Spline

Consider a set of data points $(t_j, g(t_j))$, $j = 1, \ldots, M$. Here $t_j$ are called knots. Natural cubic splines fit the solution to the following differential equation:

$$\frac{d^4 g}{dt^4} = 0$$

piecewise in the intervals between input points. The knots used in our spline are the discount factors calculated for the instruments and discount factors at swap payment dates. By necessity, a cubic spline interpolant is piecewise linear in its second derivative, i.e.,

$$g''(t) = \frac{t_{j+1} - t}{h_j} g''(t_j) + \frac{t - t_j}{h_j} g''(t_{j+1}), \quad t \in [t_j, t_{j+1}]$$

where $h_j = t_{j+1} - t_j$. The explicit equations for the interpolating cubic spline can classically be recovered by integration of the ordinary differential equation above and subsequently requiring the curve to pass through given data points as well as having continuous first derivatives across knots. A classical boundary condition for uniquely specifying the cubic spline is $g''(t_1) = g''(t_M) = 0$, leading to the natural cubic spline. While popular, cubic splines have known issues. The oscillatory nature of the basis function causes the spline to oscillate, sometimes to a large degree, around the knot points.

#### 7.7.3.3 Tension Spline

One setback of the natural cubic spline is that it has a built-in aversion to make tight turns as they cause large values of $g''$. This often leads to extraneous inflection points and nonlocal behavior, in the sense that perturbation of a single $g(t_j)$ will affect the appearance of the curve for $t$ values far from $t_j$. Another issue of the natural cubic spline is that the monotonicity and convexity of the original data set will typically not be preserved [11].

An attractive alternative is tension spline to improve upon both linear interpolation and cubic splines which allow for a parameterizations which will allow us to control the amount of oscillation in the spline function between knot points. The tension spline applies a tensile force to the end-points of the spline. The ordinary differential equation of the tension spline is

$$\frac{d^4 g}{dt^4} + \sigma^2 \frac{d^2 g}{dt^2} = 0$$

piecewise in the intervals between input points. The knots used in our spline will be the discount factors calculated for the instruments and discount factors at swap payment dates. $\sigma$ is the tension parameter and if $\sigma$ is zero the tension spline will produce a cubic spline, but if $\sigma$ is infinite it will produce linear interpolation. Thus $\sigma$ lets us control the amount of curvature in the resulting spline. The equation is solved with the following conditions: $g$, $\frac{dg}{dt}$, and $\frac{d^2g}{dt^2}$ are continuous at interval boundaries and $\frac{d^2g}{dg^2}$ is set to zero at the end points. The tension spline smooths out ringing close to the knots, especially where the knots are very close together.

$$g''(t) - \sigma^2 g(t) = \frac{t_{j+1} - t}{h_j}(g''(t_j) - \sigma^2 g(t_j)) + \frac{t - t_j}{h_j}(g''(t_{j+1}) - \sigma^2 g(t_{j+1})), \quad t \in [t_j, t_{j+1}]$$

where $\sigma > 0$ is a measure of the tension applied to the cubic spline. Instead of a piecewise linear secondary derivative, we assume quantity $g''(t) - \sigma^2 g(t)$ is linear on each sub-interval $[t_j, t_{j+1}]$. The solution to the ODE is

$$g(t) = Ae^{-\sigma t} + Be^{\sigma t} - \frac{1}{\sigma^2} C$$

where

$$\begin{aligned}
A &= \frac{e^{\sigma t_{i+1}} g''(t_j) - e^{\sigma t_i} g''(t_{j+1})}{\sigma^2 (e^{\sigma h_i} - e^{-\sigma h_i})} \\
B &= \frac{e^{-\sigma t_i} g''(t_{j+1}) - e^{-\sigma t_{i+1}} g''(t_j)}{\sigma^2 (e^{\sigma h_i} - e^{-\sigma h_i})} \\
C &= \frac{t_{j+1} - t}{h_j}(g''(t_j) - \sigma^2 g(t_j)) + \frac{t - t_j}{h_j}(g''(t_{j+1}) - \sigma^2 g(t_{j+1}))
\end{aligned}$$

**Example 20** *Construction of the discount curve using various splines*

In this example, we use both tension spline and KB spline to construct the discount curve. Datasets for this example are LIBOR rates, swap rates, and Eurodollar futures contracts at market close on February 6, 2008 as shown in the tables below.

| LIBOR rates | |
|---|---|
| term (month) | rate (%) |
| 1 | 3.1925 |
| 2 | 3.1500 |
| 3 | 3.1275 |
| 6 | 3.0025 |
| 12 | 2.7338 |

| Eurodollar futures | |
|---|---|
| contract | rate (%) |
| ED1 | 2.880 |
| ED2 | 2.440 |
| ED3 | 2.325 |
| ED4 | 2.370 |
| ED5 | 2.500 |
| ED6 | 2.700 |
| ED7 | 2.900 |
| ED8 | 3.095 |
| ED9 | 3.265 |
| ED10 | 3.445 |
| ED11 | 3.610 |
| ED12 | 3.765 |
| ED13 | 3.900 |

| Swap rates | |
|---|---|
| term (year) | rate (%) |
| 2 | 2.71626 |
| 3 | 2.96088 |
| 4 | 3.21801 |
| 5 | 3.45514 |
| 6 | 3.66813 |
| 7 | 3.85361 |
| 8 | 4.01160 |
| 9 | 4.14208 |
| 10 | 4.25507 |
| 12 | 4.43007 |
| 15 | 4.61523 |
| 20 | 4.76540 |
| 30 | 4.84073 |

In Figure 7.17(a), we display the constructed discount curves applying tension spline using three different values for $\sigma$ which are $0.1, 1, 5$ in dashed line, dotted line, and dash-dot line respectively. In Figures 7.17(b) and 7.17(c) we display the first five and ten years of the curves in Figure 7.17(a) respectively. For comparison purposes, we also plot the curves from calibration of Vasicek and CIR to these instruments which are displayed in solid line and solid line with data points respectively. In Figure 7.17(d), we display the constructed discount curves applying KB spline using three different sets for bias, tension, and continuity. As in the previous case, we also plot the curves from calibration of Vasicek and CIR to these instruments which are displayed in solid line and solid line with data points respectively. In Figures 7.17(e) and 7.17(f) we illustrate the first five and ten years of the curves in Figure 7.17(d) respectively.

In our use of the KB Hermite spline in constructing the LIBOR discount curve, we test the Catmull–Rom case where all three parameters are set to zero as a base case, and compare the different effects when varying the three different parameters, respectively. Since the algorithm requires us to access $p_{k-1}$ and $p_{k+1}$ at the boundary points when $t = 0$ and $t = n$, which are not available, we set these two points to be equal to the most adjacent points as an approximate estimation.

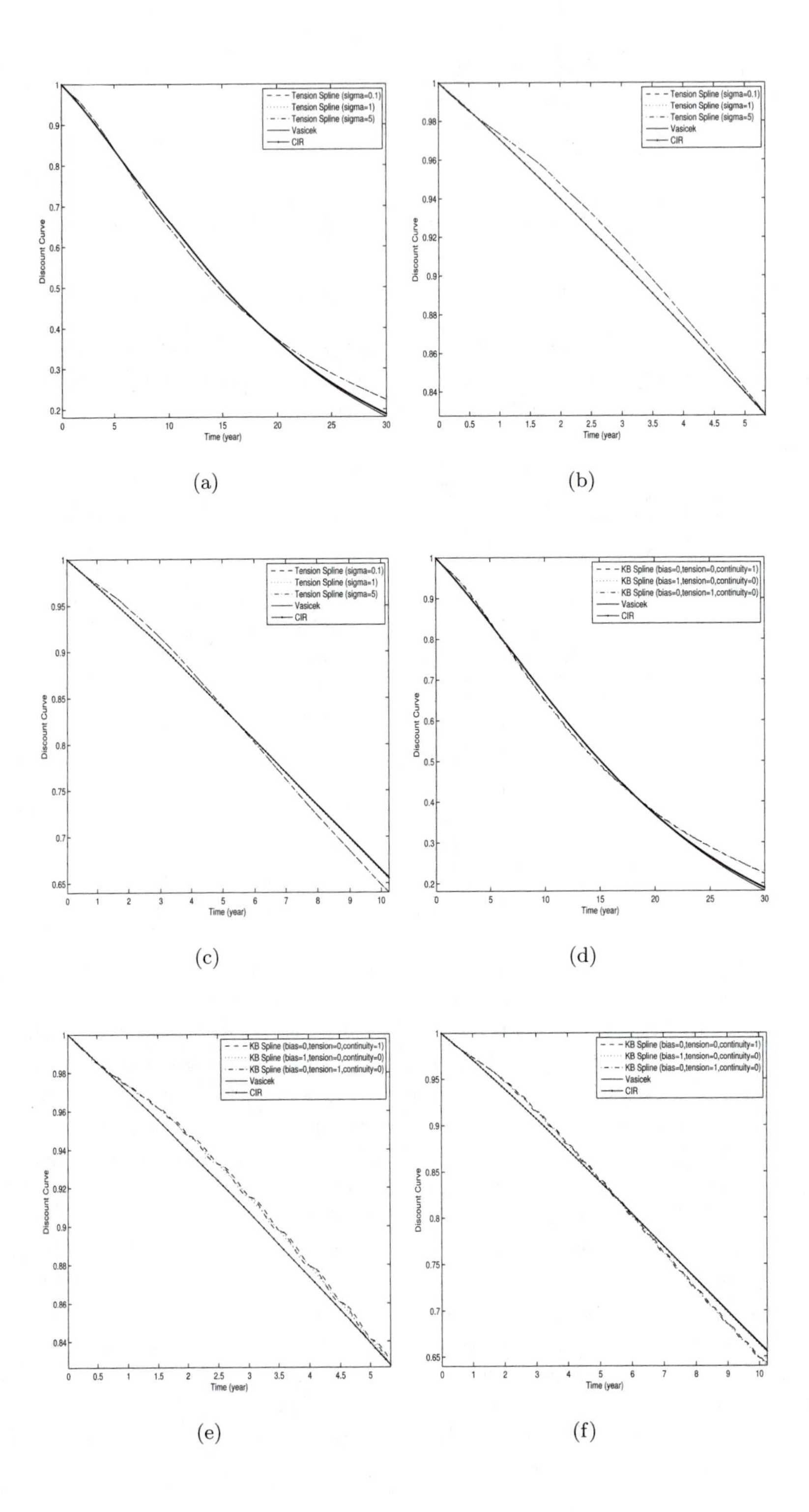


**FIGURE 7.17**: Constructed discount curves

## 7.8 Arbitrage Restrictions on Option Premiums

In this book we have presented a variety of different models and a number of different methods for solving for option premiums under those models. In this chapter we have expanded on this by presenting a number of different methods for calibrating those models to option prices in different markets. However, there are a number of different restrictions on options premia which can be derived from simple no-arbitrage arguments without the need for any modeling whatsoever. These can be used as a sanity check for options pricing and calibration procedures, as a violation of these rules would represent an immediate arbitrage opportunity. Assume strikes indexes are in an increasing order so that $K_{i+1} > K_i$ for all $i$. The rules which can be derived this way are as follows:

- Monotonicity

  Calls

$$C^{\text{ask}}(K_i) > C^{\text{bid}}(K_{i+1}) \quad \forall i$$

  Puts

$$P^{\text{bid}}(K_i) < P^{\text{ask}}(K_{i+1}) \quad \forall i$$

- Slope

  Calls

$$C^{\text{bid}}(K_i) - C^{\text{ask}}(K_{i+1}) < (K_{i+1} - K_i)e^{-r\tau} \quad \forall i$$

  Puts

$$P^{\text{bid}}(K_{i+1}) - P^{\text{ask}}(K_i) < (K_{i+1} - K_i)e^{-r\tau} \quad \forall i$$

- Convexity

  Calls

$$C^{\text{bid}}(K_{i+1}) < \lambda C^{\text{ask}}(K_i) + (1-\lambda)C^{\text{ask}}(K_{i+2}) \quad \forall i$$

  Puts

$$P^{\text{bid}}(K_{i+1}) < \lambda P^{\text{ask}}(K_i) + (1-\lambda)P^{\text{ask}}(K_{i+2}) \quad \forall i$$

where $\lambda = \frac{K_{i+2}-K_{i+1}}{K_{i+1}-K_i}$.

## 7.9 Interest Rate Definitions

In this section we present derivations of some of the rates discussed in this chapter, both simple and continuously compounded instantaneous rates.

A forward rate agreement (FRA) is a tradable contract that can be used to directly trade simple forward rates. The contract involves three time instants: the current time $t$, the expiry time $T$ where $T > t$, and the maturity time $S$ with $S > T$. The payoff of the contract at time $S$ is $1 + (S - T)F(t; T, S)$, which results in a forward investment of one dollar at time $T$. However, we can replicate this investment using the following strategy.

At $t$: sell one T-bond and buy $\frac{P(t,T)}{P(t,S)}$ S-bonds = zero net investment.

At $T$: pay one dollar.

At $S$: obtain $\frac{P(t,T)}{P(t,S)}$ dollars.

The net effect is a forward investment of one dollar at time $T$ yielding $\frac{P(t,T)}{P(t,S)}$ dollars at $S$ with certainty. Thus, by the no-arbitrage condition we are led to the following definitions:

The simple (simply compounded) forward rate for $[T, S]$ prevailing at $t$ is given by

$$\begin{aligned} 1 + (S - T)F(t; T, S) &= \frac{P(t,T)}{P(t,S)} \\ F(t,T,S) &= \frac{1}{S-T}\left(\frac{P(t,S)}{P(t,S)} - 1\right) \qquad (7.23) \end{aligned}$$

The simple spot rate for $[t, T]$ is

$$F(t,T) = F(t;t,T) = \frac{1}{T-t}\left(\frac{1}{P(t,T)} - 1\right) \qquad (7.24)$$

The continuously compounded forward rate for $[T, S]$ prevailing at $t$ is given by

$$e^{R(t,T,S)(S-T)} = \frac{P(t,T)}{P(t,S)} \qquad (7.25)$$

$$R(t,T,S) = -\frac{\log P(t,S) - \log P(t,T)}{S-T} \qquad (7.26)$$

The continuously compounded spot rate for $[t, T]$ is

$$R(t,T) = R(t,t,T) = -\frac{\log P(t,T)}{T-t} \qquad (7.27)$$

The instantaneous forward rate with maturity $T$ prevailing at time $t$ is defined by

$$f(t,T) = \lim_{T \uparrow t} R(t,T,S) = -\frac{\partial \log P(t,T)}{\partial T} \qquad (7.28)$$

The function $T \to f(t,T)$ is called the forward curve at time $t$.

The instantaneous short rate at time $t$ is defined by

$$r(t) = f(t,t) = \lim_{T \uparrow t} R(t,T) \qquad (7.29)$$

Solving Equation (7.28) for the zero-coupon bond price, $P(t, T)$, we obtain

$$P(t,T) = \exp\left(-\int_t^T f(t,u)du\right) \qquad (7.30)$$

where $P(T, T) = 1$.

## Problems

1. Assume we generated option premiums for various strikes and maturities according to the geometric Brownian motion $dS_t = (r-q)S_t dt + \sigma S_t dW_t$ with $\sigma = 40\%$. Calibrate

   (a) Heston stochastic volatility

   $$\begin{aligned} dS_t &= (r-q)S_t dt + \sqrt{v_t} S_t dW_t^1, \\ dv_t &= \kappa(\theta - v_t)dt + \lambda\sqrt{v_t} dW_t^2 \end{aligned}$$

   to these option premiums (used as calibration instruments) to obtain $\Theta = \{\theta, \kappa, \lambda, v_0, \rho\}$. What would you expect to get for the optimal parameter set $\Theta^*$? Justify your answer.

   (b) CGMY to these option premiums (used as calibration instruments) to obtain $\Theta = \{\sigma, \nu, \theta, Y\}$. What would you expect to get for the optimal parameter set $\Theta^*$? Justify your answer.

   (c) VGSA to these option premiums (used as calibration instruments) to obtain $\Theta = \{\sigma, \nu, \theta, \kappa, \eta, \lambda\}$. What would you expect to get for the optimal parameter set $\Theta^*$? Justify your answer.

2. Daily close for LIBOR rates, Eurodollar futures, and swap rates on January 19, 2007 are shown in Tables 7.14, 7.15, and 7.16 respectively (Bloomberg).

   (a) Based on that data, we have cooked the short end of the zero-coupon bond (discount factors) from 1/19/07 to 3/18/09 as displayed in Table 7.18. Having estimated (cooked) short end of the zero-coupon bond curve, estimate the 2-year swap rate and compare it with the true value given in Table 7.16 and assess your estimation.

   (b) Table 7.20 illustrates interpolated swap yields on a semi-annual basis from 01/19/2007 to 01/21/2037. Construct the discount curve from these swap rates using Equations (7.21) and (7.22).

## Case Studies

1. Repeat the calibration procedure that was done in Sections 7.3.1 and 7.3.2 using the same optimizer for the following objective functions:

   - sum of relative errors (SRE)
   - sum of the absolute errors (SAE)
   - sum of the squares of absolute errors (SSAE)

   Compare the results, write down your findings and observations, and conclude on the effect of the objective function on the results.

2. Redo the cooking process done in Example 20 to construct the discount curves.

3. It is well known that calibrating different stochastic process models to the same vanilla option surface yields different exotic option prices [137]. The goal of this case study is to provide an analysis of the effect of this model risk by considering two different models: local volatility and variance gamma with stochastic arrival (VGSA) [54]. To focus our attention we consider only the pricing of up-and-out call (UOC) options.

   **Variance Gamma with Stochastic Arrival (VGSA) Process** To obtain VGSA, as explained in [54], we take the VG process which is a homogeneous Lévy process and build in stochastic volatility by evaluating it at a continuous time change given by the integral of a Cox, Ingersoll, and Ross [82] (CIR) process. The mean reversion of the CIR process introduces the clustering phenomena often referred to as volatility persistence. This enables us to calibrate to market price surfaces across both strike and maturity simultaneously. The process has analytical expressions for its characteristic function, which will allow us to use transform based pricing for a number of different derivatives. Formally we define the CIR process $y(t)$ as the solution to the stochastic differential equation

$$dy_t = \kappa(\eta - y_t)dt + \lambda\sqrt{y_t}dW_t$$

   where $W(t)$ is a Brownian motion, $\eta$ is the long term rate of time change, $\kappa$ is the rate of mean reversion and $\lambda$ is the volatility of the time change. The process $y(t)$ is the instantaneous rate of time change and so the time change is given by $Y(t)$ where

$$Y(t) = \int_0^t y(u)du$$

   The stochastic volatility Lévy process, termed the VGSA process, is defined by

$$Z_{VGSV}(t) = X_{VG}(Y(t); \sigma, \nu, \theta)$$

   Thus $\sigma$, $\nu$, $\theta$, $\kappa$, $\eta$, and $\lambda$ are the six parameters defining the process. We define the stock process at time $t$ by the random variable

$$S(t) = S(0)\frac{e^{(r-q)t+Z(t)}}{\mathbb{E}[e^{Z(t)}]}$$

   With a closed form for the VGSA characteristic function for the log price, one can employ various techniques to price European options ([60], [66], and [111]). The resulting model may be used to estimate parameter values consistent with market option prices for vanilla options across the entire strike and maturity spectrum.

   **Local Volatility Model.** Consider the stock price process as a solution to the stochastic differential equation

$$dS_t = (r - q)S_t dt + \sigma(S_t, t)dW(t)$$

   where the function $\sigma(S, t)$ is termed the asset's local volatility function. Let $C(K, T)$ be the price of a European call option with strike $K$ and maturity $T$ under this process. It is shown in [136] that one can extend the Dupire [104] methodology to compute the local volatility function, now written as $\sigma(K, T)$, from the option prices using the following equation:

$$\sigma^2(K,T) = 2\frac{\partial C/\partial T + q(T)C + K(r(T) - q(T))\partial C/\partial K}{K^2\partial^2 C/\partial K^2}$$

   Hence the local volatility approach to pricing exotic options is to first infer local volatility functions from market option prices using calendar spread approximations for the first partial with respect to maturity and butterfly spread approximations for the second partial with respect to strike, while call spreads approximate the first partial with respect to strike. The next step is to price the exotic given the market calibrated local volatility function $\sigma(S,t)$ by either employing a finite difference solution to the underlying partial differential equation in the price of the exotic or by simulating the process. If we assume that we have a calibrated local volatility surface, we can price up-and-out call options by solving the following backward PDE:

$$\frac{\partial U_o^c}{\partial t} + \frac{\sigma(S_t,t)^2 S^2}{2}\frac{\partial^2 U_o^c}{\partial S^2} + (r(t) - q(t))S\frac{\partial U_o^c}{\partial S} = r(t)U_o^c(S,t)$$

with terminal condition

$$U_o^c(S,T) = (S-K)^+ \quad \text{for} \quad S \in [0,H)$$

and boundary conditions

$$\begin{aligned} \lim_{S\downarrow 0} U_o^c(S,t) &= U_{o\,SS}^c(S,t) = 0 \\ \lim_{S\uparrow H} U_o^c(S,t) &= 0 \end{aligned}$$

Note that in presence of rebate, the terminal condition would be

$$\begin{aligned} U_o^c(S,T) &= (S-K)^+ \quad \text{for} \quad S \in [0,H) \\ U_o^c(H,T) &= \text{rebate} \end{aligned}$$

and boundary conditions

$$\begin{aligned} \lim_{S\downarrow 0} U_o^c(S,t) &= U_{o\,SS}^c(S,t) = 0 \\ \lim_{S\uparrow H} U_o^c(S,t) &= \text{rebate} \times e^{-(T-t)} \end{aligned}$$

Equivalently we can also price up-and-out call options by solving the following forward PDE (for more details, look at [56] and [57]):

$$\begin{aligned} \frac{\sigma^2(K,T)}{2}K^2\frac{\partial^2 U_o^c}{\partial K^2} - [r(T) - q(T)]\,K\frac{\partial U_o^c}{\partial K} - q(T)U_o^c = \\ \frac{\partial U_o^c}{\partial T} + \left[\frac{\sigma^2(H,T)}{2}H^2\frac{\partial^3 U_o^c}{\partial K^3}(H,T)\right](K-H) \end{aligned}$$

with initial condition

$$U_o^c(K,0) = (S_0 - K)^+, \ \text{for} \ K \in [0,H), \ \text{and} \ S_0 < H.$$

Boundary conditions are

$$\begin{aligned} \lim_{K\downarrow 0} U_{o\,KK}^c(K,T) = 0, \ T \in [0,\bar{T}] \\ \lim_{K\uparrow H} U_{o\,KK}^c(K,T) = 0, \ T \in [0,\bar{T}] \end{aligned}$$

**Local Volatility with VGSA Calibration**. The general difficulty in implementing the process of determining a local volatility surface is that prices available in the market for a finite grid of strikes, maturities, and interpolation schemes must be invoked to infer prices for the intermediate strikes and maturities. The interpolations used may or may not be consistent with the requirements of the absence of at least static arbitrage across the strike-maturity spectrum. Even when this is accomplished, the interpolation schemes can introduce non-differentiability at various levels, leading to local volatility functions that are erratic and inspire little confidence. The task of properly interpolating the surface of option prices consistent with observed market prices is essentially the task of formulating and estimating a market consistent option pricing model. The VGSA model presented previously is one such model and delivers on such an objective. Thus our process for pricing exotic options will be to first infer prices from the calibrated VGSA model for intermediate strikes and maturities, and then derive implied local volatilities under the local volatility model using these prices. We will then compare the prices of exotic options using the original VGSA dynamics with those of the inferred local volatility model.

To complete our case study, we will do the following:

Sanity check the option quotes provided in Tables 7.21 and 7.22 by checking all the model free arbitrage restrictions discussed in Section 7.8.

Make sure you can match European prices for VGSA via simulation with those of fractional fast Fourier transform.

Obtain VGSA parameters via calibration to S&P 500 out-of-the-money options (Tables 7.21 and 7.22) using the following parameters as the initial guess for the VGSA parameter set: $\sigma = 0.20$, $\nu = 0.1$, $\theta = -0.4$, $\kappa = 2.0$, $\eta = 4.5$, $\lambda = 1.0$.

Using the calibrated VGSA model, generate a call option premium surface with sufficient strike and maturity granularity that the calendar spread, butterfly, and call spreads are sufficiently accurate.

Use this call option premium surface to construct a local volatility surface using Equation (7.3).

Use this call option premium surface to construct an implied volatility surface.

Price an up-and-out call with a strike of \$1425, an up-barrier of \$1500, a maturity of nine months, and a rebate of \$10 paid at maturity using the calibrated VGSA model via simulation (S&P 500 spot price on March 27, 2012 was \$1412.52).

Price an up-and-out call with a strike of \$1425, an up-barrier of \$1500, a maturity of nine months, and a rebate of \$10 paid at maturity using the calibrated local volatility model by solving the forward/backward PDE via finite difference.

Compare the price of each model with the exact replicated price and conclude.

Assume being short an up-and-out call option, consider various scenarios and hedging strategies to see which model does better, do not re-calibrate.

4. For this case study, use the data provided in Tables 7.21 and 7.22.

Obtain Heston stochastic parameters via calibration to S&P 500 out-of-the-money options.

Using the calibrated Heston model, generate a call option premium surface with sufficient strike and maturity granularity that the calendar spread, butterfly, and call spreads are sufficiently accurate.

Use this call option premium surface to construct a local volatility surface using Equation (7.3). Compare this surface with the one obtained in Case Study 3.

Use this call option premium surface to construct an implied volatility surface. Compare this surface with the one obtained in Case Study 3.

Draw your conclusion on comparisons.

**TABLE 7.21**: S&P 500 call option premiums on March 27, 2012

| maturity (in days) | strike | bid | ask | r | q | forward price |
|---|---|---|---|---|---|---|
| 2 | 1405 | 11.4 | 12.3 | 0.2781066 | 0.5471396 | 1411.849976 |
| | 1410 | 7.6 | 9 | 0.2781066 | 0.5471396 | 1411.75 |
| | 1415 | 5 | 6 | 0.2781066 | 0.5471396 | 1411.449951 |
| | 1420 | 3.2 | 4.1 | 0.2781066 | 0.5471396 | 1411.699951 |
| | 1425 | 2.2 | 2.4 | 0.2781066 | 0.5471396 | 1411.699951 |
| 24 | 1405 | 21.7 | 23 | 0.232443 | 1.531465 | 1410.550903 |
| | 1410 | 19.1 | 20 | 0.232443 | 1.531465 | 1410.800171 |
| | 1415 | 15.9 | 17.1 | 0.232443 | 1.531465 | 1410.549316 |
| | 1420 | 13.6 | 14.6 | 0.232443 | 1.531465 | 1410.698975 |
| | 1425 | 11 | 12.2 | 0.232443 | 1.531465 | 1410.547729 |
| 52 | 1405 | 30.8 | 32.6 | 0.3240174 | 2.108761 | 1407.851318 |
| | 1410 | 28.2 | 29.6 | 0.3240174 | 2.108761 | 1407.999023 |
| | 1415 | 25.5 | 26.8 | 0.3240174 | 2.108761 | 1407.996704 |
| | 1420 | 22.8 | 24.1 | 0.3240174 | 2.108761 | 1407.944336 |
| | 1425 | 20.6 | 21.6 | 0.3240174 | 2.108761 | 1408.092163 |
| 80 | 1405 | 39.5 | 41.4 | 0.4331506 | 2.099028 | 1406.30127 |
| | 1410 | 36.8 | 38.3 | 0.4331506 | 2.099028 | 1406.496582 |
| | 1415 | 33.8 | 35.6 | 0.4331506 | 2.099028 | 1406.291626 |
| | 1420 | 31.3 | 32.7 | 0.4331506 | 2.099028 | 1406.537109 |
| | 1425 | 28.4 | 30.1 | 0.4331506 | 2.099028 | 1406.382202 |
| 93 | 1375 | 61.8 | 66.3 | 0.4826605 | 2.081997 | 1405.738037 |
| | 1400 | 47 | 48.4 | 0.4826605 | 2.081997 | 1405.807251 |
| | 1425 | 32.8 | 34.1 | 0.4826605 | 2.081997 | 1405.82605 |
| | 1450 | 21.2 | 22.5 | 0.4826605 | 2.081997 | 1405.043945 |
| | 1475 | 12.6 | 13.9 | 0.4826605 | 2.081997 | 1405.66394 |
| 115 | 1375 | 67.4 | 69.7 | 0.5572296 | 1.981643 | 1404.051392 |
| | 1400 | 51.4 | 53.6 | 0.5572296 | 1.981643 | 1404.407959 |
| | 1425 | 37.3 | 39.2 | 0.5572296 | 1.981643 | 1404.213379 |
| | 1450 | 25.5 | 27.2 | 0.5572296 | 1.981643 | 1404.269043 |
| | 1475 | 16.2 | 17.7 | 0.5572296 | 1.981643 | 1404.124756 |
| 178 | 1375 | 80.7 | 83.2 | 0.7316633 | 2.089964 | 1400.039673 |
| | 1400 | 65.2 | 67.3 | 0.7316633 | 2.089964 | 1400.45166 |
| | 1425 | 50.7 | 53.1 | 0.7316633 | 2.089964 | 1400.311279 |
| | 1450 | 39 | 40.5 | 0.7316633 | 2.089964 | 1400.22168 |
| | 1475 | 28 | 29.7 | 0.7316633 | 2.089964 | 1400.332397 |
| 184 | 1375 | 82.2 | 84.5 | 0.7436823 | 2.098077 | 1399.140991 |
| | 1400 | 66.6 | 68.8 | 0.7436823 | 2.098077 | 1399.146729 |
| | 1425 | 52.5 | 54.5 | 0.7436823 | 2.098077 | 1399.001953 |
| | 1450 | 40.1 | 41.9 | 0.7436823 | 2.098077 | 1399.859985 |
| | 1475 | 29.5 | 31 | 0.7436823 | 2.098077 | 1399.714966 |
| 269 | 1375 | 95.5 | 98.5 | 0.8831472 | 2.154654 | 1393.822876 |
| | 1400 | 80.4 | 83.4 | 0.8831472 | 2.154654 | 1393.306274 |
| | 1425 | 66.5 | 69.1 | 0.8831472 | 2.154654 | 1393.594482 |
| | 1450 | 53.7 | 56.4 | 0.8831472 | 2.154654 | 1393.631104 |
| | 1475 | 42.4 | 44.8 | 0.8831472 | 2.154654 | 1393.516724 |

**TABLE 7.22**: S&P 500 put option premiums on March 27, 2012

| maturity (in days) | strike | bid | ask | r | q | forward price |
|---|---|---|---|---|---|---|
| 2 | 1405 | 4.4 | 5.6 | 0.2781066 | 0.5471396 | 1411.850098 |
| | 1410 | 6.1 | 7 | 0.2781066 | 0.5471396 | 1411.75 |
| | 1415 | 8.3 | 9.8 | 0.2781066 | 0.5471396 | 1411.449951 |
| | 1420 | 11.1 | 12.8 | 0.2781066 | 0.5471396 | 1411.699951 |
| | 1425 | 14.5 | 16.7 | 0.2781066 | 0.5471396 | 1411.699951 |
| 24 | 1405 | 16.2 | 17.4 | 0.232443 | 1.531465 | 1410.550903 |
| | 1410 | 18.1 | 19.4 | 0.232443 | 1.531465 | 1410.800171 |
| | 1415 | 20.3 | 21.6 | 0.232443 | 1.531465 | 1410.549316 |
| | 1420 | 22.7 | 24.1 | 0.232443 | 1.531465 | 1410.698486 |
| | 1425 | 25.3 | 26.8 | 0.232443 | 1.531465 | 1410.547729 |
| 52 | 1405 | 27.9 | 29.8 | 0.3240174 | 2.108761 | 1407.851318 |
| | 1410 | 29.9 | 31.9 | 0.3240174 | 2.108761 | 1407.999023 |
| | 1415 | 32.1 | 34.2 | 0.3240174 | 2.108761 | 1407.996704 |
| | 1420 | 34.4 | 36.6 | 0.3240174 | 2.108761 | 1407.944336 |
| | 1425 | 36.9 | 39.1 | 0.3240174 | 2.108761 | 1408.092163 |
| 80 | 1405 | 38.1 | 40.2 | 0.4331506 | 2.099028 | 1406.30127 |
| | 1410 | 40.2 | 41.9 | 0.4331506 | 2.099028 | 1406.496582 |
| | 1415 | 42.3 | 44.5 | 0.4331506 | 2.099028 | 1406.291626 |
| | 1420 | 44.6 | 46.3 | 0.4331506 | 2.099028 | 1406.537109 |
| | 1425 | 47 | 48.7 | 0.4331506 | 2.099028 | 1406.382202 |
| 93 | 1375 | 32.6 | 34.1 | 0.4826605 | 2.081997 | 1405.738037 |
| | 1400 | 41 | 42.8 | 0.4826605 | 2.081997 | 1405.807251 |
| | 1425 | 51.6 | 53.6 | 0.4826605 | 2.081997 | 1405.82605 |
| | 1450 | 64.8 | 68.7 | 0.4826605 | 2.081997 | 1405.043945 |
| | 1475 | 80.1 | 84.9 | 0.4826605 | 2.081997 | 1405.66394 |
| 115 | 1375 | 38.5 | 40.6 | 0.5572296 | 1.981643 | 1404.052002 |
| | 1400 | 47.2 | 49 | 0.5572296 | 1.981643 | 1404.407837 |
| | 1425 | 57.8 | 60.2 | 0.5572296 | 1.981643 | 1404.213013 |
| | 1450 | 70.6 | 73.4 | 0.5572296 | 1.981643 | 1404.269165 |
| | 1475 | 86.2 | 89.2 | 0.5572296 | 1.981643 | 1404.124023 |
| 178 | 1375 | 55.7 | 58.3 | 0.7316633 | 2.089964 | 1400.039673 |
| | 1400 | 64.8 | 66.8 | 0.7316633 | 2.089964 | 1400.45166 |
| | 1425 | 75.4 | 77.6 | 0.7316633 | 2.089964 | 1400.311279 |
| | 1450 | 87.7 | 91 | 0.7316633 | 2.089964 | 1400.22168 |
| | 1475 | 102 | 104.5 | 0.7316633 | 2.089964 | 1400.332397 |
| 184 | 1375 | 57.9 | 60.7 | 0.7436823 | 2.098077 | 1399.140991 |
| | 1400 | 67 | 70.1 | 0.7436823 | 2.098077 | 1399.146973 |
| | 1425 | 77.7 | 81.1 | 0.7436823 | 2.098077 | 1399.001953 |
| | 1450 | 88.5 | 93.4 | 0.7436823 | 2.098077 | 1399.859985 |
| | 1475 | 102.3 | 108.2 | 0.7436823 | 2.098077 | 1399.714966 |
| 269 | 1375 | 77.1 | 79.5 | 0.8831472 | 2.154654 | 1393.822876 |
| | 1400 | 86.8 | 90.3 | 0.8831472 | 2.154654 | 1393.306274 |
| | 1425 | 97.6 | 100.4 | 0.8831472 | 2.154654 | 1393.594482 |
| | 1450 | 109.5 | 112.6 | 0.8831472 | 2.154654 | 1393.631104 |
| | 1475 | 122.9 | 126.2 | 0.8831472 | 2.154654 | 1393.516724 |

# Chapter 8

## Filtering and Parameter Estimation

In the calibration procedure, we mostly just utilize cross-section instruments (i.e., calibration instruments) and do not bring in any time series of data into the process. In parameter estimation, however, we typically bring in a long history of prices in order to estimate the model's parameters.

In many applications, we need to generate sample paths under real world statistical measures. This is achieved by employing some filtering technique or maximum likelihood. Employing a filtering technique or maximum likelihood has many desirable properties. Maximum likelihood is consistent and guaranteed to converge as the length of the time series is increased [199]. In the case of a single noise, such as generalized autoregressive conditional heteroskedasticity (GARCH) [109],[33],[181], the likelihood function is known is an integrated form. However, for processes that the likelihood function is not known in an integrated form we will need some filtering technique for the purpose of parameter estimation and/or parameter learning.

A process that the probability density function of the process is available in an integrated form is called a fully observed process. The estimation procedure for these processes is done via maximum likelihood estimation [139]. Partially observed processes, on the other hand, are processes where the density of the process is not available in an integrated form. In partially observed processes, by conditioning on a parameter(s), conditional likelihood/density can be obtained in an integrated form. That parameter, that we condition on, is called hidden state of the process. For partially observed processes, at each day in history, the aim is to calculate the hidden state on that day to best fit that day's observation. This procedure is called *filtering*.

Assuming the model parameter set is given/known, by means of filtering we can find the time series of the hidden state that gives the best fit over time. During filtering, the parameter set is kept fixed over the entire time series of data. One can repeat this procedure by using a different parameter set for each run and find the best fit for the corresponding parameter set. The parameter set that maximizes the likelihood function or yields the smallest mean square root error or the like is the optimal parameter set. This procedure of finding the optimal parameter is called parameter estimation from historical time series of data [139].

We start this chapter with a couple of examples on fully and partially observed processes. We then cover various filtering techniques with depth analysis on their implementations with some examples on filtering and parameter estimation.

**Example 21** *An example of a fully observed process*

As previously explained, the variance gamma (VG) process $X(t;\sigma,\nu,\theta)$ is obtained by evaluating Brownian motion with drift $\theta$ and volatility $\sigma$ at a random time given by a gamma process $\gamma(t;1,\nu)$ with mean rate unity and variance rate $\nu$ as

$$X(t;\sigma,\nu,\theta) = \theta\gamma(t;1,\nu) + \sigma W(\gamma(t;1,\nu))$$

Suppose the stock price process is given by the geometric VG law with parameters $\sigma$, $\nu$, $\theta$

and the log price at time $t$ is given by

$$\ln S_t = \ln S_0 + (r - q + \omega)t + X(t; \sigma, \nu, \theta)$$

where

$$\omega = \frac{1}{\nu} \ln(1 - \theta\nu - \sigma^2\nu/2)$$

is the usual Jensen's inequality correction ensuring that the mean rate of return on the asset is risk neutral $(r - q)$. For variance gamma model, calling

$$x_h = z_k - (r - q)h - \frac{h}{\nu} \ln(1 - \theta\nu - \sigma^2\nu/2)$$

where

$$\begin{aligned} z_k &= \ln(S_k/S_{k-1}) \\ h &= t_k - t_{k-1} \end{aligned}$$

provides the following integrated density (likelihood function) of stock return:

$$p(z_k|z_{1:k-1}) = \frac{2e^{\theta x_h/\sigma^2}}{\nu^{\frac{h}{\nu}}\sqrt{2\pi}\sigma\Gamma(\frac{h}{\nu})} \left(\frac{x_h^2}{2\sigma^2/\nu + \theta^2}\right)^{\frac{h}{2\nu}-\frac{1}{4}} K_{\frac{h}{\nu}-\frac{1}{2}}\left(\frac{1}{\sigma^2}\sqrt{x_h^2(2\sigma^2/\nu + \theta^2)}\right)$$

where $K_n(x)$ is the modified Bessel function of the second kind (see [175] for more details). We see that the dependence on the gamma distribution is integrated out. Hence, there would not be any need for filtering.

To estimate the parameter set for VG via maximum likelihood, we simulate the stock price process assuming it follows VG geometric law. In our example, we assume the following parameter set: $S0 = 100$, $\sigma = 0.25$, $\nu = 0.15$, $\theta = -0.15$, $\mu = 0.01$, and $\Delta t = 1/12$, $T = 40$. Figure 8.1 displays the simulated path used in our estimation. Using starting

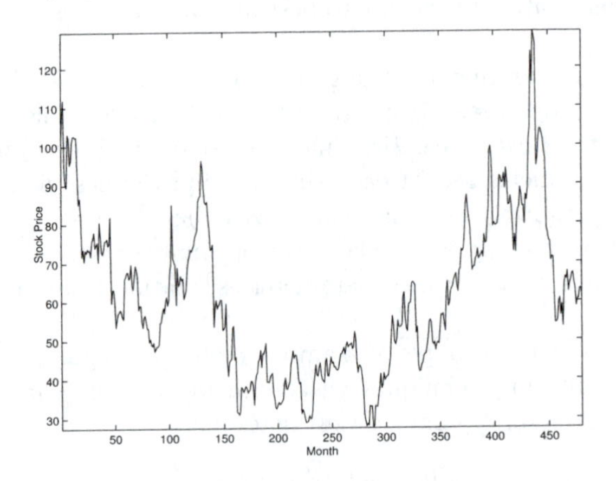


**FIGURE 8.1**: VG simulated path used for parameter estimation of the VG model via MLE

values $\Theta = \{\sigma_0, \nu_0, \theta_0, \mu_0\} = \{0.4, 0.05, 0.1, 0.1\}$, maximizing likelihood via the Nelder–Mead simplex method, we obtain the following parameter set:

$$\Theta = \{\hat{\sigma}, \hat{\nu}, \hat{\theta}, \hat{\mu}\} = \{0.2401, 0.1603, -0.1073, 0.0497\}$$

which is pretty close to the original parameter set used for simulating the path. It is important to mention that as $\Delta t$ approaches zero it becomes pretty difficult to estimate the parameters due to the instability of the Bessel function. For the Bessel function $K_{\frac{h}{\nu}-\frac{1}{2}}(x)$ to be stable, we recommend choosing $h$ such that $\frac{h}{\nu} - \frac{1}{2} > 0$. In our example, we choose a stable $h = \Delta t = 1/12 = 0.083$, since $h = 0.083 > \frac{\nu}{2} = \frac{0.15}{2} = 0.075$.

**Example 22** *An example of a partially observed process*

For VSGA, the stock process under the risk-neutral framework follows

$$d \ln S_t = (r - q + \omega)dt + X(h(dt); \sigma, \nu, \theta)$$

and

$$X(h(dt); \sigma, \nu, \theta) = B(\gamma(h(dt), 1, \nu); \theta, \sigma)$$

and the gamma cumulative distribution function

$$F_\nu(h, x) = \frac{1}{\Gamma(\frac{h}{\nu})\nu^{\frac{h}{\nu}}} \int_0^x e^{-\frac{t}{\nu}} t^{\frac{h}{\nu}-1} dt$$

and $h(dt) = y_t dt$ with

$$dy_t = \kappa(\eta - y_t)dt + \lambda\sqrt{y_t} dW_t$$

For this process the likelihood function does not exist in an integrated form; however, by conditioning on arrival rate we can find the conditional likelihood function. For a given arrival rate $dt^* = y_t dt$ we have a VG distribution and the corresponding integrated density from Equation (8.1) would be

$$p(z_k|h^*) = \frac{2e^{\theta x_h/\sigma^2}}{\nu^{\frac{h^*}{\nu}} \sqrt{2\pi}\sigma\Gamma(\frac{h^*}{\nu})} \left(\frac{x_h^2}{2\sigma^2/\nu + \theta^2}\right)^{\frac{h^*}{2\nu}-\frac{1}{4}} K_{\frac{h^*}{\nu}-\frac{1}{2}}\left(\frac{1}{\sigma^2}\sqrt{x_h^2(2\sigma^2/\nu + \theta^2)}\right)$$

where $h^* = y_t h$ and $x_h$ as in the previous example. Hence the arrival rate is the hidden state in filtering. Later in this chapter, we provide an example on estimation of parameters of the VGSA model.

---

## 8.1 Filtering

Before any formal definition or rigorous mathematical derivation, we start with a simple example to give an intuition behind filtering. Assume the hidden state evolves according to the following simple linear model:

$$x_{t+1} = a x_t + w_{t+1}$$

where $w_{t+1} \sim \mathcal{N}(0, \lambda^2)$ for some $\lambda$. Also assume the parameter set, $\Theta$, is known. Moreover, we assume that factors at time $t$, namely, $x_t$, are given. Now given an observation at time $t+1$ we want to calculate the best estimate of $x_{t+1}$, namely, $\hat{x}_{t+1}$, that is,

$$\hat{x}_{t+1} = \mathbb{E}\left(x_{t+1} \mid z_{t+1}\right)$$

where $z_{t+1}$ is the observation at time $t+1$. Assume the model price is given by $h(x_{t+1};\Theta)$ where as earlier stated $\Theta$ is the parameter set. The assumption in filtering is that the market price (observation) at time $t+1$, $z_{t+1}$, is linked to the model price via the following relationship:

$$z_{t+1} = h(x_{t+1};\Theta) + u_{t+1}$$

where $u_{t+1} \sim \mathcal{N}(0,\sigma^2)$ for some $\sigma$. Both $\lambda$ and $\sigma$ are part of the parameter set $\Theta$ and therefore are already estimated and known.

Knowing the evolution of $x_{t+1}$ we first generate $M$ samples for $x_{t+1}$.

$$x_{t+1}^{(i)} = a x_t + \mathcal{N}(0,\lambda^2)$$

for $i = 1,\ldots,M$. Having $M$ samples of $x_{t+1}^{(i)}$ we can calculate $M$ samples for the model price, namely $h(x_{t+1}^{(i)};\Theta)$. Now we can generate $M$ samples for $u_{t+1}$, having observed the market price at time $t+1$:

$$u_{t+1}^{(i)} = y_{t+1} - h(x_{t+1}^{(i)};\Theta)$$

Define $\mathcal{L}^{(i)}$ as the (conditional) likelihood function

$$\mathcal{L}^{(i)} \equiv \text{Likelihood}\left(u_{t+1}^{(i)} \mid x_{t+1}^{(i)}\right)$$

Hence $\mathcal{L}^{(i)}$ simply is

$$\mathcal{L}^{(i)} = \frac{e^{-\frac{\left(u_{t+1}^{(i)}\right)^2}{2\sigma^2}}}{\sqrt{2\pi}\sigma}$$

and therefore

$$\begin{aligned}\hat{x}_{t+1} &= \mathbb{E}\left(x_{t+1} \mid z_{t+1}\right) \\ &= \frac{\sum_{i=1}^{M} \mathcal{L}^{(i)} \times x_{t+1}^{(i)}}{\sum_{i=1}^{M} \mathcal{L}^{(i)}}\end{aligned}$$

This is the best estimate of $x_{t+1}$. For the next time step prediction what we just obtained is used as the best estimate of the current step and proceed sequentially.

Mathematically the filtering problem is solved by computing the following posterior probability $p(x_t|z_t)$, that is, given the observation at time $t$, $z_t$ what is the probability of the hidden state $x_t$? A related problem is to track $p(x_t|z_t)$ sequentially through time, although this is more difficult due to its higher dimensionality. It is very important to stress the sequential nature of the filtering problem, that is, the goal is to compute or approximate $p(x_t|\Theta, z_t)$ as new data arrives; this is different from the general smoothing problem [154].

### 8.1.1 Construction of $p(\mathbf{x}_k|\mathbf{z}_{1:k})$

Theoretically one can construct the posterior density, $p(\mathbf{x}_k|\mathbf{z}_{1:k})$, recursively in two stages:

**Time update (prediction): Chapman–Kolmogorov equation**. Given the observation up to time $t_{k-1}$, what is the best prediction for $x$ at time $t_k$, $x_k$? That is,

$$\begin{aligned} p(\mathbf{x}_k|\mathbf{z}_{1:k-1}) &= \int p(\mathbf{x}_k|\mathbf{x}_{k-1}, \mathbf{z}_{1:k-1})p(\mathbf{x}_{k-1}|\mathbf{z}_{1:k-1})d\mathbf{x}_{k-1} \\ &= \int p(\mathbf{x}_k|\mathbf{x}_{k-1})p(\mathbf{x}_{k-1}|\mathbf{z}_{1:k-1})d\mathbf{x}_{k-1} \end{aligned}$$

Here for the time update iteration we apply the Chapman–Kolmogorov equation by using the Markov property.

**Measurement update (filtering): Bayes' rule**. Now having an observation $z_k$ at $t_k$, what is the probability of $x_k$? Here we can use Bayes' rule

$$p(\mathbf{x}_k|\mathbf{z}_{1:k}) = \frac{p(\mathbf{z}_k|\mathbf{x}_k)p(\mathbf{x}_k|\mathbf{z}_{1:k-1})}{p(\mathbf{z}_k|\mathbf{z}_{1:k-1})}$$

where the probability in the denominator $p(\mathbf{z}_k|\mathbf{z}_{1:k-1})$ can be written

$$p(\mathbf{z}_k|\mathbf{z}_{1:k-1}) = \int p(\mathbf{z}_k|\mathbf{x}_k)p(\mathbf{x}_k|\mathbf{z}_{1:k-1})d\mathbf{x}_k$$

and it corresponds to the time $t_k$ likelihood function. Following the argument in [152] we can write

$$\begin{aligned} p(\mathbf{x}_k|\mathbf{z}_{1:k}) &= \frac{p(\mathbf{z}_{1:k}|\mathbf{x}_k)p(\mathbf{x}_k)}{p(\mathbf{z}_{1:k})} \\ &= \frac{p(\mathbf{z}_k, \mathbf{z}_{1:k-1}|\mathbf{x}_k)p(\mathbf{x}_k)}{p(\mathbf{z}_k, \mathbf{z}_{1:k-1})} \\ &= \frac{p(\mathbf{z}_k|\mathbf{z}_{1:k-1}, \mathbf{x}_k)p(\mathbf{z}_{1:k-1}|\mathbf{x}_k)p(\mathbf{x}_k)}{p(\mathbf{z}_k|\mathbf{z}_{1:k-1})p(\mathbf{z}_{1:k-1})} \\ &= \frac{p(\mathbf{z}_k|\mathbf{z}_{1:k-1}, \mathbf{x}_k)p(\mathbf{x}_k|\mathbf{z}_{1:k-1})p(\mathbf{z}_{1:k-1})p(\mathbf{x}_k)}{p(\mathbf{z}_k|\mathbf{z}_{1:k-1})p(\mathbf{z}_{1:k-1})p(\mathbf{x}_k)} \\ &= \frac{p(\mathbf{z}_k|\mathbf{x}_k)p(\mathbf{x}_k|\mathbf{z}_{1:k-1})}{p(\mathbf{z}_k|\mathbf{z}_{1:k-1})} \end{aligned}$$

It is obvious that it is assumed that at time step $k$, values of $\mathbf{z}_{1:k-1}$ are already known.

---

## 8.2 Likelihood Function

Having posterior density, $p(\mathbf{x}_k|\mathbf{z}_{1:k})$, at time $t_k$ we can write likelihood $l_k$

$$l_k = p(z_k|z_{1:k-1}) = \int p(z_k|x_k)p(x_k|x_{k-1}, z_{1:k-1})dx_k \tag{8.1}$$

and therefore the total likelihood is

$$\ln(L_{1:N}) = \sum_{k=1}^{N} \ln(l_k)$$

In practice, instead of trying to calculate the value of the expression in (8.1), one uses a proxy for likelihood $l_k$ as will be discussed later. We assume the following generic models/equations for state and observation:

$$\begin{aligned} x_{t+1} &= f(x_t, \Theta, u_{t+1}) \\ z_{t+1} &= g(x_{t+1}, \Theta, v_{t+1}) \end{aligned}$$

namely, state equation and measurement equation, respectively. Alternatively

$$\begin{array}{rl} p(z_{t+1}|x_{t+1}, \Theta) & \text{observation density} \\ p(x_{t+1}|x_t, \Theta) & \text{state transition} \\ p(x_0) & \text{initial distribution (prior)} \end{array}$$

In the case of a known parameter set we can write $p(x_t|z_t) = p(x_t|z_t, \Theta)$.

There are three general approaches: (a) approximates the model via either a linearization or by approximate state variables with a continuous distribution by a discrete state Markov chain; then apply a Kalman filter, (b) numerical integration routines to approximate the integrals; this could run easily into curse of dimensionality, (c) Monte Carlo, which leads to the particle filter.

**Example 23** *Filtering and parameter estimation of the discrete-time double gamma stochastic volatility model via likelihood*

In this example, we go over filtering and parameter estimation for the discrete-time double gamma stochastic volatility model proposed in [138]. We suppose the interest rate $r_t$ prevailing over the next period of time rate is a linear function of the factors $x_t$, specifically.

$$r_t = a_r + b_r^\top x_t$$

Assume the following discrete-time stochastic volatility model:

$$\begin{aligned} x_{t+1} &= x_t + (b_\gamma - Bx_t)\,\Delta t + \Sigma\sqrt{v_{t+1}}\sqrt{\Delta t} z_{t+1} \\ v_{t+1} &\sim Gamma(\lambda v_t + x, \frac{1}{\delta}) \\ x &\sim Gamma(\gamma, \frac{1}{\eta}) \end{aligned}$$

where $z_{t+1} \sim \mathcal{N}(0, I)$ and the probability density of $Gamma(\alpha, \frac{1}{\beta})$ is

$$\begin{aligned} f(t) &= \frac{\beta e^{-\beta t}(\beta t)^{\alpha-1}}{\Gamma(\alpha)} \\ &= \frac{\beta^\alpha e^{-\beta t} t^{\alpha-1}}{\Gamma(\alpha)} \end{aligned}$$

In the filtering step, the assumption is that the parameter set, $\Theta = \{\sigma, \gamma, \eta, \lambda, \delta, a_r, b_r, b_\gamma, B, \Sigma\}$, is known. We also assume $t$-time factors namely, $x_t$ and $v_t$, are given. Now given observations at time $t+1$ we wish to calculate

$$(\hat{x}_{t+1}, \hat{v}_{t+1}) = \mathbb{E}\left(x_{t+1}, v_{t+1} \mid y_{t+1}\right)$$

Assume that the model price is given by $\phi(x_{t+1}, v_{t+1}; \Theta)$ where $\Theta$ is the parameter set.

Moreover assume the market price (observation) at time $t+1$, $y_{t+1}$, is linked to the model price via the following relationship:

$$y_{t+1} = \phi(x_{t+1}, v_{t+1}; \Theta) + u_{t+1}$$

where $u_{t+1} \sim \mathcal{N}(0, \sigma^2)$ for some $\sigma$ (the assumption is $\sigma$ is part of the parameter set $\Theta$ and therefore already estimated and known).

Start by first generating $M$ samples for $x_{t+1}$ and $v_{t+1}$ as follows:

$$\begin{aligned}
x^{(i)} &\sim Gamma(\gamma, \frac{1}{\eta}) \\
v_{t+1}^{(i)} &\sim Gamma(\lambda v_t + x^{(i)}, \frac{1}{\delta}) \\
x_{t+1}^{(i)} &= \mathcal{N}\left(x_t + (b_\gamma - Bx_t)\,\Delta t, \left(\Sigma\sqrt{v_{t+1}^{(i)}}\sqrt{\Delta t}\right)^2\right)
\end{aligned}$$

for $i = 1, \ldots, M$. Having $M$ samples of $x_{t+1}^{(i)}$ and $v_{t+1}^{(i)}$ we can calculate $M$ samples for the model price, namely, $\phi(x_{t+1}^{(i)}, v_{t+1}^{(i)}; \Theta)$. Now we can generate $M$ samples for $u_{t+1}$ having observed the market price at time $t+1$:

$$u_{t+1}^{(i)} = y_{t+1} - \phi(x_{t+1}^{(i)}, v_{t+1}^{(i)}; \Theta)$$

Define $\mathcal{L}^{(i)}$ as a (conditional) likelihood function

$$\mathcal{L}^{(i)} \equiv \text{Likelihood}\left(u_{t+1}^{(i)} \mid x_{t+1}^{(i)}, v_{t+1}^{(i)}\right)$$

Hence $\mathcal{L}^{(i)}$ simply is

$$\mathcal{L}^{(i)} = \frac{e^{-\frac{\left(u_{t+1}^{(i)}\right)^2}{2\sigma^2}}}{\sqrt{2\pi}\sigma}$$

Therefore $\hat{x}_{t+1}, \hat{v}_{t+1}$ can be calculated as follows:

$$\begin{aligned}
(\hat{x}_{t+1}, \hat{v}_{t+1}) &= \mathbb{E}\left(x_{t+1}, v_{t+1} \mid y_{t+1}\right) \\
&= \frac{\sum_{i=1}^{M} \mathcal{L}^{(i)} \times (x_{t+1}^{(i)}, v_{t+1}^{(i)})}{\sum_{i=1}^{M} \mathcal{L}^{(i)}}
\end{aligned}$$

that is weighted mean where weights are conditional likelihoods.

For parameter estimation, we start with some parameter set

$$\Theta' = \{\sigma', \gamma', \eta', \lambda', \delta', b'_\gamma, B', \Sigma', \ldots\}$$

as the prior and calculate the likelihood as follows:

for $t = 0, \ldots, T-1$

for $i = 1, \ldots, M$

$$\begin{aligned}
x^{(i)} &\sim Gamma(\gamma', \frac{1}{\eta'}) \\
v_{t+1}^{(i)} &\sim Gamma(\lambda' v_t^{(i)} + x^{(i)}, \frac{1}{\delta'}) \\
x_{t+1}^{(i)} &\sim \mathcal{N}\left(x_t^{(i)} + \left(b'_\gamma - B' x_t^{(i)}\right)\Delta t, \left(\Sigma'\sqrt{v_{t+1}^{(i)}}\sqrt{\Delta t}\right)^2\right) \\
u_{t+1}^{(i)} &= y_{t+1} - \phi(x_{t+1}^{(i)}, v_{t+1}^{(i)}; \Theta')
\end{aligned}$$

$$
\begin{aligned}
\mathcal{L}(y_{t+1}, x_{t+1}^{(i)}, v_{t+1}^{(i)}) &= \frac{e^{-\frac{(u_{t+1}^{(i)})^2}{2\sigma^2}}}{\sqrt{2\pi}\sigma} \\
&\times \frac{\eta^{\gamma} e^{-\eta x^{(i)}} (x^{(i)})^{\gamma-1}}{\Gamma(\gamma)} \\
&\times \frac{\delta^{\lambda v_t^{(i)} + x^{(i)}} e^{-\delta v_{t+1}^{(i)}} (v_{t+1}^{(i)})^{\lambda v_t^{(i)} + x^{(i)} - 1}}{\Gamma(\lambda v_t^{(i)} + x^{(i)})} \\
&\times \frac{\exp\left( - \frac{\left(x_{t+1}^{(i)} - (x_t^{(i)} + (b_\gamma - B x_t^{(i)})\Delta t)\right)^2}{2\left(\Sigma \sqrt{v_{t+1}^{(i)}} \sqrt{\Delta t}\right)^2} \right)}{\sqrt{2\pi}\Sigma \sqrt{v_{t+1}^{(i)}} \sqrt{\Delta t}}
\end{aligned}
$$

endfor

endfor

Having the log likelihood, we can employ either the Nelder–Mead simplex method or the EM algorithm ([119], [96]) for estimating the parameter set. In the case of the EM algorithm we minimize the following objective function starting from $\Theta_0^*$):

$$
\begin{aligned}
\Theta_{n+1}^* &= \min_{\Theta} -\frac{1}{M} \sum_{t=0}^{T-1} \sum_{i=1}^{M} \log \mathcal{L}(y_{t+1}, x_{t+1}^{(i)}, v_{t+1}^{(i)}, \Theta | \Theta_n^*) \\
&= \min_{\Theta} -\mathbb{E}(\log \mathcal{L}(\Theta | \Theta_n^*))
\end{aligned}
$$

That is the combined $E$ and $M$ steps. The optimal $\Theta^*$ is found until $\|\Theta_{n+1}^* - \Theta_n^*\| < \epsilon$ where $\epsilon$ is a threshold. In our example, we choose $\epsilon = 1.0e-5$. Datasets used for this example are: (a) LIBOR rates with maturities of 1, 2, 3, 6 and 12 months and (b) swap rates at maturities 2, 3, 5, 10, 12, 15, 20, and 30 years. The data are daily close from December 14, 1994 through May 26, 2005. Here are our results:

(a) Using the Nelder–Mead simplex method, we obtain the following parameter set:

$$
\begin{aligned}
a_r &= 0.0304 \\
b_r &= [0.0010, 0.0044, 0.1191] \\
b_\gamma &= [0.0031, 2.0519, -1.0177] \\
B &= \begin{pmatrix} 0.0002 & & \\ 0.0531 & 0.1326 & \\ -0.0264 & -0.0236 & 0.6950 \end{pmatrix} \\
\Sigma &= \begin{pmatrix} 0.7044 & & \\ & 1.2417 & \\ & & -0.1465 \end{pmatrix} \\
\lambda &= 1.5265 \\
\delta &= 2.3504 \\
\gamma &= 9.5712 \\
\eta &= 3.0376
\end{aligned}
$$

In Figures 8.2(a) and 8.2(b) we show 1-month and 6-month LIBOR rates predictions versus actual rates, respectively. Figures 8.3(a), 8.3(b), and 8.3(c) illustrate 5-year, 15-year, and 30-year rate predictions versus actual rates, respectively. Finally we display the hidden states

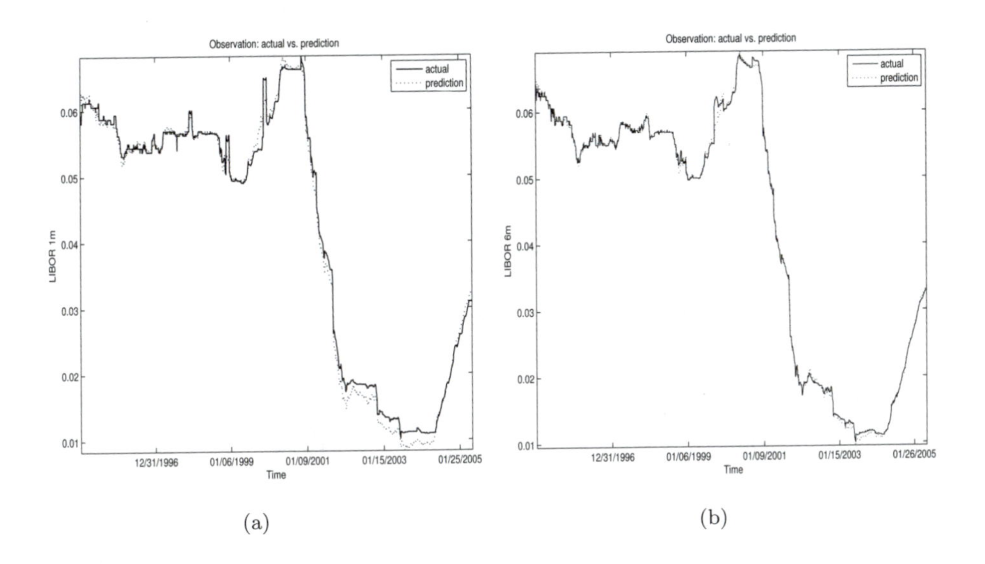


(a) (b)

**FIGURE 8.2**: (a) 1-month LIBOR rate, (b) 6-month LIBOR rate prediction versus actual, in the discrete-time double gamma stochastic volatility example using the simplex method for parameter estimation

$x_t$ and $v_t$ through time in Figure 8.4. All results are obtained using the simplex method for parameter estimation.

(b) Using the EM algorithm, we obtain the following parameter set:

$$
\begin{aligned}
a_r &= 0.03852 \\
b_r &= [0.0000139, 0.0000150, 0.1402806] \\
b_\gamma &= [0.0029114, -1.5080950, -2.3860751] \\
B &= \begin{pmatrix} 0.0001004 & & \\ -0.0556363 & 0.0311322 & \\ -0.0835994 & -0.1588748 & 2.7916005 \end{pmatrix} \\
\Sigma &= \begin{pmatrix} 0.3801258 & & \\ & 0.9065671 & \\ & & -0.0739867 \end{pmatrix} \\
\lambda &= 2.7533120 \\
\delta &= 4.1201996 \\
\gamma &= 6.7723249 \\
\eta &= 2.9314949
\end{aligned}
$$

In Figures 8.5(a) and 8.5(b) we display 1-month and 6-month LIBOR rates predictions versus actual rates, respectively. Figures 8.6(a), 8.6(b), and 8.6(c) display 5-year, 15-year, and 30-year rates predictions versus actual rates, respectively. Figure 8.7 illustrates the hidden states, $x_t$ and $v_t$ through time. All results are obtained using the EM algorithm for parameter estimation.

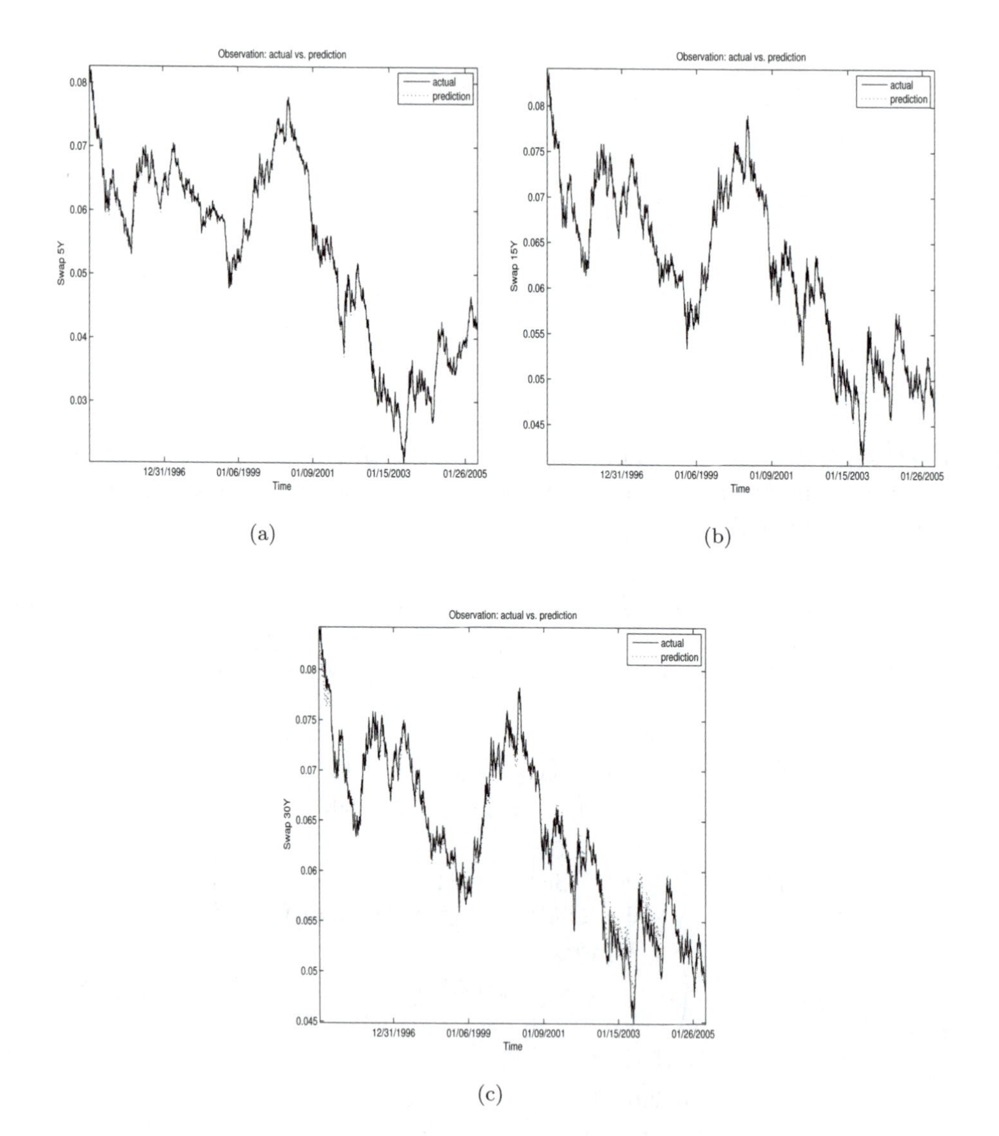


**FIGURE 8.3**: (a) 5-year swap rate, (b) 15-year swap rate, (c) 30-year swap rate prediction versus actual in the discrete-time double gamma stochastic volatility example using the simplex method for parameter estimation

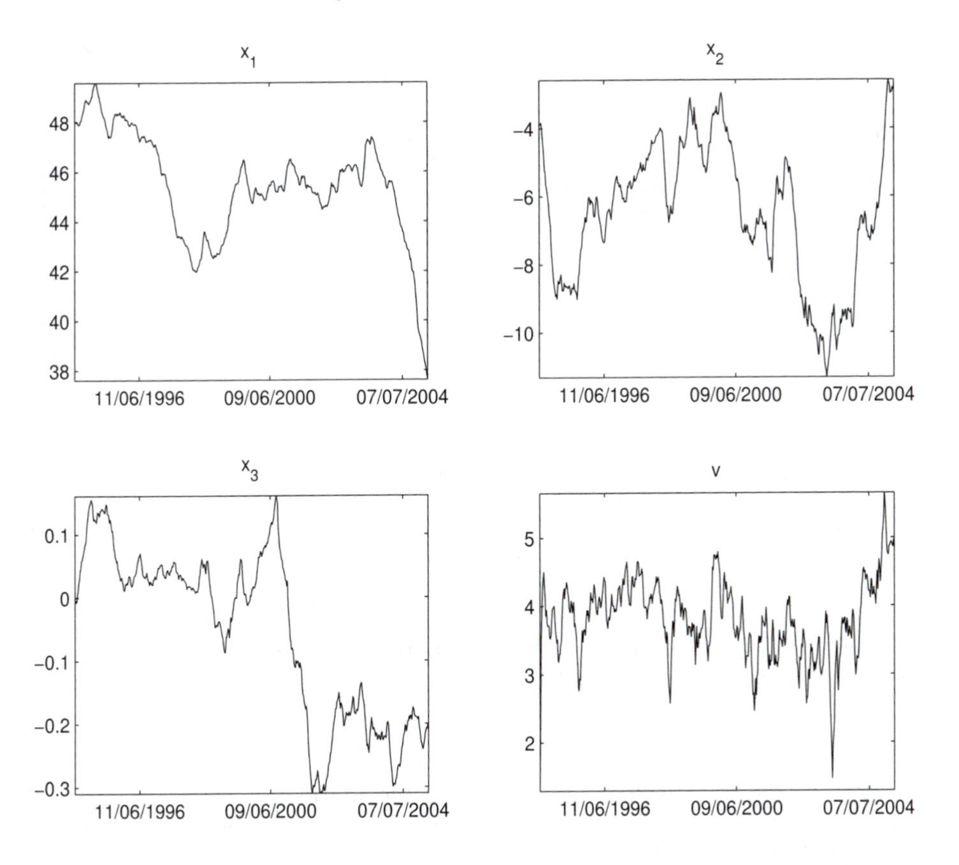


**FIGURE 8.4**: Discrete-time double gamma stochastic volatility example using the simplex method for parameter estimation: states

---

## 8.3 Kalman Filter

The Kalman filter ([129],[130]) is an efficient recursive filter[1] which estimates states of a linear dynamic system from a time series of observations. Combined with the linear-quadratic regulator the Kalman filter solves the linear-quadratic-Gaussian problem that is one of the most fundamental optimal control problems.

### 8.3.1 Underlying Model

Kalman filters are based on linear dynamical systems discretized in the time domain.[2] They are modeled on a Markov chain built on linear operators perturbed by a Gaussian noise. The state of the system is an $n \times 1$ vector of real numbers where $n$ is the dimension of the system. At each time increment, a linear operator is applied to the state to generate the

[1] In a recursive filter only the estimated state from the previous time step and the current measurement are being used to compute the estimate for the current state. Unlike batch estimation techniques, in the Kalman filter we do not use any history of observations and/or estimates.

[2] Most filters (e.g., low-pass filters) are formulated in the frequency domain and then they are transformed back to the time domain for implementation.

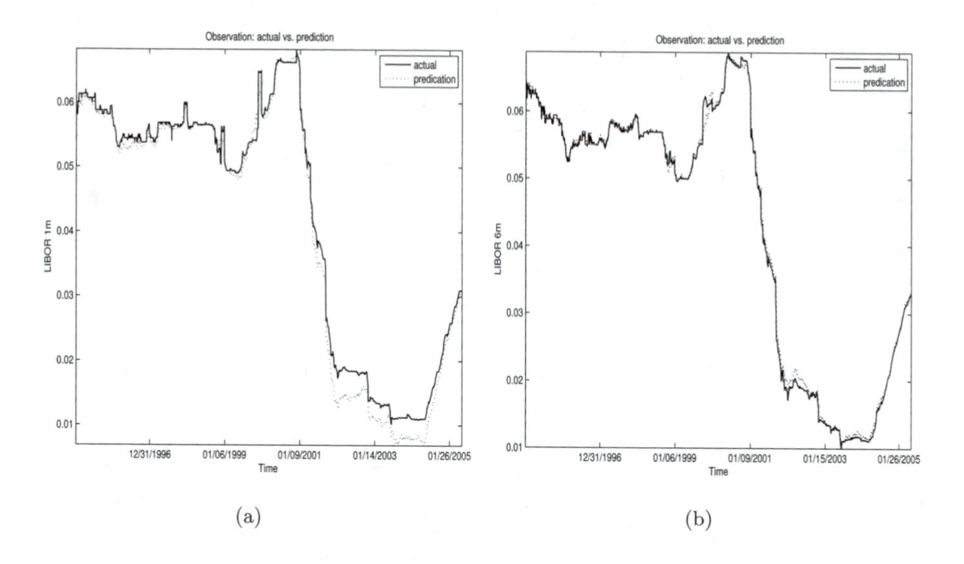


**FIGURE 8.5**: (a) LIBOR 1-month prediction versus actual, (b) LIBOR 6-month prediction versus actual, in the discrete-time double gamma stochastic volatility example using the EM algorithm for parameter estimation

new state, with some added noise, and optionally some information from the system control (if it is available and known). Then, another linear operator with more noise generates the visible outputs from the hidden state.

In [120], the authors show that there is a duality between the equations of the Kalman Filter and those of the hidden Markov model. The key difference is that the hidden state variables take values in a continuous space as opposed to a discrete state space as in the hidden Markov model. Also the hidden Markov model can represent an arbitrary distribution for the next value of the state variables, in contrast to the Gaussian noise model that is used for the Kalman filter. In order to use the Kalman filter to estimate the internal state of a process given only a sequence of noisy observations, one must model the process in accordance with the framework of the Kalman filter. In the Kalman filter, the following linear model is assumed for the evolution of the true state at time $t_k$:

$$x_k = F_k x_{k-1} + B_k u_k + w_k$$

where

$x_k$ is the (true) state at time $t_k$

$F_k$ is the state transition matrix

$B_k$ is the control-input model applied to the control vector $u_k$

$w_k$ is the process noise assumed to be a multivariate normal distribution with zero mean and covariance $Q_k$, i.e., $w_k \sim \mathcal{N}(0, Q_k)$

Here $x_k$ is an $n \times 1$ vector of real numbers and $F_k$ is an $m \times n$ matrix. At time $k$ an observation (measurement) $z_k$ of state $x_k$ is made and we assume the following measurement equation for its evolution:

$$z_k = H_k x_k + v_k$$

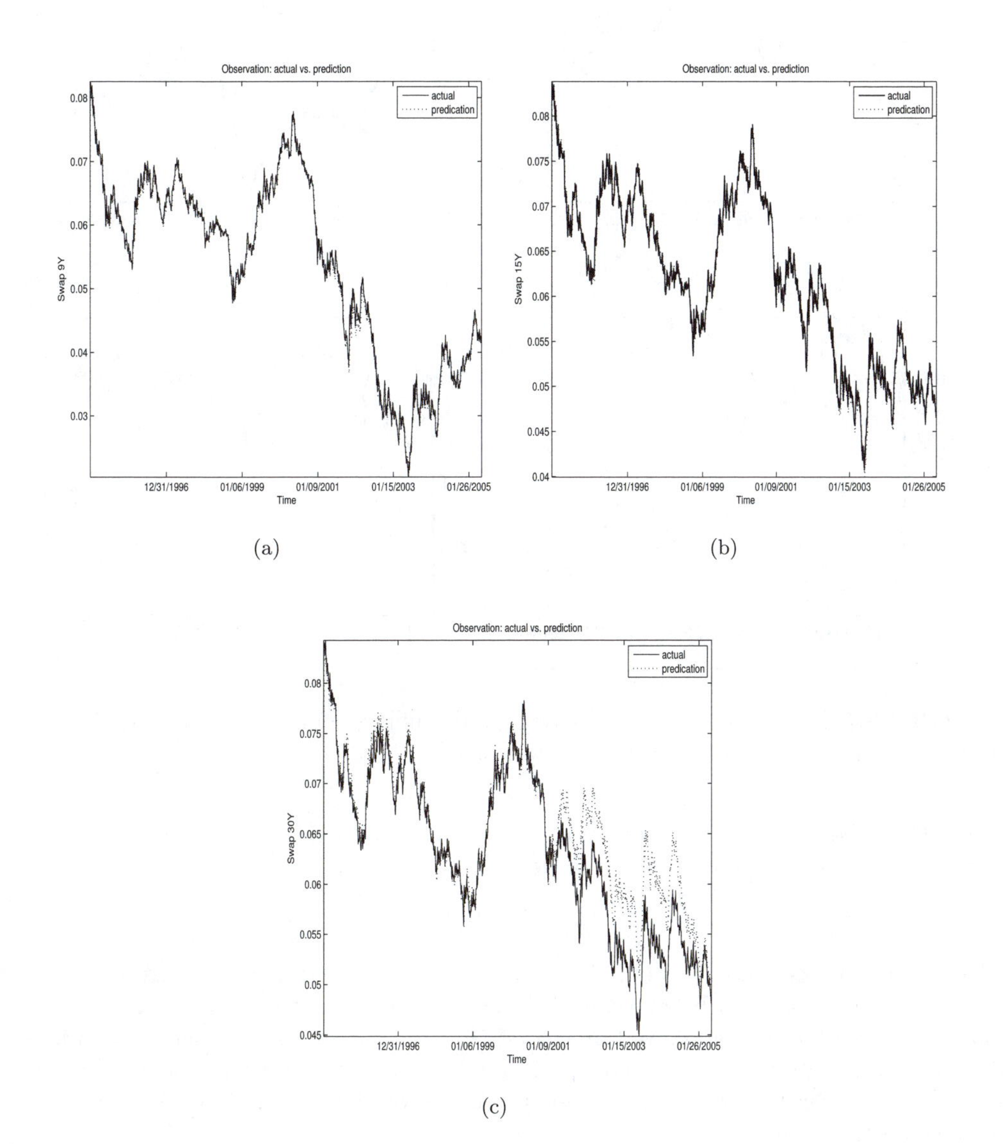


**FIGURE 8.6**: (a) 5-year swap rate prediction versus actual, (b) 15-year swap rate prediction versus actual, (c) 30-year swap rate prediction versus actual, in the discrete-time double gamma stochastic volatility example using the EM algorithm for parameter estimation

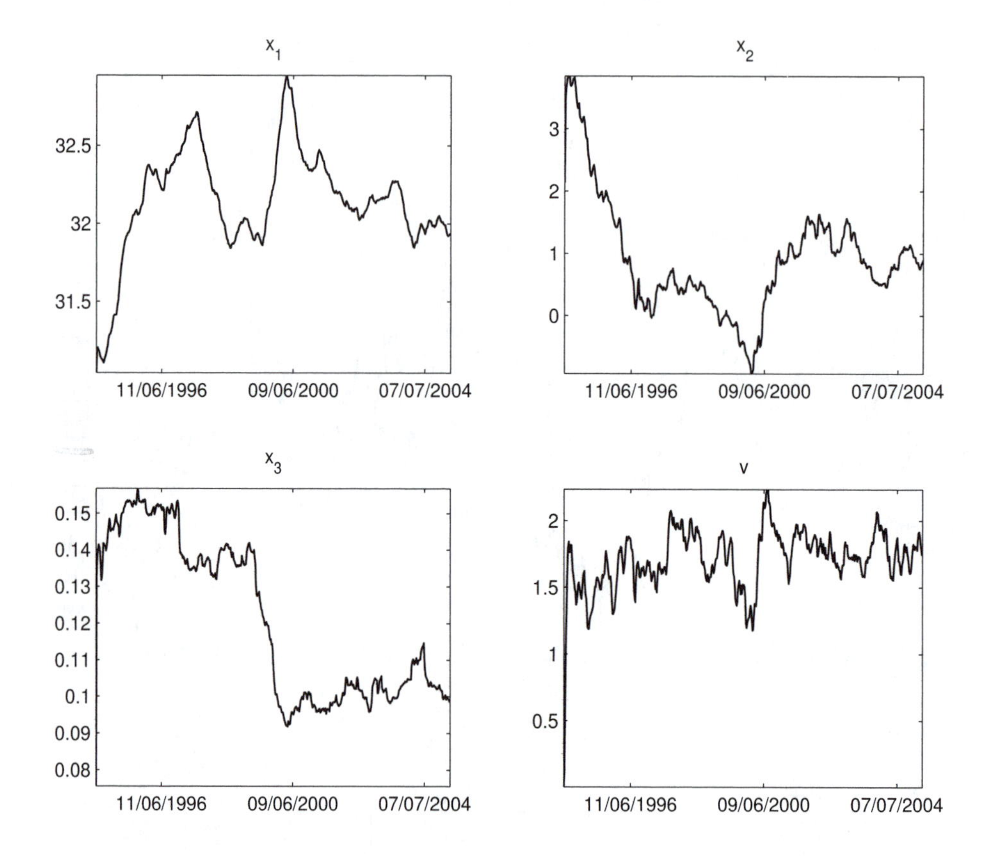


**FIGURE 8.7**: Hidden states of the discrete-time double gamma stochastic volatility model using the EM algorithm for parameter estimation

where $H_k$ is the observation matrix and $v_k$ is the observation noise assumed to be Gaussian with zero mean and covariance $R_k$, that is,

$$v_k \sim \mathcal{N}(0, R_k)$$

The initial state $x_0$ and the state noise vectors $w_1, \ldots, w_k$ and measurement noise vectors $v_1, \ldots, v_k$ are assumed to be mutually independent. There are rare cases in which a dynamical system follows the above framework. Nonetheless considering the Kalman filter is designed to operate in the presence of noise makes it a useful filter and should be a good start. We will later describe variations and extensions on the Kalman filter that would allow more sophisticated models.

The state of the filter is represented by the following variables:

$\hat{x}_{k|k-1}$ the estimate of the state at time $k$ given observations up to and including time $k-1$

$\hat{x}_{k|k}$ the estimate of the state at time $k$ given observations up to and including time $k$

$P_{k|k-1}$ the error covariance matrix (a measure of the estimated accuracy of the state estimate) at time $k$ given observations up to and including time $k-1$

$P_{k|k}$ the error covariance matrix at time $k$ given observations up to and including time $k$

As mentioned earlier, filtering is done in two steps: (a) *time update or prediction* and (b) *measurement update or filtering.* In the prediction phase, we use the state estimate from the previous time step in order to produce an estimate of the state at the current time step. In the filtering phase, we use the measurement at the current time step to refine this prediction to obtain a new and (hopefully) a more accurate state estimate for the current time step.

**Time update**

Predicted state estimate: $\hat{x}_{k|k-1} = F_k \hat{x}_{k-1|k-1} + B_{k-1} u_{k-1}$

Predicted estimate covariance: $P_{k|k-1} = F_k P_{k-1|k-1} F_k^\top + Q_{k-1}$

**Measurement update**

Measurement residual: $\hat{\delta}_k = z_k - H_k \hat{x}_{k|k-1}$

Residual covariance: $S_k = H_k P_{k|k-1} H_k^\top + R_k$

Updated state estimate: $\hat{x}_{k|k} = \hat{x}_{k|k-1} + K_k \hat{\delta}_k$

Updated estimate covariance: $P_{k|k} = (I - K_k H_k) P_{k|k-1} (I - K_k H_k)^\top + K_k R_k K_k^\top$

where $K_k$ is the Kalman gain (we do not make any assumption yet on its optimality). Before we derive the posterior estimate covariance matrix, we need to define some invariants. If the model is accurate and the values for $\hat{x}_{0|0}$ and $P_{0|0}$ accurately reflect the distribution of the initial state values, then the following invariants are preserved: (a) all estimates have mean error zero

$$\begin{aligned} \mathbb{E}(x_k - \hat{x}_{k|k}) &= 0 \\ \mathbb{E}(x_k - \hat{x}_{k|k-1}) &= 0 \\ \mathbb{E}(\tilde{\delta}_k) &= 0 \end{aligned}$$

and (b) covariance matrices accurately reflect the covariance of estimates

$$\begin{aligned} P_{k|k} &= \text{cov}(x_k - \hat{x}_{k|k}) \\ P_{k|k-1} &= \text{cov}(x_k - \hat{x}_{k|k-1}) \\ S_k &= \text{cov}(\hat{\delta}_k) \end{aligned}$$

Now, we start with the error covariance $P_{k|k}$; from our invariants we can write

$$\begin{aligned} P_{k|k} &= \text{cov}(x_k - \hat{x}_{k|k}) \\ &= \text{cov}(x_k - (\hat{x}_{k|k-1} + K_k \hat{\delta}_k)) \\ &= \text{cov}(x_k - (\hat{x}_{k|k-1} + K_k (z_k - H_k \hat{x}_{k|k-1}))) \\ &= \text{cov}(x_k - (\hat{x}_{k|k-1} + K_k (H_k x_k + v_k - H_k \hat{x}_{k|k-1}))) \\ &= \text{cov}((I - K_k H_k)(x_k - \hat{x}_{k|k-1}) - K_k v_k) \end{aligned}$$

The assumption is that measurement error $v_k$ is uncorrelated with other terms and therefore we arrive at

$$\begin{aligned} P_{k|k} &= \text{cov}((I - K_k H_k)(x_k - \hat{x}_{k|k-1})) + \text{cov}(K_k v_k) \\ &= (I - K_k H_k)\text{cov}(x_k - \hat{x}_{k|k-1})(I - K_k H_k)^\top + K_k \text{cov}(v_k) K_k^\top \end{aligned}$$

Now using the definition of $P_{k|k-1}$ and $R_k$ it becomes

$$P_{k|k} = (I - K_k H_k) P_{k|k-1} (I - K_k H_k)^\top + K_k R_k K_k^\top$$

and it holds for any arbitrary value for $K_k$. So far we have not made any assumption on $K_k$. We can reduce this expression further for the optimal Kalman gain.

### 8.3.2 Posterior Estimate Covariance under Optimal Kalman Gain and Interpretation of the Optimal Kalman Gain

The Kalman filter minimizes posterior state estimation which is equivalent to saying it minimizes mean-square error estimator. The error in the posterior state estimation is

$$\epsilon_k \equiv x_k - \hat{x}_{k|k}$$

The goal is to minimize the expected value of the square of $\epsilon_k$

$$\mathbb{E}(|x_k - \hat{x}_{k|k}|^2)$$

which is the same as minimizing the trace of the posterior estimate covariance matrix $P_{k|k}$. We get the following after expanding the terms in $P_{k|k}$:

$$\begin{aligned} P_{k|k} &= P_{k|k-1} - K_k H_k P_{k|k-1} - P_{k|k-1} H_k^\top K_k^\top + K_k (H_k P_{k|k-1} H_k^\top + R_k) K_k^\top \\ &= P_{k|k-1} - K_k H_k P_{k|k-1} - P_{k|k-1} H_k^\top K_k^\top + K_k S_k K_k^\top \end{aligned}$$

Now taking the first derivative of the trace of $P_{k|k}$ with respect to $K_k$ and setting it equal to zero, we find the optimal $k_k$:

$$\frac{\partial \mathrm{tr}(P_{k|k})}{\partial K_k} = -2(H_k P_{k|k-1})^\top + 2K_k S_k = 0$$

Solving it for $K_k$ yields

$$K_k S_k = (H_k P_{k|k-1})^\top = P_{k|k-1} H_k^\top$$

or

$$K_k = P_{k|k-1} H_k^\top S_k^{-1}$$

This Kalman gain minimizes the mean-square error estimate and is the optimal Kalman gain.

$$K_k S_k K_k^\top = P_{k|k-1} H_k^\top K_k^\top$$

Substituting back into $P_{k|k}$ yields

$$P_{k|k} = P_{k|k-1} - K_k H_k P_{k|k-1} - P_{k|k-1} H_k^\top K_k^\top + K_k S_k K_k^\top$$

Noticing that the last two terms cancel out give us

$$\begin{aligned} P_{k|k} &= P_{k|k-1} - K_k H_k P_{k|k-1} \\ &= (I - K_k H_k) P_{k|k-1} \end{aligned}$$

An interpretation of the Kalman filter could be based on linear regression. In the case of having time series of $\{\mathbf{x}_k\}$ and $\{\mathbf{z}_k\}$ the linear regression yields

$$\mathbf{z}_k = \alpha + \beta \mathbf{x}_k + \epsilon_k$$

with $\alpha$ the intercept, $\beta$ the slope, and $\epsilon_k$ the residual. Under linear regression we have

$$\beta = P_{k|k-1} H_k^\top S_k^{-1}$$

which is the expression for the Kalman gain.

In the Kalman filter, the log likelihood for each time step is $\log p(z_t|z_{1:t-1})$, which is obtained by evaluating the log of the probability density function of a multivariate Gaussian density with mean zero and covariance of $S_k$ evaluated at the values in $\hat{\delta}_k$, that is, the log likelihood of innovation

$$\begin{aligned}
\log p(z_t|z_{1:t-1}) &= \log\left(\frac{1}{(2\pi)^{d/2}|S_k|^{1/2}} \exp\left(-\frac{1}{2}(\hat{\delta}_k - 0)^\top S_k^{-1}(\hat{\delta}_k - 0)\right)\right) \\
&= -\frac{d}{2}\log(2\pi) - \frac{1}{2}\log|S_k| - \frac{1}{2}(\hat{\delta}_k - 0)^\top S_k^{-1}(\hat{\delta}_k - 0) \\
&= -\frac{d}{2}\log(2\pi) - \frac{1}{2}\log|S_k| - \frac{1}{2}(z_k - H_k\hat{x}_{k|k-1})^\top S_k^{-1}(z_k - H_k\hat{x}_{k|k-1}).
\end{aligned}$$

Calling this log likelihood $\ln(l_k)$, then the log likelihood for the entire time series is $\ln(L_{1:T}) = \sum_{k=1}^{T} \ln(l_k)$. Having the log likelihood, we can either use an optimization routine (e.g., Nelder–Mead simplex) or employ the EM algorithm ([119], [96]) to minimize the negative log likelihood for parameter estimation. It is worth mentioning that the Kalman gain does not come into the parameter estimation; it only enters into the filtering step.

**Example 24** *Parameter estimation, filtering, and prediction via MLE and EM in the Kalman Filter*

Assuming

$$\begin{aligned}
x_{t+1} &= F x_t + w_t \text{ where } w_t \sim \mathcal{N}(0, Q) \\
y_{t+1} &= H x_{t+1} + v_t \text{ where } v_t \sim \mathcal{N}(0, R)
\end{aligned}$$

The parameter set for estimation is $\Theta = \{F, H, Q, R, x_0, v_0\}$. In our simulation study we assume the following parameters: $F = 1$, $H = 2$, $Q = 0.1$, $R = 0.1$, $x_0 = 9.9355$, and $v_0 = 0.01$. Starting values for parameter estimation are

$$\{F^{start}, H^{start}, Q^{start}, R^{start}, x_0^{start}, v_0^{start}\} = [2, 1, 0.2, 0.2, 5, 0.05]$$

Implementation of the Nelder–Mead simplex method is as follows: parameter set $\Theta = \{F, H, Q, R, x_0, v_0\}$ and the objective function to minimize

$$f(\Theta) = \sum_{t=1}^{T} \left((y_{t+1} - Hx_{t+1|t})^\top S_{t+1}^{-1}(y_{t+1} - Hx_{t+1|t}) + \log|S_{t+1}|\right)$$

where

$$\begin{aligned}
x_{t+1|t} &= F x_{t|t} \\
P_{t+1|t} &= F P_{t|t} F^\top + Q \\
S_{t+1} &= H P_{t+1|t} H^\top + R
\end{aligned}$$

with $x_{0|0} = x_0$ and $P_{0|0} = v_0$. Using the Nelder–Mead simplex method to minimize the objective function we obtain the subsequent parameter set

$$\{\hat{F}, \hat{H}, \hat{Q}, \hat{R}, \hat{x}_0, \hat{v}_0\} = [0.9952, 1.7478, 0.1322, 0.0846, 11.7743, 0.0005]$$

Implementation of EM of the Kalman filter on a linear system [119] is as follows. In the *E*-step, we calculate $L = \mathbb{E}(\log P(\{x\}, \{y\})|\{y\}))$ with $\{x\} = (x_1, x_2, \ldots, x_T)$ and $\{y\} = (y_1, y_2, \ldots, y_T)$ and

$$\begin{aligned} \log P(\{x\}, \{y\}) &= -\sum_{t=1}^{T}(\frac{1}{2}(y_t - Hx_t)^\top R^{-1}(y_t - Hx_t)) - \frac{T}{2}\log|R| \\ &- \sum_{t=2}^{T}(\frac{1}{2}(x_t - Fx_{t-1})^\top Q^{-1}(x_t - Fx_{t-1})) - \frac{T-1}{2}\log|Q| \\ &- \frac{1}{2}(x_1 - \pi_1)^\top V^{-1}(x_t - \pi_1) - \frac{1}{2}\log|V_1| - \frac{T(p+k)}{2}\log 2\pi \end{aligned}$$

where $p$ is the dimension of state vector $x_t$ and $k$ is the dimension of measurement vector $y_t$ and

$$\begin{aligned} \pi_1 &= \mathbb{E}(x_1) \\ v_1 &= var(x_1) \end{aligned}$$

$L$ depends on the following smoothing of states and covariance:

$$\begin{aligned} \hat{x}_t &= \mathbb{E}(x_t|\{y\}) \\ P_t &= \mathbb{E}(x_t x_t^\top|\{y\}) \\ P_{t,t-1} &= \mathbb{E}(x_t x_{t-1}^\top|\{y\}) \end{aligned}$$

Note that the state estimate, $\hat{x}_t$, depends on both past and future observations. In that regard, it differs from the one computed in the Kalman filter. In the *M-step*, we update $F$, $H$, $Q$, and $R$ as follows

$$\begin{aligned} \frac{\partial L}{\partial F} = 0 &\Rightarrow F_{new} = (\sum_{t=2}^{T} P_{t,t-1})(\sum_{t=2}^{T} P_{t-1})^{-1} \\ \frac{\partial L}{\partial H} = 0 &\Rightarrow H_{new} = (\sum_{t=1}^{T} y_t\hat{x}_t^\top)(\sum_{t=1}^{T} P_t)^{-1} \\ \frac{\partial L}{\partial Q} = 0 &\Rightarrow Q_{new} = \frac{1}{T-1}(\sum_{t=2}^{T} P_t - F_{new}\sum_{t=2}^{T} P_{t,t-1}) \\ \frac{\partial L}{\partial R} = 0 &\Rightarrow R_{new} = \frac{1}{T}(\sum_{t=1}^{T}(y_t y_t^\top - H_{new}\hat{x}_t y_t^\top)) \\ \frac{\partial L}{\partial \pi_1} = 0 &\Rightarrow \pi_1^{new} = \hat{x}_1 \\ \frac{\partial L}{\partial V_1} = 0 &\Rightarrow V_1^{new} = P_1 - \hat{x}_1\hat{x}_1^\top \end{aligned}$$

EM maximum likelihood estimates are

$$\{\hat{F}, \hat{H}, \hat{Q}, \hat{R}, \hat{x}_1, \hat{v}_1\} = [0.9953, 2.0064, 0.1009, 0.0831, 9.9548, 4.2093e-005]$$

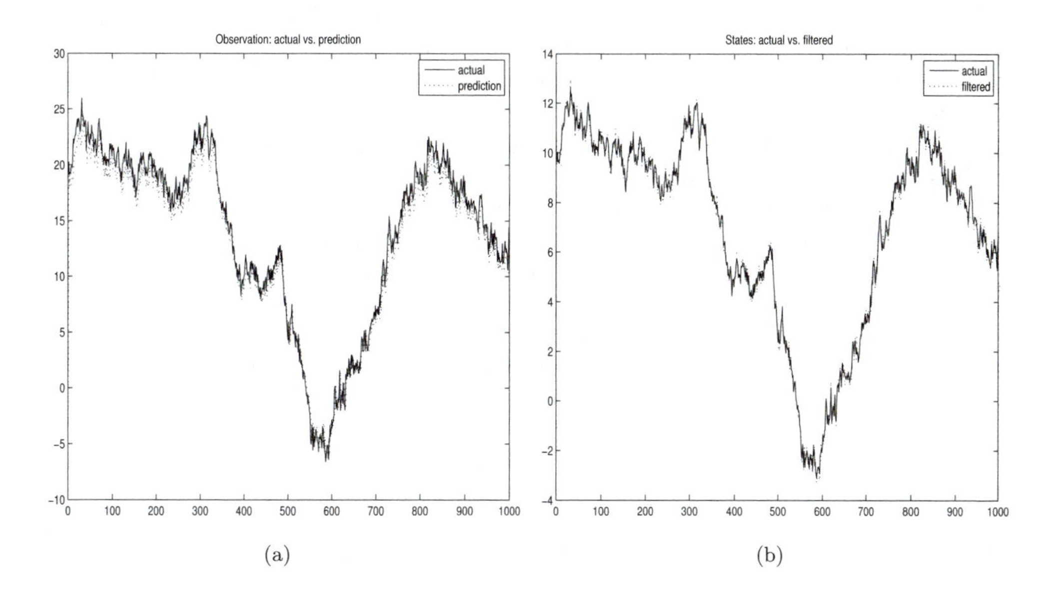


**FIGURE 8.8**: Kalman filter example: (a) observation prediction, (b) state filtering

Our observation is the EM algorithm converges faster than the MLE simplex method.

In optimization methodology it is important to note that the likelihood function may very well be non-differentiable; therefore it is wise to employ a search algorithm that does not use the gradient. The direction-set algorithm is a good choice from this point of view. The more number of parameters we have, the more difficult the convergence of the optimization process; the likelihood function becomes flat.

## 8.4 Non-Linear Filters

The basic Kalman filter is limited to a linear assumption. However, most non-trivial systems are non-linear. The non-linearity can be associated either with the process model or with the observation model or with both [149]. For details on nonlinear filtering of stochastic volatility models see [155].

## 8.5 Extended Kalman Filter

In the extended Kalman filter (EKF), the state transition and observation models need not be linear functions of the state but may instead be differentiable functions.

$$\begin{aligned} x_k &= f(x_{k-1}, u_k) \\ z_k &= h(x_k, v_k) \end{aligned}$$

The function $f$ can be used to compute the predicted state from the previous estimate and similarly the function $h$ can be used to compute the predicted measurement from the predicted state. However, $f$ and $h$ cannot be applied to the covariance directly. Instead a matrix of partial derivatives, the Jacobian, is computed. At each time step the Jacobian is evaluated with current predicted states. These matrices can be used in the Kalman filter equations. This process essentially linearizes the non-linear function around the current estimate. It uses a first-order linearization technique to allow its use with nonlinear processes.

Here the assumption as in the Kalman filter is that $w_k$ and $v_k$ are uncorrelated sequences of standard normal variables with mean zero and covariance matrices of $Q_k$ and $R_k$, respectively.

As in the Kalman filter, we call predicted state or a priori process estimates $\hat{x}_{k|k-1}$ and updated state estimate or a posterior estimate $\hat{x}_{k|k}$. Starting with our invariant on the error covariance $P_{k|k}$ as above

$$\begin{aligned} P_{k|k-1} &= \operatorname{cov}(x_k - \hat{x}_{k|k-1}) \\ P_{k|k} &= \operatorname{cov}(x_k - \hat{x}_{k|k}) \end{aligned}$$

Matrix of partial derivatives, the Jacobians, are formed as follows:

$$\begin{aligned} F_{ij} &= \frac{\partial f_i}{\partial x_j}(\hat{x}_{k|k-1}, 0) \\ U_{ij} &= \frac{\partial f_i}{\partial u_j}(\hat{x}_{k|k-1}, 0) \\ H_{ij} &= \frac{\partial h_i}{\partial x_j}(\hat{x}_{k|k-1}, 0) \\ V_{ij} &= \frac{\partial h_i}{\partial v_j}(\hat{x}_{k|k-1}, 0) \end{aligned}$$

which are evaluated at $\hat{x}_{k|k-1}$ and zero noise. Having the Jacobians, we follow what was done in the Kalman filter to write the time update equation. For the time update we have

$$\hat{x}_{k|k-1} = f(\hat{x}_{k-1|k-1}, 0)$$

and its corresponding covariance matrix

$$\hat{P}_{k|k-1} = F_k \hat{P}_{k-1|k-1} F_k^\top + U_k Q_{k-1} U_k^\top$$

The Kalman gain matrix, $K_k$ is defined in the measurement equation as

$$\hat{x}_{k|k} = \hat{x}_{k|k-1} + K_k(z_k - h(\hat{x}_{k|k-1}, 0))$$

and corresponding covariance matrix

$$P_{k|k} = (I - K_k H_k) P_{k|k-1}$$

As in the Kalman filter, the optimal Kalman gain that minimizes the mean square error over all linear estimators is given by

$$K_k = P_{k|k-1} H_k^\top (H_k P_{k|k-1} H_k^\top + V_k R_k V_k^\top)^{-1}$$

**Example 25** *Parameter estimation of the Heston stochastic volatility model via the extended Kalman filter*

In this example, we set up parameter estimation of the Heston stochastic volatility model via the extended Kalman filter. In Heston stochastic volatility, the underlying process follows the following SDE

$$\begin{aligned} dS_t &= \mu S_t dt + \sqrt{v_t} S_t dW_t^1, \\ dv_t &= \kappa(\theta - v_t)dt + \lambda\sqrt{v_t} dW_t^2, \end{aligned}$$

where the two Brownian components $W_t^1$ and $W_t^2$ are correlated with rate $\rho$ under physical measure. Define $y_t = \ln(S_t)$ and using Itô's lemma we can write it as

$$\begin{aligned} dy_t &= (\mu - \frac{1}{2}v_t)dt + \sqrt{v_t} dW_t^1 \\ dv_t &= \kappa(\theta - v_t)dt + \sigma\sqrt{v_t} dW_t^2 \end{aligned}$$

For Heston, the state equation is

$$\begin{aligned} x_k = f(x_{k-1}, u_k) &= \begin{pmatrix} y_t \\ v_k \end{pmatrix} \\ &= \begin{pmatrix} y_{t-1} + (\mu - \frac{1}{2}v_{k-1})\Delta t + \sqrt{v_{k-1}}\sqrt{\Delta t} Z_{k-1}^1 \\ v_{k-1} + \kappa(\theta - v_{k-1})\Delta t + \lambda\sqrt{v_{k-1}}\sqrt{\Delta t} Z_{k-1}^2 \end{pmatrix} \end{aligned}$$

with system noise

$$w_k = \begin{pmatrix} Z_k^1 \\ Z_k^2 \end{pmatrix}$$

and the covariance matrix

$$Q_k = \begin{pmatrix} 1 & \rho \\ \rho & 1 \end{pmatrix}$$

It is easy to see that in the extended Kalman filter, $F_k$ and $U_k$ for Heston stochastic volatility are

$$F_k = \begin{pmatrix} 1 & -\frac{1}{2}\Delta t \\ 0 & 1 - \kappa\Delta t \end{pmatrix}$$

and

$$U_k = \begin{pmatrix} \sqrt{v_{k-1}}\sqrt{\Delta t} & 0 \\ 0 & \lambda\sqrt{v_{k-1}}\sqrt{\Delta t} \end{pmatrix}$$

We assume measurement equation $y_k = \ln(S_k)$, which implies $H_k = (1\ 0)$. For a given set of parameters $\Theta = \{\kappa, \theta, \lambda, \rho, v_0\}$, we would minimize the following summation as our objective function to obtain the optimal parameter set for the model

$$\sum_{i=1}^{N} \left( \ln(A_k) + \frac{e_k^2}{A_k} \right)$$

where

$$e_k = z_k - h(\hat{x}_{k|k-1}, 0)$$

and

$$A_k = H_k P_{k|k-1} H_k^\top + V_k R_k V_k^\top$$

As mentioned earlier, minimization can be performed via the Nelder–Mead optimization algorithm. We leave the estimation and filtering as an exercise at the end of the chapter.

## 8.6 Unscented Kalman Filter

When the state transition and observation models, that is, the predict and update functions $f$ and $h$ are highly non-linear, the extended Kalman filter can give particularly poor performance. This is because only the mean is propagated through the non-linearity. The unscented Kalman filter (UKF) ([216], [91]) uses a deterministic sampling technique known as the unscented transform to pick a minimal set of sample points called sigma points around the mean. These sigma points are then propagated through the non-linear functions, and the covariance of the estimate is then recovered. The result is a filter which more accurately captures the true mean and covariance. This can be verified using Monte Carlo sampling or through a Taylor series expansion of the posterior statistics. In addition, this technique removes the requirement to explicitly calculate Jacobians, which for complex functions can be a difficult task in itself (i.e., requiring complicated derivatives if done analytically or being computationally costly if done numerically).

### 8.6.1 Predict

As with the EKF, the UKF prediction can be used independently from the UKF update, in combination with a linear (or indeed EKF) update, or vice versa. The estimated state and covariance are augmented with the mean and covariance of the process noise.

$$\begin{aligned} x^a_{k-1|k-1} &= \begin{bmatrix} \hat{x}^\top_{k-1|k-1} \\ \mathbb{E}[w^\top_k] \end{bmatrix} \\ P^a_{k-1|k-1} &= \begin{bmatrix} P_{k-1|k-1} & 0 \\ 0 & Q_k \end{bmatrix} \end{aligned}$$

A set of $2L+1$ sigma points is derived from the augmented state and covariance where $L$ is the dimension of the augmented state.

$$\begin{aligned} \chi^0_{k-1|k-1} &= x^a_{k-1|k-1} \\ \chi^i_{k-1|k-1} &= x^a_{k-1|k-1} + \left(\sqrt{(L+\lambda)P^a_{k-1|k-1}}\right)_i \qquad i = 1, \ldots, L \\ \chi^i_{k-1|k-1} &= x^a_{k-1|k-1} - \left(\sqrt{(L+\lambda)P^a_{k-1|k-1}}\right)_{i-L} \qquad i = L+1, \ldots, 2L \end{aligned}$$

where

$$\left(\sqrt{(L+\lambda)P^a_{k-1|k-1}}\right)_i$$

is the $i$-th column of the matrix square root of

$$(L+\lambda)P^a_{k-1|k-1}$$

The matrix square root should be calculated using numerically efficient and stable methods such as the Cholesky decomposition.

The sigma points are propagated through the transition function $f$.

$$\chi^i_{k|k-1} = f(\chi^i_{k-1|k-1}) \quad i = 0, \ldots, 2L$$

The weighted sigma points are recombined to produce the predicted state and covariance.

$$\begin{aligned}
\hat{x}_{k|k-1} &= \sum_{i=0}^{2L} W_i^{(m)} \chi_{k|k-1}^i \\
P_{k|k-1} &= \sum_{i=0}^{2L} W_i^{(c)} [\chi_{k|k-1}^i - \hat{x}_{k|k-1}][\chi_{k|k-1}^i - \hat{x}_{k|k-1}]
\end{aligned}$$

Weights for the state and covariance are given by

$$\begin{aligned}
W_0^{(m)} &= \frac{\lambda}{L+\lambda} \\
W_0^{(c)} &= \frac{\lambda}{L+\lambda} + (1 - \alpha^2 + \beta) \\
W_i^{(m)} &= \frac{1}{2(L+\lambda)} \qquad i = 1, \ldots, 2L \\
W_i^{(c)} &= \frac{1}{2(L+\lambda)} \qquad i = 1, \ldots, 2L
\end{aligned}$$

where

$$\lambda = \alpha^2 (L + \kappa) - L$$

### 8.6.2 Update

The predicted state and covariance are augmented as before, except now with the mean and covariance of the measurement noise.

$$\begin{aligned}
x_{k|k-1}^a &= \begin{bmatrix} \hat{x}_{k|k-1}^\top \\ \mathbb{E}[v_k^\top] \end{bmatrix} \\
P_{k|k-1}^a &= \begin{bmatrix} P_{k|k-1} & 0 \\ 0 & R_k \end{bmatrix}
\end{aligned}$$

As before, a set of $2L+1$ sigma points is derived from the augmented state and covariance where $L$ is the dimension of the augmented state.

$$\begin{aligned}
\chi_{k|k-1}^0 &= x_{k|k-1}^a \\
\chi_{k|k-1}^i &= x_{k|k-1}^a + \left(\sqrt{(L+\lambda) P_{k|k-1}^a}\right)_i \qquad i = 1, \ldots, L \\
\chi_{k|k-1}^i &= x_{k|k-1}^a - \left(\sqrt{(L+\lambda) P_{k|k-1}^a}\right)_{i-L} \qquad i = L+1, \ldots, 2L
\end{aligned}$$

Alternatively, if the UKF prediction has been used, the sigma points themselves can be augmented along the following lines:

$$\chi_{k|k-1} = [\chi_{k|k-1}^\top \mathbb{E}[v_k^\top]]^\top \pm \sqrt{(L+\lambda) R_k^a}$$

where

$$R_k^a = \begin{bmatrix} 0 & 0 \\ 0 & R_k \end{bmatrix}$$

The sigma points are projected through the observation function $h$.

$$\gamma_k^i = h(\chi_{k|k-1}^i) \quad i = 0, \ldots, L$$

The weighted sigma points are recombined to produce the predicted measurement and predicted measurement covariance.

$$\begin{aligned}
\hat{z}_k &= \sum_{i=0}^{2L} W_i^{(m)} \gamma_k^i \\
P_{z_k z_k} &= \sum_{i=0}^{2L} W_i^{(c)} [\gamma_k^i - \hat{z}_k][\gamma_k^i - \hat{z}_k]^\top
\end{aligned}$$

The state-measurement cross-covariance matrix

$$P_{x_k z_k} = \sum_{i=0}^{2L} W_i^{(c)} [\chi_{k|k-1}^i - \hat{x}_{k|k-1}][\gamma_k^i - \hat{z}_k]^\top$$

is used to compute the UKF Kalman gain.

$$K_k = P_{x_k z_k} P_{z_k z_k}^{-1}$$

As with the Kalman filter, the updated state is the predicted state plus the innovation weighted by the Kalman gain

$$\hat{x}_{k|k} = \hat{x}_{k|k-1} + K_k(z_k - \hat{z}_k)$$

And the updated covariance is the predicted covariance minus the predicted measurement covariance, weighted by the Kalman gain.

$$P_{k|k} = P_{k|k-1} - K_k(P_{z_k z_k} K_k^\top)$$

### 8.6.3 Implementation of Unscented Kalman Filter (UKF)

1. Initialize

$$\begin{aligned}
\hat{x_0} &= \mathbb{E}(x_0) \\
P_0 &= \mathbb{E}[(x_0 - \hat{x_0})(x_0 - \hat{x_0})^\top]
\end{aligned}$$

2. Calculate sigma points

$$\mathbf{x}_{k-1} = \left[ \hat{x}_{k-1} \quad \hat{x}_{k-1} + \gamma\sqrt{P_{k-1}} \quad \hat{x}_{k-1} - \gamma\sqrt{P_{k-1}} \right]$$

3. Time update

$$
\begin{aligned}
\mathbf{x}^*_{k|k-1} &= F[\mathbf{x}_{k-1}, u_{k-1}] \\
\hat{x}^-_k &= \sum_{i=0}^{2L} W_i^{(m)} \mathbf{x}^*_{i,k|k-1} \\
P^-_k &= \sum_{i=0}^{2L} W_i^{(c)} [\mathbf{x}^*_{i,k|k-1} - \hat{x}^-_k][\mathbf{x}^*_{k|k-1} - \hat{x}^-_k]^\top + Q \\
\mathbf{x}_{k|k-1} &= \left[ \hat{x}^-_k \quad \hat{x}^-_k + \gamma\sqrt{P^-_k} \quad \hat{x}^-_k - \gamma\sqrt{P^-_k} \right] \\
\mathbf{y}_{k|k-1} &= H[\mathbf{x}_{k|k-1}] \\
\hat{y}^-_k &= \sum_{i=0}^{2L} W_i^{(m)} \mathbf{y}_{i,k|k-1}
\end{aligned}
$$

4. Measurement update equations

$$
\begin{aligned}
P_{\hat{y}_k\hat{y}_k} &= \sum_{i=0}^{2L} W_i^{(c)} [\mathbf{y}_{i,k|k-1} - \hat{y}^-_k][\mathbf{y}_{i,k|k-1} - \hat{y}^-_k]^\top + R \\
P_{x_k y_k} &= \sum_{i=0}^{2L} W_i^{(c)} [\mathbf{x}_{i,k|k-1} - \hat{x}^-_k][\mathbf{y}_{i,k|k-1} - \hat{y}^-_k]^\top \\
K_k &= P_{x_k y_k} P^{-1}_{\hat{y}_k\hat{y}_k} \\
\hat{x}_k &= \hat{x}^-_k + K_k(y_k - \hat{y}^-_k) \\
P_k &= P^-_k - K_k P_{\hat{y}_k\hat{y}_k} K_k^\top
\end{aligned}
$$

where $Q$ and $R$ are the process noise covariance and measurement noise covariance matrices, respectively. For discrete time and continuous time models, we can get the analytical form of $R^v$, but we do not have the formula for $R$. We can get this matrix through a parameter estimation process, or for simplicity we assume $R = \delta I$, where $\delta$ is a small scaling value and $I$ is the identity matrix.

Some comments:

(1) The square root of $P$ is defined by $SS^\top = P$, which requires a Cholesky factorization.

(2) To get the square root of $P$, $P$ must be positive semi-definite. In implementation, if $P$ has a negative eigenvalue we modify it by adding a diagonal matrix with very small entries on the diagonal, that is, $P = P + \delta I$. We keep doing it until all eigenvalues of $P$ become non-negative. Given a proper model parameter set, the filter convergence would be fast, $P$ stays positive semi-definite, and no modification is needed.

(3) To calculate the Kalman gain, the inverse matrix of $P_{\hat{y}_k\hat{y}_k}$ is needed. If $P_{\hat{y}_k\hat{y}_k}$ is likely singular, the inverse matrix would have extreme values, consequently the new sigma points would be in an unreasonably large range. So adding a proper $R_n$ is a must.

(4) We use $\mathbf{x}_{k|k-1} = \mathbb{E}(\mathbf{x}_{k-1})$ to calculate the one step forward sigma points and $cov(\mathbf{x}_{k-1})$ is the process noise covariance $Q$.

**Example 26** *Parameter estimation of the Heston stochastic volatility model via unscented Kalman filter*

UKF implementation details for Heston model is as follow: starting from the SDE

$$\begin{aligned} dS_t &= \mu S_t dt + \sqrt{v_t} S_t dw_t^1 && (8.2)\\ dv_t &= \kappa(\theta - v_t)dt + \lambda\sqrt{v_t} dw_t^2 && (8.3) \end{aligned}$$

Define $y_t = \log S_t$, and using Itô's lemma we can rewrite (8.2) as follows:

$$dy_t = (\mu - \frac{1}{2}v_t)dt + \sqrt{v_t} dw_t^1 \quad (8.4)$$

Knowing the correlation between $dw_t^1$ and $dw_t^2$ is $\rho$ we can write

$$dw_t^2 = \rho dw_t^1 + \sqrt{1-\rho^2} dz_t \quad (8.5)$$

and substituting it into (8.3) we get

$$dv_t = \kappa(\theta - v_t)dt + \lambda\sqrt{v_t}(\rho dw_t^1 + \sqrt{1-\rho^2} dz_t) \quad (8.6)$$

From (8.4) we can write

$$dw_t^1 = \frac{dy_t - (\mu - \frac{1}{2}v_t)dt}{\sqrt{v_t}} \quad (8.7)$$

By substituting (8.7) in (8.6) we obtain

$$\begin{aligned} dv_t &= \kappa(\theta - v_t)dt + \lambda\sqrt{v_t}(\rho\frac{dy_t - (\mu - \frac{1}{2}v_t)dt}{\sqrt{v_t}} + \sqrt{1-\rho^2} dz_t)\\ &= \kappa(\theta - v_t)dt + \lambda\rho(dy_t - (\mu - \frac{1}{2}v_t)dt) + \lambda\sqrt{v_t}\sqrt{1-\rho^2} dz_t) && (8.8) \end{aligned}$$

Discretizing the state equation (8.8) we obtain

$$v_t = v_{t-1} + \kappa(\theta - v_{t-1})\Delta t + \lambda\rho(y_t - y_{t-1} - (\mu - \frac{1}{2}v_{t-1})\Delta t) + \lambda\sqrt{v_{t-1}(1-\rho^2)\Delta t}\; z_1 \quad (8.9)$$

where $z_1 \sim \mathcal{N}(0,1)$. We can write the state equation as follows: Define $x_k \triangleq v_k$ and $z_k \triangleq y_k$

$$\begin{aligned} x_k &= F_k x_{k-1} + u_k + w_k && (8.10)\\ w_k &= \sqrt{Q_k}\; e_1 && (8.11) \end{aligned}$$

where

$$\begin{aligned} F_k &= 1 - \kappa\Delta t + \frac{1}{2}\rho\lambda\Delta t && (8.12)\\ U_k &= \kappa\theta\Delta t - \lambda\rho\mu\Delta t + \lambda\rho(z_t - z_{t-1}) && (8.13)\\ Q_k &= \lambda^2(1-\rho^2)x_{k-1}\Delta t && (8.14) \end{aligned}$$

Here $U_t$ plays the role of control input vector. For the measurement equation (8.4), we use an explicit-implicit scheme to get

$$z_k = z_{k-1} + (\mu - \frac{1}{2}x_k)\Delta t + \sqrt{x_{k-1}\Delta t}\; z_2 \quad (8.15)$$

**TABLE 8.1**: Heston stochastic volatility parameters via UKF for the S&P 500 and USD/JPY

| data set | $\kappa$ | $\theta$ | $\sigma$ | $\mu$ | $\rho$ | $v_0$ |
|---|---|---|---|---|---|---|
| S&P500 | 2.1924 | 0.0133 | 0.2940 | 0.2016 | -0.6143 | 0.0233 |
| JPY/USD | 1.4785 | 0.0072 | 0.1000 | 0.0103 | -0.5656 | 0.0121 |

where $z_2 \sim \mathcal{N}(0,1)$ or equivalently

$$z_k = H_k x_k + U_k + \sqrt{R_k}\, z_2 \tag{8.16}$$

where

$$H_k = -\frac{1}{2}\Delta t \tag{8.17}$$

$$U_k = z_{k-1} + \mu\Delta t \tag{8.18}$$

$$R_k = x_{k-1}\Delta t \tag{8.19}$$

The rest are fully described in the UKF procedure for propagating state propagating, prediction, and formulation of the log likelihood.

Datasets used for this example are: (a) daily close of S&P 500 from November 11, 2001 to November 11, 2011 (b) daily close of U.S. dollar vs. Japanese yen (USD/JPY) exchange rate from November 11, 2001 to November 11, 2011. For each we use the time series to estimate Heston stochastic volatility model parameters. The parameter set for estimation is $\Theta = \{\kappa, \theta, \sigma, \mu, \rho, v_0\}$ and the hidden state for filtering is $\{v_t, 1 \leq t \leq T\}$. Table 8.1 displays Heston stochastic volatility parameters obtained from estimation via UKF for S&P 500 and USD/JPY datasets. For parameter estimation, the starting point for USD/JPY parameter sets $\Theta = [\kappa, \theta, \sigma, \mu, \rho, v_0]$ is given by $\Theta_0 = [2.3618, 0.0166, 0.3451, -0.0676, -0.0806, 0.0166]$, lower bound and upper bound for $\Theta$: $lb = [0.3, 0.00025, 0.05, -0.5, -0.99, 0.00025]$, $ub = [5, 0.64, 0.7, 0.5, 0.3, 0.64]$. For parameter estimation, the starting point for the S&P 500 parameter set $\Theta = [\kappa, \theta, \sigma, \mu, \rho, v_0]$ is given by $\Theta_0 = [1.4440, 0.0184, 0.3353, 0.0125, -0.5267, 0.0225]$, lower bound and upper bound for $\Theta$: $lb = [0.3, 0.00025, 0.05, -0.5, -0.99, 0.00025]$, $ub = [5, 0.64, 0.7, 0.5, 0.3, 0.64]$. The initial covariance matrix $P_0 = var(x_0) = 0.00001$. In Figures 8.9(a) and 8.10(a) we display time series versus its prediction for S&P 500 and USD/JPY respectively. Figures 8.9(b) and 8.10(b) show the hidden underlying volatility for S&P 500 and USD/JPY, respectively.

**Example 27** *Parameter estimation of an affine term structure with constant volatility via unscented Kalman filter*

In this example we assume a three-factor affine dynamic term structure model with constant volatility. This class belongs to the affine class of dynamics term structure models of [101], [100] and studied in [24]. Assume the short rate is affine and follows

$$r_t = a_r + b_r^\top x_t$$

where we take the vector $x_t$ to follow the matrix $OU$ equation, driven by Brownian noise under physical measure $\mathbb{P}$. Namely,

$$dx_t = -Bx_t dt + dW_t^{\mathbb{P}}$$

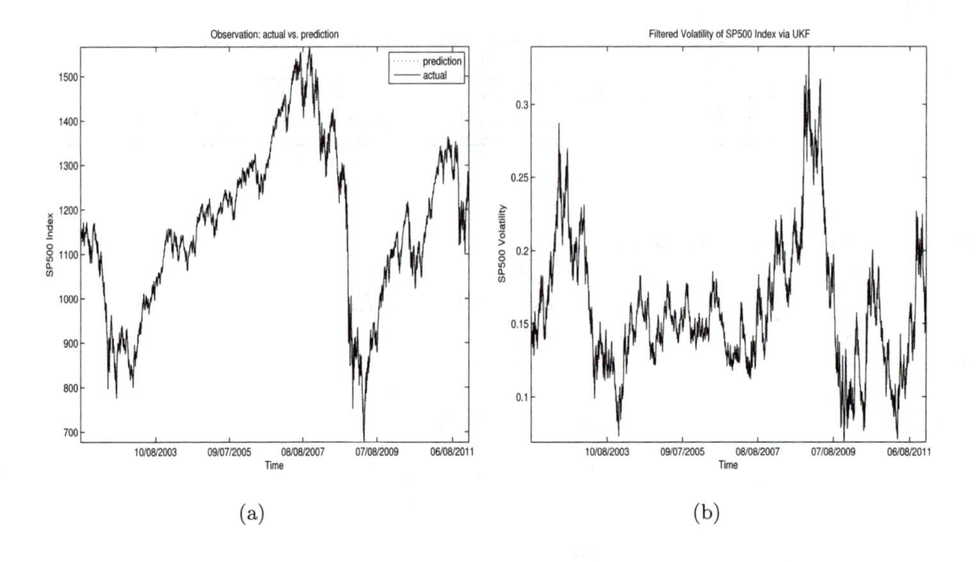


**FIGURE 8.9**: Heston stochastic volatility model UKF example for S&P 500. (a) Actual versus prediction, (b) volatility actual versus prediction

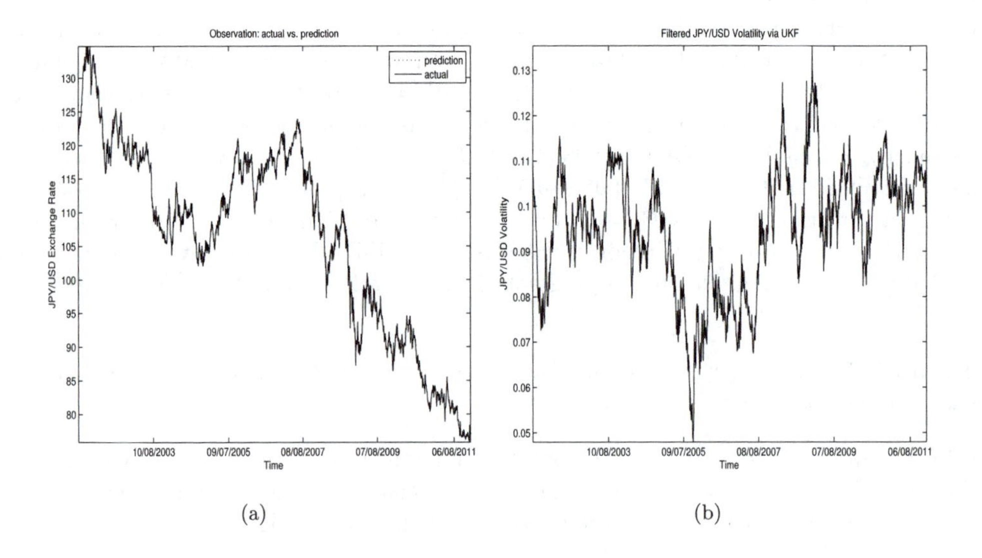


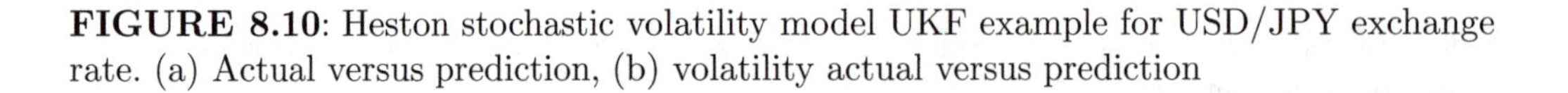
**FIGURE 8.10**: Heston stochastic volatility model UKF example for USD/JPY exchange rate. (a) Actual versus prediction, (b) volatility actual versus prediction

In general, matrix $B$ may be full, with all eigenvalues having a positive real part. We assume an affine market price of risk

$$\gamma(x_t) = b_\gamma + B_\gamma x_t$$

where $b_\gamma \in \mathbb{R}^3$ and $B_\gamma$ is a $3 \times 3$ matrix. With the market price of risk, the dynamic of factors under a risk-neutral measure are

$$dx_t = -b_\gamma dt - B^* x_t dt + \Sigma dW_t^{\mathbb{Q}}$$

where $B^* = B + b_\gamma$. Zero-coupon bond prices under this framework are exponential affine in the state vector $x_t$.

$$P(x_t, \tau) = \exp(-a(\tau) - b(\tau)^\top x_t)$$

where $\tau = T - t$ time to maturity and the loading factors $a(\tau)$ and $b(\tau)$ are determined by the following Riccati equations:

$$\begin{aligned} a'(\tau) &= a_r - b(\tau)^\top b_\gamma - b(\tau)^\top b(\tau)/2 \\ b'(\tau) &= b_r - b(\tau)^\top B^* \end{aligned}$$

subject to the initial conditions $a(0) = 0$ and $b(0) = 0$. There might be analytical solutions to this, but we solve it via a numerical procedure. The discrete version of the state equation is

$$x_{t+1} = \Phi x_t + \sqrt{Q}\varepsilon$$

where $\Phi = \exp(-B\Delta t)$ and $\varepsilon \sim \mathcal{N}(0, 1)$ and the conditional variance

$$\begin{aligned} Q &= \int_0^{\Delta t} e^{-uB}\Sigma\Sigma^\top e^{-uB^\top} du \\ &= \int_0^{\Delta t} Ue^{-uD}U^\top \Sigma\Sigma^\top U e^{-uD^\top} U^\top du \\ &= U\left(U^\top \Sigma\Sigma^\top U\right)\left(\int_0^{\Delta t} e^{-uD} e^{-uD^\top} du\right) U^\top \end{aligned}$$

The conditional variance $Q$ can be computed based on eigenvalues and eigenvectors of matrix $B$ as follows:

$$Q = U\left(U^\top \Sigma\Sigma^\top U^\top\right) D^{-1}(I - e^{-D\Delta t})U^\top$$

where $U$ and $D$ are eigenvectors and eigenvalues of $B + B^\top$ respectively, and $B + B^\top = UDU^\top$. The observation is linked to the measurement equation as follows:

$$\begin{aligned} y_t &= h(x_t; \Theta) + e_t \\ &= \begin{bmatrix} LIBOR(x_t, T_i) \\ swap(x_t, T_j) \end{bmatrix} + e_t \end{aligned}$$

with $cov(e_t) = R$ where $e_t$ denotes the measurement error at time $t$. We assume that measurement error is independent of the state vector and also mutually independent on each series with a distinct variance $\sigma_i$. Therefore $R_{ii} = \sigma_i^2$ for $i = 1, \ldots, N$ and $R_{ij} = 0$ for $i \neq j$. We use 15 years of LIBOR rates and swap rates to estimate model parameters. Time series of the interest rate factors (hidden states) are the by-product of the estimation procedure. Data consists of 1) LIBOR rates with maturities of 1, 2, 3, and 6 months, 2) swap rates at maturities 2, 3, 5, 10, 15, and 30 years. All interest rates are in U.S. dollars. The data are daily closing mid-quotes from March 3, 1997 through October 20, 2011 (3,687 observations for each instrument). The likelihood function can be constructed based on the conditional density of state variables and the pricing errors. In this example, the state propagation equation is Gaussian linear but the measurement equations in terms of LIBOR and swap rates are nonlinear in the state variables.

For the estimation, we assume that the measurement errors are independent with distinct variance $\sigma^2$. We also assume that $B$ and $B^*$ are lower triangular matrices. Thus we have the following parameter set: $\Theta = \{B, B^*, a_r, b_r, b_\gamma, \sigma\}$.

Denote the log-likelihood of each day's observation on the forecasting errors of the observed series

$$l_{k+1}(\Theta) = -\frac{1}{2}\log|P_{\hat{y}_k\hat{y}_k}| - \frac{1}{2}\left((y_{k+1} - \hat{y}_{k+1}^-)^\top P_{\hat{y}_{k+1}\hat{y}_{k+1}}^{-1}(y_{k+1} - \hat{y}_{k+1}^-)\right)$$

We define the measurement equation using LIBOR rates and swap rates assuming additive normal errors. The parameter set for estimation is $\Theta = \{a_r, b_r, b_\gamma, B, B^\star\}$. The hidden state for filtering is $\{x_t, 1 \leq t \leq T\}$.

Obtained parameters are

$$\begin{aligned}
a_r &= 0.0345 \\
b_r &= [0.000000614016347, 0.000000000001379, 0.004292529402460] \\
b_\gamma &= [0.0174, -0.0445, 0.3406] \\
B &= \begin{pmatrix} 0.0052 & & \\ -0.0668 & 0.4500 & \\ -0.0676 & -0.3796 & 0.5134 \end{pmatrix} \\
B^\star &= \begin{pmatrix} 0.0000007 & & \\ -0.1328506 & 1.1443663 & \\ -0.0547356 & -0.0892296 & 0.9696811 \end{pmatrix} \\
\Sigma &= \begin{pmatrix} 1.3769 & & \\ 1.5553 & -2.0813 & \\ -0.0590 & 0.0759 & 0.3536 \end{pmatrix}
\end{aligned}$$

In Figures 8.11(a)–8.11(d) we display one-month LIBOR actual vs. fitted, one-month LIBOR actual vs. prediction, 6-month LIBOR actual, and 6-month LIBOR actual vs. prediction respectively obtained for the affine term structure model with constant volatility using unscented Kalman filter. In Figures 8.12(a)–8.12(d) we display five-year swap actual vs. fitted, five-year swap actual vs. prediction, fifteen-year swap actual vs. fitted, and fifteen-year swap actual vs. prediction respectively. In Figures 8.13(a) and 8.13(b) we display thirty-year swap actual versus fitted and thirty-year swap actual versus prediction respectively. In Figure 8.14 we display time series of the hidden state obtained through filtering.

**Example 28** *Parameter estimation of an affine term structure with stochastic volatility via unscented Kalman filter*

In this example, we assume a three-factor affine dynamic term structure model with stochastic volatility [132]. This class belongs to the affine class of dynamics term structure models of [101], [100]. Assume the short rate is affine and follows

$$r_t = a_r + b_r^\top x_t$$

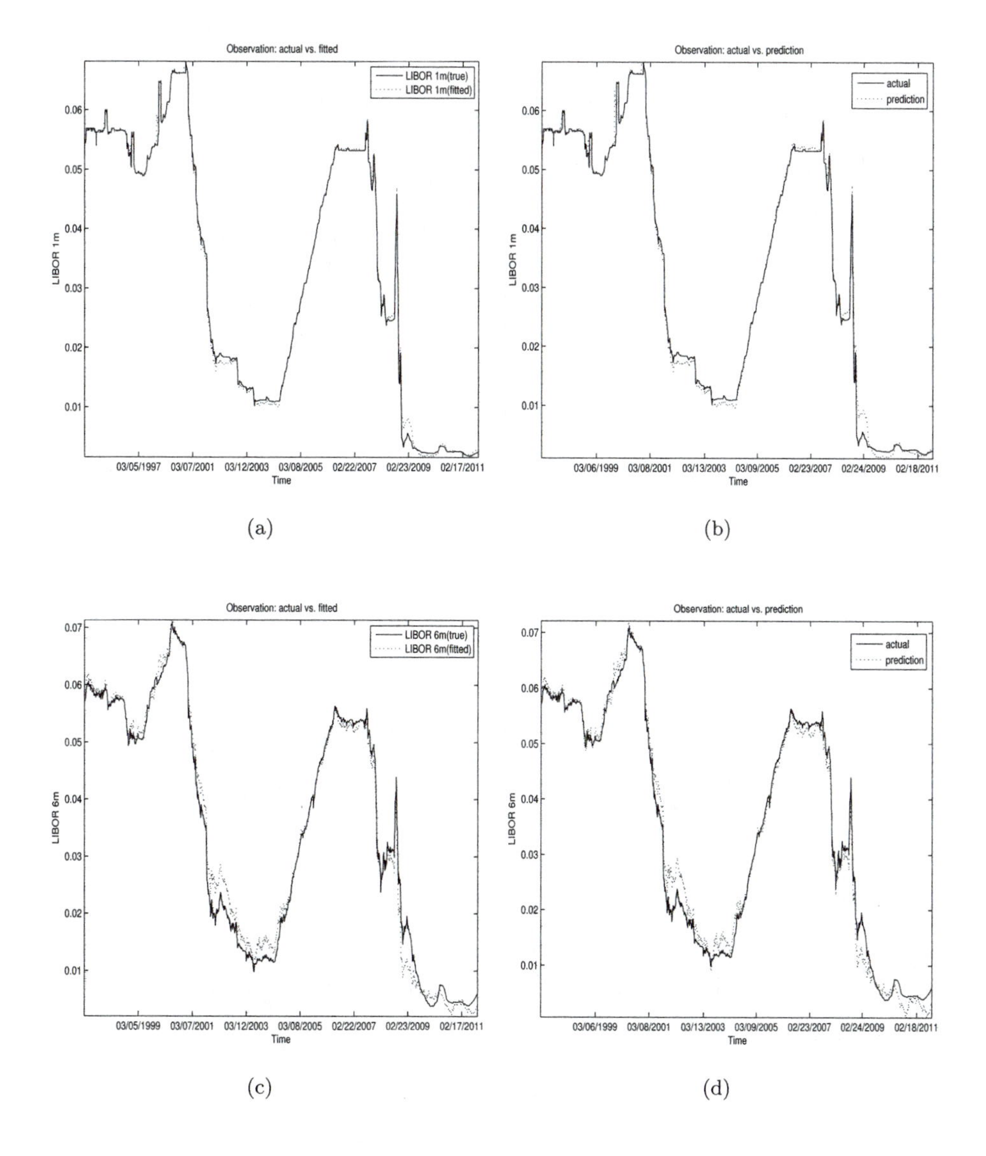


**FIGURE 8.11**: Affine term structure model with constant volatility UKF example: (a) one-month LIBOR actual vs. fitted, (b) one-month LIBOR actual vs. prediction (c) 6-month LIBOR actual vs. fitted (d) 6-month LIBOR actual vs. prediction

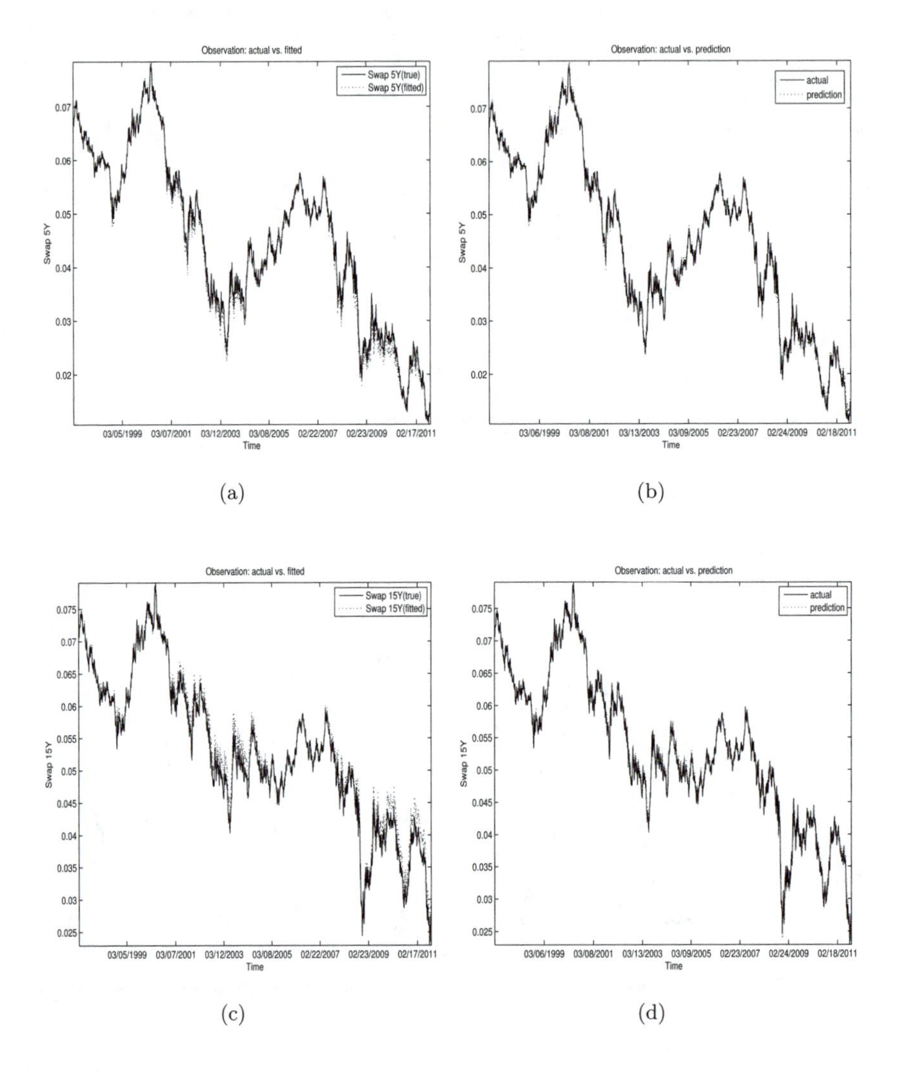


**FIGURE 8.12**: Affine term structure model with constant volatility UKF example: (a) 5-year swap actual vs. fitted, (b) 5-year swap actual vs. prediction, (c) 15-year swap actual vs. fitted, (d) 15-year swap actual vs. prediction

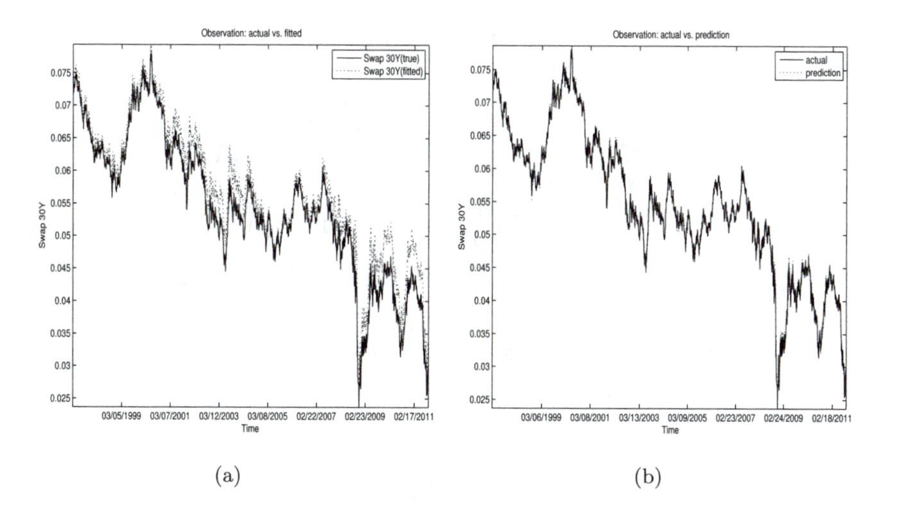


(a) (b)

**FIGURE 8.13**: Affine term structure model with constant volatility UKF example: (a) 30-year swap actual vs. fitted, (b) 30-year swap actual vs. prediction

where $x_t$ is an $n \times 1$ state factor follows the following SDE:

$$dx_t = (-b_\gamma - Bx_t)dt + \Sigma\sqrt{v_t}dW_t$$

and $v_t$ follows

$$dv_t = \kappa(\theta - v_t)dt + \lambda\sqrt{v_t}dZ_t$$

for $Z_t$ a univariate Brownian motion possibly correlated with $W_t$ with vector correlation $\rho = d < W, Z >$.

In this affine term structure framework with stochastic volatility, the bond pricing and the characteristic function of the log swap rate under the swap measure are derived as shown in [132]. Having the bond prices, one can calculate LIBOR rates and swap rates; having the characteristic function, we can employ any transform technique to price swaptions. Using LIBOR rate, swap rates, and swaption premiums, model parameters can be estimated using the unscented Kalman filter.

To utilize UKF for parameter and state estimation of this three factor affine model, we set up a state formula for this continuous-time model. For one step forward we have

$$\begin{aligned} x_{k+1} &= F(x_k) = \mathbb{E}(x_{k+1}|x_k) \\ &= x_k + (-b_\gamma - Bx_k)\Delta t \\ v_{k+1} &= F(v_k) = \mathbb{E}(v_{k+1}|v_k) \\ &= v_k + \kappa(\theta - v_k)\Delta t \end{aligned}$$

and

$$\begin{aligned} P_{xx} &= var(x_{t+1}) = \Sigma\Sigma^\top v_t \Delta t \\ P_{yy} &= var(v_{t+1}) = \lambda^2 v_t \Delta t \\ P_{xy} &= cov(x_{t+1}, v_{t+1}) = \lambda v_t \rho \Sigma dt \\ R_v(v_t) &= \begin{bmatrix} P_{xx} & P_{xy} \\ P_{xy}^\top & P_{yy} \end{bmatrix} \end{aligned}$$

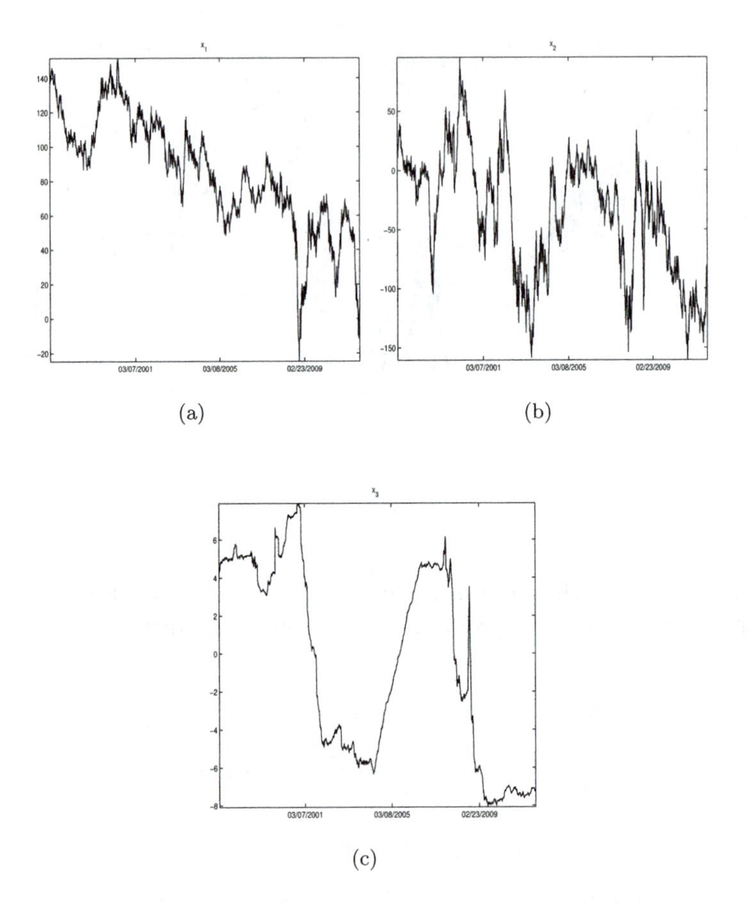


**FIGURE 8.14**: Affine term structure model with constant volatility UKF example: hidden state

We assume

$$\mathbb{E}(F(x_0)) = x_0 \tag{8.20}$$
$$\mathbb{E}(F(u_0)) = v_0 \tag{8.21}$$

Therefore

$$x_0 = -B^{-1}b_\gamma \tag{8.22}$$
$$v_0 = \theta \tag{8.23}$$

and

$$y_t = \begin{bmatrix} LIBOR(x_t, T_i) \\ swap(x_t, T_j) \\ swaption(x_t, T_i, T_j) \end{bmatrix} + e_t$$

Data used for parameter estimation consists of 1) LIBOR rates with maturities of one, two, three, six and twelve months, 2) swap rates at maturities two, three, five, ten, fifteen and thirty years and 3) at-the-money swaption premiums with maturities of one, two, five

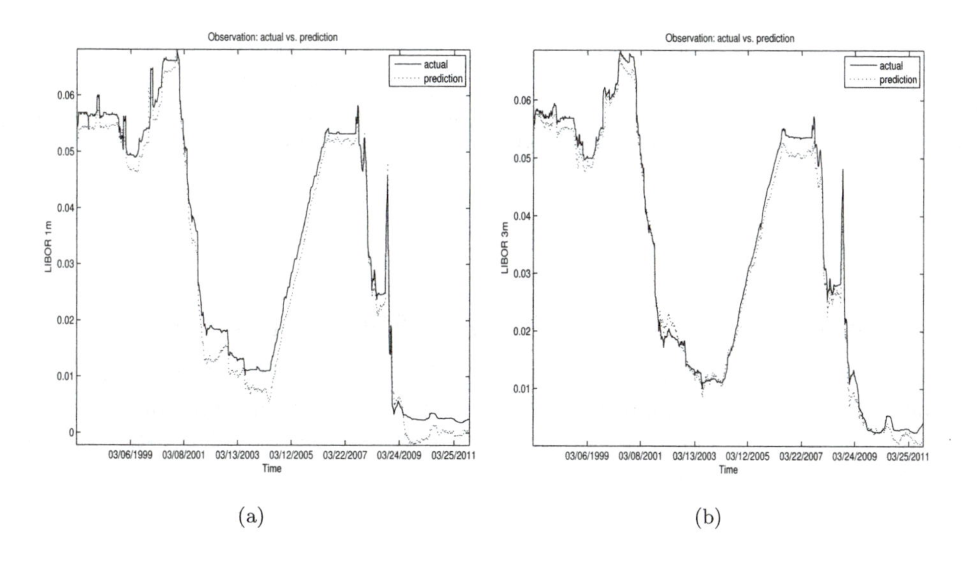


**FIGURE 8.15**: Affine term structure stochastic volatility UKF example: (a) 1-month LIBOR actual versus prediction, (b) 3-month LIBOR actual versus prediction

and ten years. At each option maturity, we have three contracts with different underlying swap maturities: two, five, and ten years. All interest rates and interest rate options are on U.S. dollars. The data are daily closing mid-quotes from March 3, 1997 through October 20, 2011 (3,687 observations for each instrument). Parameter set for estimation $\Theta = \{a_r, b_r, b_\gamma, B, B^\star, \Sigma, \kappa, \theta, \lambda\}$. Hidden State for Filtering is $\{v_t, 1 \leq t \leq T\}$. We obtain the following parameters from the estimation procedure.

$$\begin{aligned}
a_r &= 0.1098 \\
b_r &= [0.0008, 0.0009, 0.0836] \\
b_\gamma &= [0.3588, -0.1526, -0.3297] \\
B &= \begin{pmatrix} 0.3703 & & \\ -0.0582 & 0.2857 & \\ -0.2002 & 0.0074 & 1.3889 \end{pmatrix} \\
B^\star &= \begin{pmatrix} 0.0043 & & \\ -0.1403 & 0.0043 & \\ -0.8564 & -0.0269 & 0.8514 \end{pmatrix} \\
\Sigma &= \begin{pmatrix} 2.4496 & 0.0313 & 0.0074 \\ 0.1854 & 2.3478 & -0.1181 \\ 0.0127 & 0.4236 & 0.4310 \end{pmatrix} \\
\kappa &= 0.9997 \\
\theta &= 1.0014 \\
\lambda &= 1.1628
\end{aligned}$$

In Figures 8.15(a) and 8.15(b) we display one-month LIBOR actual versus prediction and three-month LIBOR actual versus prediction respectively obtained for the affine term structure model with stochastic volatility using unscented Kalman filter. In Figures 8.16(a)–8.16(d) we display five-year swap rate, ten-year swap rate, fifteen-year swap rate, and thirty-year swap rate actual versus prediction respectively. In Figure 8.17 we display states of the affine term structure with stochastic volatility obtained via unscented Kalman filter.

In [132] the authors look more closely at the relationship between model premiums and interest rate factors, as well as market premiums and interest factors to conclude that long-dated swaptions are highly correlated to the slope of the yield curve.

---

## 8.7 Square Root Unscented Kalman Filter (SR UKF)

Implementation of Cholesky factorization in UKF requires $O(L^3/6)$ computations where $L$ is the state dimension, a square root unscented Kalman filter [92] can reduce it to $O(L^2)$; and speed up the filtering process especially for the system that has a large state dimension. Here are the implementation steps.

1. Initialize

$$\begin{aligned}
\hat{x}_0 &= \mathbb{E}(x_0) \\
S_0 &= chol\left(\mathbb{E}[(x_0 - \hat{x_0})(x_0 - \hat{x_0})^\top]\right) \\
x_{t0} &= -B^{-1}b_\gamma \\
v_{t0} &= \theta \\
\hat{x}_0 &= \begin{bmatrix} x_{t0} \\ v_{t0} \end{bmatrix}
\end{aligned}$$

   and $S_0 = chol(R_v)$ or identity matrix for simplicity.

2. Calculate sigma points

$$\mathbf{x}_{k-1} = [\hat{x}_{k-1} \quad \hat{x}_{k-1} + \gamma S_k \quad \hat{x}_{k-1} - \gamma S_k]$$

3. Time update

$$\begin{aligned}
\mathbf{x}^*_{k|k-1} &= F[\mathbf{x}_{k-1}, u_{k-1}] \\
\hat{x}^-_k &= \sum_{i=0}^{2L} W_i^{(m)} \mathbf{x}^*_{i,k|k-1} \\
S^-_k &= qr\{[\sqrt{W_1^{(c)}}[\mathbf{x}^*_{1:2L,k|k-1} - \hat{x}^-_k] \quad \sqrt{R^v}]\} \\
S^-_k &= cholupdate\{S^-_k, \mathbf{x}^*_{0,k} - \hat{x}^-_k, W_0^{(c)}\} \\
\mathbf{x}_{k|k-1} &= [\hat{x}^-_k \quad \hat{x}^-_k + \gamma S^-_k \quad \hat{x}^-_k - \gamma S^-_k] \\
\mathbf{y}_{k|k-1} &= H[\mathbf{x}_{k|k-1}] \\
\hat{y}^-_k &= \sum_{i=0}^{2L} W_i^{(m)} \mathbf{y}_{i,k|k-1}
\end{aligned}$$

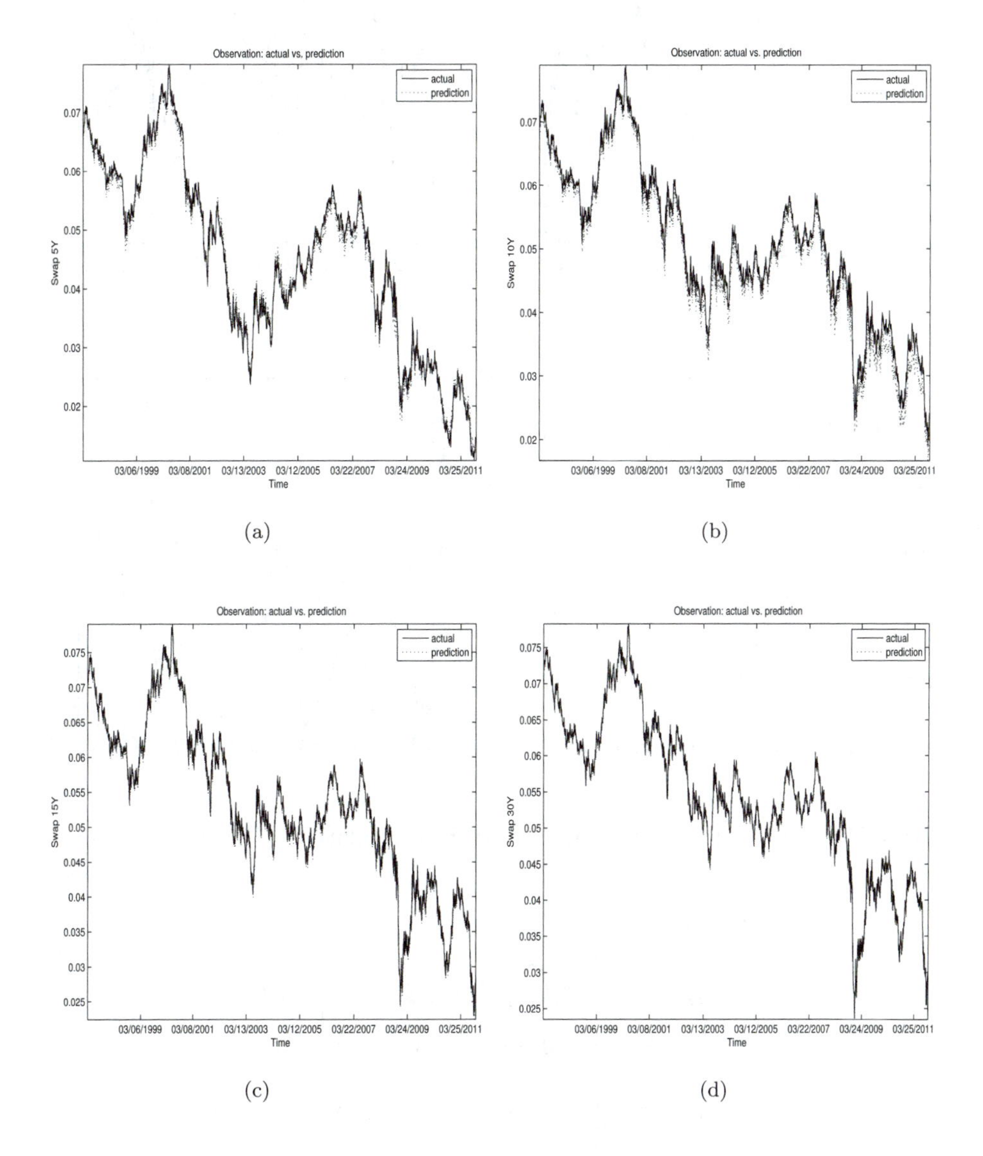


**FIGURE 8.16**: Affine term structure stochastic volatility UKF example: actual versus prediction for (a) 5-year swap rate, (b) 10-year swap rate, (c) 15-year swap rate, (d) 30-year swap rate

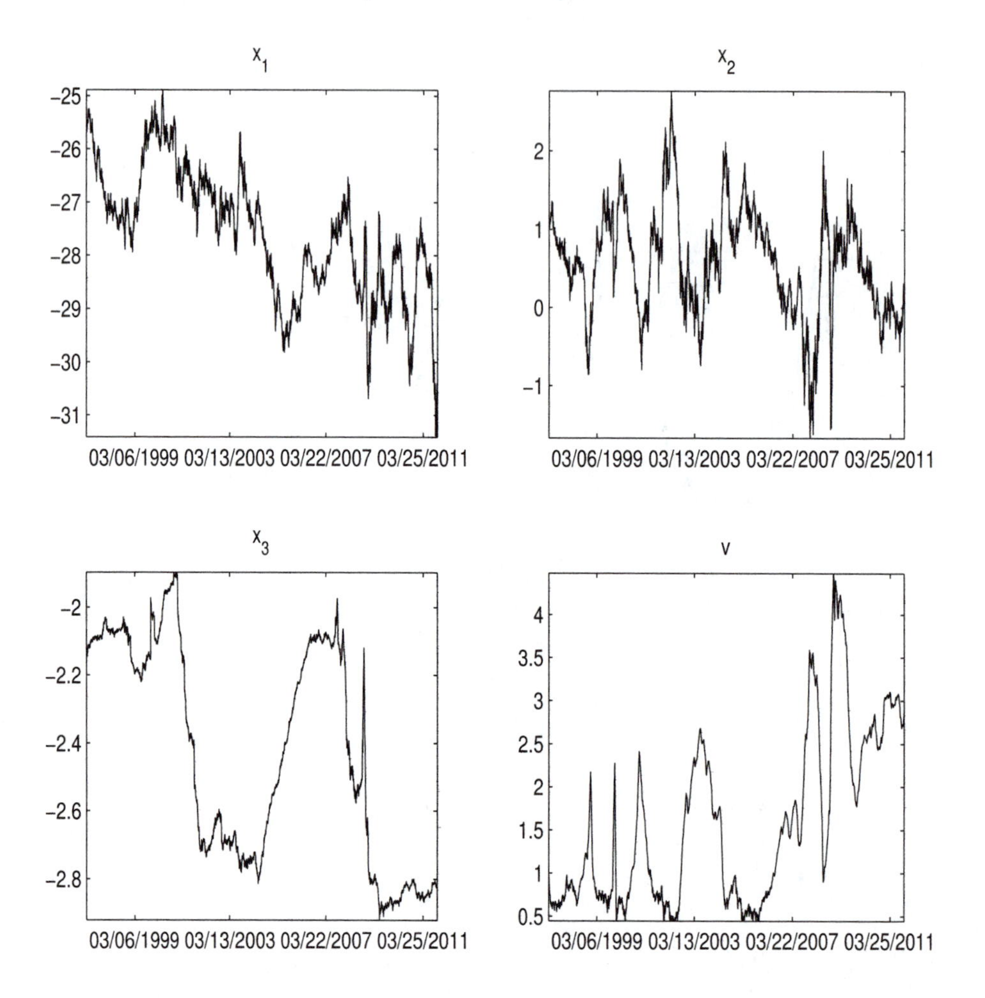


**FIGURE 8.17**: Affine term structure stochastic volatility UKF example: states

4. Measurement update equations

$$\begin{aligned}
S_{\hat{y}_k} &= qr\{[\sqrt{W_1^{(c)}}[\mathbf{y}_{1:2L,k} - \hat{y}_k] \quad \sqrt{R_k^n}]\} \\
S_{\hat{y}_k} &= cholupdate\{S_{\hat{y}_k}, \mathbf{y}_{0,k} - \hat{y}_k, W_0^{(c)}\} \\
P_{x_k y_k} &= \sum_{i=0}^{2L} W_i^{(c)} [\mathbf{x}_{i,k|k-1} - \hat{x}_k^-][\mathbf{y}_{i,k|k-1} - \hat{y}_k^-]^\top \\
K_k &= \left(P_{x_k y_k}/S_{\hat{y}_k}^\top\right)/S_{\hat{y}_k} \\
\hat{x}_k &= \hat{x}_k^- + K_k(y_k - \hat{y}_k^-) \\
U &= K_k S_{y_k} \\
S_k &= cholupdate\{S_k^-, U, -1\}
\end{aligned}$$

The variables used are defined as

$$\begin{aligned}
W_0^m &= \lambda/(L+\lambda) \\
W_0^{(c)} &= \lambda/(L+\lambda)+(1-\alpha^2+\beta) \\
W_i^{(m)} &= W_i^{(c)} = 1/\{2(L+\lambda), \quad i=1,\ldots,2L\} \\
\lambda &= \alpha^2(L+\kappa)-L \\
\gamma &= \sqrt{(L+\lambda)}
\end{aligned}$$

We recommend the following for $\alpha$, $\beta$, and $\kappa$:

$$\begin{aligned}
\alpha &= 1.0 \\
\beta &= 1.0 \\
\kappa &= 0.0
\end{aligned}$$

The definitions of $R^v$ and $R^n$ are the same as in the unscented Kalman filter.

Some comments:

(1) QR decomposition — we define $A^\top = QR$, where $A \in \mathbb{R}^{L\times N}$, and $N \geq L$. The $Q$ is an orthogonal matrix, $R$ is an upper triangular matrix, and the qr{.} in the algorithm denotes the upper triangular part of $R$.

(2) Cholesky factor updating — cholupdate$(S, u, \pm v)$ where $S = chol(A)$ is the original Cholesky factorization of $A$, returns the upper triangular Cholesky factor of $A \pm \sqrt{v}uu^\top$, where $u$ is a column vector of appropriate length. If $u$ is a matrix not a vector, then the result is $M$ consecutive updates of the Cholesky factor use the $M$ column of $u$. This algorithm is $O(L^2)$ per update.

(3) Given the same initial value and model parameters, the filtering result of UKF and SR-UKF should be the same. One exception is when $P$ is not positive semidefinite during UKF, and has been added an $I$.

In the case of using discrete time stochastic model introduced in [138] and discussed earlier, the state formula and one step forward are

$$\begin{aligned}
x_{k+1} &= F(x_k) = \mathbb{E}(x_{k+1}|x_k) \\
&= x_k + (b_\gamma - Bx_k)\Delta t \\
v_{k+1} &= F(v_k) = \mathbb{E}(v_{k+1}|v_k) \\
&= \frac{\lambda}{\delta}v_k + \frac{\gamma}{\delta\eta}
\end{aligned}$$

and

$$\begin{aligned}
P_{xx} &= var(x_{t+1}) = \frac{1}{\delta}\Sigma\Sigma^\top\Delta t\lambda v_t + \frac{1}{\delta\eta}\Sigma\Sigma^\top\Delta t\gamma \\
P_{yy} &= var(v_{t+1}) = \frac{1}{\delta^2}\lambda v_t + \frac{\gamma}{\delta^2\eta} \\
P_{xy} &= cov(x_{t+1}, v_{t+1}) = 0 \\
R_v(v_t) &= \begin{bmatrix} P_{xx} & P_{xy} \\ P_{xy}^\top & P_{yy} \end{bmatrix}
\end{aligned}$$

We assume

$$\begin{aligned}
\mathbb{E}(F(x_0)) &= x_0 \\
\mathbb{E}(F(u_0)) &= v_0
\end{aligned}$$

Therefore it is easy to see that

$$x_0 = B^{-1}b_\gamma \tag{8.24}$$

$$v_0 = \frac{\gamma}{(\delta - \lambda)\eta} \tag{8.25}$$

---

## 8.8 Particle Filter

A more recent alternative for filtering nonlinear processes is provided by a collection of techniques known as *particle filtering* [18]. Particle filters, also known as *sequential Monte Carlo*, approximate the continuous density function $p(x_t|y_t)$ by a discrete density function, effectively a histogram via simulation. The idea, based on importance sampling, is to use Monte Carlo simulation to replace the Gaussian approximation for $p(x_t|y_t)$ that was used in the Kalman filter or the extended Kalman filter [18]. Thus a better estimate of the parameters from fundamentally nonlinear processes might be obtained. Like all sampling based approaches we generate a set of samples that approximate the distribution function $p(x_t|y_t)$. For $N$ sample points we have $\{x_t^{(i)}, w_t^{(i)}\}_{i=1}^N$ where $x_t^{(i)}$ are support points to the discrete distribution called particles and $w_t^{(i)}$ is weight associated to support point $x_t^{(i)}$ and obviously sum of the weights are one. The expectation with respect to the filtering is approximated by

$$\mathbb{E}(f(x_t)) = \int f(x_t)p(x_t|y_t)dx_t \approx \sum_{i=1}^{N} w_t^{(i)} f(x_t^{(i)}) \tag{8.26}$$

Like other filtering techniques, the first step in the particle filtering algorithm is initialization. The choice of a proper initial state value plays a more crucial role in particle filtering than in the unscented Kalman filter. A bad initial value could make the sigma points (the particles) far from the actual value, and the probability of a particle to the actual state can be very small; therefore the converge process could take a long time. One solution of finding a proper initial state value is using UKF for the prior state value. The UKF method converges quickly to the true state value; then one would switch to a particle filter for more accurate estimation.

For a finite set of particles, the algorithm performance depends on the choice of the proposal distribution, $\pi(x_k|x_{0:k-1}, y_{0:k})$. The optimal proposal distribution is given by the target distribution

$$\pi(x_k|x_{0:k-1}, y_{0:k}) = p(x_k|x_{k-1}, y_k)$$

However, the transition prior is often used as the importance function, since it is easier to draw particles and perform subsequent importance weight calculations as

$$\pi(x_k|x_{0:k-1}, y_{0:k}) = p(x_k|x_{k-1})$$

As will be discussed later, resampling is used to avoid the problem of degeneracy of the algorithm, that is, avoiding the situation where all but one of the importance weights are close to zero. The performance of the algorithm can also be affected by the proper choice of resampling method. The stratified sampling proposed by [165] is optimal in terms of variance.

The probability function $p(y_k|x_k)$ that is used for weight calculation is critical to the filtering. In the case of no analytical form for this probability for the model under consideration we can alternatively assume that the measures or signal's noise has a normal distribution and associate it with $R_v$. Therefore

$$p(y_{j,k}|x_k) = \frac{1}{\sqrt{2\pi R_{v,j,j}}} e^{-\frac{(y_{j,k}-F_j(x_k))^2}{2R_{v,j,j}}} \tag{8.27}$$

At step $k$, the proxy for likelihood that is its Monte Carlo approximation is

$$l_k = \sum_{i=1}^{N_{sim}} \frac{p(y_k|x_{i,k})p(x_{i,k}|x_{i,k-1})}{\pi(x_{i,k}|x_{i,k-1}, y_k)} \tag{8.28}$$

Therefore to estimate the parameters we maximize the logarithmic of likelihood (or minimize its negative):

$$-\sum_{i=1}^{N} \log l_k \tag{8.29}$$

Note that the equivalent formulation is

$$l_k = \sum_{i=1}^{N_{sim}} w_k^{(i)} \tag{8.30}$$

with the interpretation of the likelihood as the total weight.

### 8.8.1 Sequential Importance Sampling (SIS) Particle Filtering

The sequential importance sampling (SIS) algorithm is as follows:

1. Simulate the state from the prior (or another proposal distribution) that is drawing $N$ samples according to the model

$$x_k^{(i)} = f(x_{k-1}^{(i)}, u_{k-1}^{(i)}), \quad i = 1, 2, \ldots, N \tag{8.31}$$

2. Associate to each simulated point a weight. That is done by updating the importance weights

$$w_k^{(i)} = w_{k-1}^{(i)} \frac{p(z_k|x_k^{(i)})p(x_k^{(i)}|x_{k-1}^{(i)})}{\pi(x_k^{(i)}|x_{k-1}^{(i)}, z_k)}$$

where $p(z_k|x_k^{(i)})$ is probability, $p(x_k^{(i)}|x_{k-1}^{(i)})$ is transition probability, and $\pi(x_k^{(i)}|x_{k-1}^{(i)}, z_k)$ is the proposal distribution. If we assume $\pi(x_k^{(i)}|x_{k-1}^{(i)}, z_k) = p(x_k^{(i)}|x_{k-1}^{(i)})$, the expression can be simplified as

$$w_k^{(i)} = w_{k-1}^{(i)} p(z_k|x_k^{(i)})$$

3. Normalize the weights $\widetilde{w}_k(x_k^{(i)}) = \frac{w_k^{(i)}}{\sum_i w_k^{(i)}}$

Then the best approximation of $x_k$ is its conditional expectation given $z_{1:k}$ , that is,

$$\mathbb{E}(x_k|z_{1:k}) \approx \sum_{i=1}^{N_{sims}} \widetilde{w}_k(x_k^{(i)})x_k^{(i)}$$

### 8.8.2 Sampling Importance Resampling (SIR) Particle Filtering

The variance of the weights increases over time in SIS, so the algorithm will diverge [18]. This is known as the degeneracy problem in sampling importance sampling. A solution is proposed in [18] by Arulampalam et al. that has to do with resampling. That is, regenerate particles with higher weight and eliminate those with lower weight. The resampling algorithm is done as follows: compare the cumulative distribution function (CDF) created from the normalized weights with a CDF constructed from a uniformly simulated number $\mathcal{U}[0,1]$. At time step $k$ and for $j = 1, \ldots, N_{sims}$, if

$$\frac{1}{N_{sims}}(\mathcal{U}[0,1] + j - 1) \geq \sum_{l=1}^{i} \widetilde{w}_k(x_k^{(l)})$$

then increment and *skip* $i$, otherwise take $x_k^{(i)}$ and set its weight to $\frac{1}{N_{sims}}$. Here is the pseudo-code for the resampling algorithm:

for $j = 1, \ldots, N_{sims}$

$\quad c(j) = \sum_{l=1}^{j} \widetilde{w}_k(x_k^{(l)})$

end for

$i = 1;$

for $j = 1, \ldots, N_{sims}$

$\quad u(j) = \frac{1}{N_{sims}}(\mathcal{U}[0,1] + j - 1)$

$\quad$ while $(u(j) > c(i))$

$\quad\quad i = i + 1;$

$\quad$ end while

$\quad \widetilde{x}_k^{(j)} = x_k^{(i)}$

$\quad \widetilde{w}_k^j = \frac{1}{N_{sims}}$

end for

The sampling importance resampling (SIR) algorithm is as follows:

1. Simulate the state from the prior (or another proposal distribution)
2. Associate to each simulated point a weight equal to the conditional likelihood density $w_k^{(i)} = w_{k-1}^{(i)} p(z_k | x_k^{(i)})$
3. Normalize the weights $\widetilde{w}_k(x_k^{(i)}) = \frac{w_k^{(i)}}{\sum_i w_k^{(i)}}$
4. Resample according the pseudo-code explained earlier

An alternative to that is to compute an estimate of the effective number of particles as

$$\hat{N}_{eff} = \frac{1}{\sum_{j=1}^{N} (\widetilde{w}_k^j)^2} \tag{8.32}$$

**TABLE 8.2**: Heston stochastic volatility parameters via particle filter for the S&P 500 and USD/JPY

| dataset | $\kappa$ | $\theta$ | $\sigma$ | $\mu$ | $\rho$ | $v_0$ |
|---|---|---|---|---|---|---|
| S&P 500 | 4.3758 | 0.1505 | 0.3473 | 0.0984 | -0.2541 | 0.0989 |
| JPY/USD | 2.3618 | 0.0166 | 0.3451 | -0.0676 | -0.0806 | 0.0166 |

Resample if $\hat{N}_{eff} < N_{\mathcal{T}}$, where $N_{\mathcal{T}}$ is a given threshold, and draw $N$ particles from the current particle set with probabilities proportional to their weights. Replace the current particle set with this new one. Reset weights as $w_{i,k} = 1/N$.

Compared with the unscented Kalman filter (UKF), the advantage of the particle filter (PF) is that, with sufficient samples, it approaches the Bayesian optimal estimate, so it is more accurate. However, a larger number of samples requires heavier computation for the simulation, and in many scenarios, it would be huge, while UKF only requires $2L + 1$ samples where $L$ is the state dimension.

We apply the particle filtering algorithm to Heston stochastic volatility, VGSA, and NIGSA models and estimate the optimal parameter sets for a time series via the maximization of the likelihood under the physical measure framework. We then compare these optimal parameters with those obtained from a cross-sectional fitting using options of different strike prices (under risk-neutral framework).

**Example 29** *Particle filter for Heston stochastic volatility model*

We use two different datasets for estimation of the Heston stochastic volatility model: (a) daily close of S&P 500 from November 11, 2001 to November 11, 2011 (b) daily close of the USD/JPY exchange rate from November 11, 2001 to November 11, 2011. The parameter set for estimation is $\Theta = \{\kappa, \theta, \sigma, \mu, \rho, v_0\}$ and the hidden state for filtering is $\{v_t, 1 \leq t \leq T\}$.

For parameter estimation, the starting point for USD/JPY parameter sets $\Theta = [\kappa, \theta, \sigma, \mu, \rho, v_0]$ is given by $\Theta_0 = [1.2, 0.0898, 0.3995, -0.0552, -0.0064, 0.0800]$, lower bound and upper bound for $\Theta$: $lb = [1, 0.01, 0.1, -0.1, -0.9, 0.01]$, $ub = [8, 0.09, 0.4, 0.2, 0.2, 0.09]$. For parameter estimation, the starting point for S&P 500 parameter sets $\Theta = [\kappa, \theta, \sigma, \mu, \rho, v_0]$ is given by $\Theta_0 = [5, 0.09, 0.3, 0.02, -0.4, 0.09]$, lower bound and upper bound for $\Theta$: $lb = [1, 0.00025, 0.1, -0.2, -0.9, 0.00025]$, $ub = [8, 0.49, 0.6, 0.2, 0.1, 0.49]$. The initial particles are set to be all equal to $v_0$.

For the Heston stochastic volatility model we use the following densities [5]:

$$\begin{aligned}
\pi(x_k^{(i)}|x_{k-1}^{(i)}, y_k) &= n\left(x_k^{(i)}, x_{k-1}^{(i)}, \sqrt{P_k^{(i)}}\right)\\
p(y_k|x_k^{(i)}) &= n\left(y_k,\ y_{k-1}+(\mu-\frac{1}{2}x_k^{(i)})\Delta t,\ \sqrt{x_k^{(i)}\Delta t}\right)\\
p(x_k^{(i)}|x_{k-1}^{(i)}) &= n\left(x_k^{(i)},\ x_{k-1}^{(i)}+\delta\Delta t+\rho\lambda(y_{k-1}-y_{k-2}),\ \lambda\sqrt{1-\rho^2}\sqrt{x_{k-1}^{(i)}\Delta t}\right)
\end{aligned}$$

where

$$\begin{aligned}
\delta &= \kappa\theta-\rho\lambda\mu-(\kappa-\frac{1}{2}\rho\lambda)x_{k-1}^{(i)}\\
n(x, m, s) &= \frac{1}{\sqrt{2\pi}s}\exp(-\frac{(x-m)^2}{2s^2})
\end{aligned}$$

We use two different approaches for the Heston stochastic volatility model. One approach is for parameter estimation and the other one is for prediction. In the first approach we follow the work done by Aihara et al. [5] for filtering and prediction and in the second approach we follow the work done by Mimouni [180] for parameter estimation. Implementation details of the first approach for the Heston stochastic volatility model (Heston Model-1) is as follows: start from the Heston SDE

$$dS_t = \mu S_t dt + \sqrt{v_t} S_t dw_t^1 \tag{8.33}$$

$$dv_t = \kappa(\theta - v_t)dt + \lambda\sqrt{v_t}dw_t^2 \tag{8.34}$$

Define $y_t = \log S_t$ using Itô's lemma we can rewrite (8.33) as follows:

$$dy_t = (\mu - \frac{1}{2}v_t)dt + \sqrt{v_t}dw_t^1 \tag{8.35}$$

Knowing the correlation between $dw_t^1$ and $dw_t^2$ is $\rho$, $< dZ_t, dB_t >= \rho dt$, we can write

$$dw_t^2 = \rho dw_t^1 + \sqrt{1-\rho^2}dz_t \tag{8.36}$$

and substituting it into (8.34) we get

$$dv_t = \kappa(\theta - v_t)dt + \lambda\sqrt{v_t}(\rho dw_t^1 + \sqrt{1-\rho^2}dz_t) \tag{8.37}$$

From (8.35) we can write

$$dw_t^1 = \frac{dy_t - (\mu - \frac{1}{2}v_t)dt}{\sqrt{v_t}} \tag{8.38}$$

By substituting (8.38) in (8.37) we obtain

$$\begin{aligned} dv_t &= \kappa(\theta - v_t)dt + \lambda\sqrt{v_t}(\rho\frac{dy_t - (\mu - \frac{1}{2}v_t)dt}{\sqrt{v_t}} + \sqrt{1-\rho^2}dz_t) \\ &= \kappa(\theta - v_t)dt + \lambda\rho(dy_t - (\mu - \frac{1}{2}v_t)dt) + \lambda\sqrt{v_t}\sqrt{1-\rho^2}dz_t) \end{aligned} \tag{8.39}$$

Rewriting the discrete version of the state equation (8.39)

$$v_t = v_{t-1} + \kappa(\theta - v_{t-1})\Delta t + \lambda\rho(y_t - y_{t-1} - (\mu - \frac{1}{2}v_{t-1})\Delta t) + \lambda\sqrt{v_{t-1}(1-\rho^2)\Delta t}\ z_1 \tag{8.40}$$

where $z_1 \sim \mathcal{N}(0,1)$. Now we use explicit-implicit discretization to discretize 8.35 as follows:

$$y_t = y_{t-1} + (\mu - \frac{1}{2}v_t)\Delta t + \sqrt{v_{t-1}}\sqrt{\Delta t}\ z_2 \tag{8.41}$$

and plug (8.41) into (8.40) to get

$$\begin{aligned} v_t &= v_{t-1} + \kappa(\theta - v_{t-1})\Delta t + \lambda\rho((\mu - \frac{1}{2}v_t)\Delta t - (\mu - \frac{1}{2}v_{t-1})\Delta t) \\ &+ \lambda\sqrt{v_{t-1}(1-\rho^2)}\Delta z_2 + \lambda\rho\sqrt{v_{t-1}}\sqrt{\Delta t}z_1 \end{aligned} \tag{8.42}$$

Notice that if we did not use explicit-implicit discretization the terms in (8.42) would have canceled each other. By doing this we make the scheme stable. From (8.40) we can construct the optimal proposal (importance) function, $\pi(v_t|v_{t-1}, y_t)$, that is,

$$\pi(v_t|v_{t-1}, y_t) = n(v_t, m_I, \sigma_I) \tag{8.43}$$

where

$$\begin{aligned} m_I &= v_{t-1} + \kappa(\theta - v_{t-1})\Delta t + \lambda\rho(y_t - y_{t-1} - (\mu - \frac{1}{2}v_{t-1})\Delta t) \\ \sigma_I &= \lambda\sqrt{v_{t-1}(1-\rho^2)\Delta t} \\ n(x,m,s) &= \frac{1}{\sqrt{2\pi}s}\exp(-\frac{(x-m)^2}{2s^2}) \end{aligned}$$

From (8.41), we have the likelihood function $p(y_t|v_t, v_{t-1}, y_{t-1})$

$$p(y_t|v_t, v_{t-1}, y_{t-1}) = n(y_t, m_L, \sigma_L)$$

where

$$\begin{aligned} m_L &= y_{t-1} + (\mu - \frac{1}{2}v_t)\Delta t \\ \sigma_L &= \sqrt{v_{t-1}}\Delta t \end{aligned}$$

From (8.42) we have the transition density $p(v_t|v_{t-1})$

$$p(v_t|v_{t-1}) = n(v_t, m_T, \sigma_T) \tag{8.44}$$

where

$$\begin{aligned} m_T &= (1+\frac{1}{2}\lambda\rho\Delta t)^{-1}(v_{t-1} + \kappa(\theta - v_{t-1})\Delta t + \frac{1}{2}\lambda\rho v_{t-1}\Delta t) \\ \sigma_T &= (1+\frac{1}{2}\lambda\rho\Delta t)^{-1}\lambda\sqrt{v_{t-1}\Delta t} \end{aligned}$$

Then weights are updated as follows: initial weights are all equal to $\frac{1}{N}$, then

$$w_t^i = w_{t-1}^i \frac{p(y_t|v_t, v_{t-1}, y_{t-1})p(v_t|v_{t-1})}{\pi(v_t|v_{t-1}, y_t)}$$

Then the likelihood is approximated by $\sum_{t=1}^{T}(\log\sum_{i=1}^{N} w_t^i)$ where $T$ is number of days and $N$ is number of particles used.

**Implementation of the prediction step**

$$\begin{aligned} p(y_t|y_{1:t-1}) &= \int p(y_t, x_{t-1}|y_{1:t-1})dx_{t-1} \\ &= \int p(y_t|x_{t-1})p(x_{t-1}|y_{1:t-1})dx_{t-1} \\ &= \int\int p(y_t, x_t|x_{t-1})p(x_{t-1}|y_{1:t-1})dx_t dx_{t-1} \\ &= \int\int p(y_t|x_t)p(x_t|x_{t-1})p(x_{t-1}|y_{t-1})dx_t dx_{t-1} \end{aligned}$$

The above equation suggests the following approximation for predictive density $p(y_t|y_{1:t-1})$. Suppose $x_{t-1}^{(i)}$ is the $i$-th sample from $p(x_{t-1}|y_{1:t-1})$ and $x_{t-1}^{i,j}$ is the $j$-th sample from $p(x_t|x_{t-1}^{(i)})$, then we can estimate each $p(y_t|y_{1:t-1})$ via

$$p(y_t|y_{1:t-1}) = \frac{1}{N \times M}\sum_{i=1}^{N}\sum_{j=1}^{M} p(y_t|x_t^{i,j})$$

We implement the prediction as follows: given time $t-1$ (today), particles $v_{t-1}^{(i)}$ is as approximation for posterior samples $p(v_{t-1}|y_{1:t-1})$, doing one step propagation of the state equation to generate $v_t^i$ via the following equation

$$v_t^{(i)} = v_{t-1}^{(i)} + \kappa(\theta - v_{t-1}^{(i)})\Delta t$$

Given $v_t^{(i)}$, we predict $y_t^i$ using the following equation:

$$y_t^i = y_{t-1} + (\mu - \frac{1}{2}v_t^i)\Delta t + \sqrt{v_{t-1}^i}\Delta w^1(t)$$

Then given $y_t^k, k = 1, \ldots, N$, we calculate the predictive density of $y_t^k$ by

$$p(y_t^k|y_{1:t-1}) = \frac{1}{NM}\sum_{i=1}^{N}\sum_{j=1}^{M} p(y_t^k|x_t^{i,j})$$

Simplify it by using $M = 1$, then $p(y_t^k|y_{1:t-1}) = \frac{1}{N}\sum_{i=1}^{N} p(y_t^k|x_t^i)$, then the prediction of log stock price at time $t$ is given by

$$y_t^{pred} = \sum_{k=1}^{N} p(y_t^k|y_{1:t-1})y_t^k$$

**Parameter estimation**

The simulation study indicates that the above formulation of likelihood has difficulty in identifying $\kappa$ and $\rho$. We observe that in estimation, we constantly hit the boundary of $\kappa$ or $\rho$. Instead we switch to another discretization. The Heston model second approach has to do with parameter estimation, which is more stable than the Heston Model-1 described earlier. In this approach, we follow the work done by Mimouni [180]. The second approach is as follows (Heston Model-2):

$$\begin{aligned} dS_t &= \mu S_t dt + \sqrt{v_t} S_t dw_t^1 && (8.45) \\ dv_t &= \kappa(\theta - v_t)dt + \lambda\sqrt{v_t}dw_t^2 && (8.46) \end{aligned}$$

Define $y_t = \log S_t$ and by means of Itô's lemma we obtain the following for (8.45):

$$dy_t = (\mu - \frac{1}{2}v_t)dt + \sqrt{v_t}dw^1(t) \quad (8.47)$$

and as previously mentioned, knowing the correlation, we can write

$$dw_t^2 = \rho dw_t^1 + \sqrt{1-\rho^2}dz_t \quad (8.48)$$

Substituting (8.48) into (8.46) and using (8.47) and then discretizing it using Euler's scheme to get

$$\begin{aligned} v_t &= v_{t-1} + \kappa(\theta - v_{t-1})\Delta t + \lambda\rho(y_t - y_{t-1} - (\mu - \frac{1}{2}v_{t-1})\Delta t) \\ &+ \lambda\sqrt{v_{t-1}(1-\rho^2)}\sqrt{\Delta t}z_1 && (8.49) \end{aligned}$$

where $z_1 \sim \mathcal{N}(0, 1)$. Also, discretization of (8.47) yields

$$y_t = y_{t-1} + (\mu - \frac{1}{2}v_{t-1})\Delta t + \sqrt{v_{t-1}}\Delta t z_2$$

where $z_2 \sim \mathcal{N}(0,1)$. Following the argument in [180], we use Equation (8.49) to propagate the state from $t-1$ to $t$:

$$\begin{aligned} v_t^{(i)} &= v_{t-1}^{(i)} + \kappa(\theta - v_{t-1}^{(i)})\Delta t + \lambda\rho(y_t - y_{t-1} - (\mu - \frac{1}{2}v_{t-1}^{(i)})\Delta t) \\ &+ \lambda\sqrt{v_{t-1}^{(i)}(1-\rho^2)\Delta t} z_1 \end{aligned} \tag{8.50}$$

and use $v_t^{(i)}$ obtained from Equation (8.50) to predict $y$ at $t+1$ and compute $y_{t+1}$ to calculate likelihood and update the weights. Initial weights are all equal to $\frac{1}{N}$.

$$\begin{aligned} w_t^{(i)} &= w_{t-1}^{(i)} L(t) \\ L(t) &= n(v_t^{(i)}, \mu_L, \sigma_L) \end{aligned}$$

where

$$\begin{aligned} \mu_L &= y_{t+1} - y_t + (\mu - \frac{1}{2}v_t^{(i)})\Delta t \\ \sigma_L &= \sqrt{v_t^{(i)}\Delta t} \end{aligned}$$

The rationale for the second approach working for estimation is the fact that we use future data $y_{t+1}$ to smooth $v_t$, that is, $v_t|y_{1:t+1}$ since we discretize

$$y_{t+1} = y_t + (\mu - \frac{1}{2}v_t)\Delta t + \sqrt{v_t}\Delta w_t^1$$

while in the first approach we only estimate posterior $v_t|y_{1:t}$. Also in this approach proposal distribution and transition distribution are the same. We plot results of Heston stochastic volatility particle filter example in Figures 8.18(a)–8.18(d). In Figures 8.18(a)–8.18(d) we illustrate filtered volatility of USD/JPY, USD/JPY actual versus prediction, filtered volatility of S&P 500, and S&P 500 actual versus prediction respectively.

**Example 30** *Parameter estimation of variance gamma with stochastic arrival (VGSA) via particle filtering*

Under the VGSA model, the stock price process (in the risk-neutral framework) follows

$$d\ln S_t = (r - q + \omega)dt + X(h(dt); \sigma, \nu, \theta)$$

with $\omega = \frac{1}{\nu}\ln(1 - \theta\nu - \sigma^2\nu/2)$ and

$$X(h(dt); \sigma, \nu, \theta) = B(\gamma(h(dt), 1, \nu), \theta, \sigma)$$

and the gamma cumulative distribution function

$$F_\nu(h, x) = \frac{1}{\Gamma(\frac{h}{\nu})\nu^{\frac{h}{\nu}}} \int_0^\infty e^{-\frac{t}{\nu}} t^{\frac{h}{\nu}-1} dt$$

and $h(dt) = y_t dt$ with

$$dy_t = \kappa(\eta - y_t)dt + \lambda\sqrt{y_t}dW_t$$

The Euler discretized VGSA process could be written via the auxiliary variable

$$y_k = y_{k-1} + \kappa(\eta - y_{k-1})\Delta t + \lambda\sqrt{y_{k-1}}\sqrt{\Delta t} Z_{k-1}$$

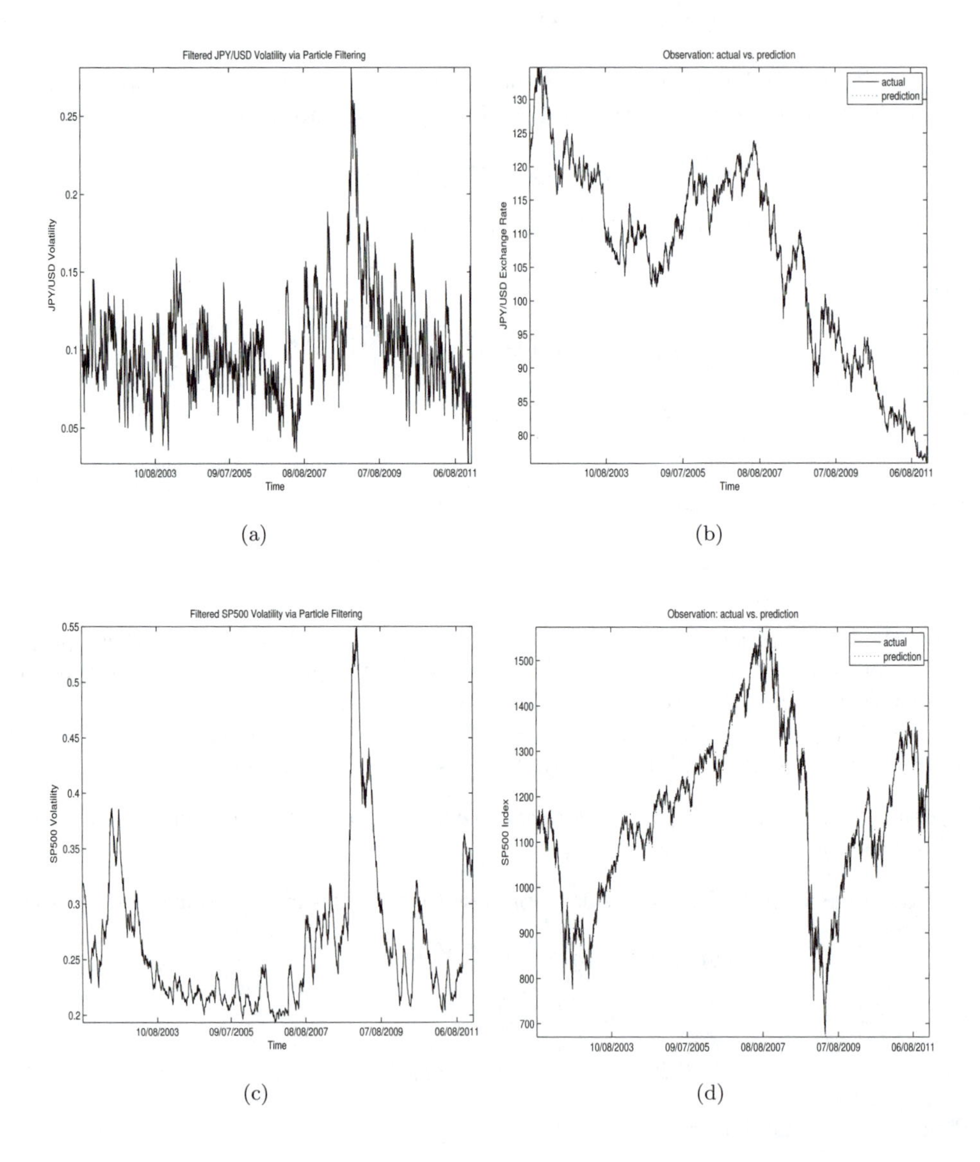


**FIGURE 8.18**: Heston stochastic volatility particle filtering example: (a) filtered volatility of USD/JPY, (b) actual versus prediction USD/JPY, (c) filtered volatility of S&P 500 and (d) actual versus prediction S&P 500

and the state

$$x_k = F_\nu^{-1}(y_k \Delta t, \mathcal{U}[0,1])$$

as well as the observation $z_k = \ln S_k$

$$z_k = z_{k-1} + (r_{k-1} - q_{k-1} + \omega)\Delta t + \theta x_k + \sigma\sqrt{x_k} B_k$$

with $\omega = \frac{1}{\nu}\ln(1 - \theta\nu - \sigma^2\nu/2)$ and $W_{k-1}$ and $B_k$ are $\mathcal{N}(0,1)$.
The particle filter algorithm could therefore be written as follows (the first approach):

- initialize the arrival rate $y_0^{(j)}$, the state $x_0^{(i)}$, and the weight $w_0^{(i)}$ for $j$ between 1 and $M_{sims}$, and $i$ between 1 and $N_{sims}$
- while $1 \leq k \leq N_{sims}$
  - simulate the arrival-rate $y_k$ for $j$ between 1 and $M_{sims}$

$$y_k^{(j)} = y_{k-1}^{(j)} + \kappa(\eta - y_{k-1}^{(j)})\Delta t + \lambda\sqrt{y_{k-1}^{(j)}}\sqrt{\Delta t}\mathcal{N}^{-1}\left(\mathcal{U}^{(j)}[0,1]\right)$$

  - simulate the state $x_k$ for each $y_k^{(j)}$ and for $i$ between 1 and $N_{sims}$

$$\tilde{x}_k^{(i|j)} = F_\nu^{-1}\left(y_k^{(j)}\Delta t, \mathcal{U}^{(j)}[0,1]\right)$$

  - compute the unconditional state

$$\tilde{x}_k^{(i)} = \int \tilde{x}_k^{(i)}(y_k)p(y_k|y_{k-1}dy_k \sim \frac{1}{M_{sims}}\sum_{j=1}^{M_{sims}} \tilde{x}_k^{(i|j)}$$

  - calculate the associated weights for each $i$

$$w_k^{(i)} = w_{k-1}^{(i)} p(z_k|\tilde{x}_k^{(i)})$$

    with

$$p(z_k|\tilde{x}_k^{(i)}) = n(z_k, m, s)$$

    is the normal density with mean of $m = z_{k-1} + (r_k - q_k + \omega)\Delta t + \theta\tilde{x}_k^{(i)}$ and standard deviation of $s = \sigma\sqrt{\tilde{x}_k^{(i)}}$
  - normalize the weights

$$\tilde{w}_k^{(i)} = \frac{w_k^{(i)}}{\sum_{j=1}^{N_{sims}} w_k^{(i)}}$$

  - resample the points $\tilde{x}_k^{(i)}$ and get $x_k^{(i)}$ and reset $w_k^{(i)} = \tilde{w}_k^{(i)} = \frac{1}{N_{sims}}$
- end of while loop

We can also take advantage of the fact that VG provides an integrated density of stock return. Calling $z = \ln(S_k/S_{k-1})$ and $h = t_k - t_{k-1}$ and posing $x_h = z - (r-q)h - \frac{h}{\nu}\ln(1 - \theta\nu - 2\sigma^2\nu/2)$ we have

$$p(z|h) = \frac{2\exp(\theta x_h/\sigma^2)}{\nu^{\frac{h}{\nu}}\sqrt{2\pi}\sigma\Gamma(\frac{h}{\nu})}\left(\frac{x_h^2}{2\sigma^2/\nu + \theta^2}\right)^{\frac{h}{2\nu}-\frac{1}{4}} K_{\frac{h}{\nu}-\frac{1}{2}}\left(\frac{1}{\sigma^2}\sqrt{x_h^2(2\sigma^2/\nu + \theta^2)}\right)$$

As we can see, the dependence on the gamma distribution is *integrated out* in the above. For the VGSA for a given arrival rate $dt^* = y_t dt$ we have a VG distribution and

$$d\ln S_t = (r - q + \omega)dt + B(\gamma(dt^*, 1, \nu); \theta, \sigma)$$

and the corresponding integrated density becomes

$$p(z|h, h^*) = \frac{2\exp(\theta x_h/\sigma^2)}{\nu^{\frac{h^*}{\nu}}\sqrt{2\pi}\sigma\Gamma(\frac{h^*}{\nu})}\left(\frac{x_h^2}{2\sigma^2/\nu + \theta^2}\right)^{\frac{h^*}{2\nu}-\frac{1}{4}} K_{\frac{h^*}{\nu}-\frac{1}{2}}\left(\frac{1}{\sigma^2}\sqrt{x_h^2(2\sigma^2/\nu + \theta^2)}\right) \quad (8.51)$$

Hence the idea of using the arrival rate as the state and use the following alternative algorithm (the second approach) for particle filtering:

- initialize the state $x_0^{(i)}$ and the weight $w_0^{(i)}$ for $i$ between 1 and $N_{sims}$
- while $1 \leq k \leq N$
  - simulate the state $x_k$ for $i$ between 1 and $N_{sims}$

$$y_k^{(i)} = x_{k-1}^{(i)} + \kappa(\eta - x_{k-1}^{(i)})\Delta t + \lambda\sqrt{x_{k-1}^{(i)}}\sqrt{\Delta t}\mathcal{N}^{-1}\left(\mathcal{U}^{(i)}[0,1]\right)$$

  - calculate the associated weights for each $i$

$$w_k^{(i)} = w_{k-1}^{(i)} p(z_k|y_k^{(i)})$$

    with $p(z_k|y_k^{(i)})$ as defined in (8.51) where $h$ will be set to $\Delta t$ and $h^*$ to the simulated state $\tilde{x}_k^{(i)}$ times $\Delta t$
  - normalize the weights

$$\widetilde{w}_k^{(i)} = \frac{w_k^{(i)}}{\sum_{j=1}^{N_{sims}} w_k^{(j)}}$$

  - resample the points $\tilde{x}_k^{(i)}$ and get $x_k^{(i)}$ and reset $w_k^{(i)} = \tilde{w}_k^{(i)} = \frac{1}{N_{sims}}$.
- end of while loop

The advantage of this method is that there is one simulation process instead of two and we *skip* the gamma distribution altogether. However, the dependence of the observation $z_k$ on $x_k$ is highly nonlinear, which makes the convergence more difficult. The log-Likelihood for particle filtering (PF) algorithm ought to be maximized is

$$\ln(L_{1:N}) = \sum_{k=1}^{N} \ln\left(\sum_{i=1}^{N_{sims}} w_k^{(i)}\right)$$

**TABLE 8.3**: VGSA parameters via particle filter for USD/JPY and S&P 500

| | USD/JPY | USD/JPY | S&P500 |
|---|---|---|---|
| $\sigma$ | 0.1308 | 0.1578 | 0.3573 |
| $\nu$ | 0.0726 | 0.1053 | 0.3288 |
| $\theta$ | -0.1762 | -0.2718 | -0.0735 |
| $\kappa$ | 3.8542 | 3.8283 | 2.2940 |
| $\eta$ | 6.3822 | 4.3537 | 4.0748 |
| $\lambda$ | 8.5093 | 8.2152 | 0.1205 |
| $\mu$ | -0.0451 | -0.0394 | -0.0078 |

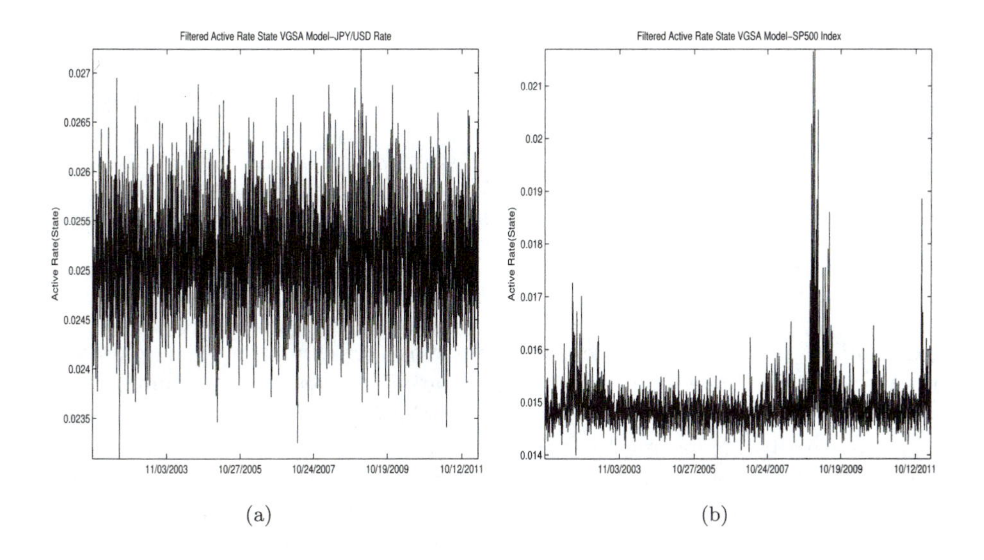


**FIGURE 8.19**: VGSA particle filtering example arrival rate: (a) USD/JPY, (b) S&P 500

We use two different data sets for this example: (a) USD/JPY from November 8, 2001 to November 8, 2011 daily close, (b) S&P 500 from November 11, 2001 to November 11, 2011 daily close. The parameter set for estimation is $\Theta = \{\sigma, \nu, \theta, \kappa, \eta, \lambda, \mu\}$. The hidden state for filtering is $\{y(t), 1 \leq t \leq T\}$. The idea is to find the optimal parameter set via the maximization of likelihood. The maximization takes place over the parameter set $\Theta$. Using the second approach, the parameter set obtained for each data set is tabulated in Table 8.3. In Figures 8.19(a) and 8.19(b), we illustrate the hidden state, arrival rate for USD/JPY exchange rate and S&P 500.

**Example 31** *Comparison of VGSA parameters obtained from option premiums versus stock prices*

In this example [139], we use five years of daily close prices for the S&P 500 Index from January 2, 1998 to January 2, 2003 to estimate the VGSA model parameter set following the same approach as in Example 30. This will provide us with the sequences for the spot price $S_t$ for $t = 1, \ldots, T$ and the dates $t = 1, \ldots, T$. We can also obtain the corresponding overnight LIBOR rates using Bloomberg which gives the drift rate $r_t$ for $t = 1, \ldots T$. As for

**TABLE 8.4**: S&P empirical results, VGSA statistical parameters estimated for the period January 2, 1998 to January 2, 2003 via the particle filter (SIR) versus the risk-neutral parameters

| | Physical measure | Risk-neutral measure |
|---|---|---|
| $\sigma$ | 0.087 | 0.2063 |
| $\theta$ | -0.025 | -0.4160 |
| $\nu$ | 0.002 | 0.0635 |
| $\kappa$ | 5.1319 | 2.17711 |
| $\eta$ | 6.4996 | 5.71047 |
| $\lambda$ | 4.36 | 5.67908 |
| $\mu$ | -0.009 | 0.0538 |

the dividend yields $q_t$ for $t = 1, \ldots, T$, we can write for any date $t$ and option maturity $T$

$$q_t = r_t - \frac{1}{T-t} \ln\left(\frac{F_t}{S_t}\right)$$

where

$$F_t = K + e^{r_t(T-t)} \left(C_t(S_t, K) - P_t(S_t, K)\right)$$

where $C_t$ and $P_t$ are the S&P 500 call and put close prices at time $t$ and $K$ could be a near-the-money strike price for liquidity considerations. Given that the number of option maturities are limited we need to choose an interpolation scheme. Then the question begs, what can we learn by comparing the two parameter sets? How can we take advantage of having both $\mathbb{P}$ and $\mathbb{Q}$? The research in [59] is an example of optimal positioning in derivative pricing. In [6], the authors compare the risk-neutral density estimated in complete markets from the cross-section of S&P 500 option prices to the risk-neutral density inferred from the time series density of the S&P 500 index. If investors are risk-averse, the latter densities are different from the actual density that could be inferred from the time series of S&P 500 returns. Naturally, the observed asset returns do not follow the risk-neutral dynamics, which are therefore not directly observable. In contrast to the existing literature, they avoid making any assumptions on investors' preferences, by comparing two risk-adjusted densities, rather than a risk-adjusted density from option prices to an unadjusted density from index returns. Our only maintained hypothesis is a one-factor structure for the S&P 500 returns.

**Example 32** *Normal inverse Gaussian with stochastic arrival (NIGSA)*

It is the same process as VGSA except we replace $\gamma(h(dt))$ with $\phi(h(dt))$ with $\phi$ the inverse Gaussian distribution with the cumulative distribution function

$$F_\nu(h, x) = \mathcal{N}\left(\frac{\nu x - h}{\sqrt{x}}\right) + e^{2h\nu} \mathcal{N}\left(-\frac{\nu x + h}{\sqrt{x}}\right)$$

It is same exact algorithm except we change the definition of $F_\nu^{-1}(h, x, \nu)$, as previously stated.

### 8.8.3 Problem of Resampling in Particle Filter and Possible Panaceas

The problem of resampling when doing SIR is as follows:

Sample impoverishment, high weight particles get selected several times, which lead to particle become the same after some time step. The bad consequence is that particles are no longer independent and identically distributed, so that the law of large numbers is vanished and the approximation of posteriors no longer justified.

During the resampling step, empirical CDF of weights are used to eliminate the low weights. However, the empirical CDF of the weights is the step function and is not smooth. The consequences of a non-smooth CDF will lead to discontinuity of the likelihood function, which is $\sum_{t=1}^{T}(\log \sum_{i=1}^{N} w_t^{(i)})$.

The remedy for these two issues we use for implementation: we use effective sample size to prevent over resampling which will lead to more and more repeated particles. However, sample impoverishment is not a major issue in our implementation.

The major issue is non-smoothness of the likelihood function when resampling which makes the optimizer either not converge [159], [117] or converge very slowly. We use the common random number technique to tackle this issue, as suggested in [170] and [188]. The common random number technique suggests that for different parameter sets $\Theta_1$ and $\Theta_2$, the particles share the same common random number. As an example for the Heston stochastic volatility model the way we apply the common random number is as follows:

$$v_t^{(i)} = v_{t-1}^{(i)} + \kappa(\theta - v_{t-1}^{(i)})\Delta t + \lambda\rho(y_t - y_{t-1} - (\mu - \frac{1}{2}v_{t-1}^{(i)})\Delta t) + \lambda\sqrt{v_{t-1}^{i}(1-\rho^2)\Delta t}\ z$$

Here we need $N \times T$ standard normal samples as the common random number for each complete procedure of SIR particle filtering for a given parameter set $\Theta$ and these $N \times T$ random numbers should be kept the same even when using another parameter set $\Theta^{'}$ for optimization. Here $T$ is number of data points and $N$ is the number of observation days. When doing resampling, $u(i) = \mathcal{U} + (i-1)/N$, $i = 1, \ldots, N$, where $U \sim \mathcal{U}(0,1)$ and as in the common random number case, we need $U$ as another common random number for each complete procedure of SIR particle filtering for a given parameter set $\Theta$ and these uniform random numbers should be kept the same even when the optimizer begins using another parameter set $\Theta^{'}$ for optimization. The justification for using the common random number is given in [170]. For example, Michael K. Pitt [188] uses piecewise linear CDF to replace empirical CDF. However, as Flury and Shephard [115] point out all these smoothing techniques are essentially jittering or perturbation of the particles after the original resampling step. They constructed an optimal way to jitter the particles to improve the SIR algorithm.

---

## 8.9 Markov Chain Monte Carlo (MCMC)

The Bayesian solution to any inference problem is the following simple rule: compute the conditional distribution of the unobserved variable given the observed data. Characterizing the posterior distribution; however, is often difficult. In most settings $p(\Theta, x|y)$ is complicated and high dimensional, implying that standard sampling methods either do not apply or are prohibitively expensive in terms of computing time. Markov chain Monte Carlo (MCMC) provides a simulation based method for sampling from these high dimensional distributions and is particularly useful for analyzing financial time series models that commonly incorporate latent variables. For more detail on Markov chain Monte Carlo, we refer readers to [121], [156], and [157].

## Problems

1. In Example 25 in this chapter, we lay out the procedure for parameter estimation of the Heston stochastic volatility model. Parameter set for estimation is $\Theta = \{\kappa, \theta, \sigma, \mu, \rho, v_0\}$. Assume that S&P 500 Index and USD/JPY spot currency follow Heston stochastic volatility, use ten years of data to estimate the parameter set for each time series using the extended Kalman filter. Datasets used for this problem: (a) daily close of S&P 500 from November 11, 2001 to November 11, 2011 (b) daily close of the USD/JPY exchange rate from November 11, 2001 to November 11, 2011.

2. In Example 26 in this chapter, parameters of Heston stochastic volatility were estimated via unscented Kalman filter. Use the same time series, estimate the Heston stochastic volatility model via square root unscented Kalman filter.

3. In Example 27 in this chapter, parameters of affine term structure model with constant volatility were estimated via unscented Kalman filter. Use the same time series, estimate the Heston stochastic volatility model via square root unscented Kalman filter.

4. In Example 28 in this chapter, parameters of affine term structure model with constant volatility were estimated via unscented Kalman filter. Use the same time series, estimate the Heston stochastic volatility model via square root unscented Kalman filter.

5. In Example 32 in this chapter, we lay out the procedure for parameter estimation of normal inverse Gaussian with stochastic arrival (NIGSA). Use the same dataset used in Example 30 to estimate the parameter set for NIGSA.

# References


[1] Marwan I. Abukhaled and Edward J. Allen. A class of second-order Runge–Kutta methods for numerical solution of stochastic differential equations. *Stochastic Analysis and Applications*, 16(6):977–991, 1998.

[2] Yves Achdou and Olivier Pironneau. Volatility smile by multilevel least square. October 2001.

[3] Peter John Acklam. An algorithm for computing the inverse normal cumulative distribution function. http://home.online.no/ pjacklam/notes/invnorm/, June 2002.

[4] Joachim H. Ahrens and Ulrich Dieter. Computer generation of Poisson deviates. *ACM Transactions on Mathematical Software*, 8(2):163–179, 1982.

[5] Shin Ichi Aihara, Arunabha Bagchi, and Saikat Saha. On parameter estimation of stochastic volatility models from stock data using particle filter — Application to AEX Index. *International Journal of Innovative Computing, Information and Control*, 5(1):17–27, January 2009.

[6] Y. Aït-Sahalia, Y. Wang, and F. Yared. Do option markets correctly price the probabilities of movement of the underlying asset? *Journal of Econometrics*, 2001.

[7] H. Albrecher, S. Ladoucette, and Wim Schoutens. A generic one-factor Lévy model for pricing synthetic CDOs. In Michael C. Fu, Robert A. Jarrow, Ju-Yi J. Yen, and Robert J. Elliott, editors, *Advances in Mathematical Finance*, chapter 12. Birkhauser Verlag AG, Boston, MA, USA, 2007.

[8] Ariel Almendral and Cornelis W. Oosterlee. Accurate evaluation of european and american options under the cgmy process. *SIAM Journal of Scientific Computing*, 29(11):93–117, 2007.

[9] Ariel Almendral and Cornelis W. Oosterlee. On American options under the variance gamma process. *Applied Mathematical Finance*, 14(2):131–152, May 2007.

[10] Leif Andersen. Efficient simulation of the Heston stochastic volatility model. Banc of America Securities, December 2006.

[11] Leif Andersen. Discount curve construction with tension splines. *Review of Derivatives Research*, 10(3):227–267, June 2008.

[12] Leif B. G. Andersen and Jseper Andreasen. Jumping smiles. *Risk*, 12:65–68, November 1999.

[13] Leif B. G. Andersen and Rupert Brotherton-Ratcliffe. The equity option volatility smile: an implicit finite-difference approach. *The Journal of Computational Finance*, 1(2):5–37, 1998.

[14] Leif B.G. Andersen and Vladimir V. Piterbarg. *Interest Rate Modeling. Volume 1: Foundations and Vanilla Models.* Atlantic Financial Press, 2010.

[15] Torben G. Andersen, Luca Benzoni, and Jesper Lund. An empirical investigation of continuous time equity return models. *Journal of Finance*, 57:1239–1284, June 2002.

[16] Jesper Andreasen. Implied modelling, stable implementation, hedging, and duality. Manuscript, University of Aarhus, 1998.

[17] Jesper Andreasen and Peter Carr. Put call reversal. Manuscript, University of Aarhus.

[18] S. Arulampalam, S. Maskell, N. Gordon, and T. Clapp. A tutorial on particle filters for on-line nonlinear/non-Gaussian Bayesian tracking. *IEEE Transactions on Signal Processing*, 50(2), 2002.

[19] Marco Avellaneda. The minimum-entropy algorithm and related methods for calibrating asset-pricing models. *In Proceedings of the International Congress of Mathematicians, Berlin*, 3:545–563, 1998.

[20] Marco Avellaneda, Craig Friedman, Richard Holmes, and Dominick Samperi. Calibrating volatility surfaces via relative entropy minimization. *Applied Mathematical Finance*, 4:37–64, 1997.

[21] David H. Bailey and Paul N. Swarztrauber. The fractional Fourier transform and applications. *SIAM Review*, 33(3):389–404, 1991.

[22] David H. Bailey and Paul N. Swarztrauber. A fast method for the numerical evaluation of continuous Fourier and Laplace transforms. *SIAM Journal on Scientific Computing*, 15(5):1105–1110, 1994.

[23] Gurdip Bakshi, C. Cao, and Z. Chen. Empirical performance of alternative options pricing models. *Journal of Finance*, 52(5), 1997.

[24] Turan Bali, Massoud Heidari, and Liuren Wu. Predictability of interest rates and interest-rate portfolios. *Journal of Business and Economic Statistics*, 27(4):571–527, 2009.

[25] G. Barles, J. Burdeau, M. Romano, and N. Samsoen. Critical stock price near expiration. *Applied Mathematical Finance*, 5:77–95, 1995.

[26] Ole E. Barndorff-Nielsen and Neil Shephard. Non-Gaussian Ornstein–Uhlenbeck-based models and some of their uses in financial economics. *Royal Statistical Society* 63, Part 2, 2001.

[27] Ole E. Barndorff-Nielssen. Normal inverse Gaussian distributions and stochastic volatility modelling. *Scandinavian Journal of Statistics*, 24(1):1–13, 1997.

[28] Ole E. Barndorff-Nielssen. Processes of normal inverse Gaussian type. *Finance and Stochastics*, 2(1):41–68, 1998.

[29] David S. Bates. Post-'87 crash fears in S&P 500 options. *Journal of Econometrics*, 94:181–238, 2000.

[30] Richard Bellman. *Eye of the Hurricane.* World Scientific Publishing Company, Singapore, 1984.

[31] Jean Bertoin. *Lévy Processes.* Cambridge University Press, 1996. Cambridge Tracts in Mathematics 121.

[32] Fisher Black and Myron Scholes. The pricing of options and corporate liabilities. *Journal of Political Economy*, 81(3):637–654, 1973.

[33] Tim Bollerslev. Generalized autoregressive conditional heteroskedasticity. *Journal of Econometrics*, 31:307–327, 1986.

[34] Lennart Bondesson. *Generalized Gamma Convolutions and Related Classes of Distributions and Densities.* Springer-Verlag, 1992. Lecture Notes in Statistics v. 76.

[35] H. Peter Boswijk. Volatility mean reversion and the market price of volatility risk. Tinbergen Institute and Department of Quantitative Economics, Universiteit van Amsterdam, August 2001.

[36] Svetlana I. Boyarchenko and Sergei Z. Levendorskiĭ. Non-Gaussian Merton–Black–Scholes theory. *Journal of Business*, 75(2):305–332, April 2002.

[37] P. P. Boyle, Mark Broadie, and Paul Glasserman. Simulation methods for security pricing. *Journal of Economic Dynamics and Control*, 21:1267–1321, 1997.

[38] Allan Brace, Dariusz Gatarek, and Marek Musiela. The market model of interest rate dynamics. *Mathematical Finance*, 7:127–155, 1997.

[39] D. Britz, Ole Østerby, and J. Strutwolf. Damping of Crank–Nicolson error oscillations. *Computational Biology and Chemistry*, 27(3):253–263, 2003.

[40] Mark Broadie and J. Detemple. American option valuation: New bounds, approximations, and a comparison of existing methods. *The Review of Financial Studies*, 9(4):1211–1250, Winter 1996.

[41] Mark Broadie and J. Detemple. Recent advances in numerical methods for pricing derivative securities. In L. C. G. Rogers and D. Talay, editors, *Numerical Methods in Finance.* Cambridge University Press, 1997.

[42] Mark Broadie and J. Detemple. The valuation of American options on multiple assets. *Mathematical Finance*, 7:241–286, 1997.

[43] Mark Broadie and Paul Glasserman. Estimating security price derivatives using simulation. *Management Science*, 42:269–285, 1996.

[44] Mark Broadie and Paul Glasserman. Monte carlo methods for pricing high-dimensional American options: An overview. *Journal of Economic Dynamics and Control*, 3:15–37, 1997.

[45] Mark Broadie and Paul Glasserman. Pricing American-style securities using simulation. *Journal of Economic Dynamics and Control*, 21(8–9):1323–1352, 1997.

[46] Mark Broadie and Paul Glasserman. A stochastic mesh method for pricing high-dimensional American options. Working Paper, 1998.

[47] Mark Broadie, Paul Glasserman, and G. Jain. Enhanced monte carlo estimates for American option prices. *Journal of Derivatives*, 5:25–44, 1997.

[48] Mark Broadie and Özgür Kaya. Exact simulation of stochastic volatility and other affine jump diusion models. *Operations Research*, 54(2), 2006.

[49] A. Buraschi and B. Dumas. The forward valuation of compound options. *Journal of Derivatives*, 9:8–17, Fall 2001.

[50] Kevin Burrage and E. Platen. Runge–Kutta methods for stochastic differential equations. *Annals of Numerical Mathematics*, 1:63–78, 1994.

[51] R. Caflisch, W. Morokoff, and A. Owen. Valuation of mortgage-backed securities using Brownian bridges to reduce effective dimension. *Journal of Computational Finance*, 1(1):27–46, 1997.

[52] Andrew Cairns. *Interest rate models: an introduction.* Princeton University Press, 2004.

[53] Peter Carr, Hélyette Geman, Dilip B. Madan, and Marc Yor. The fine structure of asset returns: An empirical investigation. *Journal of Business*, 75(2):305–332, April 2002.

[54] Peter Carr, Hélyette Geman, Dilip B. Madan, and Marc Yor. Stochastic volatility for Lévy processes. *Mathematical Finance*, 13(3):345–382, July 2003.

[55] Peter Carr, Hélyette Geman, Dilip B. Madan, and Marc Yor. From local volatility to local Lévy models. *Quantitative Finance*, 4(5):581–588, 2004.

[56] Peter Carr and Ali Hirsa. Why be backward? *Risk*, 16(1):103–107, January 2003.

[57] Peter Carr and Ali Hirsa. Forward evolution equations for knock-out options. In Michael C. Fu, Robert A. Jarrow, Ju-Yi J. Yen, and Robert J. Elliott, editors, *Advances in Mathematical Finance*, chapter 12. Birkhauser Verlag AG, Boston, MA, USA, 2007.

[58] Peter Carr, Robert A. Jarrow, and Ravi Myneni. Alternative characterization of American put options. *Mathematical Finance*, 2:87–106, 1992.

[59] Peter Carr and Dilip Madan. Optimal positioning in derivative securities. *Quantitative Finance*, 1, 2001.

[60] Peter Carr and Dilip B. Madan. Option valuation using the fast Fourier transform. *The Journal of Computational Finance*, 2(4):61–73, 1999.

[61] Peter Carr and Dilip B. Madan. Saddlepoint methods for option pricing. *The Journal of Computational Finance*, 13(1), Fall 2009.

[62] Peter Carr and Liuren Wu. The finite moment logstable process and option pricing. *Journal of Finance*, 58(2):753–770, April 2003.

[63] Andrew P. Carverhill. A simplified exposition of the Heath, Jarrow and Morton model. *Stochastics and Stochastic Reports*, 53(3–4):227–240, 1995.

[64] E. Catmull and R. Rom. A class of local interpolating splines. In *Computer Aided Geometric Design,* R. E. Barnhill and R. F. Reisenfeld, Eds. Academic Press, New York, pages 317–326, 1974.

[65] A. De Cezaro, O. Scherzer, and Jorge P. Zubelli. A convex-regularization framework for local volatility calibration in derivative markets: The connection with convex risk measures and exponential families. 6th *World Congress of the Bachelier Finance Society*, 2010.

[66] Kyriakos Chourdakis. Option valuation using the fast Fourier transform. *The Journal of Computational Finance*, 31(2):826–848, 2008.

[67] Neil Chriss. Transatlantic trees. *Risk*, 9(7):45–48, July 1996.

[68] W. J. Cody. Rational Chebyshev approximations for the error function. *Math. Comp.*, 23(107):631–637, July 1969.

[69] Thomas F. Coleman, Yohan Kim, Yuynig Li, and Arun Verma. Dynamic hedging with a deterministic local volatility function model. *Journal of Risk*, 4(1):63–89, 2001.

[70] Thomas F. Coleman and Yuynig Li. An interior trust region approach for nonlinear minimization subject to bounds. *SIAM Journal on Optimization*, 6(2):418–445, 1993.

[71] Thomas F. Coleman, Yuynig Li, and Arun Verma. Reconstructing the unknown local volatility function. *Journal of Computational Finance*, 2:77–102, 1998.

[72] P. Concus and Gene H. Golub. *A Generalized Conjugate Gradient Method for Non-symmetric Systems of Linear Equations in R. Glowinski and J. L. Lions, Editors.* Springer-Verlag, 1976. Lecture Notes in Economics and Mathematical Systems 134.

[73] Rama Cont. Model uncertainty and its impact on the pricing of derivative instruments. Centre de Mathématiques Appliquées CNRS Ecole Polytechnique, F-91128 Palaiseau, France, June 2004.

[74] Rama Cont. Model calibration. In Rama Cont, editor, *Encyclopedia of Quantitative Finance*, volume 3, pages 1210–1218. John Wiley and Sons Ltd, Southern Gate, Chichester, West Sussex, England, 2010.

[75] Rama Cont, Romain Deguest, and Yu Hang Kan. Default intensities implied by cdo spreads: inversion formula and model calibration. Columbia University Financial Engineering Report 2009-04, www.ssrn.com, 2010.

[76] Rama Cont and Peter Tankov. Calibration of jump-diffusion option pricing models: A robust non-parametric approach. Rapport Interne CMAP Working Paper No. 490, September 2002.

[77] Rama Cont and Peter Tankov. *Financial Modelling with Jump Processes.* Chapman & Hall/CRC Financial Mathematics Series, 2003.

[78] James W. Cooley and John W. Tukey. An algorithm for the machine calculation of complex Fourier series. *Mathematics of Computation*, 19(90):297–301, 1965.

[79] F. Costabile and A. Napoli. Economical Runge–Kutta method for numerical solution of stochastic differential equations. *BIT Numerical Mathematics*, 48(3):499–509, 2008.

[80] R. Courant, K. Friedrichs, and H. Lewy. On the partial difference equations of mathematical physics. *IBM Journal of Research and Development*, 11(2):215–234, March 1967.

[81] John C. Cox. Notes on option pricing i: Constant elasticity of variance diffusions. *Journal of Portfolio Management*, (22):15–17, 1996.

[82] John C. Cox, Jonathan E. Ingersoll, and Stephen A. Ross. A theory of the term structure of interest rates. *Econometrica*, 53(2):385–407, March 1985.

[83] Ian J. D. Craig and Alfred D. Sneyd. An alternating-direction implicit scheme for parabolic equations with mixed derivatives. *Computers & Mathematics with Applications*, 16(4):341–250, 1988.

[84] John Crank and Phyllis Nicolson. A practical method for numerical evaluation of solutions of partial differential equations of the heat-conduction type. *Advances in Computational Mathematics*, 6(1):207–226, December 1996.

[85] S. Crépey. Calibration of the local volatility in a trinomial tree using Tikhonov regularization. *Institute Of Physics Publishing*, (19):91–127, December 2002.

[86] Alan C. Curtis and M. R. Osborne. The construction of minimax rational approximations to functions. *The Computer Journal*, 9:286–293, 1966.

[87] Zhi Da and Ernst Schaumburg. The price of volatility risk across asset classes. November 2011.

[88] A. d'Aspremont. Risk-mangement methods for the LIBOR market model using semidefinite programming. *Journal Of Computational Finance*, 8(4):77–99, 2005.

[89] Dmitry Davydov and Vadim Linestsky. Pricing options on scalar diffusions: An eigenfunction expansion approach. October 2000.

[90] Dmitry Davydov and Vadim Linestsky. The valuation and hedging of barrier and lookback options for alternative stochastic processes. August 2000.

[91] R. Van der Merwe, A. Doucet, N. de Freitas, and E. Wan. The unscented particle filter. Oregon Graduate Institute, 2000.

[92] R. Van der Merwe and E. A. Wan. The square-root unscented Kalman filter for state and parameter estimation. *IEEE International Conference on Acoustics, Speech, and Signal Processing*, 6:3461–3464, 2001.

[93] Emanuel Derman. Model risk. *Quantitative Strategies Research Notes*, Goldman Sachs, April 1996.

[94] Emanuel Derman. Laughter in the dark — The problem of the volatility smile, May 2003.

[95] Emanuel Derman and Iraj Kani. The volatility smile and its implied tree. *Risk*, 7(2):32–39, February 1994.

[96] V. Digalakis, J. R. Rohlicek, and M. Ostendorf. Maximum likelihood estimation of a stochastic linear system with the EM algorithm and its application to speech recognition. *Speech and Audio Processing, IEEE Transactions on*, 1(4):431–432, 1993.

[97] J. Douglas and Jr. H. H. Rachford. On the numerical solution of the heat conduction problem in two and three space variables. *Transactions of the American Mathematical Society*, 82:421–439, 1956.

[98] Jefferson Duarte and Christopher S. Jones. The price of market volatility risk. October 2007.

[99] D. Duffie. *Dynamic Asset Pricing Theory.* Princeton University Press, Princeton, NJ, second edition, 1996.

[100] Darrell Duffie and Rui Kan. A yield-factor model of interest rates. *Mathematical Finance*, 6:379–406, 1996.

[101] Darrell Duffie, Jun Pan, and Kenneth Singleton. Transform analysis and asset pricing for affine jump diffusions. *Econometrica*, 68:1343–1376, 2000.

[102] Daniel J. Duffy. *Finite Difference Methods in Financial Engineering: A Partial Differential Equation Approach.* John Wiley and Sons Ltd, Southern Gate, Chichester, West Sussex, England, 2006.

[103] B. Dumas, J. Fleming, and R. Whaley. Implied volatilities: Empirical tests. *Journal of Finance*, (53):2059–2106, 1998.

[104] Bruno Dupire. Pricing with a smile. *Risk*, 7(1):18–20, January 1994.

[105] E. Eberlein, U. Keller, and K. Prause. New insights into smile, mispricing, and value at risk: The hyperbolic model. *Journal of Business*, 71:371–406, 1998.

[106] David Eberly. Derivative Approximation by Finite Differences. Magic Software, Inc, January 2003.

[107] David Eberly. Kochanek–Bartels cubic splines (TCB splines). Magic Software, Inc, March 2003.

[108] Herbert Egger and Heinz W. Engl. Tikhonov regularization applied to the inverse problem of option pricing: Convergence analysis and rates. Johann Radon Institute for Computational and Applied Mathematics and Johannes Kepler University Linz, Altenbergerstr. 69, A-4040 Linz, Austria, 2008.

[109] Robert E. Engle. Autoregressive conditional heteroskedasticity with estimates of the variance of U.K. inflation. *Econometrica*, 50:987–1008, 1982.

[110] A. Esser and C. Schlag. A note on forward and backward partial differential equations for derivative contracts with forwards as underlyings in foreign exchange risk. In Jurgen Hakala and Uwe Wystup, editors, *Foreign Exchange Risk: Models, Instruments and Strategies*, chapter 12. Risk Books, 2002.

[111] Fang Fang and Cornelis W. Oosterlee. A novel pricing method for European options based on Fourier-cosine series expansions. *SIAM Journal on Scientific Computing*, 8(2):1–18, Winter 2004.

[112] Fang Fang and Cornelis W. Oosterlee. Pricing early-exercise and discrete barrier options by Fourier-cosine series expansions. University of Netherlands, June 2009.

[113] K.-T. Fang and Y. Wang. *Number Theoretic Methods in Statistics.* Chapman & Hall, New York, USA, 1994.

[114] H. Faure. Discrépance de suites associées à un système de numération (en dimension s). *Acta Arithmetica*, 41(4):337–351, 1982.

[115] Thomas Flury and Neil Shephard. Learning and filtering via simulation: smoothly jittered particle filters. Oxford-Man Institute, University of Oxford, Eagle House, Walton Well Road, Oxford OX2 6ED, UK & Department of Economics, University of Oxford, December 2009.

[116] Roland W. Freund and Nöel M. Nachtigal. QMR: A quasi-minimal residual method for non-Hermitian linear systems. *SIAM Journal: Numererical Mathematics*, 60:315–339, 1991.

[117] Andras Fulop. Particle filtering with applications in finance. ESSEC, the Risk Management Institute, NUS, November 2007.

[118] Helyette Geman, Dilip B. Madan, and Marc Yor. Time changes for Lévy processes. *Mathematical Finance*, 11:79–96, 2001.

[119] Zoubin Ghahramani and Geoffrey E. Hinton. Parameter estimation for linear dynamical systems. University of Toronto, Department of Computer Science, *Technical Report CRG-TR-96-2*, February 1996.

[120] Zoubin Ghahramani and Sam Roweis. A unifying review of linear Gaussian models. *Neural Computation*, 11(2):77–99, 1999.

[121] W. R. Gilks, S. Richardson, and David J. Spiegelhalter. *Markov Chain Monte Carlo in Practice.* Chapman & Hall/CRC, London, UK, 1996.

[122] Paul Glasserman. *Monte Carlo Methods in Financial Engineering.* Springer, 2003.

[123] Gene H. Golub and Charles F. Van Loan. *Matrix Computations.* Johns Hopkins Studies in Mathematical Sciences, Baltimore, MD, USA, third edition, 1996.

[124] Geoffrey R. Grimmett and David R. Stirzaker. *Probability and Random Processes.* Oxford University Press Inc, New York, USA, second edition, 1992.

[125] Chandrasekhar R. Gukhal. Analytical valuation of American options on jump diffusion processes. *Mathematical Finance*, (11):97–115, 2001.

[126] István Gyöngy. Mimicking the one-dimensional marginal distributions of processes having an Itô differential. *Probability Theory Related Fields*, 71:501–516, 1986.

[127] J. H. Halton. On the efficiency of certain quasi-random sequences of points in evaluating multi-dimensional integrals. *Numerische Mathematik*, 2(1):84–90, 1960.

[128] Sana Ben Hamida and Rama Cont. Recovering volatility from option prices by evolutionary optimization. *Journal of Computational Finance*, 8(4):1–34, Summer 2005.

[129] A. C. Harvey. *Forecasting, Structural Time Series Models and the Kalman Filter.* Cambridge University Press, Cambridge, UK, 1989.

[130] Simon Haykin, editor. *Kalman Filter and Neural Networks.* John Wiley & Sons, New York, USA, 2001.

[131] David Heath, Robert A. Jarrow, and Andrew Morton. Bond pricing and the term structure of interest rates: a discrete time approximation. *Journal of Financial and Quantitative Analysis*, 25:419–440, 1990.

[132] Massoud Heidari, Ali Hirsa, and Dilip B. Madan. Pricing of swaption in affine term structures with stochastic volatility. In Michael C. Fu, Robert A. Jarrow, Ju-Yi J. Yen, and Robert J. Elliott, editors, *Advances in Mathematical Finance*, chapter 12. Birkhauser Verlag AG, Boston, MA, USA, 2007.

[133] Vicky Henderson, David Hobson, Sam Howison, and Tino Kluge. A comparison of option prices under different pricing measures in a stochastic volatility model with correlation. October 2004.

[134] Steve Heston. A closed-form solution for options with stochastic volatility with applications to bond and currency options. *Review of Financial Studies*, 6:327–343, 1993.

[135] Ali Hirsa. *Numerical Algorithms for Option Pricing and Convection Diffusion Equation.* PhD thesis, University of Maryland at College Park, MD, USA, December 1997.

[136] Ali Hirsa, Georges Courtadon, and Dilip B. Madan. Local volatility reengineers semimartingle models. March 2001.

[137] Ali Hirsa, Georges Courtadon, and Dilip B. Madan. The effect of model risk on the valuation of barrier options. *Journal of Risk Finance*, 4(2), Spring 2003.

[138] Ali Hirsa, Massoud Heidari, and Dilip B. Madan. Swaption pricing in discrete-time double gamma affine term structure models. Working Paper in Progress, 2009.

[139] Ali Hirsa and Alireza Javaheri. Parameter estimation for partially observed processes. Working Paper in Progress, 2003.

[140] Ali Hirsa and Dilip B. Madan. Pricing American options under variance gamma. *Journal of Computational Finance*, 7(2):63–80, Winter 2003.

[141] Ali Hirsa, Dilip B. Madan, and Li Bao. Cap pricing in continuous-time and discrete-time affine term structure models with stochastic volatility. In Fabio Mercurio, editor, *Modelling Interest Rates*, chapter 11, pages 265–278. Risk Books, Incisive Media, 2009.

[142] Willem Hundsdorfer and Laura Portero. A note on iterated splitting schemes. *Journal of Computational and Applied Mathematics*, 201(1):146–152, 2007.

[143] Ronald L. Iman and Jon C. Helton. A comparison of uncertainty and sensitivity analysis techniques for computer models. *Technical Report SAND84-1461*, Sandia National Laboratories, Albuquerque, NM, USA, 1985.

[144] Ronald L. Iman, Jon C. Helton, and James E. Campbell. An approach to sensitivity analysis of computer models: Part II — ranking of input variables, response surface validation, distribution effect and technique synopsis. *Journal of Quality Technology*, 13(4):232–240, 1981.

[145] Ronald L. Iman, Jon C. Helton, and James E. Campbell. An approach to sensitiviy analysls of computer models: Part I — introduction, input variable selection and preliminary assessment. *Journal of Quality Technology*, 13(3):174–183, 1981.

[146] Karel J. in 't Hout and S. Foulon. ADI finite diference schemes for option pricing in the Heston model with correlation. *International Journal of Numerical Analysis and Modeling*, 7(2):303–320, 2010.

[147] Jonathan E. Ingersoll. *Theory of Financial Decision Making.* Rowman & Littlefield Publishers, Inc, Lanham, Maryland, USA, 1987.

[148] Ken iti Sato. Basic results on lévy processes. In Ole E. Barndorff-Nielsen, Thomas Mikosch, and Sidney I. Resnick, editors, *Lévy processes Theory and Applications*, pages 3–37. Birkhauser, Boston, MA, USA, 2001.

[149] K. Ito and K. Xiong. Gaussian filters for nonlinear filtering problems. *IEEE Transactions on Automatic Control*, 45(5):910–927, 2000.

[150] Jean Jacod and Albert N. Shiryaev. *Limit Theorems for Stochastic Processes.* Springer-Verlag, Berlin, New York, 1987.

[151] Wolfgang Jank. Quasi-Monte Carlo sampling to improve the efficiency of Monte Carlo EM. Department of Decision and Information Technologies, University of Maryland College Park, MD, USA, wjank@rhsmith.umd.edu, November 2003.

[152] Alireza Javaheri. *Inside Voaltility Arbitrage: The Secrets of Skewness.* John Wiley & Sons, Hoboken, NJ, USA, 2005.

[153] Mark Jex, Robert Henderson, and Robert Henderson. Pricing exotics under the smile. *Derivatives Reaserch J.P. Morgan Securities Inc. London*, J.P. Morgan, September 1999.

[154] Michael Johannes and Nicolas Polson. *Bayesian Filtering.* Princeton, 2011.

[155] Michael Johannes, Nicolos Polson, and Jonathan Stroud. Nonlinear filtering of stochastic differential equations with jumps. Working Paper, Columbia University, University of Chicago and University of Pennsylvania, October 2002.

[156] Michael S. Johannes and Nicolas Polson. Mcmc methods for continuous-time financial econometrics. In Yacine Ait-Sahalia and Lars Hansen, editors, *Handbook of Financial Econometrics*, volume 2, pages 1–66. 2009.

[157] Michael S. Johannes and Nicolas Polson. Mcmc methods for continuous-time financial econometrics. In Torben Andersen, Richard David, Jens-Peter Kreiss, and Thomas Mikosch, editors, *Markov Chain Monte Carlo*, pages 1001–1013. 2009.

[158] Christian Kahl and Peter Jäckel. Fast strong approximation Monte Carlo schemes for stochastic volatility models. *Quantative Finance*, 6(6), 2006.

[159] N. Kantas, A. Doucet, S. S. Singh, and J. M. Maciejowski. An overview of sequential Monte Carlo methods for parameter estimation in general state-space models. Cambridge University Engineering Dept., Cambridge CB2 1PZ, UK and The Institute of Statistical Mathematics, Tokyo 106-8569, Japan.

[160] Ioannis Karatzas and Steven E. Shreve. *Brownian Motion and Stochastic Calculus.* Springer-Verlag, New York, USA, second edition, 1991.

[161] Ioannis Karatzas and Steven E. Shreve. *Methods of mathematical finance,* vol. 39 of Applications of Mathematics. Springer-Verlag, New York, USA, 1998.

[162] Toshiyasu Kato and Toshinao Yoshiba. Model risk and its control. *Monetary and Economic Studies*, December 2000.

[163] Ajay Khanna and Dilip B. Madan. Non Gaussian models of dependence in returns. Working Paper, Robert H. Smith School of Business, University of Maryland at College Park. MD, USA, November 2009.

[164] Manabu Kishimoto. On the Black–Scholes equation: Various derivations. Management Science and Engineering 408 Term Paper, May 2008.

[165] G. Kitagawa. Monte carlo filter and smoother for non-Gaussian nonlinear state space models. *Journal of Computational and Graphical Statistics*, 5(1):1–25, 1996.

[166] S. G. Kou. A jump-diffusion model for option pricing. *Management Science*, 48:1086–1101, 2002.

[167] Pavol Kútik and Karol Mikula. Finite volume schemes for solving nonlinear partial differential equations in financial mathematics. In Jaroslav Fořt, Jiří Fürst, Jan Halama, Raphaèle Herbin, and Florence Hubert, editors, *Finite Volumes for Complex Applications VI Problems & Perspectives Springer Proceedings in Mathematics volume 4*, pages 643–651. Springer, June 2011.

[168] Damien Lamberton and Bernard Lapeyre. *Introduction to Stochastic Calculus Applied to Finance.* Chapman & Hall, first edition, 1996.

[169] Jean-Paul Laurent and Jon Gregory. Basket default swaps, cdos and factor copulas. *Journal of Risk*, (4):1–20, Summer 2005.

[170] Anthony Lee. Towards smooth particle filters for likelihood estimation with multivariate latent variables. Master thesis, The University of British Columbia, August 2008.

[171] Alexander Lipton. The vol smile problem. *Risk*, 15(2):61–65, February 2002.

[172] Francis A. Longstaff and Eduardo S. Schwartz. Valuing American options by simulation: A simple least-squares approach. *The Review of Financial Studies*, 14(1):113–147, 2001.

[173] Roger Lord, Remmert Koekkoek, and Dick van Dijk. A comparison of biased simulation schemes for stochastic volatility models. *Journal of Quantitative Finance*, 10(2):177–194, 2010.

[174] R. Lugannani and S. O. Rice. Saddlepoint approximation for the distribution of the sum of independent random variables. *Advances in Applied Probability*, 12:475–490, 1980.

[175] Dilip B. Madan, Peter Carr, and Eric C. Chang. The variance gamma process and option pricing. *European Finance Review*, 2:79–105, 1998.

[176] M. D. McKay, R. J. Beckman, and W. J. Conover. A comparison of three methods for selecting values of input variables in the analysis of output from a computer code. *Technometrics*, 21(2):239–245, May 1979.

[177] Robert C. Merton. Theory of rational option pricing. *Bell Journal of Economics and Management Science*, 4(1):141–183, 1973.

[178] Robert C. Merton. Option pricing when underlying stock returns are discontinuous. *Journal of Financial Economics*, 3:125–144, 1976.

[179] Sergei Mikhailov and Ulrich Nögel. Hestons stochastic volatility model implementation, calibration and some extensions. *Wilmott magazine*, 6:74–79, July 2003. Fraunhofer Institute for Industrial Mathematics, Kaiserslautern, Germany, Mikhailov@itwm.fhg.de, Noegel@itwm.fhg.de.

[180] Karim Mimouni. Variance filtering and models for returns data. *Journal of Money, Investment and Banking*, 1:133–141, 2011.

[181] Daniel B. Nelson. Conditional heteroskedasticity in asset returns: A new approach. *Econometrica*, 59(2):347–370, March 1991.

[182] H. Niederreiter. *Random Number Generation and Quasi-Monte Carlo Methods.* SIAM, Philadelphia, 1992.

[183] Bernt Øksendal. *Stochastic Differential Equations. An Introduction with Applications.* Springer, New York, USA, fifth edition, 2000.

[184] Ole Østerby. Five ways of reducing the crank–nicolson oscillations. *BIT Numerical Mathematics*, 43(4):811–822, 2003.

[185] Antonis Papapantoleon. An introduction to Lévy processes with applications in finance. Lecture notes taught at the University of Piraeus, University of Leipzig, and at the Technical University of Athens, 2008.

[186] Navroz Patel. Ther evolving art of pricing cliquets. *Risk magazine*, 15(7), July 2002.

[187] D. W. Peaceman and Jr. H. H. Rachford. The numerical solution of parabolic and elliptic differential equations. *Journal of the Society for Industrial and Applied Mathematics*, 3(1):28–41, March 1955.

[188] Michael K. Pitt. Smooth particle filters for likelihood evaluation and maximisation. Department of Economics, University of Warwick, Coventry CV4 7AL, M.K.Pitt@warwick.ac.uk, July 2002.

[189] W. H. Press, S. A. Teukolsky, W. T. Vetterling, and B. P. Flannery. *Numerical Recipes in C: The Art of Scientific Computing.* Cambridge University Press, second edition, 1992.

[190] E. Reiner and M. Rubinstein. Breaking down the barriers. *Risk Magazine*, 4(8):28–35, 1991.

[191] Yong Ren, Dilip B. Madan, and Michael Q. Qian. Calibrating and pricing with embedded local volatility models? *Risk*, 20(9):138–143, September 2007.

[192] R. Richtmeyer and K. W. Morton. *Difference Methods for Initial Value Problem.* Wiley, New York, USA, 1967.

[193] L. Rogers and Z. Shi. The value of an Asian option. *Journal of Applied Probability*, 32:1077–1088, 1995.

[194] Mark Rubinstein. Implied binomial trees. *The Journal of Finance*, 49(3):771–818, July 1994.

[195] Y. Saad and M. H. Schultz. GMRES: A generalized minimal residual algorithm for solving nonsymmetric linear systems. *SIAM Journal on Scientific and Statistical Computing*, 7:856–869, 1986.

[196] Paul A. Samuelson. Rational theory of warrant pricing. *Industrial Managment Review*, 6(2):13–31, Spring 1965.

[197] Wim Schoutens, Erwin Simnos, and Jurgen Tistaert. Model risk for exotic and moment derivatives. In Wim Schoutens Andreas E. Kyprianou and Paul Wilmott, editors, *Exotic Option Pricing and Advanved Lévy Models*, chapter 4. John Wiley & Sons, 2005.

[198] Louis O. Scott. Pricing stock options in a jump-diffusion model with stochastic volatility and interest rates: Applications of Fourier inversion methods. *Mathematical Fiance*, 7(4):413–426, 1997.

[199] Steven G. Self and Kung-Yee Liang. Asymptotic properties of maximum likelihood estimators and likelihood ratio tests under nonstandard conditions. *Journal of the American Statistical Association*, 82(398):605–610, June 1987.

[200] Rudiger Seydel. *Tools for Computational Finance.* Springer, 2nd edition, 2003.

[201] Karl Sigman. Acceptance-rejection method. Columbia University, 2007.

[202] Dennis Silverman. Solution of the Black Scholes eqaution using the Green's function of the diffusion eqation. Department of Physics and Astronomy, University of California, Irvine, CA, USA, August 1999.

[203] Ashok K. Singh and R. S. Bhadauria. Finite difference formulae for unequal subintervals using Lagrange's interpolation formula. *International Journal of Mathematical Analysis*, 3(17):813–827, 2009.

[204] Robert D. Smith. An almost exact simulation method for the Heston model. *Journal of Computational Finance*, 11(1):115–125, 2008.

[205] I. M. Sobol. On the distribution of points in a cube and approximate evaluation of integrals. *U.S.S.R. Computational Mathematics and Mathematical Physics*, 7(4):86–112, 1967.

[206] J. C. Strikwerda. *Finite Difference Schemes and Partial Differential Equations.* Pacific Grove, CA: Wadsworth and Brooks, 1989.

[207] K. Stüben and U. Trottenberg. *Multigrid MInt. Journal of Math. Analysisethods: Fundamental algorithms, model problem analysis and applications, in Multigrid Methods, pp. 1–176, W. Hackbusch and U. Trottenberg editors.* Springer–Verlag, Berlin, Germany, 1982.

[208] Rangarajan K. Sundaram. Equivalent martingale measures and risk-neutral pricing: An expository note. *Journal of Derivatives*, 5(1):85–98, Fall 1997.

[209] Grigore Tataru and Travis Fisher. Stochastic local volatility. *Quantitative Development Group*, Bloomberg, February 2010.

[210] Domingo Tavella and Curt Randall. *Pricing Financial Instruments, The Finite Difference Method.* John Wiley & Sons, first edition, 2000.

[211] A. Tocino and J. Vigo-Aguiar. Weak second order conditions for stochastic Runge–Kutta methods. *SIAM Journal on Scientific Computing*, 24(2):507–523, 2002.

[212] Jürgen Topper. *Financial Engineering with Finite Elements.* Wiley Finance Series, 2005.

[213] A. van Haastrecht and A. A. J. Pelsser. Efficient, almost exact simulation of the Heston stochastic volatility model. Netspar Network for Studies on Pensions, Aging and Retirment, Discussion Paper 09/2008 - 044, September 2008.

[214] Jan Vecer. A new PDE approach for pricing arithmetic average Asian options. *Journal of Computational Finance*, 4(4):105–113, 2001.

[215] Anatoly Vershik and Marc Yor. Multiplicativity of the gamma process and asymptotic study of stable laws of index $\alpha$, as $\alpha$ converges to zero. Prépublication No. 289 Du Laboratoire de Probabilités de L'Université Paris VI., June 1995.

[216] E. Wan, R. van der Merwe, and A. T. Nelson. Dual estimation and the unscented transformation. *Neural Information Processing Systems*, 12:666–672, 2000.

[217] Eric W. Weisstein. Halley's method. From MathWorld–A Wolfram Web Resource. http://mathworld.wolfram.com/HalleysMethod.html.

[218] Paul Wilmott, Jeff Dewynne, and Sam Howison. *Option Pricing: Mathematical Models and Computation.* Oxford Financial Press, 1993.

[219] Andrew T.A. Wood, James G. Booth, and Ronald W. Butler. Saddle point approximations to the CDF of some statistics with nonnormal limit distributions. *Journal of the American Statistical Association*, 88(422):680–686, June 1993.

[220] R. Zvan, P. A. Forsyth, and K. R. Vetzal. A finite volume approach for contingent claims valuation. *IMA Journal of Numerical Analysis*, 21(3):703–731, 2001.

# Index